Catalysis by
Polymer-Immobilized
Metal Complexes

Catalysis by Polymer-Immobilized Metal Complexes

A.D. Pomogailo

Institute of Chemical Physics
Moscow, Russia

Translated from the Russian Edition by

N.M. Bravaya and A.V. Kulikov

Gordon and Breach Science Publishers

Australia • Canada • China • France • Germany • India
Japan • Luxembourg • Malaysia • The Netherlands
Russia • Singapore • Switzerland

Originally published in Russian in 1988 as Polimernye Immobilizovannye Metallokompleksnye Katalizatory by Nauka, Moscow.

Amsteldijk 166
1st Floor
1079 LH Amsterdam
The Netherlands

British Library Cataloguing in Publication Data

Pomogailo, Anatolii D. (Anatolii Dmitrievich)
   Catalysis by polymer-immobilized metal complexes
   1. Metal complexes 2. Catalysis 3. Metal catalysts
   I. Title
   541.3'95

   ISBN: 90-5699-130-2

*I dedicate this book to my little children,
daughter Dasha and son Dima.*

Contents

Foreword to English Edition

The field of catalysis by polymer-bound metal complexes has grown immensely. My interest in this field began in 1966 while investigating the applicability of incorporating the burning rate catalyst, butyl ferrocene, into rocket propellent binders. Polymer immobilization of ferrocene derivatives insured dispersion and prevented phase separation. Over the following 25 years I maintained an active research program in this area and wrote several reviews covering this topic. This experience led me to feel that I had a good command of the literature in the field. However, while reading this book by A.D. Pomogailo this quaint notion was quickly dispelled, as I continually encountered references of which I was unaware. This was true over and over again in each chapter.

Professor Pomogailo has provided the Western chemist with an impressive collection of references and summaries of research carried out in the former Soviet Union and Eastern Europe which is woven together with work emanating from the rest of the world. This text will, for the first time, give the English language reader a broad view of the work of Russian chemists, in particular, as well as efforts of Chinese and Japanese scientists which have not been readily available in the West. During the Cold War our contact with most Soviet scientists was severely limited. This, plus language differences, led to a substantial separation between East and West in the evolution of this area of catalysis. The relative isolation of these two communities from one another has begun to crumble and this monograph is a valuable contribution in that direction.

The reader will find a rather different organization and thrust in this monograph compared to many of the existing reviews which have appeared previously in the English language. These differences further point to a separate evolution of this chemistry. The extent of the work carried out in the (former) Soviet Union makes it clear that the field of polymer-immobilized catalysts was considered to be an area where important technology might arise. The highly interdisciplinary nature of research on this topic suggests this book will be a valuable resource to scientists from several fields of science. The completion of this text constitutes a valuable contribution to the unification of two communities which were artificially separated from each other for too long. Let us hope that this trend continues to accelerate in the years ahead.

Professor Charles U. Pittman, Jr

Foreword to Russian Edition

This book deals with a new and promising field developed over the last two decades on the boundary between homogeneous and heterogeneous catalysis. Due to the vigorous efforts of scientists of many countries, catalysis by immobilized complexes has developed so rapidly that it has become clear that a new interdisciplinary scientific field has been born. This field is characterized by the following specific features: it deals with numerous new and interesting objects having common features of their structure; new methods of investigation and controls of these systems have been developed; and optimal fields of their practical applications have been found.

Therefore, a review of the numerous experimental studies in this field is very timely. The author of this book works intensively in this field and knows his subject well along with the history of its development.

Common features of catalysis of a wide range of organic reactions are considered: hydrogenation, oxidation, polymerization of monomers, dimers and oligomers of various types, syntheses with participation of carbon monoxide. Special attention has been paid to electro- and photochemical stimulation of catalytic processes in the presence of immobilized catalysts.

The survey shows that the activity and selectivity of such systems are determined not only by the usual factors considered for homogeneous catalysis, but also by consideration of the specific effects caused by the properties of the macroligand and the chemical bond of the transition metal with the macroligand, the surface density of metal complexes on supports, the effect of temperature on the catalyzed reaction and also on relaxational transitions in the macroligand, etc.

Further progress in this field could be substantially accelerated by more intensive investigation of bi- and polyfunctional cooperative catalysis: macroligands actually offer limitless opportunities to immobilize active centers of different types on the same support. In this sense, there is a relation and analogy between catalysis by immobilized complexes and enzymatic catalysis. The main modern trends in these fields are also considered in this book.

Consideration of the different features of catalysis by immobilized complexes is also interesting. The polymeric structure of immobilized complexes inhibits redox processes and ligand exchange; in these systems reactions of electron transfer, as well as transformations of mononuclear complexes into polynuclear ones and cluster structures, also proceed in specific ways. As a rule, the polymeric structure of immobilized systems leads to their higher thermodynamic and kinetic stability in comparison with homogeneous analogues.

A large number of references facilitate the more detailed acquaintance of readers with almost all problems of catalysis by immobilized complexes.

Academician M.E. Vol'pin

Preface to English Edition

It is a great pleasure for me to introduce this book to English-speaking readers. Publication of a book in the English language, which most chemists speak and use to write their works, is a remarkable event for any scientist. I consider publication of my second book in English as a particular honor.

This book is not just a reproduction of the Russian edition. Substantial additions, including new literature data, have been included; and some concepts have been considered from new points of view.

One of my aims was to deliver to English-speaking readers a substantial contribution from scientists of Russia and the former USSR on catalysis by polymer-immobilized complexes because, as a rule, most Russian publications are inaccessible to scientists of other countries.

Professor M.E. Vol'pin, a well-known Russian chemist, was an editor of the Russian edition of this book, and it is a special pleasure that it was his initiative to publish the book with Gordon and Breach Science Publishers. Unfortunately, he didn't see the English edition of this book: he died on September 28th, 1996, at the age of 74. His death was a severe loss for chemists worldwide. His work initiated many trends in organic and metalorganic chemistry, metalcomplex catalysis and chemistry of coordination compounds, and was characterized by splendid ideas and deep insight into the essence of problems. He will remain in the memory of his numerous colleagues, including the author of this book, as a responsive, charming and extremely erudite person.

At present, catalysis by polymer-immobilized complexes is a combination of different fields of chemical science; therefore it is quite possible that many scientists have met for the first time within the pages of this book and have found that they work on common problems.

I am grateful to Professor C.U. Pittmann Jr, a well-known specialist in catalysis by polymer-immobilized complexes and one of the pioneers of this field, who read the book and wrote a foreword to the English edition, as well as to many scientists who sent reprints of their latest papers.

Great work was done by the translators and my colleagues Professor A.V. Kulikov (Chapters 2 and 4) and Dr N.M. Bravaya (Chapters 1, 3 and 5) who have done their best to deliver to readers the main ideas of this book.

I hope that the monograph will be useful for specialists of a broad profile and will facilitate further development of this fascinating field of science.

Preface to Russian Edition

The main reason for writing this book was the author's desire to continue the consideration and generalization of numerous experimental studies of catalysis by immobilized complexes started earlier (Pomogailo A.D. (1988) *Polimernye Immobilizovannye Metallokompleksnye Katalizatory* (Polymeric Immobilized Metal Complex Catalysts), Moscow: Nauka). Indeed, using an analogy between homogeneous, heterogeneous and enzymatic catalysis, the driving force behind the idea of metal complex immobilization has created highly effective and technological catalysts. A reliable scientific basis was created due to the rapid development of this field over the last two decades, and researchers discussed what is possible and what is impossible in catalysis by immobilized metal complexes. A substantial contribution was made toward notions and principles which are of a common character and important for theoretically choosing catalysts and in the prediction of their properties.

In the present monograph I analyze physicochemical aspects of the main reactions of organic synthesis in the presence of immobilized complexes of transition metals. This book is addressed not only to researchers who actively use immobilized catalysts in their work, but also to students and researchers who have just started investigations in this field, and therefore almost every section of this book has a brief introduction to the subject considered.

I tried to consider different opinions in an unbiased manner, nevertheless in many cases my own attitude is evident. Moreover, the problems considered are often many-sided and studied to different extents, therefore it was difficult for me as a specialist in one part of this complex field to be equally competent in all subjects.

However, my task was eased by fruitful discussions of a number of sections of the book with scientists R.I. Gvozdev, E.T. Denisov, N.T. Denisov, O.N. Efimov, P.A. Ivanchenko, E.I. Karpeiskaya, E.I. Klabunovskii, A.P. Moravskii, I.N. Meshkova, Yu.N. Nizel'skii, V.L. Rubailo and V.Z. Sharf. I am sincerely grateful to them. Discussions and advice from foreign scientists J. Sheats, C. Carraher Jr and C. Pittman (USA), D. Sherrington and F. Hartley (UK), G. Braca and G. Sbrana (Italy), K. Takamoto (Japan), Y.L. Li (China) and J. Garnett (Australia) were very useful.

I used in the monograph materials of two All-Union Symposiums on catalysis by supported complexes (organized by Yu.I. Ermakov in 1977 and 1980) and the last two International Symposiums on the relation between homogeneous and heterogeneous catalysis (Novosibirsk, 1986, and Pisa, 1989).

I am also grateful to G.I. Dzhardimaliyeva for technical assistance.

I hope that this book will be useful and encourage further progress in this field of science.

Abbreviations

-gr-	grafted	EB	ethylbenzene
4VP	4-vinylpyridine	EDTA	ethylenediaminetetraacetate
AcAc	acetylacetone	ESCA	electron spectroscopy for
AIBN	azobis (isobutyronitryle)		chemical application
AP	acetophenone	EXAFS	extended X-ray absorption
BD	branching degree		fine structure
CAAA	copolymer of acrylic acid	HDPE	high density polyethylene
	and acrylamide	HMFA	hexamethyl(phosphorous)
CDT	cyclododecatriene		triamide
CHHP	cyclohexyl hydroperoxide	HP	hydroperoxide
CHP	cumene hydroperoxide	LDLPE	low density linear
CMPS	chloromethylated poly-		polyethylene
	styrene	LGC	lamellar graphite compounds
COD	cyclooctadiene	LMCE	liquid metal complex
Cp	cyclopentadiene		exchange
CS	copolymers of styrene	MAA	methyl acetoacetate
CS3FS	copolymer of styrene with	MAc	methylacrylate
	α,β,β'-trifluorostyrene	MMAc	methylmethacrylate
CSAAc	copolymer of styrene with	MMD	molecular-mass distribution
	acrylic acid	MPC	methylphenylcarbinol
CSBr-S	copolymers of styrene with	P2VP	poly(2-vinylpyridine)
	α-bromostyrene	P2VPr	poly-N-(vinyl-2-
CSDVB	copolymer of styrene with		pyrrolidone)
	divinylbenzene	P4VP	poly(4-vinylpyridine)
CSMS	copolymer of styrene and 2-	PA	polyacetylene
	(methylsulphynyl)ethyl-	PAA	poly(allyl amine)
	methacrylate	PAAc	poly(acrylic acid)
CTC	charge transfer complex	PAAl	poly(allyl alcohol)
DAPh	dialkylphenole	PAAO	polyacrylamide oximes
Dipy	dipyridyle	PAN	polyacrylonitryl
DMFA	dimethyl formamide	PAP	protective ability of the
DMPC	dimethyl phenyl carbinol		polymer support
DMSO	dimethyl sulfoxide	PB	polybutene
DPSP	dimethyl-*p*-styrylphosphine	PBD	polybutadiene
DVB	divinyl benzene	PBIm	polybenzimidazole

PBMVP	poly(butyl methacrylate-co-4-vinylpyridine)	PPhO	polyphenylene oxide
		PS	polystyrene
PC	phthalocyanine	PTFE	poly(tetrafluoro-ethylene)
PD	polymerization degree		
PE	polyethylene	PTFE-gr-PAA	polytetrafluoro-ethylene grafted polyacrylic acid
PE-gr-P4VP	polyethylene grafted poly(4-vinylpyridine)		
		PU	polyurethane
PE-gr-PAA	polyethylene grafted poly(allyl amine)	PVA	polyvinylamine
		PVAc	poly(vinyl acetate)
PE-gr-PAAc	polyethylene grafted poly(acrylic acid)	PVAl	polyvinylalcohol
		PVC	polyvinylchloride
PE-gr-PAAl	polyethylene grafted poly(allyl alcohol)	PVIm	poly(vinylimidazole)
		PVP	poly(vinylpyridine)
PE-gr-PDAA	polyethylene grafted poly(diallyl amine)	Py	pyridine
		SAS	surface-active substance
PE-gr-PMVK	polyethylene grafted poly(methyl vinyl ketone)		
		SCA	specific catalytic activity
PEG	polyethylene glycol	TBHP	tert-butyl hydroperoxide
PEI	polyethyleneimine		
PEO	polyethylene(oxide)	TCEPD	triple copolymer of ethylene, propylene and diene
PETP	polyethylene terephthalate		
		TETA	triethylenetetramine
PhA	phenyl acetylene	T_g	glass transition temperature
PHAr	polyheteroarylene		
PhFO	phenol–formaldehyde oligomers	THF	tetrahydrofuran
		TMC	titanium magnesium catalyst
PI	polyisoprene		
PMAA	poly(methylaceto-acetate)	TMED	tetramethyl ethylene diamine
PMAc	poly(methylacrylate)		
PP	polypropylene	TN	turnover number
PP-gr-PMAc	polypropylene grafted poly(methylacrylate)	TPP	tetraphenylporphyrin
		UDT	undecatriene
PPC	polymer–polymer composites	VAc	vinylacetate
		VCH	vinylcyclohexene
PPG	polypropylene glycol	VIm	vinyl imidazole
PPhA	poly(phenylacetylene)	VP	vinylpyridine

Introduction

The wish to combine the advantages and exclude the shortcomings of homo-geneous and heterogeneous catalysts has led to the creation of 'hybrid-phase' catalytic systems. They feature a substantial restriction of the mobility of the metal complexes (MX_n) by chemically binding them to a polymeric or mineral support.

The main principles of metal complex immobilization and the properties of immobilized catalysts were considered in my previous book [1]. However, until now the vast amount of data on the catalysis of numerous reactions by these systems has not been generalized in a systematic manner, although such attempts have been made many times, from the first reviews [2, 3] to several later books [4–9]. More-over, in spite of the wide consideration of features of catalysis by immobilized complexes in the literature, the fundamental problems of this field remained unanalyzed. A number of monographs on homogeneous catalysis include sections on immobilization of metal complexes (see, for example, [10 –12]).

For the last two decades homogeneous catalysts have been widely used in industry [13] (21 Mton/year of organic products are produced by means of these catalysts [14]). Nevertheless, heterogeneous catalysts are used substantially more widely because of their economic advantages. The cost of products, produced by catalytic processing of oil, is above $300 million [15], whereas the cost of catalysts is much lower (0.2% of the cost of products). The economic parameters of catalysis by noble metal complexes are substantially lower. For instance, although rhodium carbonyl has a high catalytic activity in hydroformylation (rhodium is used at concentrations of 10–50 ppm, or 0.001–0.005% [16], comparable with the con-centrations of Fe and Ni released into the reaction mixture from the walls of equip-ment), rhodium tends not to be used because of its high cost (rhodium is ~ 1000 times more expensive than cobalt); and also because of its shortage. Therefore, the loss of 1 mg of metal per kg of oxoproduct, which is admissible for cobalt, is very high for rhodium. In other fields of homogeneous catalysis the situation is even more complex.

It seems that this problem can be solved only by implantation of such complexes into firm, elastic granules, fibers, membranes and powders, that can be easily isolated from the reaction mixture and used many times in the same way as a typical heterogeneous catalyst. Catalysts of this type are very attractive from a technological viewpoint, because it is possible to predict the main properties of new catalytic systems on the basis of complexes immobilized on polymers, and to improve the selectivity and kinetic parameters of catalytic reactions by optimization of the chemical properties and structure of the macroligands. Finally, immobilized catalysts are universal with respect to the combination of the products obtained.

However, it is not only the prospect of valuable applications of these catalysts that is important. From a theoretical viewpoint it is important that for these catalysts it is possible to isolate, at any step of a catalytic reaction, different intermediates (products of metal complex activation, active centers formed in the course of reaction, products of association of active centers and deactivation of the catalyst), to study these intermediates by various physicochemical methods including methods of non-destructive control, and to monitor changes of content and structure of the surface layer in the course of the reaction *in situ*. In many cases this allows more details of the mechanisms of catalyzed reactions to be determined.

It should be mentioned that evolution transformations of the system catalyst-reaction medium is a rather complex problem even for relatively simple heterogeneous systems [17]. For immobilized catalysts the study of these transformations is complicated by the effect of reaction medium not only on the supported metal center, but also on the polymeric support. This effect causes a number of polymer transformations (swelling, destruction, cross-linking, etc.). In many cases this effect leads to the appearance of free complexes of transition metals; in some cases catalysis by these free complexes can disguise catalysis by immobilized complexes and complicates its study.

It is important that immobilized complexes can be considered as models of biological catalysts because they can carry out multicenter activation of a substrate; this type of activation is characteristic for metal enzyme catalysis [18, 19]. The energy of conformational strains in the macroligand acquired in the course of reaction is used for activation of the catalytic center and, finally, for acceleration of substrate transformation. In enzymatic catalysis this function, recuperation of energy, is carried out by protein molecules. Probably, immobilized catalysts will be very useful for a new scientific field, physicochemical modeling of natural processes. Methodological approaches developed in catalysis by immobilized complexes can be used for these purposes: consistent investigation of the content and structure of the coordination sphere of transition metal ions, ideas of specific spatial structure of the polymeric matrix with incorporated metal complex and the study of dynamics of changes of the system in the course of reaction.

Macrocomplexes, components of immobilized systems, are also used in electronics, medicine, ecology, as materials for electroconductive elements, devices for energy conversion and information storage, and spaceships and artificial satellites, as well as in processes of separation and hydrometallurgy [20].

This book can be considered as an attempt to generalize the most important, in my opinion, catalytic processes. Perhaps, at present, detailed analysis of all types of reactions is impossible. The book deals mainly with pioneering studies that resulted in new concepts and results that were important for applications. Special attention is paid to comparison of immobilized and free catalysts; for this purpose almost every section has a brief review of the modern state of the subject considered. The characteristics of catalysts caused by immobilization of metal complexes are also considered.

The book includes results obtained by myself and my colleagues during almost 20 years of investigation of catalysis by immobilized complexes at the Institute of Chemical Physics, Academy of Sciences of the USSR, in the Department of Polymers and Composites.

References

1. Pomogailo, A.D. (1988) *Polimernye Immobilizovannye Metallokompleksnye Katalizatory* (Polymeric Immobilized Metal Complex Catalysts). Moscow: Nauka.
2. Manassen, J. (1970) *Isr. J. Chem.*, **8**, 5; (1971) *Platinum Metals Rev.*, **15**, 142.
3. Michalska, Z.M. and Webster, D.E. (1974) *Platinum Metals Rev.*, **18**, 65.
4. Ermakov, Yu.I. and Zakharov, V.A. (1980) *Zakreplennye Kompleksy na Okisnykh Nositelyakh v Katalize* (Catalysis by Complexes Supported on Oxides). Novosibirsk: Nauka.
5. Lisichkin, G.V. and Yuffa, A.Ya. (1981) *Geterogennye Metallokompleksnye Katalizatory* (Heterogeneous Metal Complex Catalysts). Moscow: Khimiya.
6. Hodge, P. and Sherrington, P.C. (1983) *Polymer-Supported Reactions in Organic Systems*. Chichester: Wiley.
7. Kopylova, V.D. and Astanina, A.N. (1987) *Ionitnye Kompleksy v Katalize* (Catalysis by Ionite Complexes). Moscow: Khimiya.
8. Bekturov, E.A. and Kudaibergenov, S. (1988) *Kataliz Polimerami* (Catalysis by Polymers). Alma-Ata: Nauka.
9. Hartley, F.R. (1985) *Supported Metal Complexes. A New Generation of Catalysts*. Dodrecht: Reidel.
10. Dickson, R.S. (1985) *Homogeneous Catalysis with Compounds of Rhodium and Indium*. Dodrecht: Reidel.
11. Hegedus, L.S. (1987) *Catalyst Design Progress and Perspectives*. New York: Wiley.
12. Mortreaux, A. and Petit, F. (1988) *Industrial Application of Homogeneous Catalysis*. Dodrecht: Reidel.
13. Keim, W. (1984) *Fundamental Researches of Homogeneous Catalysis*, N.Y., **4**, 131.
14. (1988) *Platinum Metals Rev.*, **32**, 122.
15. Burke, D.P. (1979) *Chem. Week*, **3**, 28.
16. Imyanitov, N.S. (1976) *Sintezy na Osnove Okisi Ugleroda* (Synthesis on the Basis of Carbon Dioxide). Leningrad: Khimiya.
17. Rozovskii, A. Ya. (1988) *Katalizatory i Reaktsionnaya Sreda* (Catalysts and Reaction Medium). Moscow: Nauka.
18. Likhtenstein, G.I. (1988) *Chemical Physics of Redox Metalloenzyme Catalysis*. Heidelberg: Springer.
19. Bertini, I., Drago, R.S. and Luchinat, C. (1983) *The Coordination Chemistry of Metalloenzymes*. Dodrecht: Reidel.
20. Sherrington, P.C. and Hodge, P. (1988) *Syntheses and Separation Using Functional Polymers*. New York: Wiley.

CHAPTER 1

HYDROGENATION PROCESSES CATALYZED BY METAL-CONTAINING POLYMERS

Catalytic hydrogenation is a key reaction in organic chemistry. Therefore, both theoretical studies of the reaction stages and technological investigations are very important areas of research.

Recently the basic concepts of catalyst selection have been defined and also numerous problems of hydrogenation in solutions have been solved. This applies in equal measure to both homogeneous catalysis by metal complexes [1,2] and heterogeneous catalysis by group VIII transition metal ions [3]. Thus, data on the counteraction of both hydrogenated substrate and hydrogen with metal complexes, as well as on adsorption of unsaturated compounds, hydrogen and solvent molecules on the surface of highly dispersed metals, allow us to determine the main stages of the hydrogenation process and to develop reaction kinetics models. Moreover catalytic hydrogenation has been shown to be a complex process involving by sequence of redox counteractions involving metal complexes, supports, solvents, substrates, and reagents.

Heterogenization of homogeneous catalysts provides additional possibilities for varying catalytic activity and for selectivity regulation. It is often possible to determine, in detail, both the structure of the active centers participating and the mechanism of their formation. Such catalytic properties as oxidative state and coordinative number of transition metal ion, together with the ligands surrounding it, the geometry of the metal complex as well some side reactions (cluster formation, aggregation, submicrocrystal formation and their growth, etc.) are to a certain extent controlling features in catalysis by supported metal complexes.

Unlike other catalytic reactions, which will be considered below, the hydrogenation by immobilized metal complexes is more generalized. In the many reviews, papers and monographs listed in the Introduction one can find descriptions of the specific features of hydrogenation by immobilized complexes. Sometimes hydrogenation alone is used as a test reaction for comparison of the activities of homogeneous and heterogenized metal complexes. A number of supported metal catalysts have also found uses in industry.

A more complete summary of hydrogenation by immobilized metal complexes is presented in the monograph [4] and the preprint [5]. A generally similar approach to such investigations of catalysts is found in almost all original papers and reviews. This includes the following aspects: the preparation of the polymer support, the study of metal complex fixation and subsequent investigation of the mechanism of substrate hydrogenation in the presence of this catalyst. To some extent this approach makes it difficult to determine the contribution of the polymer nature of these catalysts to the catalytic properties and therefore hinders their optimization. It is also difficult to estimate the influence of some controlling factors, such as concentration ratios, temperature, solvent nature, etc. on the hydrogenation rate and selectivity.

The methodology employed below (without consideration of the nature of both the hydrated substrate and the fixed metal ion) allows us to concentrate on the more significant influences that control the catalytic properties of metal-containing polymers.

Such an approach has been found possible due to the availability of numerous and comparative experimental data on these hydrogenations by different classes of substrate although there still remain some gaps. The main problem is the difficulty of determining the individual contributions of the different factors on the reaction.

1.1 Principal Information Concerning the Mechanism of Homogeneous Hydrogenation

We will first consider the main features of catalytic hydrogenation.

This process is rather complicated and many questions have remained unanswered up to now. We will point out here only those problems that are important in gaining insight into specific features of immobilized catalysts. Detailed analyses of various aspects of catalytic hydrogenation both by homogeneous and heterogeneous catalysts can be found in specialized monographs [1–3,6].

The bond energy of H_2 is rather high, 437 kJ/mol [7]. Consequently activation of the hydrogen molecule in the intermediate complexes is necessary. Substrate activation may be the controlling reaction factor depending on the catalyst type, substrate nature, and reaction conditions. In the situation where the catalyst surface is filled with hydrogen molecules the value of the activation energy is calculated to be 21 kJ/mol [3]. When the adsorption ability of substrate increases, there are simultaneously molecules of H_2, substrate and solvent arranged on the catalyst surface (adsorption activation energy for H_2 molecules being 30–38 kJ/mol). A further rise of adsorption capacity of the substrate results in the reaction becoming limited by H_2 adsorption (46–63 kJ/mol). In some cases the interaction of substrate with catalyst may be such great in value (as it is for nitrobenzene) that it is accompanied by the formation of anion-radicals, which are capable of interaction with any form of hydrogen. The case of irreversible adsorption of substrate or reaction products leading to the limitation of the hydrogenation process is also possible.

A change in the value of hydrogen–surface bond energy, even by 4–8 kJ/mol, may lead to a significant change in hydrogenation selectivity. The specific energy values for heterogeneous catalysts of Pt-group metals are [3]: H_2 sorption heat (Q_{ads} 12–40 kJ/mol), H_2–surface bond energy (Q_{H-Ct} 220–260 kJ/mol), electronic work function (4.5–5.3 eV) and the maximum substrate sorption capacity (e.g. for allyl alcohol $(0.4–5.1) \times 10^{-10}$ g-eq/cm^2 at 293 K). The value of H_2 sorption heat and the bond energy with the surface are related by $Q_{ads} = Q_{H-H} + 2Q_{H-Ct}$.

Without detailed scrutiny of these processes let us notice only that for step-by-step consecutive reactions $A \xrightarrow{H_2, K_1} B \xrightarrow{H_2, K_2} C$ where K_1 and K_2 are the constant rate values; A – substrate; B and C are intermediate and end product, respectively. The selectivity coefficient, S, depends on both adsorption (b_1 and b_2 are adsorption coefficients for A and B, respectively) and kinetic parameters: $S = K_1 b_1 / K_2 b_2$.

Thus, if an intermediate containing double bonds is adsorbed less strongly than an initial substrate with triple bonds, then it will be displaced from the catalyst surface during the reaction, and selective hydrogenation of the unsaturated compound is observed. Otherwise the full reduction $A \xrightarrow{2H_2} C$ takes place.

Hydrogenation follows Rideal's mechanism via adsorption of H_2 on the catalyst surface which becomes fully occupied by the substrate (when reaction is not diffusion controlled) and proceeds on the catalyst surface which is energetically uniform for each substrate [6]. The reaction rate is controlled by the adsorption of H_2, the substrate or the solvent molecules, or their desorption rate, as well as the rate of interaction.

In homogeneous catalysis by transition metal complexes (MX_n) it is necessary to activate H_2 and/or substrate molecules to initiate the reaction. The hydrogen activation follows one of the mechanisms given below: ligand substitution which is not accompanied with M oxidation; homolytic cleavage of H_2 molecules and metal ion oxidation; oxidative addition of H_2 molecules to MX_n. Unsaturated substrate is activated via formation of olefin-hydride complexes:

$$MX_n \xrightarrow{\quad H_2 \quad} X_n MH_2 \xrightarrow{\quad \text{substrate} \quad} MX_n + \text{Product}$$
$$MX_n \xrightarrow{\quad \text{substrate} \quad} X_n M - \text{Substrate} \xrightarrow{\quad H_2 \quad} MX_n + \text{Product}$$

Scheme 1.1

For realization of the above reactions the transition metal ion must possess unsaturated valences. Sometimes these are present from the outset but more often they are formed during catalysis as a result of ligand dissociation or substitution by H_2 and/or substrate molecules.

At the present time a lot of evidence demonstrates the similarity of action of homogeneous and heterogeneous hydrogenation catalysts (e.g. poisoning with the same catalytic contaminants, the reduction of olefin hydration rate in the presence of acetylene strongly bound with the metal ion or conjugated diene substrates, compounds of M–H type being detected as intermediates, multigraded hydrogen addition, etc.).

It will be shown below that immobilization of metal complexes leads to the recognition of further similarities between homogeneous and heterogeneous catalysis.

1.2 Relationships between the Preparation Conditions for Metal-Containing Polymers and their Catalytic Properties

The metal complexes of the Pt-group and those of first transition series with polymers are often used in hydrogenation of olefins, dienes, alkyne and aromatics, as well in the reduction of different functional groups reduction. Both linear and cross-linked polymers, grafted macromolecules and different bifunctional compounds, which are the products of condensation, are used as macroligands.

1.2.1 Main routes for preparation of immobilized hydrogenation catalysts

Almost all types of macroligands, such as linear, soluble polymers with monodentate or chelate-type groups, three-dimensional polyligands and substances of mixed type (mineral-polymer), etc., are widely used as supports for hydrogenation catalysts.

Platinum-group metal ions both free and with ligands (acetylacetone (AcAc), benzonitrile, carbonyl, phosphyne, etc.), as well those of the transition metal series, are selective catalysts for hydrogenation. Among a number of important catalytic processes we shall consider particularly the followings [8]: hydrogenation of nitro groups, aromatics and heteroaromatics (e.g. phenol to cyclohexanone), semihydrogenation of fats and unsaturated fatty acids, saccharide hydrogenation to polyols, transformation of alkynes into *cis*- and *trans*-forms, etc.

The structure parameters of cross-linked polymer supports influence the hydrogenation process because of different factors. Specific surface area, pores sizes and distribution, surface selectivity in aromatic and aliphatic substances, the state of gel reaction, effect of solvation, diffusion limitation, etc., are the most important. A review [9] is devoted to these correlations.

The data in [10] illustrate the connection between the structure of the microenvironment and the reactivity of fixed complexes. The structure of chloromethylated or aminated styrene copolymer with divinylbenzene (CSDVB) with different amounts of cross-linking agent significantly affect the activity of fixed boron hydride in the hydrogenation of cinnamonic aldehyde into cinnamonic alcohol (Table 1.1).

The catalyst derived from the support **I** is a gel-like polymer, so the rate of hydrogenation is controlled by the rate of substrate diffusion through the gel. The reaction rate increases in the sequence of supports as IV≥III≥II and correlates with characteristics of their microstructure. In the case of **IV**, the reagent is fixed on the surface of the microspheres, so that the availability of active centers is high.

Here, there are some examples of highly active catalysts derived from soluble oligomer ligands. Thus, ClRh(PPh$_3$), a Wilkinson catalyst attached to ethylene oligomers, M_n~80–1000 (produced by anionic polymerization of ethylene or butadiene with the following traditional modification: bromination→LiPPh$_2$→ClRh(PPh$_3$)$_3$ [11–13]), is effective in the hydrogenation of olefins at 363–383 K. The same catalyst attached to the highly developed surface of phosphynated polyethylene (PE) monocrystall[1] (S_{sp}=200 m^2/g) [5] under similar conditions shows a 6-fold activity increase in comparison with the catalyst immobilized by cross-linked polystyrene (PS), which is easily swelled under the reaction conditions. Wilkinson's catalysts immobilized by macroporous PE fibers[2] (pore diameter 400 nm, 60% porosity) and treated as mentioned above also find good use [14]. Pore sizes in such three-dimentional fibers are rather large for reagent penetration and the process is not controlled by substrate diffusion rate as it is in the case of cross-linked PS granules where the diffusion of substrate into the polymer is hindered. The related rates of hydrogenation by catalysts on some supports are as follows:

Wilkinson catalyst without support   1.00
Wilkinson catalyst on PS   0.12
Wilkinson catalyst on PE oligomers   0.89
Wilkinson catalyst on PE crystal   0.75
Wilkinson catalyst on PE-spaced fibers   0.69

The hydrogenation rate decreases also with an increase in the size of the unsaturated compound (in going from linear to cyclic olefins). This correlation is more significant in the case of PE fibers than that for PS. The transport of a polar substrate (allyl alcohol) into the non-polar PS spheres is hindered and thus decreases the hydrogenation rate

---

[1] Monocrystals of PE are prepared from 0.1% HPPE xylol solution at 360 K with followed by corresponding treatment.

[2] Produced by Mitsubishi Rayon Co.

(Table 1.2). In unswollen PE supports it increases with the rise in the polarity of the medium. Additionally the latter catalysts easily eliminate reagents after the reaction.

Table 1.1 Characteristics of supports on the base of CSDVB [10].

Support	DVB (%)	v_{pores}, (cm^3/g)	S (m^2/g)	r_{pores}, (nm)
I	10	0.040	14	10^2–10^4
II	2.0	0.192	182	10^2–10^4
III	30	0.330	222	10^2–10^4
IV	40	0.392	357	10^2–10^4

Table 1.2 A comparison of related hydrogenation rates on different substrates by a Wilkinson catalyst, immobilized by supports [14].

Substrate	Wilkinson catalyst without support	PS-catalyst	PE (crystal) catalyst	PE (spaced fibers)-catalyst
Cyclohexene	1.00	1.00	1.00	1.00
1-Dodecene	1.23	0.60	0.95	0.91
Cyclododecene	1.29	0.23	0.78	0.74
Allyl alcohol	1.09	0.62	0.92	0.84
Cholesterole	1.20		0.68	0.58

Table 1.3 Hexene hydrogenation in the presence of $ClRh(PPh_3)_3$ heterogenized on irradiated-grafted copolymers [17].

Polymer support	Grafting degree	Hexene transformation extent
PVCl-gr-poly(styryl)-diphenylphosphyne	23	3
PP-gr-PS, phosphynated	50	88
CSDVB (2% DVB)-gr-poly(styryl)diphenylphosphyne	16	25
CSDVB (2% DVB)	0	0

Thus, the replacement of reactive centers in the outermost layer of a polymer support may give some advantages. An effective approach to such catalyst construction is the grafting polymerization of monomers with different functional groups to the surface of inert polymer supports such as PE, polypropylene (PP), PS, CSDVB, etc. The main features of gas-phase grafting polymerization, induced by ionizing γ-irradiation of ^{60}Co, accelerated electrons, low-temperature gas discharge plasma, etc. are listed in [15, 16]. To prepare the hydrogenation catalysts, o-, m- and p-styryldiphenylphosphyne, vinyldiphenylphosphyne were grafted to the surface of PP films or powder followed by immobilization of Rh^+ complexes [17–19]. The high level of activity of these catalysts is caused by the occurrence of reactions within a thin layer of the catalyst[3]. These catalytic properties are dependent on a number of parameters: the nature of the polymer layer, the support, the grafted monomer, and the grating degree (Table 1.3).

The shape of the support particles (pellet, powder, film, fiber) also significantly affects the catalytic properties. Diphenylphosphyne–palladium complexes fixed on carbonic filaments show a high level of hydrogenation activity [21]. Catalysts fixed on polymer powders with large surface areas possess high levels of activity. For example, palladium complexes fixed on nylon-66 powder are more active than those immobilized by fibers [22]. Palladium complexes fixed on PE-gr-polyacrylic acid (PAAc) exhibit a 2-fold activity rise in comparison with those fixed on polyamide fibers [23]. Apart from reasons of geometry, these effects may also be connected with a different mechanism of complex bonding. Specifically, if metal ions in parent complexes are in the Pd^{2+} state[4], immobilization on polymer powder may be accompanied by Pd^0 formation [22].

It has already been shown that catalyst porosity and the value of specific surface significantly affect the catalyst activity. The S_{sp} value of CSDVB is a function of the amount of cross-linking agent used. Even simple grinding of CSDVB granules (20% DVB) with fixed titanocene [24] leads to a many-fold rise in catalyst activity (Table 1.4).

Apart from CSDVB, PE, PP and their modifications, practically all known functionalized macroligands are used as polymer supports [25]. Active centers may be either attached to backbone of polymer chain (in the case of coordinative polymers being located in the chain) or bonded to polymer molecule side groups.

We will consider here only those that are most often used in hydrogenation.

Polycarboxylate matrices (including cross-linked materials) are widely used as supports [26], in particular for the fixation of $RuH_2(PPh_3)_4$ heterogenized derivatives [27]. Rhodium complexes with amino acids fixed on anionic-exchange resins possess the same activity in hydrocarbon hydrogenations as the corresponding homogeneous catalysts [28]. Complexes of $PdCl_2$ with homo- and copolymers of 2- and 4-vinyl-pyridine (reduced by $NaBH_4$) are active in hydrogenation of allyl alcohol and dimethylethynylcarbinol [29,30]. Complexes of Pt-group metal ions with anionites are widely used in hydrogenation both of individual compounds [31] and of the C_4 acetylene fraction [32]. Metal complexes with heterochain polyethyleneimine (PEI), polydentate polymer ligand, after preliminary reduction with hydrogen, were shown to be active in the hydrogenation of cyclohexene, cyclohexadiene, etc., (e.g. see [33]).

The descending order of activity of the following metals is: Pd>Rh>Ru>Co>Ni. For immobilized Ru complexes the kinetics of the hydrogenation reaction exhibit zero order behaviour with respect to substrate concentration, and is first order in hydrogen, and is a linear function of complex concentration; E_a=29.3 kJ/mol. The product of the reaction of $PdCl_2$ with the other heterocyclic polymer, oligomer PEG (M_n=200–600), catalyzes phenyl- and diphenylacetylene hydrogenation to *cis*-alkenes (in CH_2Cl_2 at 383 K) [34].

The products of polycondensation of for example, polyamides, which possess enhanced thermal stability, are also used as macroligands. Thus, Pd/poly-*p*-phenyl-

---

[3] Catalysts of this type are crust-like substances. In particular the Pd crust thickness which coveres the support pellets is 0.1–0.4 mm; it effectively ctalyzes hydrogenation of high molecular mass oils [20].

[4] The fast oxidation of Pd^0 to Pd^{2+} promoted by CuI used as a stabilizer of polymer fibers may be responsible for this process.

terphthalamide [35], S_{sp}=50 m^2/g shows activity and selectivity in hydrogenation of unsaturated acids chloroanhydrides to the respective aldehydes.

Complexes of Pd^0 immobilized on polyamides are active and selective catalysts for both olefin hydrogenations and nitro-compound reductions [36]. Hydrogenations of aliphatic, cyclic and aromatic alkenes and alkynes are catalyzed by Pd^0- and Pt^0-coordinative polymers derived from 4,4′-isocyan biphenyl. The latter show higher activity than polymers incorporating Pd^{2+} and Pt^{2+} complexes [37]. Polybenzimidazole (PBI), sulfonated PBI, and CSDVB modified with benzimidazole groups, are the reagents for preparation of Pd^0-containing catalysts. These catalysts are used for selective reduction both of nitro-compounds and alkenes with keto and aldehyde groups [39]. Palladium-containing catalysts of benzimidazole row polyheteroarylenes are promising products due to the high thermal and chemical stability of the supports [39–41]. PdCl$_2$ adsorbed by PBI is reduced to PBI–Pd^0 by hydrogen, producing anchored Pd clusters for which the hydrogenation activity decreases in the sequence: dienes>allyl alcohol>alkene-1>alkene-2>2-methylalkene-2. The sorption ability of substrates follows the same order. The polymer support probably has an additional effect on surface Pd atoms through moderation of the metal-substrate bond strength.

Table 1.4 Initial rates of olefin hydrogenation, w_0 (ml of H$_2$/mmol of Ti), by cyclopentadiene (Cp) Ti^{4+} compounds.

Catalyst	Substrate		
	Hexene	Cyclohexene	Cyclooctene
CpTiCl$_3$	2.9	2.0	0.5
CSDVB-CpTiCpCl$_2$ (granules)	42	30	7.8
CSDVB-CpTiCpCl$_2$ (powder)	390	194	60.5

Another effect is used in gel-immobilized hydrogenation catalysts. Thus, Pd^{2+} immobilized by a triple copolymer of ethylene, propylene and uncojugated diene (norbornene ethylidene) with grafted poly-4-vinylpyridine (P4VP) shows both high selectivity and activity in cyclopentadiene hydrogenation to cyclopentene in benzene and benzene ethanol mixtures [42].

Special attention is paid to supports prepared from chelate-like polymers that lead to strong bonding of MX$_n$. Polymers with pentane 2,4-dione groups [43], ethylenediimine [44], dipyridyl (Dipy) [45], 8-aminoquinoline [46], iminodiacetate [47], etc., are often used for these purposes. Attached metal complexes, MX$_n$, are Pd(OAcAc)$_2$, PdCl$_2$, RhCl(PPh$_3$)$_3$, [RhCl(CO)]$_2$, MCl$_2$(PhCN)$_2$ (M=Pt or Pd), etc. A complete analysis of the subject can be found in the monograph [48]. These catalysts are stable, the recycling number in hydrogenation reach 300–1000 or more.

Organosilica polymers are also used for these purposes [49–51].

In homogeneous catalysis, rhodium complexes with unusual ligands such as η^5-C$_2$B$_9$H$_{11}$$^{2-}$ are used for hydrogenation. Similar polymer complexes produced by copolymerization of methylmethacrylate (MMA) with 1-isopropenyl-(3)-1,2-dicarbaundecaborate salts followed by counteraction with $RhCl(PPh_3)_3$ [52] turned out to be effective catalysts for olefin and diene hydrogenations. These complexes were also immobilized by supports derived from styrene, divinylbenzene and 4-(1'-o-carboranyl)styrene via the p-phenylene group [53]. Immobilization of Rh, Ir, and Ru complexes by polymer supports carrying CHB_9H_9 groups [54] as well dicarbaundecaborate-containing polyamides [55] are also used for the preparation of hydrogenation catalysts.

The broad range of macroligands used is much wider than reviewed above and includes such polymers as functionalized cellulose derivatives, oligo-   and polysaccharides, gelatin, etc. Thus, Pd and Rh complexes, bonded to modified cellulose containing nitrogen and phosphorus atoms, show a hydrogenation activity comparable with that of analogous homogeneous catalysts [56]. $PdCl_2$ complexes immobilized by Indian silk exhibit high activity levels in the reduction of isoprene, hexyne-1, cyclopentadiene, and nitrobenzene [57,58]. The reduced Pd^0 atoms coordinated with protein donor groups ($-NH_2$, $-OH$, $-CONH-$, $COOH$) are also active centers of hydrogenation. Metal atoms are retained in the micelles of silk fibers after reduction. Such a catalyst may be recovered a few tens of times without the loss of activity. Effective catalysts may also be prepared by fixation of group VIII metal complexes on chitin and chitozane granules or powder [57].

In practice almost all metals of the transition series, immobilized on polymers, as well as the f-orbital elements, such as the lanthanum series [59] and uranium compounds [60] are used in hydrogenation.

Immobilized catalysts of the Ziegler–Natta type also find their use although on a limited scale. Complexes of Ti, Zr, Hf, V, Mo, Co, Ni, Cr, etc., immobilized on polymers in combination with organometallic compounds (usually organoaluminium) are active catalysts for unsaturated substrate reductions. The earliest investigations are generalized in monograph [61] and more recent coverage is to be found in reviews [62,63]. Nickel catalysts[5] are among the most active. They provide selective diene hydrogenations to monoenes. Thus, the heterogenized complex $Ni(AcAc)_2$–$Al_2Et_3Cl_3$–phosphinated polystyrene in molar ratio $Ni/Al/P=1:10:5$ leads to selective transformation of 1,4-cyclohexadiene into cyclohexene under mild conditions (313 K, hydrogen pressure 0.1 MPa, ethylbenzene solvent) [66]. Ni^{2+} naphthenehydroxamates, hydroxamates, stearates, benzoates, and naphthenates, immobilized on macroligands with π-donor ability, such as polyacetylene (PA), are active catalysts for phenylacetylene (PhA) hydrogenation [61]. The interaction of transition metal ions with PA leads, probably, to unpairing of PA electrons therby increasing the free radical concentration, and as a consequence there is partial reduction of the transition metal at the stage of its bonding with the support[6]. This occurs also in the Fe^{3+}/Fe^{2+} reduction on PA. The transformations of hydrated substrates on such catalysts probably pass through a

---

[5] Catalysts of this type are widely used fixed on inorganic supports. For example, the effective catalyst of alkenes hydrogenation [64] is prepared by counteraction of Ni(butylsalycylaldiminate)$_2$ with Al_2O_3/SiO_2 followed by treatment with $AlEt_3$. Some bound organometallic compounds can also be useful for' this purpose. Particularly supported organoactinide complexes ($CpTh(benzyl)_3/\gamma$-Al_2O_3) are effective in benzene and p-xylene hydrogenation at 90°C [65].

[6] The use of conducting polymers (polythiophen, poly-p-phenylene) as metal-free catalysts for alkene (octene-1) cis-trans-isomerization is known [67].

dissotiative chemosorption stage [68]. The processes proceeding on organometal catalysts are similar to those observed on corresponding metal blacks.

The formation of Zr–H bonds active in olefin transformations has been revealed [69] in the reaction of $NaAlH_2(OCH_2CH_2OOCH_3)_2$ with $CpZrCl_3$ fixed on CSDVB modified with Cp groups; their existence was also assumed for organozirconium compounds [70].

Homo- and copolymers produced from metal-containing monomers have found applications as hydrogenation catalysts [71]. The strong cross-linking PS-Dipy-Pd catalyst produced by polymerization of palladium-containing dipyridyl monomer [72] should be noted in addition to the above ruthenium acrylate [35].

The methods of immobilization of highly despersive metal catalysts on activated carbon with the use of transparent polymer – gel-like membranes, including thermal decomposition of cellulose-nickel hydrogel membranes have been elaborated elsewhere [73] (see section 1.9.3).

1.2.2 Hydrogenation catalysts bonded to mixed-type supports

New types of hydrogenation catalysts, i.e. metal complexes immobilized on supports of mixed polymer-oxides nature, have been prepared. These are produced, as a rule, by grafting polymerization of different monomers to the surface of oxides, SiO_2, MgO, Al_2O_3, etc., polymerization of monomers in the presence of oxides, polymer-analogous transformations of grafted polymer cover, and so on.

For illustration let us consider copolymerization of methacrylate with a mixture of m- and p-DVB in the presence of SiO_2. As a result a cross-linked support SiO_2-polymethacrylate is produced. Following fixation of $PdCl_2 \cdot 2H_2O$ or H_2PtCl_6 leads to immobilized Pd^{2+} and Pt^{2+} complex formation, and they reveal high levels of activity in hydrogenation of unsaturated hydrocarbons, nitro compounds, aldehydes and ketones. The activity maximum is observed at a [COOH]/[Pd] ratio equal to 10, Pt-containing macrocomplexes being the more active with a rather high turnover number (above 2000) (Figure 1.1). Polymer-coated silica with anchored palladium complex (obtained by copolymerization of 4-vinylpyridine with DVB in the presence of SiO_2) is effective in alkene and alkyne hydrogenations [74]. Catalysts based on mixed supports are usually more active and stable than the metal-containing polymer. Thus, Rh complexes polyvinylchloride modified with $-NMe_2$ groups supported on SiO_2 particles show higher activity in reduction of benzene and alkenes under mild conditions [75] than similar complexes with polyamides. The maximum rates of hexene-1, heptene-1, decene-1, cyclohexene, styrene and toluene hydrogenations (at 303 K, H_2 pressure 0.1 MPa and N/Ph=2) are 48.0; 26.1; 9.2; 44.1; 10.1 and 3.2 H_2 ml/min respectively.

The composite PEI/SiO_2 derived by PEI sorption on SiO_2 surface effectively binds Pd^{2+} ions which are readily reduced by $NaBH_4$ to fixed Pd^0 atoms and show a high activity for the hydrogenation of nitrobenzene in CH_3OH [76]. Such catalysts show high stability, are not flammable in air, and can be decanted from the solution for multiple recycling. Heterogenized palladium catalysts for the hydrogenation of conjugated dienes and acetylenes to olefins have been produced by fixation of prearranged active complexes by supports [77], e.g. by reduction with $Al(iso$-$Bu)_2H$ of $PdCl_2$ complexes with primary amines. It is appropriate to mention the series of catalysts formed through $PdCl_2$ immobilized on covered by poly-γ-(aminoethyl-amine)propylsiloxane SiO_2, Al_2O_3, Si–Mg sorbents, etc. [78]. Their activity diminishes in the sequence $Al_2O_3 > Si-Mg > SiO_2$. A variety of the method is based on photoinduced substitution or labilization of CO groups in $ArCr(CO)_3$ by silica-containing polymer with isonitrile ligands, which is the product of 3-isocyanpropyl-triethoxysilane condensation with SiO_2 (boiling in solution) [79].

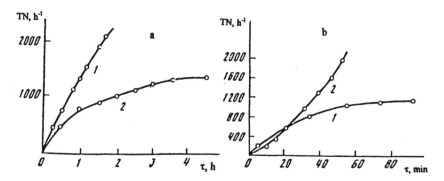

Figure 1.1 The dependence of turnover number (TN) on time of hydrogenation of nitrobenzene (a) and heptene-1 (b) by catalysts SiO_2-PMAc-Pd (1) and SiO_2-PMAc-Pt (2) ([COOH]/[M]=10; [catalyst]=0.02 mmol; [substrate]=0.05 mol).

The catalysts immobilized on flaky compounds, e.g. GFC, are also within the scope of this consideration. $[(1,5\text{-}COD)(PPh_3)_2]^+$ fixed between layers of fluoride tetrasilica cellulose [80] reveals a 4-fold rise in selectivity in 4-vinylcyclohexene hydrogenation compared with the homogeneous analogues. Another example is the catalyst of hydrogenation or hydrocracking, which involves the transition metal complexes with oligomer ligands as an interlayer binder of smectic structure (montmorillonite, hectorite, fluorihectorite or their mixture). The specific surface for such a catalyst is about 200–700 m^2/g. It is important that smectite-based catalysts retain their structure, lattice constant, 1.7–2.5 nm, and interlayer distance, 10–35 nm.

The combination of polymer and inorganic support has advantages in catalysts of mixed-type and leads usually to improvement of the catalytic properties of fixed metal complexes.

1.2.3 The influence of the conditions of synthesis on the catalytic properties of the metal complexes formed

Let us consider in outline the influences concentration, immobilization duration and reaction temperature, together with the solvent character on the activity of prepared catalysts. Depending on the MX_n concentration initially considering the ratio $[M_0]/[L_0]$, where $[L_0]$ is the concentration of polymer functional groups, catalysts of different activities may be formed. Such an effect can be seen more clearly in the example of phosphynated CSDVB. It has been reported [81] that an excess of $[CODRhCl]_2$ leads to formation of dimer-like Rh^+ complexes having little activity for hydrogenation which, however, can be activated by PPh_3 addition according to the scheme:

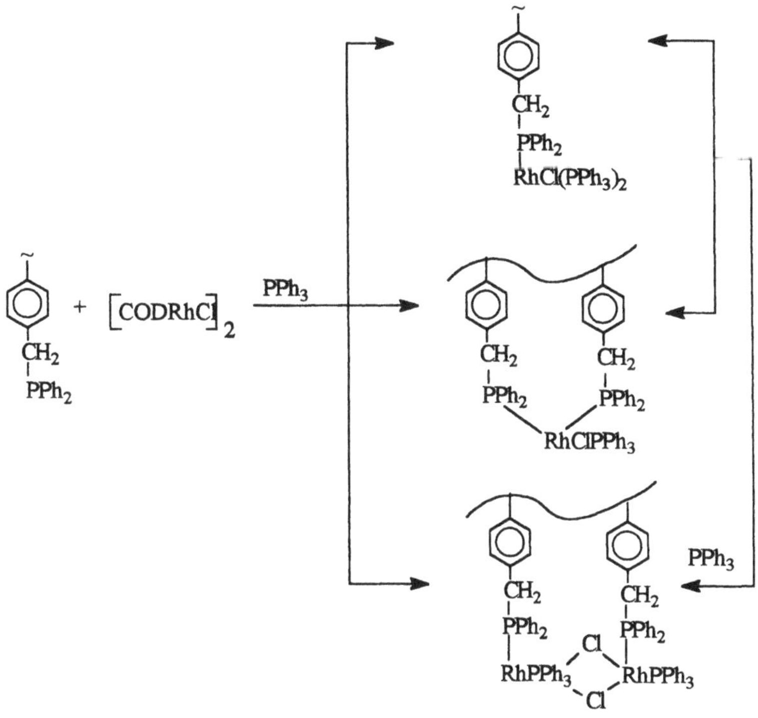

Scheme 1.2

The formation of dimer complexes has been confirmed by X-ray analysis [82] for MX_n immobilized on the polymers with low extents of cross-linking (up to 2% DVB) and flexible polymer chains. Functional groups in strongly cross-linked polymers (20% DVB) show less mobility, and thus prevent dimer complex formation. It is very probable that rhodium fixation by more than two phosphyne groups of macroporous PS support is hindered due to steric reasons [83]. Unsaturated valences of MX_n are necessary for H_2 or substrate activation. Consequently an excess of functional groups hinders their formation thereby decreasing the hydrogenation rate. There is an optimal ratio [81] $[P]/[Rh] \approx 2$ (Figure 1.2) characteristic both for homogeneous complexes and those heterogenized on linear [84], low-linked [81] and strongly cross-linked (25% DVB) [84] phosphynated PS. To demonstrate, RhL_3Cl is bonded to macroporous phosphynated CSDVB (2% DVB) by two $-PPh_2-$ groups of the polymer providing labile groups [85]. RuL_3Cl_2 shows a tendency to coordinate three PPh_2 groups of the polymer matrix, the product being active if $[P]/[Ru] < 3$ and there are no free $-PPh_2-$ groups. Moreover, in strongly cross-linked polymers there are also opportunities for bonding unsaturated compounds. This trend has been demonstrated [86] by the example of $[CODRhCl]_2$ immobilized by CSDVB.

The nature of the macroligand has an impact on the processes of monomerization of dimer complexes and dimerization of monomer complexes. Thus, polymeric amines and thiols interact with the dimer $[Rh(CO)_2Cl]_2$ without its cleavage [87].

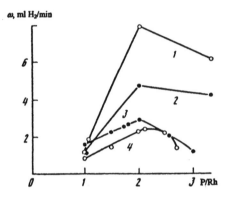

Figure 1.2 The dependence of hexene-1 (*1,4*) and cyclohexene (*2,3*) hydrogenation rates by rhodium complexes with triphenylphosphyne (*1,2*) and by its heterogenized analogues (*3,4*) on the ratio [P]/[Rh].

Concentration ratios can not only affect the constitution and the structure of immobilized complexes, but also affect their topochemistry. For example, at low MX_n concentrations fixed complexes are arranged only on the surface or in a thin surface layer of weakly linked phosphynated PS [88]. The excess of MX_n and prolonged reaction time leads to complex fixation throughout the whole polymer. The increase in counteraction of polyamide-66 suspension in H_2PtCl_6 solution causes a linear rise of platinum content in the polymer. The temperature rise also favors the internal penetration of metal ions into the polymer [89]. It stands to reason also that the solvent character affects metal complex fixation because it is responsible for the flexibility of chains and the range of action of the reagents. In solvents that do not result in polymer swelling, complex fixation proceeds only in the surface layer of the support and this determinates their higher specific catalytic activity (SCA).

Among the numerous examples which corroborate this conclusion we shall pick up the following [90]. Pd-containing strongly cross-linked phosphynated PSs show extremely high activity in hydrogenation of aromatics and reduction of different functional groups of the complex substrate. This is because over 91% of catalytic metal particles are on the surface of the support. The dependence of polymer swelling ability on solvent nature was used for the development of the physical (caging) method for catalyst heterogenization in the polymer matrix. Some part of the metal complex passing from solution into the polymer matrix became fixed there due to contraction of the polymer network under the action of another solvent. The procedure does not decrease catalyst activity but increases its selectivity with respect to the size of the reagents. For instance hexene-1 hydrogenation has been observed on Rh complexes incorporated into cross-linked (1%) PS [91].

1.2.4 Formation of active centers in immobilized catalysts

Sometimes immobilized complexes need no additional activation for hydrogenation reactions. But more frequently they are activated by reduction (as a rule to a zero valent state) with hydrogen, sodium borohydride or hydrazine, formaldehyde as well as organolithium, -magnesium or -aluminium compounds. The method of formation of the metal particles influences their size and as a consequence the catalytic properties of the polymer catalyst. Thus, Pd^0 particles on macroporous ion-exchange resin, Amberlist-15, may be produced [92] by reduction of fixed $Pd(NH_3)_4^{2+}$ either by heating in a vacuum followed by partial volatilization of the ammonia produced or by H_2 treatment at elevated temperatures (>353 K) with complete ammonia volatilization. In the latter case stable Pd^0 particles are located within the inner sorbent surface. Both the size (4–11 μm) and distribution of the particles formed are dependent on the engaged sorbent capacity. The catalyst shows high activity levels in ethylene hydrogenation and is resistant to heating. Larger particles are also formed and some palladium hydride is detected during reduction with formaldehyde.

It is important that activation processes are not usually accompanied by rupture of metal-polymer bonds. This is especially characteristic of macromolecular metal chelates. Thus, the reduction of Rh^{3+} fixed on polyamide modified with dipyridyl groups provides Rh^+ ions according to the scheme [93]:

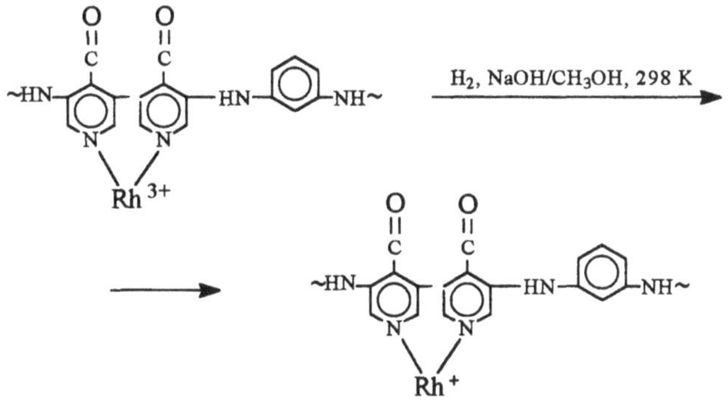

Scheme 1.3

This reaction proceeds in two steps. The first stage leads to the formation of Rh^+ and the second is accompanied by autocatalytic Rh^{3+} reduction with participation of the formed Rh^+ ions formed according to the equation:

$$d[Rh^+]/dt = K[Rh^{3+}]^2[Rh^+]$$

The immobilized Rh^+ ions are stable in H_2 atmosphere. A similar route for the reduction was observed for Pd^{2+} ions [94]:

$$■\text{-Dipy} + Pd(OCOCH_3)_2 \longrightarrow ■\text{-Dipy}\cdot Pd(OCOCH_3)_2 \xrightarrow{LiAlH_4} ■\text{-Dipy}\cdot Pd^0$$

where ■– represents the rest of polymer.

An interesting sequence of transformations should also be mentioned in this respect [95]. This includes a Pd^{2+} coupling reaction with toluene diisocyanate followed by reduction of the palladium complex formed to Pd^0.

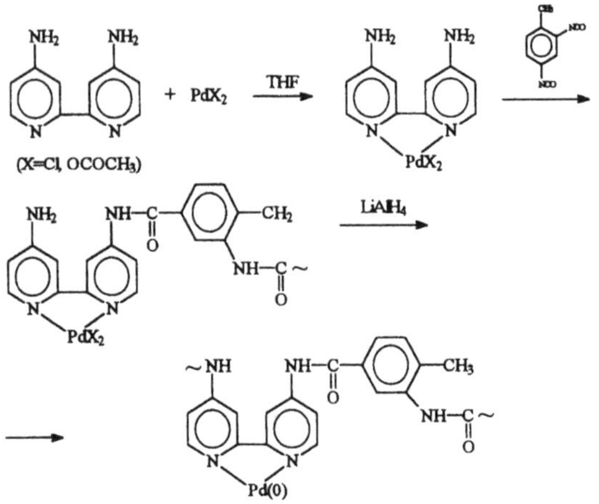

Scheme 1.4

There are many such examples. However, in the case of monodentate metal complex reduction, for example the reduction of Pd^{2+} complexes fixed on poly-2-vinylpyridine (P2VP) by $NaBH_4$, the polymer support contains both bonded Pd^0 atoms and free pyridine groups [30] after the reaction is completed. Special studies of Pd losses in the hydrogenation of 1,5,9-cyclododecatriene (100°C, hydrogen pressure 1.96 MPa) by polymer-immobilized Pd catalysts have been carried out [96]. The Pd loss depends on the type of polymer support (functional groups and even pore size). The lowest palladium losses were recorded in the case of palladium chelation with the polymer.

1.3 Composition, Structure and Activity of Immobilized Metal Complexes

The catalytic properties of immobilized metal complexes are controlled by a number of factors, such as the nature and distribution of attached transition metal ions and the character of the polymer support, together with unreacted functional groups of the polymer after fixation and activation of the metal complexes.

The nature of the transition metal affects the catalytic activity of immobilized catalysts in a similar way to that found in homogeneous and heterogeneous catalysis. In some cases it is possible to carry out a comparative analysis of activities for such systems. The effect of macroligands, if you neglect the macrochain mobility and polymer swelling, is not direct and is transmitted either by the bound metal complexes or by neighboring free functional groups (L) of the polymer. The functional groups of the polymer support may not only take part in MX_n binding but also influence the catalytic reaction by promoting a rise in the local concentration of active centers due to side reactions or by the protection of active centers from the action of impurities or catalytic poisons.

Let us consequently examine the role of L, the nature of the polymer support and the influence of immobilized complex ligands, MX_n, in the catalytic hydrogenation reaction.

1.3.1 The role of the nature of polymer support functional groups

As it was mentioned above (see Section 1.2.4), the stability of the M^0 metal particles produced and the strength of their bonds with the polymer support after the preliminary metal complex activation with reducing reagents, including H_2, depends to a great extent on the nature of both L and the ligand dentate number. Catalytic properties also depend on the stability of coordinate centers in the polymer particles. Numerous examples of such correlations are known. Thus, immobilized by functionalized PE, palladium complexes show different activities and selectivity in hydrogenation of p-nitrochlorobenzene and under comparable conditions are dependent on the character of the ligand (Table 1.5).

Table 1.5 Hydrogenation of p-nitrochlorobenzene by complexes immobilized with grafted PE[a] [97].

Grafting monomer	$[L_0]$, (mol/g) $\times 10^4$	$[Pd_0]$, (mol/g) $\times 10^4$	SCA^b	Reaction products, %[c]			
				Nitro-benzene	Aniline	p-Amino-chloro-benzene	Phenyl-hydroxyl-amine
Vinyl-imidazole	15.3	0.76	1.27	7.5	2.8	1.4	–
Allylamine	3.6	0.70	6.41	4.3	49.5	11.1	–
4-Vinyl-pyridine	12.6	2.26	5.10	4.6	60.7	10.1	–
Vinyl-carbazole	10.1	0.20	2.09	4.4	26.9	8.5	–
Vinyl-pyrrolidone	19.0	0.95	11.31	7.4	63.4	13.0	–
Acrylonitrile	12.0	0.91	4.01	2.0	53.0	18.0	3.0

[a] Reaction conditions: P_{H_2} =0.1 MPa, 323 K, 10 ml of 2% $NaBH_4$ in isopropanol, 1 mmol p-nitrochlorobenzene, [Pd]=$1 \cdot 10^{-5}$ mol.

[b] SCA: units mol of H_2/(mol of Pd min).

[c] The remainder is unreacted substrate.

In the general case SCA correlates with the effective complex formation rate constant, K_0 (Figure 1.3) and increases with the rise of the share of the bonded functional groups of the ligand, f. The metal bond strength with the polymer support probably also determines the effect of catalyst working off during hydrogenation reaction. The different stabilities of Pd and Pt complexes with PVP and PMAc is most probably the reason for some specific features of the heptene-1 and/or nitrobenzene hydrogenations (Figure 1.4). The gradual formation of Pd^0 or Pt^0 occurs during the reaction and the higher mobility of carboxyl complexes than for vinylpyridine results in comparatively higher levels of activity. The different activities of Rh-containing carboxylate polymers in hydrogenation and isomerization of pentene-1 (Figure 1.5) is the result [98] of an unequal stabilizing effect by the macroligand on the coordinative centers $Rh^{2+} \cdot O_2^-$ that are supposed to be responsible for the catalytic reactions. But the presence of such complexes in reductive media seems to be questionable as Rh^{2+} complexes in a H_2 atmosphere are expected to give Rh^{3+} hydride [1]:

$$L_nRh^{2+} \xrightarrow{H_2} 2HRh^{3+}L_n$$

An increase in the distance between the polymer backbone and L significantly influences both the MX_n bonding process and the activity of the active centers so formed.

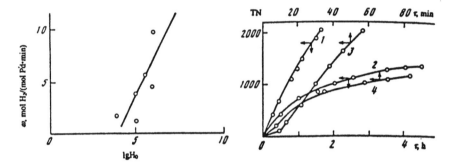

Figure 1.3 The dependence of immobilized palladium complex activity in p-nitrochlorobenzene hydrogenation on the lg K_0.

Figure 1.4   The influence of properties of fixed metal complexes on their activity in hydrogenation of nitrobenzene (1,2) and heptene-1 (3,4). Metal-containing polymers: 1, Pd-PMAc; 2, Pd–PVP; 3, Pt-PMAc; 4, Pt–PVP, COOH/M=10, N/M=18, [substrate]=0.05 mol, [catalyst]=0.02 mmol.

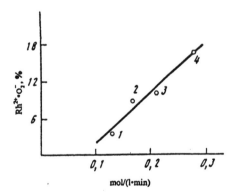

Figure 1.5 The dependence of the initial rates of both hydrogenation and isomerization of pentene-1 on the content of rhodium triphenylphosphyne complex, immobilized on polycarboxylate matrices. 1, polyacrylic acid (PAAc); 2, copolymer of vinylbenzyl ester with maleic acid; 3, tercopolymer of vinyl alcohol, vinylbenzene ester and maleic acid; 4, copolymer of ethylene and maleic acid. Solvent, benzene; P_{H_2} =0.1 MPa, [pentene-1]=0.83 mol, [pentene-1]/[Rh]=365.

The last is the result of an increase in active center mobility facilitating the interaction with substrates. Thus, the SCA of a Rh catalyst in the hydrogenation of octene-1 is close to that for a homogeneous catalysts [99]. This can be attributed to the presence of a rather long "anchor"-like chain between the backbone and the fixed reactive site.

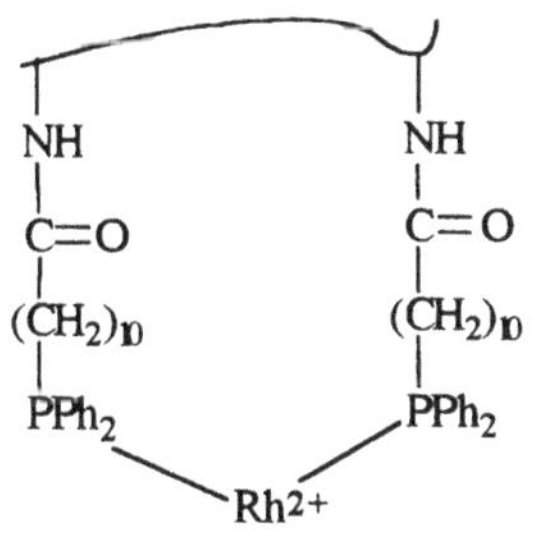

Scheme 1.5

The activity of palladium-containing aminated CSDVB in hydrogenate and/or hydrogenizing amination rises with an increase in the polymer amine fragment length [23].

A wide variety of functional groups of ionites allows one to reduce the influence of L on the catalytic properties of immobilized metal complexes. Thus, the activity of palladium catalysts immobilized on ion-exchange resins decreases in the sequence [100]: strong basic anionite $-N^+Me_3$>strong acid cationite $-SO_3$ or weak basic anionite$-NM_2J \approx$ weak acid cationite $-COOH$> chelate resin $-N(CH_2COO)_2$. The reduction of $PdCl_4^{2-}$ attached to such supports results in partial Pd^0 formation, the particles being coordinated with N^+Me_3. Complexes of different stabilities are probably formed including those of high stability, incapable of generating unsaturated vacancies for reagent activation even under treatment with activated agents.

The mechanism of the influence of ligand properties on the catalytic process is rather varied. One of the results is the influence on the rate of active center formation and their concentration. Another feature has been observed in hydrogenization processes. This is discussed in detail in Section 2.3.3 and concerns cyclohexene oxidation by Co^{2+} complexes with carboxyl-containing polymer matrices: the vacant carboxylic acid groups of the support are capable of absorbing reagents to high concentrations by, for example, hydrogen bonding. A similar effect has been observed in styrene hydrogenation by Pd, immobilized on functionalized CSDVB (4–6% DVB) [101]. It has been noticed that the reaction rate increases for supports with functional groups capable of π-complexes formation with styrene, thus increasing the lifetime of the absorbed substrate complexes. In allylacrylate hydrogenation, the catalyst activity correlates with both π-acceptor and polar properties of functional groups of the polymer matrix. The groups capable of associating with substrates are $-NO_2$, and $-OMe$; others may affect the substrate behavior by polar factors. Polymers with a hydrophobic surface cause a decrease in the hydrogenation rate of non-polar substrates. Even a small amount (7%) of introduced 2,4-dinitrophenyl groups leads to a significant fall in the 4-vinylcyclohexene hydrogenation rate.

The rather high activity of Rh^+ complexes immobilized on phosphynated CSDVB in hydrogenation of polynuclear heteroaromatics may be caused by enhanced substrate concentration near the metal site together with the promotion of spatial and electronic features. In the case of strongly cross-linked polymer supports such support-substrate interactions play a meaningful role. The number of available groups in these catalysts is often significantly higher than that taking part in MX_n bonding. Moreover strong adducts of L and exchanged polymer ions may form on the surface of ionites with high exchange capacity. Basically these effects are connected with strong-weak acid-base interactions.

Interaction of L with components of the reaction mixture is often accompanied by different processes, such as protonation, ionization, macroligand alkylation and sometimes quaternization, oxidation or reduction. For example, in systems with fixed palladium phosphyne complexes the catalytic "quaternization" of a coordinated phosphyne group by substrate with the formation of the salt of the phosphonium base $(EtPPh_3)^+$ proceeds in the course of ethylene dimerization or hydrogenation [102]. These processes influence the catalytic activity and the selectivity due, at least, to changes in the polymer surface properties (hydrophobicity, dielectric permeability, etc.). Thus, metal redox ability, surface hydrophobity and the size of the ether units significantly affect the catalytic properties of rhodium or rhutenium polymer contacts [103].

1.3.2 About the influence of support nature on catalytic properties

MX_n fixation on macroligands with similar functional groups but different structures (the value of cross-linking is one of most important parameters) may yield catalysts with rather different catalytic properties, activities and selectivities. Moreover, the nature of polymer support may specify the route of the hydrogenation reaction. Rhodium complexes immobilized on phosphynated polydiacetylene or silica gel show activity for the hydrogenation of aromatics [104] but loose this activity when immobilized on phosphynated PS. This may be connected with the ease of polymer ligand cleavage, so that the formation of Rh^0 particles due to $Rh^{3+} \rightarrow Rh^0$ reduction is complicated by cross-linked PS. The rate of allyl alcohol, dimethylethynylcarbynol and nitrophenol hydrogenation slows down in the following sequence of supports [105]: polyvinyl alcohol (PVAl)>polyacrylonitrlile (PAN)>polyhexamethylene adipinamid>polyethylene terephthalate (PETP). The higher activity of Pd-PVAl is connected with the presence of polar hydroxy groups on the polymer surface.

The complexes of Ir, Pt, Rh and Pd attached to different polyamides [106–107] are active in benzene hydrogenation (Ir>Pt>Rh>Pd). At the same time the rate of olefinic substrate hydrogenations increases as follows: 1-methylcyclohexene<cyclohexene octene-1<hexene-1.

The influence of the cross-linkage level on the catalyst activity and selectivity is confirmed by numerous experimental data. The activity of polymer-immobilized catalysts is as a rule similar to that of the homogeneous molecular complexes and sometimes even above it for weakly cross-linked polymer as supports [108] (Figure 1.6). The relatively high catalytic activity mentioned is a specific feature of polymer catalysis. However, a high mobility of polymer segments within weakly linked polymer matrices may lead to the formation of binuclear complexes that are less active than immobilized mononuclear species. There are however some exceptions to the rule: immobilized polyoxim–Pd^{2+} complexes containing the bond Pd–Pd were found to be highly effective in hydrogenation of a number of substrates [109]. At the same time, their molecular analogue, bis(dimethylglycoximato)Pd^{2+} does not catalyze these processes. This distinction is probably because immobilized complexes are capable of providing coordinative vacancies by the rupture of Pd–Pd bonds, unlike the rigid structure of homogeneous, saturated Pd^{2+} complexes.

Polymer supports often hinder the process of metal complex dimerization and association [25]. This is the main reason for the rise in immobilized metal complex activity and selectivity. Numerous examples confirm this generalization. Here we will only look at olefinic hydrogenation in the presence of rhodium–polymer complexes, [98,110] together with acetylene hydrogenation by molybdenum-containing polymers [111,112].

The nature of the support also affects the processes of catalyst formation. Thus Rh^0 generation in a reducing atmosphere at elevated temperatures proceeds significantly more slowly in gel-like systems. Metal centers in such systems are located within the total volume and not only at the surface of the support [113].

Finally, let us especially point to a classic example of polymer matrix employment as a means of providing coordinate vacancies (at the stage of catalyst synthesis or easily formed in the course of synthesis) for olefin, diene and acetylene hydrogenation in the presence of immobilized metallocenes [24,114–116]. Homogeneous titanocene systems experience fast deactivation due to complex dimerization, whereas a polymer matrix (PS modified with Cp groups) prevents this process. Heterogenized catalysts therefore show not only 10–20-fold rise in activity but are also significantly more stable than homogeneous analogues (see Table 1.4).

Substrate diffusion into the polymer matrix is not a limiting step in the catalytic process, although some examples of reaction rate increase with the use of preliminary ground catalyst particles are known.

1.3.3 The effect of transition metal properties and the ratio M/L

Comparisons of the activities of metal–polymer catalysts on the basis of different MX_n complexes attached to the same polymer matrix is not a very simple job. There are many reasons for this. Firstly, it is rather difficult to provide the same metal complex content on the support. Moreover, activation of such catalysts yields different concentrations of active sites, which may alter catalyst properties, their distribution on the polymer support and accessibility to the substrate. Finally, metal centers are very specific for each substrate so that experimental data are not consistent generally and are attributed to the overall content of transition metal.

The catalyst activity depends in a complex way on both the metal nature (M=Pd, Pt) and the metal content (Figure 1.7). Thus, the extremal activity for octyne-1 hydrogenation by Pd complexes immobilized on nylon-66 is reached at a Pd content of 0.4 wt.% [117] whereas, at such a content, Pt-containing catalysts are inactive. An increase of metal content leads to SCA diminution for Pd catalyst in contrast with a rise for Pt-containing active species. In the sequence of transition metal complexes (Rh^{3+}, Ru^{3+}, Ni^{2+}, Cu^{2+}, Co^{2+}, Pd^{2+} and Mo^{6+}) immobilized on polymer supports with isonitrile groups, $PdCl_2$ complexes show the greatest activity in phenylacetylene and benzaldehyde hydrogenation [118]. At the same time $Pd(OAc)_2$, coordinated to cross-linked isocyanines [119], results in substrate hydrogenation followed by isomerization; the more selective hydrogenation of cyclohexene proceeds with the use of polymer-supported ruthenium (III) complexes as catalysts [120].

The SCA of benzene hydrogenation with different metal complexes attached to nylon-66 increases in the sequence of the transition metals (the catalytic turnover number, TN, molecules of substrate converted per catalyst molecule, is given in parenthesis) [106]: Pd(0.6)< Pt, Rh (13.2 both)<Ir(33). However the activities of the same metal-containing polymers in nitrobenzene hydrogenation follow the reverse

order [107]: Rh< Pt<Pd. This activity sequence reversal seems to provide evidence that benzene ring hydrogenation proceeds at different active sites.

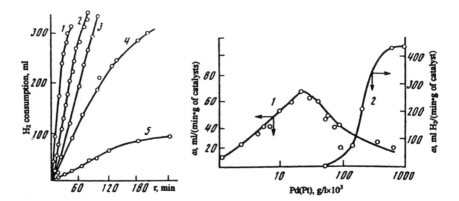

Figure 1.6 Changes in diphenylacetylene hydration rate in the presence of metal-containing polymers: 1, Dipy-Pd^{2+} acetate on CSDVB (2% DVB); 2, Pd^0 on CSDVB (2% DVB); 3, 10% Pd/C; 4, dipyridyl palladium acetate on CSDVB (20% DVB); 5 – soluble analog, dipyridyl palladium(II) acetate.

Figure 1.7 The dependence of catalysts on the polymer base of nylon-66 activity, in hydrogenation of octene-1, on the metal content in the catalyst: 1, Pd-nylon-66; 2, Pt-nylon-66.

Nitrobenzene hydrogenations have already been shown to proceed effectively under mild conditions (323 K, P_{H_2}=0.1 MPa) in the presence of Pd-containing ionites [121]. In contrast, under the same conditions, Pt-, Co-, Ni- and Fe-containing catalysts show no activity at all and Rh- even less. Such differences are apparently connected with different coordination site stabilities in the polymer matrix. Only palladium complexes can reduce to Pd^0 under these conditions. It was demonstrated [122] that Rh complexes fixed on PAAc are more active in furfural hydrogenation than Pt and Pd analogues, but that Rh-PAAc and Pd-PAAc catalysts show similar activity in hexene-1 hydrogenation [123].

It would be interesting to compare the reactivities of similar active sites, for example transition metal hydrides. Such centers are probably formed in pentene-1 hydrogenation by Rh and Ru complexes immobilized on copolymers of maleic acid, vinyl ethyl ether or hexyl 2-ethyl ether [103]. Their activities are (TN) 29, 15.7 and 4 h^{-1}, respectively.

The formation of coordinatively unsaturated complexes is the important factor determining the catalytic activity of immobilized catalysts [84,124,125]. Optimization of the metal-to-ligand ratio, M/L, is one of the ways of reaching this either at the stage of catalyst preparation or during the catalytic reaction (dynamic effect). It has been noticed already that optimal ratio is P/Rh~2 for catalysts of the type Rh-CSDVB [81] (Figure 1.2). It should also be noted that the constant catalytic action that has been observed for the majority of immobilized catalysts is probably evidence of similar composition of the active forms. Thus, in selective reduction of polynuclear heteroaromatics by Rh^{1+}, anchored to phosphynated CSDVB, this P/Rh ratio lowers slightly, even after several thousands of catalytic cycles [126].

Stable coordinatively unsaturated metal centers are formed during heterogenization of Ir(CO)Cl(PPh$_3$)$_2$ (Vaska's complex) on CSDVB (1% DVB) due to steric hindrance and some specific features of the mobility of the macrochains [127]:

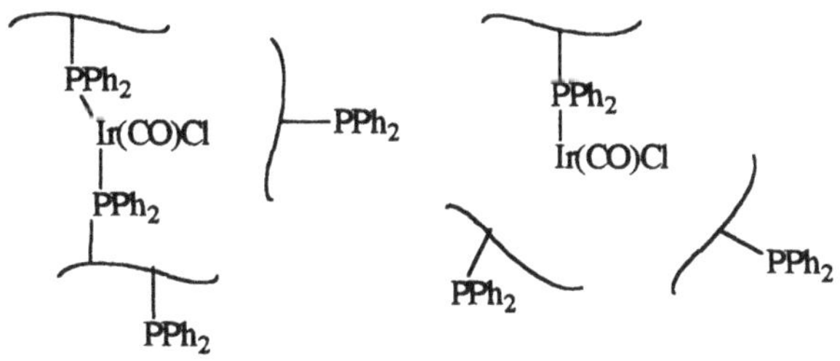

Scheme 1.6

A rise of hydrogenation rate was observed for catalysts with low P/Ir ratio (Table 1.6). The increase of numbers of phosphyne groups leads to a diminution in the fraction of coordinative unsaturated metal centers. Accordingly, such catalysts are comparable in activity with the homogeneous materials. The introduction of low molecular mass analogues, such as triphenylphosphyne, into the reactive system leads to retardation of the hydrogenation reaction [103]. Accordingly, not only ligands of the polymer support but also low molecular mass additives may affect the structural organization of immobilized catalysts. It is also important, in this connection, that supported macrocomplexes are dynamic systems, and that unsaturated vacancies of metal centers may either appear or disappear.

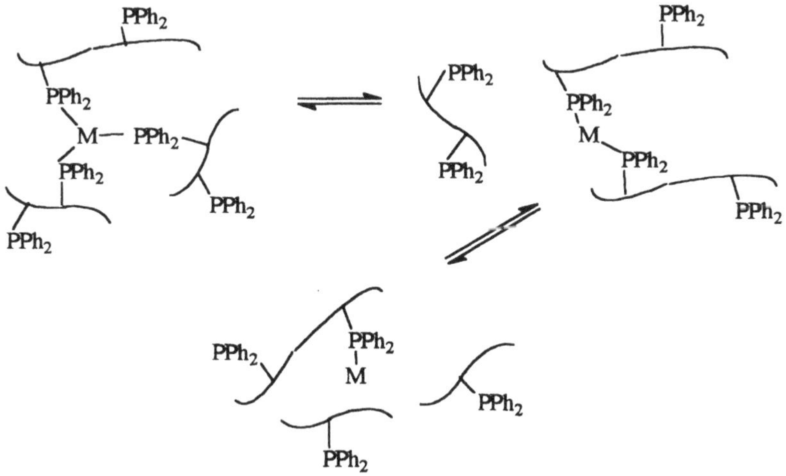

Scheme 1.7

Table 1.6 Comparison of Ir(CO)Cl(PPh$_3$)$_2$ and the immobilized complex activities in hydrogenation of 1,5-cyclooctadiene [127].

Catalyst	P/Ir	Reaction time (h)	Yield (%)		Selectivity^a (%)
			cyclooctene	cyclooctane	
Ir(CO)Cl(PPh$_3$)$_2$	3	33.0	37	3	92.5
]^b–(PPh$_2$)$_2$Ir(CO)Cl	3	0.25	57	4	93.4
Ir(CO)Cl(PPh$_3$)$_2$	4	48.0	44	3	93.6
]–(PPh$_2$)$_2$Ir(CO)Cl	4	0.25	69	5	93.2
Ir(CO)Cl(PPh$_3$)$_2$	22	72.0	15	1	93.8
]–(PPh$_2$)$_2$Ir(CO)Cl	22	93.0	18	1.5	92.3

^{a} Selectivity is equal to the ratio of concentrations [cyclooctene]/([cyclooctene]+[cyclooctane]).
^{b} Open squares represent the "rest of polymer".

The dispersivity of Rh^0 particles formed on the surface of phosphynated CSDVB (2% DVB) also depends on the P/Rh ratio [128–130] and consequently affects catalyst activity in 1,5-cyclooctadiene (1,5-COD), acetylene and butadiene hydrogenation [131]. The relations between these parameters are summarized in Table 1.7. By use of extended X-ray absorption fine structure (EXAFS) the average size of the rhodium particles produced (Rh$_s$) on the support surface has been determined and their fractions in relation to the total metal content has been estimated.

The rate of butadiene hydrogenation depends in a complex way both on the content of phosphyne groups and the ligand-to-metal (P/Rh$_s$) ratio (Figure 1.8). The general tendency is as follows. The decrease of phosphorus content in the polymer as well as P/Rh$_s$ ratio leads to an increase of Rh$_s$ particle size at low phosphyne group contents and P/Rh$_s$ values approaching the Rh$_s$ particle size formed on the unfunctionalized polymer (CSDVB).

An increase of functional group content leads to retardation of the decomposition of rhodium complexes to free metal particles and at P/Ph=40 metal crystallites have not been revealed even during heating at 398 K for some days [113].

1.3.4 The influence of the immobilized MX$_n$ ligands character on catalytic properties

The initial ("own" or "frame") ligands which are retained in the inner coordination sphere of MX$_n$ after immobilization, together with interchanged ions and/or counterions of ionites, etc., may affect complex stability, the process of active site formation and their activity and selectivity in hydrogenation reactions.

Analysis of extensive data shows that interaction of halide metal complexes of the platinum group (especially K$_2$PdCl$_4$) with anionites may lead to the formation of complexes with a halide deficiency, although their oxidation state is (+2) or (+1), as in palladium complexes (Table 1.8). Such metal centers more easily transform into Pd^0 than, say, in [PdCl$_4$]$^{2-}$ under the action of reducing agents. However the nature of macrocomplex halide, PE-gr-P4VP·PdX$_2$, may significantly affect the catalytic activity of the complexes formed. Consequently, nitrobenzene hydrogenation rates diminish as follows: Cl>Br>I [23]. The halide ligands appear to modify both the reaction of Pd^{2+} to Pd^{0} reduction (see Table 1.8) and/or hydride formation

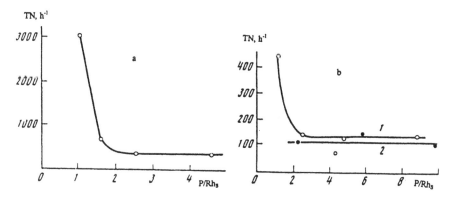

Figure 1.8 The dependence of 1,3-butadiene hydrogenation turnover number on the molar ligand-to-metal ratio, P/Rhs. Phosphorus content in polymer support (%): (a) 0.5; (b) 1, 1.2; 2, 1.7.

In addition to the halides, carbonyls, nitrosyls, phosphines, AcAc, CH$_3$COO, as well as pyridine and Dipy, can also be used as frame ligands or MX$_n$. It is very difficult to estimate the influence of the ligands characteristics because these are often substituted by reagents during the catalytic process. However, the phosphine basicity and phosphorus substituent size have been shown to be important parameters affecting activity, selectivity and stability of metal-containing polymer catalysts with phosphine ligands [134]. The rate of cyclohexene hydrogenation by rhodium-containing immobilized catalysts decreases with an increase of phosphine basicity [134,135]. Polymer-immobilized organometallic compounds are rarely used for hydrogenation. The activities of the resulting catalysts are only modest (for example compared with that of 6,10,14-trimethylpentadeca-3,5-diene-2-one) [136,137].

An excess of phosphine ligands is known to inhibit olefin hydrogenation by Wilkinson's catalysts. One of the ways of producing activity stabilization is through catalyst modification by polymer matrices capable of effective adsorption of the phosphine redundant from the solvent, which then promotes rhodium-coordinative unsaturation [138]. A most effective sorbent in this case has been found to be the silver salt of polystyrene sulfonic acid.

The problem of elucidating composition–structure–property relationships is thus far from solution for immobilized catalysts and in particular for hydrogenation processes. Significant aspects of the problem have been studied only for some special cases. Although it seems that the progress in homogeneous hydrogenation in combination with the possibility of separation of intermediates from immobilized ones will provide a more precise characterization of the active forms, these are additional problems intrinsic to immobilized catalysts. Some of these may be solved by analysis of dynamic factors of liquid-phase hydrogenation such as the solvent nature, reaction temperature, concentration ratios, etc.

Table 1.7 Some characteristics of rhodium complexes attached to phosphynated CSSDVB (2% DVB) and their activities in butadiene and acetylene hydrogenation [131].

P (%)	Rh (%)	Rh$_s$, (%)	Rh particles size (nm)	P/Rh$_s$	$N^a \times 10^2$	
					butadiene	acetylene
1.6	1.1	1.0	1.5	5.8	1.3	1.0
1.7	0.6	0.5	1.7	10.7	1.0	12
1.7	1.6	1.3	1.9	4.3	0.8	–
1.2	0.5	0.5	1.2	8.4	1.4	–
1.2	1.0	0.9	1.6	4.5	1.3	0.8
1.2	2.0	1.7	1.8	2.3	1.5	–
1.2	5.0	3.3	2.9	1.2	4.2	5.0
1.7	3.0	2.4	1.9	2.3	1.1	–
0.5	0.5	0.4	2.6	4.6	3.9	–
0.5	1.0	0.7	2.9	2.4	3.8	21
0.5	2.0	1.1	4.4	1.6	6.3	–
0.5	5.0	1.7	6.7	0.9	30	–
CSDVB	1.0	0.5	4.8	–	28	42

^{a} N – number of substrate molecules /(atm [Rh$_s$] s); reaction conditions 348 K; P_{H_2} =67 kPa; $P_{C_4H_6}$ =0.11 kPa, $P_{C_2H_2}$ =0.11 kPa (368 K).

Table 1.8 Pd$_{3d5/2}$ and Cl$_{2p}$ electron binding energies of some palladium-containing catalysts.

Catalyst	Electron binding energy (eV) (± 0.2 eV)		Pd/Cl ratio	Ref.
	Pd$_{3d5/2}$	Cl$_{2p}$		
PdCl$_2$	383.3	199.4	2.0	[132]
K$_2$PdCl$_4$	338.9	199.3	4.0	[132]
PdCl$_2$(PPh$_3$)$_2$	338.3	198.6	–	[133]
PdCl$_2$-phosphynated PS	338.6	199.1	–	[133]
5% Pd/C	336.1		–	[132]
PdCl$_2 \cdot$(4-vinylpyridine)$_2$	338.6	198.5	2.0	[132]
PdCl$_2 \cdot$P4VP	338.2	198.3	1.9	[132]
PdCl$_2 \cdot$PE-gr-P4VP	338.1	198.2	1.8	[132]
PdCl$_2 \cdot$PHAr^a	338.2		–	[39]
Pd$^0 \cdot$PHAr^a	335.7	–	–	[39]

^{a}PHAr, polyheteroarylene – poly(benzo-bis(benzylimidazole)phenanthroline).

1.4 The Influence of Solvent Properties on the Rate of Hydrogenation

The role of the solvent is one of the foremost factors affecting the hydrogenation process, so it has been widely investigated (e.g. see [84,103,135,139–142]). In general the effects of solvent on the rate and mechanism of hydrogenation are manifested by numerous factors: bond energies both of the substrate and hydrogen with the catalyst surface; adsorption ability of solvent molecules; substrate and product concentration distribution between catalyst surface and solution; the ratio of components; activation and desorption rates; hydrogen solubility in the solvent and diffusion rate at the gas-liquid boundary; solvation of the reactive components; polar impurities capable of selective sorption; redox interactions between solvent, catalyst and especially the support.

The influences of these factors are not equal in strength. Firstly in immobilized systems the solvent causes support swelling and chain flexibility and is responsible for both long-range interactions and the formation of complexes of various compositions. Also, solvent often determines the condition of participating metal-containing polymers. Thus, supports based on modified PS are glass-like species in n-hexadecane but resin-like in m-xylene at 343 K [143]. The form of the macromolecules carrying the metal complexes is also of significant importance. Thus, Rh and Ru immobilized by alternating copolymers of maleic acid and vinyl ethers, which are rather compact in n-octane, produce less activity [144].

A decrease in activity is often observed for polymer species with modest swelling ability. The linear dependence of SCA on the P/Ru ratio is connected with coordinative unsaturation of the transition metal ion. The high-level activity of Rh-containing polymers may be accounted for by partial decomposition at the stage of catalyst activation followed by formation of coordinative unsaturated sites due to partial release of PPh$_3$ groups. An analogous pattern of reactivity has been observed [145] for catalysts of hydrogenation of $trans$-cinnamic acid (298 K, 0.1 MPa of H$_2$) when Pd^{2+} was fixed on P4VP or a copolymer of styrene with 4VP. Stable particles of Pd^0 bound to the polymer have been generated in the course of the reaction (Pd leaching was not above 1% within 10 turnovers). The catalyst activity decreases in the solvent sequence AcOH>EtOH>etyl acetate>benzene>acetone, being higher for copolymer complexes than for homopolymer. The rate of octene-1 hydrogenation in the presence of macromolecular Rh$^+$ chelates in tetrahydrofuran (THF), dioxane and methanol is equal to ($w\times10^6$): 1.07, 1.91, 0.58 mol/min, respectively, and in solvent in which the polymers do not swell (benzene), absorption of hydrogen has not been observed even after 120 min [44].

Nevertheless any correlation between metal polymer swelling and activity is rather indefinite. Thus, the rate of allylbenzene hydrogenation depends in a complex way on the swelling of the gel-immobilized catalytic system [146,147]. At the start a gradual swelling is accompanied by a rise in activity (Figure 1.9) due to the increase of active sites accessibility. A further increase in the amount of swelling does not result in further hydrogenation rate increase. Similar regularities have been observed [148] in the hydrogenation of allyl alcohol by Pd^{2+} complexes immobilized on P2VP. The reaction rate decreases in the solvent sequence methanol>ethanol>propanol>butanol (Figure 1.10), the smallest in hexane being dependent on the component ratio in water/alcohol mixtures. The intrinsic viscosity of the polymers also follows the above sequence. Swelling ability of reduced catalysts is higher in alcohols than in water or in hexane. In the latter case palladium particles are usually located within the polymer and do not take part in catalytic reaction. Only those located on the surface of the polymer coil can catalyze the reaction.

Palladium macrocomplexes with ampholytes containing iminodiacetate functional groups are selective catalysts for cyclopentadiene hydrogenation to cyclopentene [149]. The reaction rate increases with a rise in solvent polarity (Figure 1.11).

When analyzing a physical factor such as swelling one must bear in mind that it may be different for "own" polymer matrix and immobilized macrocomplexes as immobilization may be followed by macrochain linkage. Moreover, the effect of solvent may be different for non-swelling, weakly and strongly swelling polymer matrices.

Weakly cross-linked gel-like polymers are known to absorb amounts of solvent that are several times their own mass. Essentially polymer gel with a cross-linkage value of 2% consists of mostly solvent with only a small part of its own mass. All the internal volume of the polymer is accessible to both solvent and reagents. A rise in the extent of cross-linkage results in a diminution in the ability of the polymer to expand in compatible solvent preventing solvent diffusion to the core of polymer coil [150].

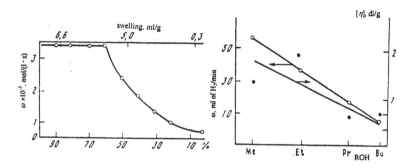

Figure 1.9 The dependence of the initial allylbenzene hydrogenation rate on the swelling of metal polymer catalyst Ni(AcAc)₂ fixed on a copolymer of butylacrylate, acrylonitrile, acrylic acid and methylmethacrylate (96:2.1:1 mol%). The abscissa gives the volume percentage of benzene in the mixture benzene/heptane.

Figure 1.10 The dependence of the hydrogenation rate of allyl alcohol (0.17 mol) by Pd (0.03 g) fixed on P2VP, and the values of P2VP intrinsic viscosity, on the nature of alcohols.

Figure 1.11 The influence of solvent nature on the rate of cyclopentadiene hydrogenation by macrochelate Pd complexes: 1, CH₃OH; 2, C₂H₅OH; 3, HCON(CH₃)₂; 4, CH₃(C₂H₅)CO; 5, CH₃O(C₂H₅)CO; 6, C₆H₆; 7, n-C₆H₁₄.

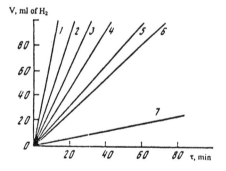

The swelling ability of PS and polyolefins decreases with a rise of solvent polarity, which consequently reduces the size of the transport channels of catalysts and increases the diffusion hindrances. The last effect is most important for solvated substrate molecules. Similar processes also take place in non-swelling and weakly swelling polymers. A rise of solvent polarity favors the diffusion of non-polar substrates to the active centers and concurrently acts to prevent the diffusion of polar reagents [84, 140]. This fact is of great importance for hydrogenation catalyzed by insoluble metal-containing polymers. Thus, the rate of hydrogenation in alcohol and benzene decreases for substrates in the order cyclopentene<styrene<acrylonitrile, which corresponds to

their polarities [84] (rhodium catalyst complexes immobilized on phosphynated CSDVB; 25% DVB). In spite of this, the cyclohexene hydrogenation rate (mol/min$\times 10^3$) increases with a rise of solvent polarity in the order ethanol(0.36)>benzene(0.32)>n-heptane(0.15) under similar hydrogen solubility conditions in these solvents (The Henri constant is the same for all the above solvents). The relative rates of cyclohexene hydrogenation in benzene, a benzene/THF mixture and THF are 1:1.66:3.26, respectively.

It is worth noticing that a similar effect has been observed [84] during cyclopentene hydrogenation in the presence of soluble metal-containing catalysts: the reaction rate increased twice in the mixture of benzene and ethanol compared with that in benzene alone. Conversely, acrylonitrile hydrogenation is accelerated in benzene, producing 1:1 complexes with the substrate (complexes of formula 1:2 have been found for the same substrate with ethyl alcohol).

The high level of polymer support swelling may also promote immobilized catalyst deactivation. The reasons of this effect are connected with the flexibility of polymer matrices in assisting the association of active sites and permitting available ligand sites (L) to take part both in intra- and intermolecular reactions and provide additional contacts with active sites which can lead to the loss of their coordinative unsaturation. This enables the sharp decrease in activity of Rh- and Ru-containing polycarboxylates, even after the first catalytic cycle [103], to be explained. Deactivation of the catalysts may be retarded or prevented by neutralization or etherification of free carboxylic acid groups by the use of metal-containing polymers produced by polymerization of the respective metal-containing monomers.

Another effect of protonic solvents on the rate of hydrogenation should be noted here. This concerns the possible participation of alcohols in transition metal hydride formation which may (or may not) be accompanied by changes of metal ion oxidation state. Much available experimental data relate to this fact. For example, the formation of hydride complexes of rhodium [84, 139] and ruthenium [151] has been observed when the corresponding metal-containing polymers have interacted with protonic solvents. In contrast, Pd derivatives do not produce hydrides under the same conditions, as shown by the use of deuterated compounds. This may be explained on the premise of $Pd^{2+} \rightarrow Pd^0$ reduction by the action of alcohol [124, 152]. Similar reactions take place at the stage of metal complex immobilization (see Table 1.8) perhaps by an autocatalytic mechanism (see section 1.2.4).

It is interesting to note that the homogeneous system $PdCl_2(PPh_3)_2$ is not activated by alcohols, a result that may be connected with higher electron density of the Pd–Cl bond than that in immobilized complexes [124]. Unfortunately there are unsufficient data to contend that substrate activation proceeds by its protonation according to the reduction-protonation sequence that is characteristic of homogenous catalytic hydrogenation [1].

It should be noted that the rate of furfural hydrogenation by $RhCl_3$ fixed on PAA also depends on the nature of the solvent following the sequence [153]: H_2O/iso-$C_3H_7OH < H_2O/C_2H_5OH < H_2/iso$-$C_5H_{11}OH$ (at volume ratio 1:1), whereas in phenol hydrogenation the nature of the alcohol used does not affect the rate of the reaction [154], only the presence of alcohol in solvent mixture being important. These experimental data may also be connected with formation of immobilized Rh^0 particles following substrate coordination by these active sites. Another illustration is the 1.5–8-fold increase in palladium polymer catalyst activity on hydrogenation of dienes following substitution of aprotic solvents by protonic ones [155]. The rate of diene hydrogenation by Pd–PEI complexes in protonic solvents is some orders of magnitude higher than that in benzene or THF [33,156].

One of the ways in which to avoid to some extent the solvent influence is to use of macroporous polymers with 6–75 nm pore size instead of gel-like polymers.

The solvent properties significantly affect the hydrogenation reaction catalyzed not only by immobilized mono-metal complexes but also polynuclear, cluster-type and different metal complexes anchored to supports of mixed type (oxide-polymers). Thus polymer complexes PdCl$_2$ with poly-γ-(aminoethylamine)-propyleneoxane supported on SiO$_2$, Al$_2$O$_3$, etc., are in the order of activities in different solvents MeOH>DMFA>*iso*-PrOH>THF>benzene [80].

Finally, solvent molecules may act as specific ligands coordinated by active sites (or located in the vicinity of their action) thereby altering their activity. The direction of these changes (activation or deactivation) depends on many factors and particularly on the solvent–metal center bond strength. Thus, the greatest rate of styrene hydrogenation has been found [124,133] in some oxygen-containing solvents (see Table 1.4).

Table 1.9 Some characteristics of Pd catalyst activities in hydrogenation of styrene[a] in different solvents.

Solvent	Initial rate, (ml of H$_2$/min)	Solvent	Initial rate, (ml of H$_2$/min)
DMFA	8.04	Ethylacetate	2.08
Ethanol	7.74	Methylene chloride	1.68
THF	6.13	Acetic acid	1.45
Benzene	2.84	Nitromethane	0.10
Chloroform	2.45	Dimethyl sulfoxide	0.09

[a]Reaction conditions: [Pd]=3.19 mmol/l, [styrene]=0.67 mol/l. hydrogen pressure 0.1 MPa, 298 K, 12 ml of solvent.

The presence of the unshared electron pair on the heteroatom in the solvents in Table 1.9 appears to favor the activation of the metal center, but strong solvent coordination (dimethyl sulfoxide, nitromethane) hinders the subsequent reagent recoordination. Similar results have been obtained [121] in hydrogenation amination of aldehydes with nitrobenzene by palladium-containing ionites. The rate of these reactions decreases in the order alcohols>ketones>aliphatic hydrocarbons, with N-acetamide and acetonitrile completely inhibiting the reaction.

The examples mentioned above show the multiple influences of the solvent in hydrogenation by immobilized catalysts which are much more versatile than those in homogeneous catalysis. Selection of solvent may be used as a reaction control. In many cases the substrate plays the role of the solvent. In homogeneous catalysis the choice of solvents is restricted by solubility and stability of the complexes. These restrictions may often be removed in catalysis by immobilized metal complexes. One can make high concentrations of such catalysts in practically any reaction mixture, even to the use of slurries. To strengthen the effect one can also use polymer catalysts with metal complexes distributed on the surface of the polymer support. The formation of active sites at the stage of metal–polymer complex immobilization allows a reduction of the dependence of SCA on the degree of swelling and possibly on the properties of the support. Different polymer supports show different swelling capacity with different concentrations of available functional groups and can also be produced by copolymerization of corresponding monomers, including those that contain metals.

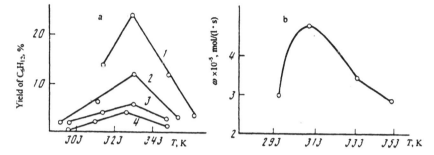

Figure 1.12 The influence of reaction temperature on the catalytic activity of metal polymer catalysts in hydrogenation of benzene (*a*) and allylbenzene (*b*). (*a*) 1, Pt-nylon-66, 2, Rh-nylon-66, 3, Ir-nylon-66, 4, Pd-nylon-66. (*b*) Ni^{2+} copolymer of butylacrylate, acrylonitrile, acrylic acid and methylolacrylamide 96:2.1:1 (mol%).

Figure 1.13 Arrhenius plot for hydrogenation of gaseous ethylene by Rh^+ complexes immobilized on phosphynated CSDVB (2% DVB). The catalyst was evacuated before each experiment at 343 K (1) and 363 K (2); $P_{C_2H_4} = P_{H_2} = 27$ kPa.

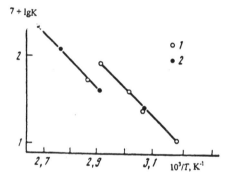

1.5 The Influence of Reaction Temperature on the Catalytic Hydrogenation

The rates of hydrogenation on both homogeneous and heterogeneous catalysts are as a rule well defined by the reaction temperature. This increases with temperature within an active range. Influence of temperature on the rate of hydrogenation by heterogenized metal complexes is more complex. This arises not only because the catalyzed reaction is controlled by the temperature but also by the state of macroligand–MX_n system. The curve $w=f(T)$ is often of "bell-shaped" form. In particular, these dependences have been revealed (Figure 1.12) in the hydrogenation of benzene by platinum-group metal complexes immobilized by nylon-66 [106], allylbenzene hydrogenation by gel-immobilized nickel complexes [146], in hydrogenating amination of 2-methyl propionic aldehyde with nitrobenzene by Pd-polytrimethylolmelamine [157], etc.

There are reasons for this atypical temperature dependence. These include the possible decomposition of polymer matrices, especially at elevated temperatures which can be initiated or accelerated by metal complexes. It may be that the curves drop downward because of metal center aggregation; this process becomes especially significant at elevated temperatures.

However, more objective is the temperature influence on phase changes in the metal polymer matrices. Polymer reactions are known to be coupled with mobility of chain segments controlling catalytic properties of immobilized metal complexes. A rise in chain mobility at a characteristic temperature may lead to changes in coordinative unsaturation of the transition metal ion of the active center, as has been shown for $Pd(PPh_3)_4$ immobilized by phosphynated weakly cross-linked CSDVB [125] (see Scheme 1.4).

A particularly interesting phenomenon is the conformation of the polymer matrix providing the favored configuration for the catalytic center. To illustrate this point let us describe the abnormal Arrhenius dependence of the ethylene hydrogenation rate in the gas phase by rhodium complexes immobilized on phosphynated polystyrene [158]. A sharp increase of hydrogenation rate has been observed when the temperature increases across the glass transition point, T_g, of the polymer (341 K; Figure 1.13). The value of the activation energy of the reaction at $T>T_g$ was practically unchanged (33 kJ/mol for the higher temperatures and 35 kJ/mol below T_g). Consequently all changes in the rate constant value for this reaction are connected with revisions in the entropy factor. It is most probable that the effect is the result of polymer segments unfreezing and promoting for the correct configuration of the active centers for catalytic action. Alternatively such a promotion of lability can also cause the aggregation of metal particles of different sizes composed of zero-valent metal atoms with an activity much higher than that of isolated metal atoms[7]. Another possible reason of the discontinuity in Arrhenius plot is existence of limitation for ethylene coordination. A comparison of olefin adsorption with their rates of hydrogenation has revealed that both enthalpy and entropy of coordinated ethylene molecules correlate with changes in polymer chain mobilities (this phenomenon will be discussed in detail in Section 5.4).

It is very difficult to establish the quantitative relation between segment mobility and catalytic activity, especially as there are many examples of the normal Arrhenius dependence of hydrogenation rate. For example, octene-1 hydrogenation by macrochelate rhodium complexes across the total operating temperature range (303–323 K) gives a value of the activation energy of 23 kJ/mol [44]. It is worth noticing that sometimes it is possible to connect such dependences with translatory movements of active centers [159] (e.g. in ethylene polymerization; see Chapter 3).

In such a way, the complex dependence $w=f(T)$ in hydrogenation by immobilized complexes may be explained by different reasons. The rising branch of the curves (see Figure 1.12) is the Arrhenius plot while the curves drop downward results from a decrease of hydrogenation rate possibly diminishing the response of the number of active sites. The target-oriented analysis of the response of polymer-immobilized catalysts to the action of temperature may lead to creation of

---

[7] It is interesting that similar regularities have been observed in C_2H_4 hydrogenation by metallic Pd dispersed on Chromosorb containing a polymer phase, a mixture of poly(ethylene) oxide and PS in 1:1 proportion by mass [160]. The temperature increase above T_g for PS (~368 K) led to the catalyst activity rise.

hydrogenation catalysts operating at elevated temperatures and also optimization of their usage.

Another phenomenon promoting the creation of temperature-regulated systems should also be noted. Poly(alkene oxides), block copolymers of ethylene oxide, propylene oxide modified with $SOCl_2/LiPPh_2$, treated with $Rh_2Cl_2(C_8H_{12})_2$, or with $RhCl(PPh_3)$ that are soluble at low temperature ($0°C$), however, demonstrate phase separation at higher (room) temperatures. Such temperature-dependent solubility permits a combination of the advantages of homogeneous catalysis and heterogeneous (catalyst separation); such catalyst action has been demonstrated in allyl alcohol hydrogenation [160].

1.6 The Dependence of Hydrogenation Rate on Concentration Ratios

If we exclude from consideration the processes of active site formation, the general scheme of hydrogenation by immobilized catalysts may be represented by two main stages. The first one is the interaction of substrate molecule with a fixed center containing activated hydrogen. This is the reversible step. In the second stage the adduct irreversibly reacts with hydrogen yielding the product with subsequent regeneration of the catalytic active center (the Michaelis-Menten mechanism):

$$]-MH + S \underset{k_{-1}}{\overset{k_1}{\Longleftrightarrow}}]-M-SH$$

$$]-MSH + H_2 \overset{k_2}{\longrightarrow}]-MH + SH_2$$

The rate of hydrogenation follows the general equation[8]:

$$w = \frac{dS}{dt} = \frac{k_1 k_2 [S][H_2][]-MH]}{k_{-1} + k_1[S] + k_2[H_2]}$$

(Adsorption coefficients of substrate and adduct enter in the rate constant values).

Under conditions of liquid phase hydrogenation at a low (as a rule at barometric) H_2 pressure:

$$k_1[S] \gg (k_1 + k_2[H_2]) \qquad \text{then}$$

$$w = -dS/dt = k_2[H_2][]-MH]$$

The mechanism of 1-octene hydrogenation by $RhCl(PPh_3)_3$ immobilized on chloromethylated modified with ethylenediamine groups CSDVB (2 and 5% DVB) follows the scheme:

---

[8] This equation is easily transformed into Michaelis-Menten form applied for homogeneous fermentation catalysis. The rate of the reaction rises with the rise of substrate concentration tending to the constant value w_{max}:

$$w = w_{max}[S]/(K_m + [S])$$

where K_m is the Michaelis factor corresponding to the substrate concentration value at which the reaction rate is one-half of w_{max}.

$$]-RhCl(PPh_3)_2 \xrightarrow{\text{solvent}}]-RhClPPh_3 + PPh_3,$$

$$]-RhClPPh_3 + H_2 \underset{k_{-1}}{\overset{k_1}{\longleftrightarrow}}]-RhCl(H)_2 PPh_3,$$

$$]-RhCl(H)_2 PPh + S \xrightarrow[K]{\text{slow}}]-[RhPPh_3(H)_2 - S]^+ + Cl^-,$$

$$]-[RhPPh_3(H)_2 - S]^+ \xrightarrow{\text{fasr}}]-[Rh(PPh_3)_2 + HSH \text{ (product)}$$

The rate of the reaction follows the equation:

$$w = \frac{Kk_1[]-ML][H_2][S]}{k_{-1} + K[S]}$$

The rate of cyclooctene hydrogenation by rhutenium complexes immobilized on carboxylated polymers [161] increases proportionally both with H_2 pressure and catalyst concentration. A plot of $1/w - 1/[S]$ being a straight line. The general equation for this

$$w = \frac{K[]-Ru][H_2][S]}{1 + [S]}$$

However, the reaction often follows two parallel routes. For example, cyclohexene hydrogenation by palladium acetate supported on PS modified with dipyridyl groups at high substrate concentrations is described by the scheme [161]:

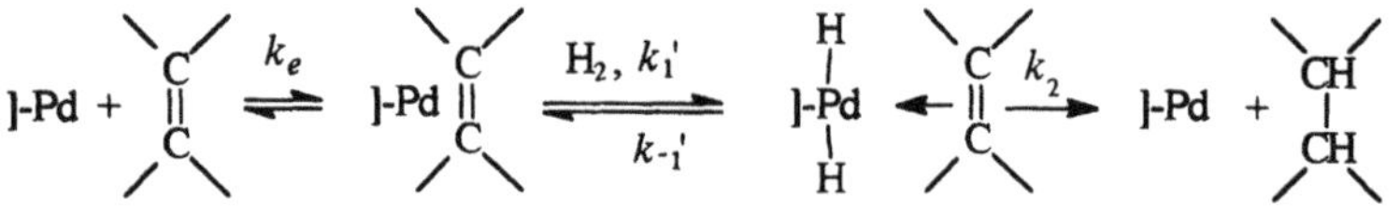

Scheme 1.8

This reaction is described by the equation:

$$w = \frac{k_1 k_2 []-Pd][H_2]}{k_1[H_2]+k_{-1}+k_2}$$

The second alternative first involves interaction with molecular hydrogen:

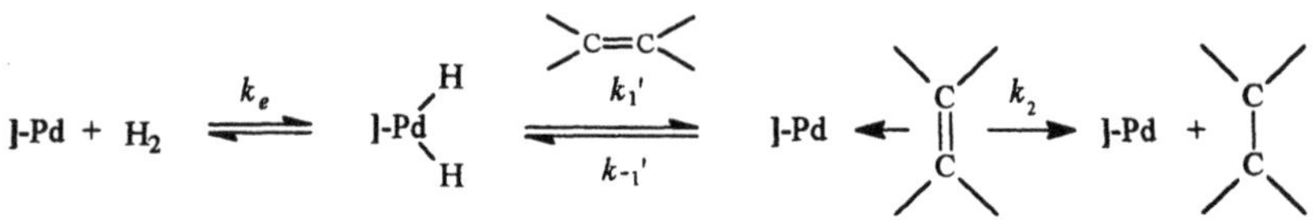

Scheme 1.9

and is expressed by the equation:

$$w = \frac{k_1' k_2' k_e' \, []-Pd][H_2][S]}{k'[S]+k_1'+k_2'}$$

Proceeding from a kinetic analysis of the most frequently observed dependences of hydrogenation rate on the concentration of catalyst, substrates of different type and H_2 pressure one may reach the following conclusions.

1. At low H_2 pressure the rate of hydrogenation increases directly in proportional to the catalyst concentration. This is the most frequently observed experimental dependence. Typical examples are: the hydrogenation of cyclohexene, styrene and 1,3-cyclooctadiene by palladium immobilized on phosphynated PS; the similar reactions of styrene and crotonic acid by rhutenium-containing ionites [124,133,162] (Figure 1.14); aromatic nitrocompounds in the presence of palladium polymer catalysts [157], etc. It should be noted that a study of the hydrogenation rate dependence on the immobilized catalyst concentration is often used as a test for diffusion or kinetic control reaction rate.

2. The dependence of hydrogenation rate may be either linear in H_2 concentration [23,124,133,162,163] or not [124,133,162,163] (Figure 1.15). The latter dependence has been observed for the case when $k_1[S]>>(k_{-1}+k_2[H_2])$ and becomes linear when $1/w = f(1/P_{H_2})$ (Figure 1.15,b). This procedure is often employed for calculation of k_1 and k_2 values.

3. The hydrogenation rate is ordinarily proportional to substrate concentration. However some examples are known where reaction rate is independent of [S], i.e. it is zero order with respect to substrate concentration, [124,133,164] (Figure 1.16). This is probably connected with the processes of substrate interaction with the polymer support (complex formation with its functional groups, polymolecular adsorption, etc.).

Some examples of a non-linear dependence of hydrogenation rate on substrate concentration are also known [124,133,162] (Figure 1.17). A plot of $1/w$ against $f(1/[S])$ yields a straight line. The values of k_1 and k_{-1} can be calculated from this plot.

While avoiding detailed comparison of immobilized catalysts with others, we shall only note that sometimes the parallels reveal kinetic features that are intrinsic in homogeneous hydrogenation catalysts, and at times resemble heterogeneous catalysts. However, a specific feature of metal polymer catalysts is the increase of local reagent concentration in the polymer phase. This is a result of higher substrate concentration in the polymer phase than throughout the total reaction volume due to the differences between polymer and solvent polarities. A non-polar substrate in a polar solvent tends to penetrate into the polymer volume across a polarity gradient. As a result one can observe a rise of the hydrogenation rate. Experimental data for the hydrogenation of aromatic compounds with different functional groups (especially with $>C=O$, $-C\equiv N$, $-NO_2$) are rather contradictory due to a number of competitive substrate actions in the course of the reaction.

1.7 The Role of the Metal Complex Surface Density in the Process of Active Site Formation

Not only the total amount of the bound metal complex but also the character of their distribution in the polymer matrices (topography) can have a profound effect on the rate of hydrogenation. As a rule, catalysts in which all active centers are accessible to the substrate show the largest activity. Such MX_n distributions can be specified either at the stage of catalyst synthesis or sometimes in the course of catalyzed reaction. Thus, Creamer's complex [Rh(C₂H₄)₂Cl]₂ bound to phosphynated CSDVB with different amounts of cross-linking agent (2–60%) under conditions which exclude polymer swelling [165] gave a catalyst that was 1.5–4 times more active than that produced from polymers that swell easily (Table 1.10).

Figure 1.14 The dependences of hydrogenation rates of different substrates on the catalyst concentration (a) 1, styrene; 2, 1,3-cyclooctadiene; 3. cyclohexene (palladium on phosphynated PS); (b) crotonic acid (rhutenium bound to ionite); (c) nitroanthraquinone (Pd bound to anionite).

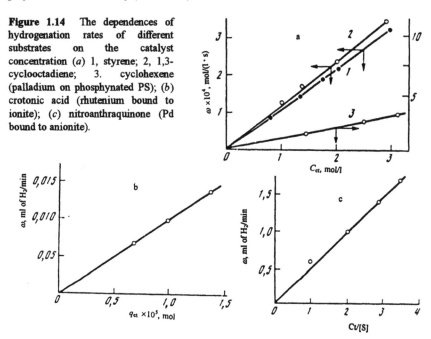

The specific catalytic activity depends in a complex way on the bound metal complex content (or density of the metal complex on the polymer support). As a rule, the SCA increases (Figure 1.18), reaches its largest value and then diminishes with the rise in the amount of metal complex.

Such a bell-like form of SCA dependence on the surface density of transition metal ions has also been observed in other catalytic reactions (see Chapters 2 and 3), probably being one of the specific features of catalysis by immobilized metal complexes. If there is no well-founded explanation of the rising branch of the plot, the diminishing trend may be connected with the formation and further growth of low- and/or zero-valent transition metal ion associations diminishing the catalyst efficiency.

Some ideas about the effect of active site distribution have been developed for di- and polymerization processes (see Chapter 3) [166] by extensions and models for hydrogenation catalysis. Active centers of immobilized catalysts are localized on the boundaries of cluster-like substances with stabilization by their electron systems. With the rise of metal complex surface density the number of such sites suitable for catalysis decreases. It is important that these associations are dynamic; their number may change in the course of bound complexes activation. Thus the reactivity of rhodium complexes bound on phosphynated CSDVB (8 or 40% of DVB) in hexene-1 hydrogenation was much greater after preliminary treatment with hydrogen [167] (Figure 1.19). The time of treatment affects in a complex way the activity of the catalysts so formed, SCA having the highest value for catalysts with small amounts of bound rhodium ([Rh]=0.51 and 1.22%). Obviously in catalytic contacts the highest proportion of Rh is localized on the surface of the support and its cross-linked structure prevents rhodium particle aggregati-

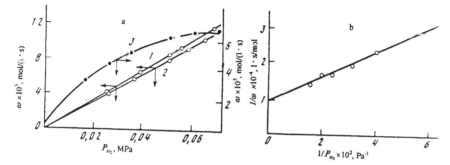

Figure 1.15 The dependence of substrate hydrogenation rate on the hydrogen pressure: (a) 1, styrene; 2, 1,3-cyclooctadiene; 3, cyclohexene; (b) linearization of hexene-1 hydrogenation rate dependences.

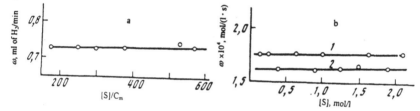

Figure 1.16 The dependence of hydrogenation rate on substrate concentration (a) nitroquinone; (b) 1, styrene; 2, 1,3-cyclooctadiene.

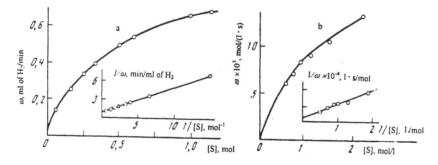

Figure 1.17 Non-linear dependence of hydrogenation rate on the substrate concentration and its linearization (a) styrene; (b) cyclohexene.

on both at the stage of activation and in the course of hexene-1 hydrogenation. The reactivity of such catalysts in benzene hydrogenation testifies (in combination with measurements of particle surface values) to the formation of metal centers composed of Rh^0 particles capable of enlargement in the course of the reaction [168].

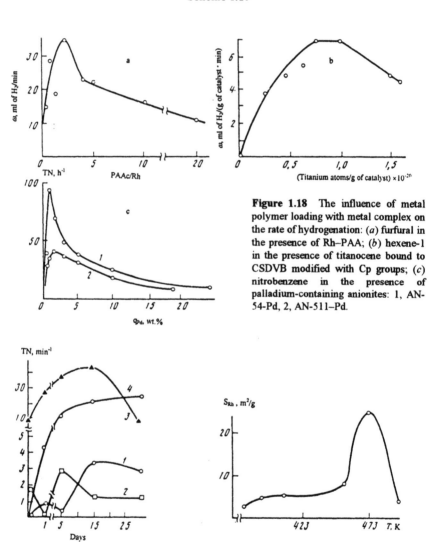

Scheme 1.10

Figure 1.18 The influence of metal polymer loading with metal complex on the rate of hydrogenation: (*a*) furfural in the presence of Rh–PAA; (*b*) hexene-1 in the presence of titanocene bound to CSDVB modified with Cp groups; (*c*) nitrobenzene in the presence of palladium-containing anionites: 1, AN-54-Pd, 2, AN-511–Pd.

Figure 1.19 The dependence of metal polymer catalyst activity in hydrogenation of hexene-1 on the time of treatment with hydrogen. DVB content (%) in phosphynated CSDVB: 1, 3 – 8; 2,4 – 40; 1,2 – catalysts with high metal content, 3,4 – catalysts with low metal content.

Figure 1.20 The influence of the temperature of Rh–CSDVB treatment with hydrogen on the surface of zero-valent rhodium particles. 60% DVB; P/Rh=3; Ssp=300 m^2/g.

The activity of metal polymer catalysts rose with an increase in the degree of decarbonylation. Fully decarbonylated catalyst (Rh^0 particles) has been shown to be the most active one. The temperature of catalyst pretreatment with hydrogen influences both the rate of Rh^0 formation and the value of the catalyst specific surface (S_{sp}) [165] (Figure 1.20). A sharp increase of S_{sp} in the range 453–473 K is probably the result of autocatalytic reduction and the diminishing branch of the plot is connected with Rh^0 aggregation.

Table 1.10 Hexene-1 hydrogenation in the presence of bound Creamer's complex [154].

DVB content in the support (%)	P/Rh in the catalyst		Activity TN, s^{-1}	
	Unswelled	Swelled	Unswelled	Swelled
2	25.0	6.5	2.35	0.49
25	5.6	3.0	1.02	0.60
60	2.0	3.0	0.66	0.40

Electron spectroscopy for chemical applications (ESCA) has shown that after the reduction of palladium-containing polymers with $NaBH_4$ or H_2 all the metal in the catalyst is in the Pd^0 state (see Table 1.8). The degree reduction of $Pd^{2+} \rightarrow Pd^0$ and consequently the catalyst activity for olefin hydrogenation changes significantly even for macroligands of the same type (for example, for polyheteroarylenes) [39–41] (Table 1.11). Two different types of Pd^0 particles have been detected in such reduced forms. The value and the sign of the electron bond energy shift for relatively substantial metallic palladium ($E^b_{Pd\ 3d\ 5/2}$=335.6 eV) in combination with other known data [133] allow us to conclude that Pd^0 particles in such catalysts are in the size range 2–3 nm on average. X-ray analysis has shown the presence of clusters composed of Pd^0 particles in such catalysts of 3.5±1 nm diameter. Such small particles are capable of strong interactions with polymer matrices. Palladium atoms assigned a positive charge, $Pd^{\delta+}$, due to the electron transfer, service the coordination of substrate and consequently the activity of such catalysts. The presence of a π-conjugated system of the polymer support which transfers into an ion-radical state also benefits the effect. The process of strong metal ion interactions (in the course of reduction with $NaBH_4$) with the support is followed by electron transfer and metal reduction up to leuco-base state.

Because the selectivity of the reactions is practically the same for all the above metal polymers, one may conclude that hydrogenation proceeds by Pd^0 immobilized on the polymer matrices. Analysis of the hydrogenation rate dependences on the value of catalyst potential shift $\Delta\phi$ into anode range supports this conclusion. The value of $\Delta\phi$ to a first approximation is assumed to be a measure of Pd^0 particle formation as during the reduction of bound Pd^{2+} ions the released vinylpyridine groups increases the basicity of the supports and accordingly the saturation potential of the catalysts.

Metal centers composed of Pd^0 particles have been detected in many other metal polymer systems. Thus the character of the SCA changes (Figure 1.22) and the induction period of hydrogenation only by Pd^{2+} species shows that the activity of palladium bound to phosphynated CSDVB in 4VP hydrogenation is connected to the Pd^0 content [134]. The degree of the metal palladium dispersion changes within rather broad limits (from 0.05 nm to 1 μm) during hydrogenation and depends on the Pd^0 content in starting catalyst. This coincides with the expectation of a catalytic activity decrease with the fall of metal dispersion due to metal aggregation. ESR spectra of such catalysts show no peaks corresponding to Pd^{+1}. It is important that Pd^0 particles have been formed already at the stage of metal complex immobilization by the phosphynated polymer matrix. The Pd^0 fraction increases with a rise of the initial concentration of metal complexes. These processes have already been discussed in Section 1.2.1.

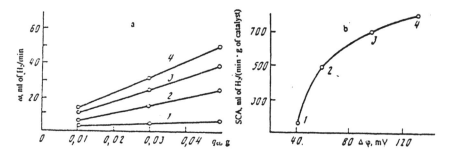

Figure 1.21 The dependence of allyl alcohol hydrogenation rate on catalyst amount, its nature (a) and the value of the catalyst potential shift (b): 1, Pd^{2+}–P2VP; 2, Pd^{2+}–P4VP; 3, Pd^0–P2VP; 4, Pd^0–P4VP.

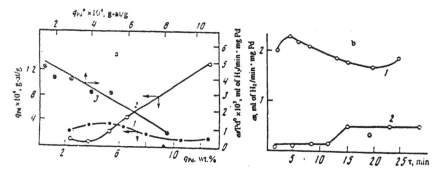

Figure 1.22 The relationship between Pd^{2+} (1) and Pd^0 (2) in $PdCl_2$-phosphynated CSDVB catalysts and their SCA in 4VP hydrogenation (related to Pd^0) (3) (a) and 4VP hydrogenation in the presence of metal polymer with a mixture of Pd^{2+} and Pd^0 (1) and only Pd^{2+} (2) (b).

Palladium catalysts bound to polyvinylpyridines may be arranged by relative activities in the sequence [29,30] (the selectivity in % is given in parenthesis): P4VP·Pd^0 (86) >P2VP·Pd^{2+} (86) >P4VP·Pd^{2+} (85) >P2VP·Pd^{2+} (89) (Figure 1.21) It is most probable that the active metal centers in metal polymer systems are used for hydrogenation (as in the majority of heterogeneous catalysts) combine M^0 species. Their formation proceeds both at the stage of catalyst formation and during catalyst activation as well as during the course of the catalyzed reaction. Polymer matrices probably promote stabilization metal centers (including transition metal ions of unusual valence states) in particular preventing their aggregation into large particles. These processes depend both on the nature of the polymer support and its functional groups; amino groups hinder aggregation more effective than phosphyne groups [169]. Polymer ligands appear to exert a double function. Primarily they must show electron donor properties being coordinated with transition metal ions in high oxidation state and also act as electron acceptors to stabilize metal particles in the low oxidation state.

1.8 Catalysis by Dispersed Colloid Particles and Transition Metal Clusters

A great number of studies on dispersion in the polymer of small metal particles as a means of immobilized catalyst production have appeared for the reason that formation of metal complex aggregates composed of zero-valent metal species bound to the polymer supports has often been observed during catalysis. The main role of the polymer matrix in such catalysts is the stabilization of particles by prohibiting further aggregation. There are many approaches to the preparation of such catalysts that show different activities and selectivities.

Thus, effective hydrogenation catalysts containing metal particles of Pd, Ru, Rh, Ag, Os, Ir, Pt, Au, Ni, etc., of different sizes have been produced by boiling the transition metal salts with macroligands in protonic solvents [170–174]. In particular, the following transformations in reaction systems have been observed by boiling the mixture $RhCl_3$-poly(vinyl)alcohol (PVAl)–methanol–water.

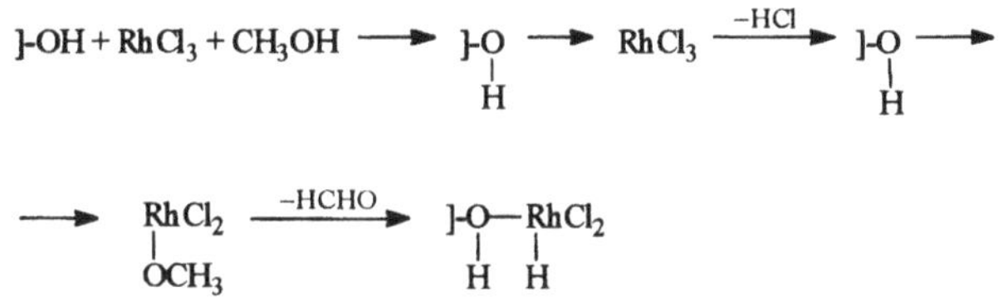

$$]-OH + RhCl_3 + CH_3OH \longrightarrow]-\underset{\underset{H}{|}}{O} \longrightarrow RhCl_3 \xrightarrow{-HCl}]-\underset{\underset{H}{|}}{O} \longrightarrow$$

$$\longrightarrow \underset{\underset{OCH_3}{|}}{RhCl_2} \xrightarrow{-HCHO}]-\underset{\underset{H}{|}}{O}-\underset{\underset{H}{|}}{RhCl_2}$$

Scheme 1.11

This sequence includes the coordination of $RhCl_3$ with PVAl followed by oxonium compound formation which further transforms into a bound hydride complex via an intermediate alcoxyde. A series of these reactions is necessary for the formation of homogeneous colloid because rhodium hydrides disproportionate [1] and this causes metal particle to become coarser.

$$PVA-Rh^{3+} \xrightarrow{CH_3OH} PVA \cdot Rh_3 \text{ (particles of diameter 0.8 nm)} \longrightarrow$$

$$\xrightarrow{growth} PVA \cdot Rh_{13} \text{ (particles of diameter 1.3 nm)}$$

Scheme 1.12

Thirteennuclear rhodium clusters packed in face centered cubic crystals (rhodium coordinative number 12) of 4 nm diameter are the end product of these transformations. These particles are bound to protective colloid either by electrostatic attraction or by physical adsorption and also possibly by coordinative bonds[9].

---

[9] It has already been shown [165] that 25 metal atoms are sufficient for adsorbed particle properties to become the same as those of larger volume metal particles.

Table 1.11 The dependence of the olefin hydrogenation rate constant on the electron bond energy of $Pd_{3d\,5/2}$ in palladium-containing heteroarylenes [39–41].

	Pd^{2+}		$Pd^{\delta+}$		Pd^{0}		$k \times 10^2$, min^{-1}			
	E_b (eV)	Content (%)	E_b (eV)	Content (%)	E_b (el')	Content (%)	Octene-1	Octene-2	cis-octene-2	Cyclohexene
Pd·PNBI[a]										
	338.0	40.0	336.0	60	–	–	50.0	5.3	6.6	8.1
Pd·PNBI-X[a]										
	338.0	32.5	336.3	57	335.7	10.5	18.4	–	–	3.3
Pd·PNT[a]										
	337.0	65.0	–	–	335.5	35.0	2.8	0	0	0

[a] PNBI, polynaphthoylenebenzimidazole; PNBIX, polynaphthoylenebenzimidazole–Cl; PNT, polynaphthoylene-S-triazole.

Cluster particles of different sizes incorporated in polymer matrices (Table 1.12) can be produced either by variation of reaction conditions or changes of alcohol and polymer (poly-N-(vinyl-2-pyrrolidone) (P2VPr), copolymer of MMA with N-vinyl(2-pyrrolidone), poly(acrylamide) gel, etc.). The activity of such catalysts in olefin hydrogenation (303 K, EtOH/H_2O (1:1), H_2 pressure 0.1 MPa) is 2–22 times higher than that of standard 5% Rh/C. Whilst the olefin hydrogenation rate is fairly insensitive to the particle size[10] the rate of cyclohexene hydrogenation decreases with increasing particle size.

Colloids of Pd^0 (1.8–5.6 nm) detected by electron microscopy, have been shown to be active and to be selective catalysts for cyclopentadiene reduction to cyclopentane (methanol, 303 K, H_2 pressure 0.1 MPa), as well as for 1,5-, 1,4-, 1,3- and 4,4-hexadiene hydrogenation. The reduction of conjugated dienes proceeds either at the 1,2 and 1,4 positions depending on the nature of the diene. Palladium particles (diameter 1.8 nm) in PVPr-protected colloid are effective catalysts for hydrogenation of the methyl ester of linolenic acid as well as for partial hydrogenation of soybean oil under ambient conditions [172].

The specific features of such catalysts are a narrow distribution of particle size, colloid stability within a large range of pH values (from 2 to 13), the ease of catalyst separation and long-term stability after many times usage.

Some other examples of such catalysts are as follows. Nickel boride sol, an effective catalyst for olefin hydrogenation has been produced by $NiCl_2$ reduction with $NaBH_4$ in ethanol in the presence of PVPr as a protective colloid [175]. β-cyclodextrin may be used as a stabilizing agent for Rh, Pt and Pd colloid particles (1–100 nm of the size). Such systems are effective catalysts for olefin hydrogenation as well as for semihydrogenation of dienes, etc.

Photoreduction of K_2PdCl_4 solution encapsulated in polymer bubbles [176] gave active colloid palladium particles of ≤5–6 nm diameter; their activity in ethylene hydrogenation rose with decreasing Pd species size.

Table 1.12 Sizes of Rh^0 colloid particles and their activity in olefin[a] hydrogenation [170].

System	Average Rh particle size (nm)	Activity (mol H_2/g-at of Rh·s)	
		Hexene-1	Cyclohexene
RhCl$_3$/PVA/CH$_3$OH/H$_2$O	4.0	15.2	3.1
RhCl$_3$/polymethyl ester/CH$_3$OH/H$_2$O	4.3	22.5	9.6
RhCl$_3$/PVPr/CH$_3$OH/H$_2$O	3.4	15.8	5.5
RhCl$_3$/PVPr/C$_2$H$_5$OH	2.2	15.5	10.3
RhCl$_3$/PVPr-iso/C$_3$H$_7$OH	2.4	–	9.3
RhCl$_3$/PVPr-CH$_3$OH/NaOH	0.9	16.9	19.2

[a]Reaction conditions: 303 K, H_2 pressure 0.1 MPa, [olefin]=25 mmol; [Rh]=0.01 mmol.

---

[10] As will be shown in Section 4.8, the activity of colloidal Pt in H_2O photoconversion depends not on dispersity but mainly on the protective ability of polymer support. The quantitative measure of this value [177] is the protective ability for some gold mass (in g) applied to 1 g of the polymer to the action of a 1% solution of NaCl.

Photoreduction is carried out [177] in the presence of capillary-active substances in the medium of water-soluble polymers. The colloid particles formed can be coated with micelles or microcapsulated [178]. Ultradispersed Ni particles within the size range 5–20 nm (effective catalysts for 1,3-cyclohexadiene reduction) have been produced [171] by diffusion of $Ni(NO_3)_2$ through gel membranes of cellulose acetate with subsequent decomposition by H_2.

Noble metal clusters dispersed in macroreticular polymer supports are active catalysts for hydrogenation. These may be produced by reduction of metal ions in the polymer medium [101], evaporation of Pd onto graphite (particles size 0.8–2.5 nm), or by sputtering of isolated or solvated metal atoms under vacuum at 77 K [179], etc. Even PS and CSDVB dopes for Rh- and Pt-blacks yield a 1.5–2.0 rate increase for the hydrogenation of nitrobenzene and dimethylethylcarbinol [180].

Other cluster-containing polymers produced by immobilization of active cluster complexes are also of great interest. Such catalysts provide the possibility of permanent cluster-structure control. It is often possible to correlate the changes in catalyst activity with changes in cluster complex structure. Thus, the rate of ethylene hydrogenation promoted by tetrairidium or tetrahutenium carbonyl clusters bound to phosphynated polymers [181,182] decreases with an increase of the number of donor phosphyne ligands in the cluster environment (Table 1.13). This effect is also accompanied by a rise of ethylene–catalyst bond strength.

Table 1.13 The influence of P/Ir ratio on the ethylene hydrogenation[a] rate [183].

Catalyst	P/Ir	$w \times 10^3$ (mol of C_2H_4/g-at of Ir·s)
]–PPh_2Ir_4(CO)_{11}	4.5	21.3
]–PPh_2Ir_4(CO)_{11}	14.3	25.3
]–PPh_2)_2Ir_4(CO)_{10}	3.4	3.76
]–PPh_2)_2Ir_4(CO)_{10}	6.9	3.60

[a]Reaction conditions: ethylene pressure 20 kPa, H_2 pressure 80 kPa, 313 K.

Hydrogenation is fully inhibited in the presence of traces of carbon monooxide but the activity is restored after removal of CO. IR–spectroscopy data show no M–CO bond rupture during hydrogenation. So, coordinative vacancies arise as a result of L–M bond rupture or most probably breaking of the M–M bond in (Ru–Ru, Rh–Rh, Os–Os, etc.). Active catalysts for both hydrogenation and isomerization of hexene-1 have been produced [183] by immobilization of $H_4Ru_4(CO)_{12}$ or $H_4Ru_4(CO)_{10}(PPh_3)_2$ on polymers or the mixed-type supports. Such factors as solvent character, polymer reticulation and L/M ratio significantly influence the activity of immobilized cluster complexes.

Reduction processes in solutions catalyzed by simple cluster complexes (e.g. of styrene with tetranuclear osmium clusters [179]) also follow the mechanism of cluster fragmentation accompanied by the formation of highly active particles with low nuclearity, possibly mononuclear. The concentration of such species, which are probably active sites of the reaction, is rather low. Cluster Fe_4S_4, activated with LiPh, in hydrogenation reactions splits under H_2 action and gives rise to hydride cluster complexes[11] [184]. This system is a catalyst for double bond hydrogenations of monoenes and dienes. It is also very selective in hydrogenation of terminal bonds [186] and in hydrogenation of norbornadiene to norbornene. Substrate selectivity may be controlled through the molar ratio of components.

Another interesting example of cluster structure transformation is the heterometallic $RhOs_3$-cluster (derived from $H_2RhOs_3(AcAc)(CO)_{10}$) that segregates in the model double reaction of ethylene hydrogenation and butene-1 isomerization [187]. This provides the bond separated trinuclear osmium clusters responsible for isomerization and the metallic rhodium particles 1.0 nm in diameter, that are active in hydrogenation[12]

1.9 Selectivity in Hydrogenation Processes

The practical importance of immobilized catalysts is characterized not only by their high catalytic activity and long-term stability but also, with selectivity, i.e. the ability of the reaction to yield one or more desirable products. Four main requirements may be imposed for a heterogeneous catalyst to provide the best selectivity [191]. These are: (i) a balance between bond strengths; (ii) the ability to supply a specific substrate coordination; (iii) the possibility of active ensemble formation; (iv) a template effect. All these parameters depend on both electronic and geometric properties of the metal particles. It is rather difficult however to detect the influence of these separate effects in many component reactions.

The main processes that accompany the hydrogenation reaction and decrease its selectivity are isomerization of substrate, of intermediate or of desired product, alkylation and di- or trimerization, etc. The knowledge of the reaction mechanism and possibilities of reaction control at the stage of catalyst synthesis and upon changes in reaction conditions enable the control of reaction products.

1.9.1 The influence of composition and complex structure on the catalyst selectivity

In essence those factors that are significant for catalyst activity also affect catalyst selectivity. These include the nature and topography of immobilized complexes, the nature of the polymer matrix and functional groups, the L/M ratio and reaction conditions. Thus, the nature of the functional groups (in particular their π-acceptor properties) of modified CSDVB has a very profound influence on 4-vinylcyclohexene hydrogenation [101]. The degree of catalyst selectivity (the ratio of 4-vinylcyclohexene yield to other products formed) being equal to 0.4 may reach 0.97–0.98 after 4-ethyl-cyclohexene separation. Immobilized catalysts show high selectivity in hydrogenation of

---

[11] Dialkylidine cluster $Os_3(CO)(\mu_3\text{-CPh})(\mu_3\text{-CCH}_3O)$ is capable of hydration with H_2 in toluene at 383 K [185]; the process is accompanied with alkylidine ligand coupling, thus the product is a mixture of the two complexes.

[12] The heterometallic cluster $CpNiOs_3$ anchored to $\gamma\text{-Al}_2O_3$, activated at 373-573 K, is more active than the catalyst produced by reduction and calcination of the separate metal salts. One may relate this observation [188] with the formation of a more highly dispersed metallic catalyst. Heterometallic vinylidene cluster $HFe_3Rh(CO)_{11}$ (C=CHR) (R=H or Ph) catalyzes olefin hydrogenation [189]. At the begining of the reaction it isomerizes in preference the terminal double bonds but segregates in the course of the reaction producing some $H_3Fe_3(CO)_9(\mu\text{-CCH}_2Ph)$, which becomes a catalyst for the hydrogenation. This is also characteristic of the $NiRu_3$ cluster active in selective hydrogenation of conjugated dienes and isomerization of unconjugated dienes and cyclic olefins [190].

aromatics and heteroaromatics, for example in hydrogenation of phenol to cyclohexanol and saccarides to polyols, and alkenes isomerization to *cis-* and *trans*-alkenes [8]. Hydrogenation of allylbenzene and 1,5-cyclohexadiene with cyclopentadienyl derivatives of zirconium and hafnium bonded to polymers [67] causes isomerization with double bond migration and the formation of the more stable *cis-* and *trans*-propenylbenzenes and 1,4-cyclooctadiene.

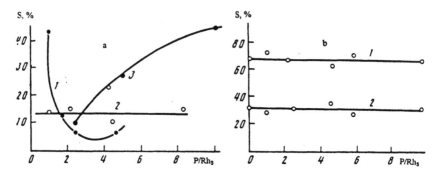

Figure 1.23 The selectivity of phosphynated CSDVB immobilized rhodium catalysts in hydrogenation of 1,3-butadiene to butene-1 (*a*) and acetylene to butene-1 (*b*) as functions of P/Rh$_s$ ratios. *a*: phosphorus content in polymer, %, 1, 0.5; 2, 12; 3, 1.7. *b*: 1, ethane; 2, ethylene.

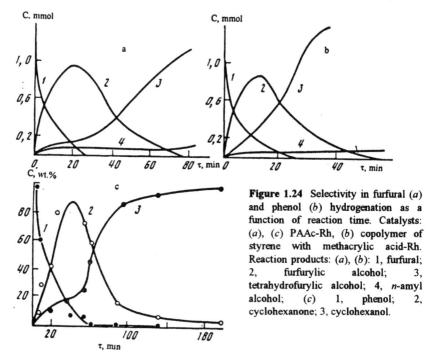

Figure 1.24 Selectivity in furfural (*a*) and phenol (*b*) hydrogenation as a function of reaction time. Catalysts: (*a*), (*c*) PAAc-Rh, (*b*) copolymer of styrene with methacrylic acid-Rh. Reaction products: (*a*), (*b*): 1, furfural; 2, furfurylic alcohol; 3, tetrahydrofurylic alcohol; 4, *n*-amyl alcohol; (*c*) 1, phenol; 2, cyclohexanone; 3, cyclohexanol.

By analogy with homogeneous catalysis [1], the substrate can isomerize if the active complex formation step is reversible:

$$MX_n \xrightarrow{H_2,\,\text{substrate}} X_nM\text{-substrate} \underset{+H}{\overset{-H}{\rightleftharpoons}} X_nM\text{-substrate}^I$$

$$X_nM\text{-substrate}^I \xrightarrow{H_2} X_nM + \text{product}^I$$

Scheme 1.13

The appearance of isomerization products testifies to the stepwise addition of hydrogen to substrate.

The following examples demonstrate the influence of immobilized catalyst composition on selectivity.

The greater the bulk surrounding Rh (in $[Rh(COD)Cl]_2$ catalyst) the greater the catalyst selectivity in 1,3-butadiene hydrogenation to butene-1 [131] (Figure 1.23). A reciprocal dependence has been observed at low P/Rh_s ratios (for similar supports containing only 0.5% of phosphorus) when more large rhodium particles formed. Thus, selectivity as the function of P/Rh_s ratio experience a combined effect of both steric hindrances and particle size with dominant role of the first. For the catalyst series with phosphorus content equal to 1.2% these factors counterbalance each other and selectivity does not depend on P/Rh_s ratio. As the diminishing of particle size is accompanied with both geometry changes of surface species and electronic state of metal particles it is rather difficult to determine which of these factors is responsible for selectivity changes. Meanwhile it is obvious that appearance of additive coordinative vacancies of the active sites, free of solvent molecules on the contrary to homogeneous systems, must favor olefin isomerization. The mechanism of hydrogenation and isomerization with immobilized Wilkinson's complex is shown by the scheme [192]:

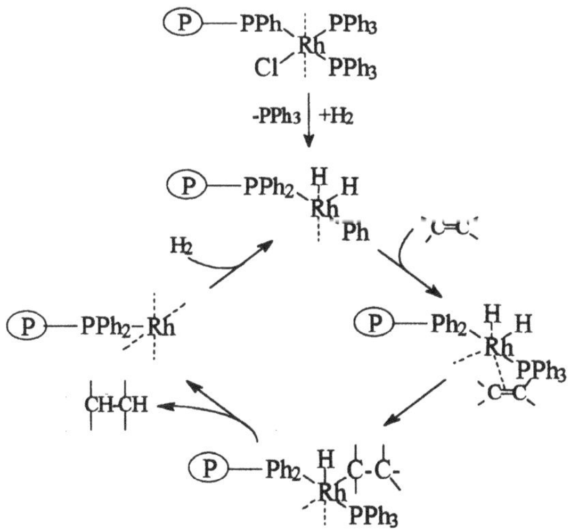

Scheme 1.14

In hydrogenation of phenyl- and diphenylacetylene (PdCl$_2$·PEI) the main reaction products are *cis*-alkenes with a slight amount of *trans*-alkenes and unsaturated compounds [193]. The rates of hydrogenation of 1,3- and 1,4-cyclopentadienes (323 K, H$_2$ pressure 0.5 MPa, catalyst PdCl$_2$-PEI) are equal to 3.24 and 0.87 mol/l·s. Catalyst is highly selective: only 1,3-diene reacts in the mixture of dienes from initial stages up to 95% conversion and after that 1,4-diene undergoes the reaction. Diene disproportionation is inhibited with acetylene additives.

However, the rate of acetylene hydrogenation (rhodium catalyst on phosphynated CSDVB) does not depend on the P/Rh$_s$ ratio below 10.7. This indicates that both ethylene desorption in the intermediate state and acetylene redesorption [131] are not controlled by steric factors. In other words, the amount of phosphyne does not influence the equilibrium constant value:

$$Rh(C_2H_2) \xrightarrow{H_2} Rh(C_2H_4),$$

$$Rh(C_2H_4) + C_2H_2 \underset{}{\overset{K}{\rightleftharpoons}} Rh(C_2H_2) + C_2H_4,$$

$$Rh(C_2H_4) \xrightarrow{H_2} Rh + C_2H_6$$

Scheme 1.15

Moreover, K does not depend on an increase of electron density of the metal particles.

Palladium complexes bound to gel-like and macroreticular ion-exchange resins are selective catalysts for the hydrogenation of vinylacetylene to 1,3-butadiene [32], but their selectivity decreases sharply with increasing conversion. Pt^0, Pd^0 as well as Pt^{2+} and Pd^{2+} complexes are active in the semihydrogenation of 1,5-cyclopentadiene with a selectivity factor of 93–94% and that for acetylenes, 76–84%. These are also effective in selective hydrogenation of alkynes with internal triple bonds [37]. Both at low cross-linking degree and modest phosphyne loadings the system Ni(AcAc)$_2$/Al$_2$Et$_3$Cl$_3$/phosphynated PS shows selectivity in the hydrogenation of 1,4-cyclohexadiene to cyclohexene analogously to that for the homogeneous system, Ni(AcAc)$_2$/Al$_2$Et$_3$Cl$_3$/PPh$_3$ [66], whereas nickel complexes bound to acid-nafion are only selective in olefin isomerizations [66].

Furfural is reduced to furfurylic and tetrahydrofuryl alcohols [153]:

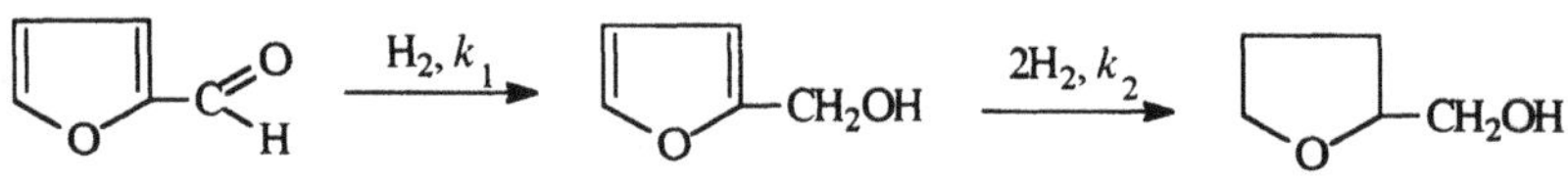

Scheme 1.16

The dependence of these product yields on the reaction time is demonstrated in Figure 1.24 (*a*) and (*b*); the values of rate constant, k_2, are equal to 0.04 min^{-1} and 0.01 min^{-1} for the metal complex of the copolymer of styrene with methyl acrylate

(MAc) and with polyacrylic acid (PAAc), respectively. The former catalyst is preferable for tetrahydrofuryl alcohol production and the second for the furfurylic series. Palladium-rhodium catalysts immobilized on sugar coal [194] are effective for furfural hydrogenation; the yield of furfurylic alcohol increases with a rise of the Rh content.

Selectivity can be promoted by variations of reaction conditions. Thus, phenol hydrogenation with Rh-PAA [154] can be controlled to result in the production of cyclohexanol (practically quantitatively) or cyclohexanone (yield not less than 80%) formation (Figure 1.24c). The shapes of the product yield plots reveal the following stepwise reactions:

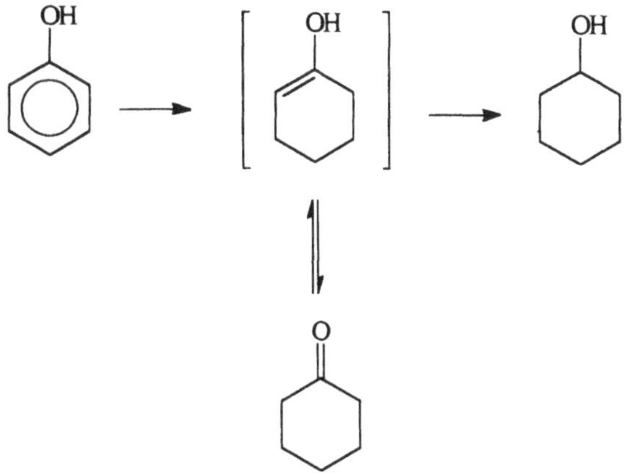

Scheme 1.17

Complexes of PEI with palladium and rhodium chlorides are highly selective catalysts for hydrogenation of multiple bonds of allyl residues of phenol derivatives [195] (allylphenol, allylphenylether, etc.).

Hydrogenation of tetrahydrofurylacetylene alcohols proceeds at a fast rate and high selectivity in the presence of supported Pd catalysts. Under complete substrate conversion, selectivity reaches 98–99%, their activity being equal to 17 000–13 000 substrate mol/g-at of Pd per h at 293 K and H_2 pressure of 0.1 MPa [196,197]. Hydrogenation of benzaldehyde proceeds completely under ambient conditions (298 K, H_2 pressure 0.1 MPa) with PAAc-Rh catalyst [198]. By monitoring the reaction conditions it is possible to change the degree to which the hydrogenation reaction proceeds, i.e. to reduce the aldehyde residue to alcohol, or to carry out the reduction of the alcohol residue and even the benzene ring. A reduction in catalytic activity is caused by inhibition of the reaction either by the substrate or the reaction products formed. Selective hydrogenation of carbonic acid chlorides (with a selectivity factor with respect to aldehydes of 60-100%) by palladium complexes bound to poly-*p*-(phenyleneterephthamide) proceeds under the action of metallic palladium at elevated temperatures (393 K), being controlled by H_2 diffusion to the catalyst surface. Chelated complexes of Cu^+ supported on phosphynated CSDVB show greater selectivity (91–94% depending on the nature of R) [199] for this process. Below is a general reaction scheme.

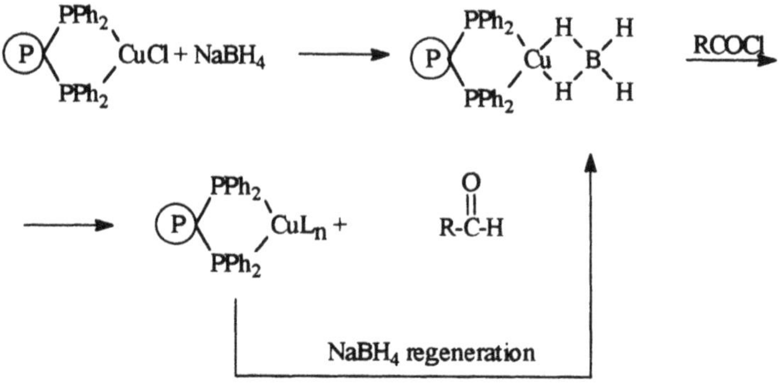

$(R = CH_3(CH_2)_5, C_6H_{11}, C_6H_5)$

Scheme 1.18

As a rule, selectivity is the result of the specific adsorption of one of the reagents or the reaction products to the catalyst surface. This may be the reason of selective hydrogenation of soybean oil by dispersed Pd particles stabilized with PVPd [200]. Unsaturated aliphatic acids themselves can also be reduced with high regio- and stereo-selectivity by immobilized metal complexes such as Pd- and Ni-complexes bound to carboxymethyl cellulose [201] or $PdCl_2$ on ion-exchange resins [202]. Colloidal Pt and Pd particles stabilized with PVPd are highly effective selective catalysts for hydrogenation of double bonds in both 10-undecylenic and 2-undecylenic acids as well their sodium salts [177]. Regioselectivity testifies that hydrogenation follows substrate dispersion in the polymer due to hydrophobic repulsion. In this connection the selective hydrogenation of benzocrown esters (in particular dibenzo-18-crown-6) by rhodium on carboxylic polymers [203]) should also be noted.

Comparisons of the activities of homogeneous and heterogenized Wilkinson's catalyst and the selectivity in hydrogenation of polynuclear heteroaromatics [126] showed that the SCA of the heterogenized catalyst was 10–20 times higher. Quinoline transformed to 1,2,3,4-tetrahydroquinoline with initial rates of 0.013 for the homogeneous and 0.29% min^{-1} (353 K) for the heterogenized catalyst. Under the action of metal-polymer complexes acridine reduced only to 9,10-dihydroacridine while up to 50% of 1,2,3,4-tetrahydropyridine was produced by $RhCl(PPh_3)_3$.

Another type of catalyst selectivity has been observed in hydrogenation of 4-vinyl-(cyclohexene) [80]. This proceeded into the side chains but not into the ring (independently of the nature of solvent) under action of both homogeneous and heterogenized (by modified cellulose) catalytic systems derived from $Rh(1,5-COD)(PPh_3)_2^+$. However, the selectivity of the bound complexes is 4 times greater than that of the homogeneous system. The same relations have been observed in both selective hydrogenation of 4-vinylpyridine [204] by Pd complexes and selective hydrogenation of benzene to cyclohexene by Rh/nylon [205].

A fall in selectivity may be caused by some intramolecular reactions of free functional groups such as hydrolysis, esterification, reesterification, amination, etc. These reactions may involve not only the substrate but also the solvent molecules, especially with alcohols. In addition, dissociation of some functional groups may cause a change in pH which can also influence both activity and selectivity of the metal-polymer catalysts.

The numbers of free functional groups can also influence the selectivity. Thus, the activity of $[RuCl_2(CO)_3]_2$ bound to polymers with carboxylic groups increases by 25 times with a decrease of functional group loading with metal complexes from 73 to 6% [206]. This may also be connected with the decrease in the ability of the polymer to swell [207]. On the other hand, hydrogenation of pentene-1 by $RhHCO(PPh_3)_3$ bound to a polymer of maleic acid proceeds until PPh_3 groups are released [144]. After all the PPh_3 groups have been removed, the catalyst becomes more active in isomerization than in hydrogenation, similar to the behaviour observed in the homogeneous system.

Other examples of intramolecular reactions that are capable of selectivity reuction include the following. The disproportionation of cyclohexadiene to benzene and cyclohexene (catalyst $PdCl_2$-polybenzimidazole) [151] and aromatization (hydroformic process) of hydrocarbons, more often proceeding by the action of immobilized technetium catalysts [208].

The very interesting mechanism of isomerization which accompanies the model reaction of butene-1 hydrogenation by heterometallic cluster $RhOs_3$ bound to CSDVB has already been mentioned above. In the course of the reaction this cluster segregates into a stable, bound, triosmium carbonyl cluster complex, active in isomerization of butene-1 and metal rhodium particles (diameter of <1 nm) capable of butene-1 hydrogenation.

The character of the metal center distribution can also influence the catalyst activity and selectivity. This has been demonstrated by comparison of both activity and selectivity (hydrogenation of butyne-2) of trinuclear Pd clusters bound to phosphynated PS and incorporated in the volume of the polymer bead or located only in the presurface catalyst layer. Catalytic turnover number for the last catalyst was 35 times greater than that for the first. However, the selectivity (with respect to cis-butene-2 formation) was 1.4–3.7 times higher for the first catalyst. This seems to be the result of differences in the structures of the active metal centers, i.e. differences in coordinate unsaturation of palladium ions in these two cases, or in the P/Pd ratio responsible for selectivity.

1.9.2 The geometrical selectivity in hydrogenation by immobilized metal complexes

The dependence of the hydrogenation rate on the size of the substrate molecule (geometrical or matrix selectivity) is an intrinsic feature of catalysis by immobilized metal complexes. The ratio of hydrogenation rates by immobilized and analogous homogeneous systems, S_M, may serve as the measure of the effect. Both experimental data concerning the dependence of the hydrogenation rate on the substrate molecule size and theoretical approaches used to explain the effect (influence of steric factors on both the absorption process and hydrogenation rate, electronic properties of substituents in the substrate molecule, some specific interactions between substrate and support and so on) are given in a number of works (for example, [24,39,114,134,147,192,209–213]). Table 1.14 gives an idea of the order of S_M value. The main experimental data on the hydrogenation of different substrates by rhodium [170,209,210], rhutenium [144], palladium [124,133,214], titanium [115], nickel [215] immobilized complexes are reported in Table 1.15.

The rate of alkene hydrogenation decreases with elongation of the olefin chain and a rise of cycloolefin size, and diminishes for substrates with inter-chain or inter-ring double bond. The polymer pore diameter is also of great importance (selective hydrogenation of 1,5,9-cyclododecatriene to cyclododecene in the presence of immobilized complexes of Rh and Ru [216]).

The rise of the number of substituents in the vicinity of the double bond in both linear and cyclic olefins, as well in alkenes with heteroatoms, also causes a fall in the hydrogenation rate. Electronic properties of substituents located close to the double bond are also of great importance. Donor substituents accelerate alkene reductions while acceptors retard (Table 1.15).

The values of both S_M and the hydrogenation rate are sensitive to such parameters as the properties of the polymer support, the ligand and the solvent, as well as reaction temperature, etc. Geometrical selectivity, however, is a general phenomenon intrinsic not only to hydrogenation but also to other catalytic reactions in the presence of immobilized catalysts. Thus, in contrast to the homogeneous system, elongation of the olefin chain leads to a decrease of a hydrogenation rate by immobilized $RhCl(PPh_3)_3$ with preservation of other specific features for this process (e.g., the hydrogenation rates follow the sequence: cis-olefins>trans-olefins, α-olefins>β-olefins); quantitative relationships depend on the properties of the support (Table 1.16).

Under other similar conditions, the S_M value depends on the character of the transition metal ion. Thus, the rate of hexene-1 hydrogenation is 18 times greater than that of hexene-2 or cyclohexene by Pt bound to PS modified with Dipy. However, the same reactions, catalyzed with similar Pd catalysts, are less sensitive to the substrate nature: the rate of hexene-1 hydrogenation is only 1.7 times greater than that of cyclohexene and 2 times greater than that of hexene-2 [108]. Generally speaking, the data concerning the hydrogenations of geometrical isomers are rather discrepant. Thus, if trans-isomers are reduced more rapidly in the presence of nickel borane stabilized by PVPd [215], under the action of other immobilized catalysts [40,217], the on the cis-isomers are reduced more rapidly.

Table 1.14 Relative rates of olefin hydrogenation by Wilkinson's complex and that immobilized by polymers [192, 209–213].

Olefin	Homo-geneous	Support		Support	
		J–C6H4CH2PPh2	S_M	J–C6H4PPh2	S_M
Hexene-1	1.00	1.00	1.00	1.00	1.00
Cyclohexene	0.71	0.38	0.54	1.00	1.41
Cyclooctene	0.71	0.15	0.21	0.50	0.70
Cyclododecene (70% cis-, 30% trans-isomers)	0.50	0.08	0.16	0.30	0.60
Octadecene (the mixture of isomers)	0.50	0.19		0.38	0.80	1.60
D^2-cholestene	0.50	0.01	0.02	0.10	0.20

To a considerable degree the phenomenon of selectivity is exhibited in hydroge-nation of aromatics and substrates containing functional groups, such as ≡N, . –C=O, –NO2, –C≡N, etc. Thus, the rate of hydrogenation decreases with introduction of bulky substituents (e.g. *tert*-butyl) in the benzene ring. The hydrogenation rates (catalyst rhutenium immobilized on PS) of benzene, toluene, *o*-xylene, *tert*-butylbenzene and hexamethylbenzene are in the propotions [218] 1.00:0.62:0.27:0.01:0.

Table 1.15 The influence of substrate structure on the hydrogenation rate in the presence of different metal polymer catalysts[a].

Substrate	The relative rate in the presence of catalysts					
	I	II	III	IV	V	VI
Hexene-1	1.00	1.00		1.00	1.00	1.00
2-Methylhexene-1					0.04	
Hexene-2	0.26					
cis-Hexene-2						0.04
trans-Hexene-2						0.05
Heptene-1			1.0			
Heptene-2			0.15			
Octene-1		0.93		0.99		
Decene-1		0.79				
Dodecene-1		0.76		1.77		
Hexadecene-1		0.64				
Octadecene-1		0.59				
Cyclopentene						0.34
Cyclohexene	0.35		0.21	0.63	0.42	0.03
Cycloheptene						0.28
Cyclooctene	0.07		0.02		0.16	0.05
Styrene	0.16		2.06		1.14	2.35
α-Methylstyrene						1.38
Allylbenzene			1.29			1.06
Allyl alcohol			1.48	1.14		3.04
Allyl chloride			0.64			
Acrolein			0.48			
Acrylic acid				0.63		
Acrylamide						0.23
Acrylonitrile	0.05					
Methylacrylate	1.12		1.42			1.06
Methylmethacrylate	0.96			0.78		0.74
Allylacetate				1.30		
Vinylacetate				0.84		

[a]Catalysts: I, RhCl$_3$ stabilized with PVPd [169]; II, RhCl(PPh$_3$)$_3$ bound to phosphynated PS [141]; III, PdCl$_2$ bound to phosphynated PS [124]; IV, palladium-containing polyurethane [95,214]; V, Ti^{4+} bound to CSDVB modified with Cp groups [115]; VI, nickel borane stabilized with PVPd [215].

In contrast with olefin substrates and aromatics, the structure of carbonyl compounds is only slightly dependent on the rate of C=O group reduction as in the case of hydrogenation in the presence of nickel borane stabilized with PVPd [213], but the general features remain the same. Cyclic ketones are more easily reduced than linear ones. This is most probably connected with steric factors.

Nitro-compounds with different functional groups are reduced to the corresponding aminocompounds in high yield, the reaction rates increasing for substrates with donor substituents (e.g. polyamide complexes of palladium [36], Pt·PEI [219], Pt^{2+} and Pd^{2+} bound to P4VP [220]). Both the hydrogenation rates of substituted nitrobenzenes and those of aldehydes hydrogenatic aminanation catalyzed by these catalysts ($PdCl_2$·PE-gr-P4VP) correlate with substituents Hammet's constants[13] (Figure 1.25) [97, 157]. The values of initial rates of hydrogenation of these substrates [222] are presented in Table 1.17.

Table 1.16 The comparison of some properties of different rhodium complexes[a] [192].

Property	$RhCl(PPh_3)_3$	Rh/SiO_2–PS	Rh/phosphynated CSDVB	Rh/CSDVB
Catalyst pretreatment	H_2	Evacuation at 393 K; H_2 at 373 K	Evacuation at 393 K; H_2 at 373 K	Evacuation at 373 K; H_2 at 373K
The order of activities	$C_5>C_4>C_3>C_2$	$C_2>C_3>C_4$	$C_2>C_3>C_4$	$C_2>C_3>C_4$
	$cis \Rightarrow trans$	$cis \Rightarrow trans$	$cis \Rightarrow trans$	$cis \Rightarrow trans$
Selectivity	No isomerization	0.7	318–363 K	313–363K
S_h^i			0.3–0.9	
S_1^t		5.8	10.0–7.0	7.3–11
Activation energy, kJ/mol				
of hydrogenation		25.2	59.6	45.8
of isomerization		30.6	47.0	30.6

[a] Rh/SiO_2PS–SiO_2 with adsorbed PS impregnated with $RhCl(PPh_3)_3$; Rh/phosphynated CSDVB–$RhCl(PPh_3)_3$ bound to copolymer of styrene, p-styryldiphenylphosphyne and divinylbenzene (93:5:2 mol%); Rh/CSDVB–$RhCl(PPh_3)_3$ bound to strongly cross-linked CSDVB; S_h^i, the ratio of initial rates of hydrogenation and isomerization product formation ($[C_4H_{10}]/([trans$-β-$C_4H_8]+[\alpha$-$C_4H_8])$); S_1^t – the ratio $[trans$-β-$C_4H_8]/[\alpha$-$C_4H_8]$.

On the one hand, a rise of both heterogenized catalysts (Pd^{2+}·PE-gr-P4VP) activity and selectivity has been reported [97] for p-nitrochlorobenzene reduction; on the other hand, hydrogenation of aromatics and heteroaromatics may be inhibited by the substrate [223] due to their action on the metal complex, as has been observed for homogeneous catalysts. Consequently the variety of such catalysts is rather scant. Macromolecular chelates providing high catalyst stability are supposed to be prospective catalysts in this connection [224]. These includes rhodium complexes with N-phenylanthranilic acid bound to polymer support [224], immobilized metal chelates of the platinum group with oxyquinolines [225], supported Pt^{2+} complexes with 1-phenyl-2-azonaphthole in DMFA [226].

The nature of vinyl pyridine isomers in modified CSDVB (2-, 3- or 4-VP) can also affect the rate of hydrogenation by immobilized palladium complexes [204]. The most effective are metal complexes bound to 2-VP, the least active are the complexes with 3-VP.

---

[13] The rate of alifatic nitriles hydrogenation in the presence of nickel catalysts also decreases with a rise of the number of carbon atoms in the substituent, i.e. with the decrease of σ^* values (Tapht's constants) [221].

Table 1.17 The initial rates of different nitro-compounds hydrogenation in the presence of Pd-poly(methyl methacrylate) (PdPMMA) [222].

Substrate	Initial rate, (ml/min)	Ratio of the initial rates of substituted substrate and nitrobenzene hydrogenation	Substrate	Initial rate, (ml/min)	Ratio of the initial rates of substituted substrate and nitrobenzene hydrogenation
Nitrobenzene	46.1	1.00	o-Nitrotoluene	34.3	0.74
p-Nitroaniline	54.5	1.18	o-Nitrophenol	48.0	1.04
m-Nitroanilin	50.0	1.08	Nitromethane	9.6	0.21
o-Nitroanilin	28.8	0.63	Nitroethane	4.6	0.10

The effect of geometrical selectivity may be usefully explicited in hydrogenation of one substrate in the presence of another or in the hydrogenation of a mixture of unsaturated compounds.

Figure 1.25 The influence of benzene ring substituent on the rate of hydrogenation of aromatic nitrocompounds, $XC_6H_4NO_2$, in the presence of $PdCl_2$·PE-gr-P4VP. X: 1, p-NH_2; 2, p-OH; 3, p-CH_3; 4, m-CH_3; 5, H; 6, p-Cl; 7, p-CHO; 8, m-Cl; 9, m-COOH.

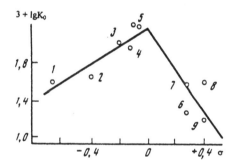

1.9.3 The membrane catalysis of hydrogenation

In heterogeneous catalysis for making semihydrogenated products which form active intermediate products with H_2, special catalysts fabricated of palladium and its alloys [227,228] are used. A number of different reactors of special design have been developed for this purposes. Recently, an alternative to these catalysts has been developed through special composite membrane catalysts. These usually consist of inert polymer support permeable to H_2 with a supported metal layer. Cation-exchange resins are convenient supports for such contacts. These have been used [229] for nitro-compound reduction (e.g. sodium p-nitrophenoxide). In such films (thickness of 0.4 μm) hydrogen ions show abnormally high mobility. They migrate in the membrane by both a relay-race mechanism and through vacancies. Thus, the value of the proton diffusion coefficient D in such cation-exchange membranes is by an order of magnitude greater than that for metallic palladium membranes [230]: 3.2×10^{-9} and $(0.4-4.0) \times 10^{-10}$ m^2/s, respectively; activation energies are 12–20 kJ/mol and 23.9–28.4 kJ/mol, respectively. Such catalysts are able to promote the reactions in thin films[14], and control the rate at which reagents penetrate through the polymer membrane.

Membranes used as catalysts for phenylacetylene (PhA) hydrogenation have been produced [232] by treatment of deuteropoly(acetylene) with a mixture of NiBr$_2$/dimethoxyethane in a medium of THF/hexamethyl(phosphorus)triamide with subsequent metal ion reduction to Ni^0; the selectivity of the reaction to styrene formation (the ratio [styrene]/([PA]+[styrene]+[ethylbenzene] (EB)) is 0.88. Use of multifold films has led to some decrease of the reaction rate without a change in selectivity. It is important that the relative rate of PhA hydrogenation to styrene and styrene to EB are dependent on the amount of Ni^0. The rate of EB formation increases with an increase of Ni^0 content. It is probably possible to control the rate of substrate migration through the membrane and in such a way as to enhance the selectivity, or in other words, separate zones of substrate and reagent.

There are different procedures for the preparation of membranes, from simple polymer flow-over to more complicated methods. Thus, an alternative composition of the hydrophobic films of PTFE–CSDVB (50–100 µm thickness), with a sputtered platinum covering (1–1.4%) and hydrophilic layers of 50–100 µm thickness, has been produced [185]. Both hydrophobic and hydrophilic layers of the catalyst are twisted to a spirals and spaced from each other with a hydrophilic network to increase the turbulence of the streams. An ultrafine dispersion of metal particles in the membrane, shows high activity in 1,3-cyclooctadiene (COD) hydrogenation to cyclooctene (Table 1.18). These catalysts have been obtained by impregnation of gel-like celluloseacetate membranes (200 µm thickness) with both Ni(NO$_3$)$_2$ and NaOH followed by further reduction with H$_2$ at 673 K [73,233].

Figure 1.26 Composition of the reaction mixture (mmol) in hydrogenation of 1,3-COD by Ni/C catalyst in ethanol at 298 K Products: 1, 1,3-cyclooctadiene; 2, cyclooctene; 3, cyclooctane.

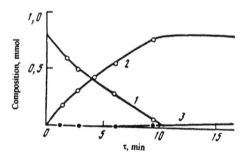

As demonstrated in Figure 1.26, the reduction of COD proceeds at the initial stage with sa electivity of 100% in respect to cyclooctene. The accompanying reduction of cyclooctene to cyclooctane proceeds at a low rate and with significant delay.

Optimization of membrane catalysts for both hydrogenation and oxidation (as will be shown in Chapter 2) as well as the main problems of their usage have yet to be developed and solved.

---

[14] The technology of membrane production method for natural and synthetic polymers and their main characteristics have been described previously [231].

Table 1.18 Characteristics and catalytic properties of thermal decomposition of gellular membranes [168].

Diameter of particle (nm)	S_{sp}, (m^2/g)	Ni content (mass%)	The rate of COD hydrogenation (mol/g-at of Ni min)	Selectivity^a (%)	
				A^b	B^c
5	220	39	300	100	100
17	210	29	105	100	97
19	95	74	90	100	98
21	120	80	75	100	98
Reney nickel	–	–	67	95	85

^{a}Selectivity is equal to the ratio [cyclooctene]/([cyclooctane]+[cyclooctene]).
^{b}Conversion 50%.
^{c}Conversion 90%.

Table 1.19 Hydrogenation of prochiral amino acids $\begin{smallmatrix} R \\ \diagdown \\ \end{smallmatrix} C = C \begin{smallmatrix} NHAc \\ \diagup \\ \end{smallmatrix}$ in the presence of homogeneous and immobilized rhodium complexes [189,190].

R in substrate	Optical yield (%)			
	M–R	M–S	J–R	J–S
Ph	90(S)	91(R)	91(R)	90(S)
AcO	87(S)	87(R)	83(R)	87(S)
HO–Ph	93(S)	–	33(R)	–
CH₃O / AcO benzene ring	88(S)	86(R)	88(R)	88(S)

1.10   Enantioselective Hydrogenation in the Presence of Immobilized Metal Complexes

Enantioselective catalysis is of vital importance for all animate nature. All life-support substances are generated in tissues of people, animals and plants by this mechanism. Many enantioselective processes that proceed under the action of homogeneous catalysts such as hydrogenation, hydrosilylation, cyclopropanation, epoxidation, Grignard cross-coupling, etc., are known at present. These systems provide significant advantages for the separation of racemates [234–236].

Great progress has been made in the study of asymmetric hydrogenation in the presence of immobilized, especially chiral rhodium phosphyne polymer catalysts. Considerable success has been achieved with catalysts involving a range of macroligands including optically active residues such as 4,5-bis(diphenylphosphyne)methyl-1,3-dioxolane (I) (DIOP) or dibenzophosphonic (II) groups as well polymers formed from (2S, 4S)–N-acryloil-4-diphenylphosphyno-2-

[(diphenylphosphyno)-methyl)]pyrrolidone (**III**) (derivatives of natural *L*–hydroxy-proline):

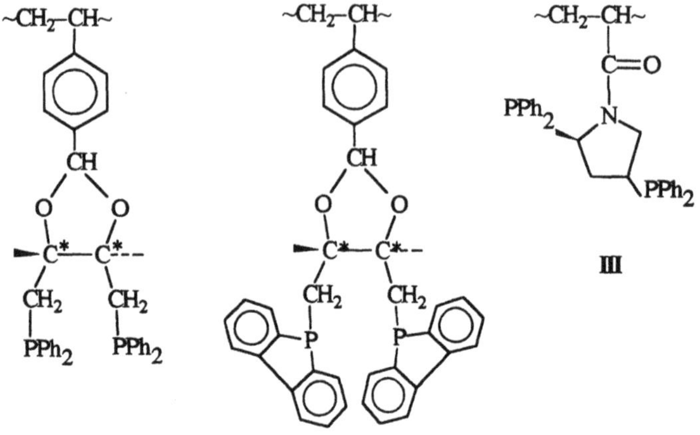

Scheme 1.19

Immobilized chiral centers for metal bonding can also be produced by radical poly- or copolymerization of *N*-methacryloyl-*L*-alanine, alkaloids of quinine bark and unsaturated amino acids and their derivatives. Reviews of these works can be found in ([25] pp.85–89; [4] pp.193–203; [237–240]). These polymers or their copolymers with styrene, DVB, hydroxyethylacrylate are swollen by EtOH or THF and capable of interaction in particular with μ-dichloro-bis(1,5-cyclooctadiene)di(rhodium) (**I**).

The efficiency of an asymmetric catalytic reaction is characterized either by access of the predominantly formed enantiomer, for example of the *R*–configuration and its optical yield (%), $P=([R]-[S])/([R]+[S])$, or by the value of the enantioselectivity, $E=[R]/[S]$.

Among the substrates hydrogenized with higher optically active product yields, amino acids (prochiral enamides) are of a special interest. Progress in studies of the hydrogenation of these substrates has stimulated the search for both new chiral ligands and immobilized metal complexes.

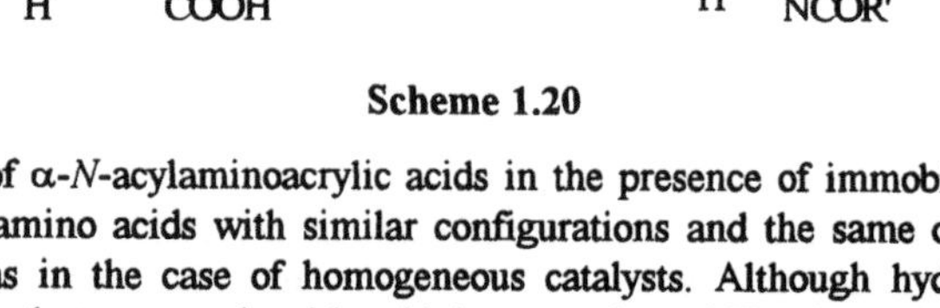

Scheme 1.20

Hydrogenation of α-*N*-acylaminoacrylic acids in the presence of immobilized catalysts gives the same amino acids with similar configurations and the same optically active product yields as in the case of homogeneous catalysts. Although hydrogenation by homogeneous catalysts proceeds with a higher rate, immobilized metal complexes can be easily separated after the reaction and reused without loss of optical purity [241]. Very interesting results have been obtained with chiral palladium-containing catalysts (macroligands was the product of copolymerization of optically active monomers *S*– and

R–1-(4-vinylphenyl)ethylamine with styrene with some additives of DVB in the reaction of reductive aminolysis of azalactone [242,243].

At the same time there are numerous examples where products with the same optical yields but opposite configurations (in comparison with homogeneous systems) have been formed under the action of immobilized catalysts. The alternate chirality may be connected with a change of active site structure in the immobilized metal complexes in contrast with the homogeneous catalyst. To illustrate the effect let us consider some examples of dehydroamino acid hydrogenation by both homogeneous and heterogenized catalysts with asymmetric groups: [2*R*, 4*R*] or [2*S*, 4*S*] *N-tert*-butoxycarbonyl)[4-diphenylphosphyno)methyl pyrrolidone (M–*R*, M–*S* and]–*R*,]–*S* [189,190], respectively). The reaction product (α-*N*-acylamino acid), has been obtained in practically the same yield from both systems (Table 1.19). However this product obtained with homogeneous system showed an absolute configuration opposite to that of obtained with a catalyst with optically active ligands, whereas heterogenized catalysts generated with product which are similar to the bound chiral center configuration, i.e. in the presence of]–*R*-catalysts *R*–acid is generated and the *S*–acid in the case of the]–*S* ligand.

Enantiomeric excess depends on many factors including both the nature of polymer the support and the solvent. Thus, 2-acetamidoacrylic acid hydrogenation proceeds in ethanol with *P*=67% and *P*=59% in THF. The optical yield is slightly decreased after repeated use of a catalyst, e.g. it falls from 64 to 60% after 3-fold catalyst usage [244]. The cationic chelate, phenyl-4,6-*O*(*R*)-benzyliden-2,3-*O*-bis(diphenylphosphyno)-β-*D*-dextrosepyranozid Rh$^+$, bound to an ion-exchange resin, showed comparatively high activity in asymmetric hydrogenation of enamides [245]:

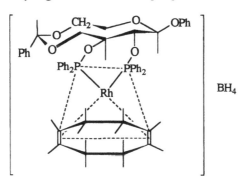

Scheme 1.21

The enantioselectivity of this catalyst was 3% greater than that of the catalyst in solution. The catalyst was also effective in repeated employment, it showed geometrical selectivity, and the active component almost did not leach in the course of the reaction.

The rise of optical yields is caused [246] by changes in the complex chiral ligand due to solvolysis of dioxane rings catalyzed by protons. In comparative hydrogenation of *E*– and *Z*–isomers of α-*N*-benzoilamino-2-butenic acid methyl ester, or their mixture, the rhodium phosphyne catalyst showed higher selectivity in hydrogenation of the *Z*–isomer [247]. This fact is of practical interest because the reaction product can be further reduced to aminobutanol which is an optically active component of the anti-tuberculosis drug, (+)–ethambutanol.

Immobilized rhodium complexes with chiral diphenylphosphynes are active in asymmetric hydrogenation of prochiral ketones [248]

$$RCOR' \xrightarrow{\text{H}_2, \text{ catalyst}} (S)\text{-RCH(OH)R'}$$

An increase of H_2 pressure leads to a rise of alcohol yield and optical purity. The reaction conditions[15] (which were analyzed in detail for the example of acetophenone reduction in the presence of $[Rh(norbornadiene)Cl]_2$+DIOP in [249]) may be the main reason for differences in enantioselectivity of homogeneous and immobilized complexes.

Some other immobilized metal complexes are also active in asymmetric hydrogenation. Thus, chiral metal polymers derived from $NiCl_2$ and $CoCl_2$, bound to polymer containing NH_2- , EtOCO- , $CH(OPPh_2)$- , or $CH(OPh_2)$ neighboring groups, are effective in asymmetric hydrogenation of α-acetylamidocinnamic acid with optical yields of 29.9 and 25.4%, respectively [250]. Skeleton nickel catalyst, modified with tartaric acid and NaBr reduced methylacetoacetate to methyl-3-oxybutyrate with a yield 80% but lost activity after three runs [251]. Incorporation of the catalyst into vulcanized silastic, which is well permeable to H_2, led to significant catalyst stabilization[16]. It is important to note the effect of changes of metal complex dispersion, metal content (at fixed dispersion) and conditions of catalyst preactivation on the process of asymmetric hydrogenation [253]. Catalysts for asymmetric hydrogenation on Ti^+ with chiral menthyl- or neomenthylcyclopentadiene ligands are also known [254].

Cluster complexes with the asymmetric ligand DIOP, derived from $Rh_4(CO)_{12}$ and $Rh_6(CO)_{16}$, are to be mentioned particularly as catalysts for hydrogenation of enamides [255]. The ratio of DIOP/Rh are of great importance for such catalysts (the highest asymmetric induction has been observed at DIOP/Rh$\geq$1). Monomeric species $HRh(CO)_2 \cdot [(-)DIOP]$ produced by oxidative addition of H_2 and following cluster segregation ,are supposed to be the active centers in these catalysts.

Complete clarity in the subject of asymmetric hydrogenation by immobilized metal complexes has not, as yet, been achived. Each of three aspects has to be understood in this connection: (i) coordination chemistry of the triple complexes, (ii) the mechanism of catalytic hydrogenation and (iii) the reason for enantioselectivity.

A mechanism involving correlation between stability of the alter-ligand complex on the catalyst surface containing chiral ligand (X^l) and catalyst enantioselectivity[17] [256] appears to be the most aaceptable explanation.

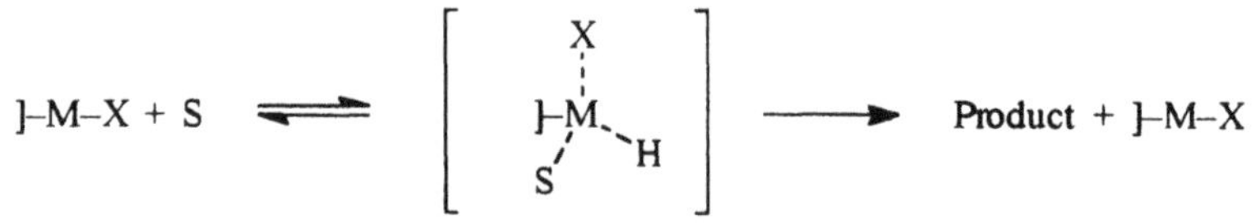

$$]-M-X + S \rightleftharpoons \left[\begin{array}{c} X \\ | \\ ]-M \diagdown_{H} \\ S \diagup \end{array} \right] \longrightarrow Product +]-M-X$$

Scheme 1.22

---

[15] Asymmetric ketones hydrosilylation will be discussed in Section 1.13.

[16] For comparison we may mention that immobilization of chiral rhodium complexes on SiO_2 led to the rise of their stability and enantioselectivity in liquid phase asymmetric hydrogenation of organic acids [252]. Their catalytic properties depend on the length of anchor chain separated complex from the support.

[17] This phenomenon will be analyzed in detail in Chapter 2.

This is the "lock and key" mechanism when enantioselectivity is specified by the first prochiral substrate bonding with chiral catalysts. There are two principal points specifying the reaction tendency: the stability of the triple complex and the influence of the chiral ligand on the route of the reaction in the transition state. Some approaches to account for the correlation between asymmetric activity and strength of intermediate complexes have been discussed [257]. Both chiral–component–substrate and substrate–metal complex interactions are considered.

The influence of the support containing the asymmetric center in the backbone on the asymmetric activity has been found to be negligible in hydrogenation of 2-acetamido-acrylic acid to N-acetylalanine [258]. The reaction was controlled by oxidative addition of H_2 to a rhodium-olefin complex. The influence of the intermediate optical center proceeds not by coordination with the metal complex but by a slower interaction with the medium (solvent). However the high enantioselectivity of $CuCl \cdot (1R, 3R)$-bis(diphenylphosphynooxy)-1,3-diphenylpropan in asymmetric hydrogenation of dibenzoylmethane has been attributed [259] to steric factors, i.e. the formation of a stressed chelate configuration after substrate entrance due to repulsion of phenyl groups in chiral ligand.

Another approach to the reaction mechanism [260] is based not on consideration of initial triple complex formation but takes into account the high reactivity of the lesser diastereomeric adduct, catalyst-substrate, corresponding to less favorable substrate bouding.

To resume consideration of experimental data we should notice that the rise of stereospecifity is reached by restriction of the conformation variety. Factors promoting high optical yields are: chelation by bidentate ligands; formation of a five-membered chelate; an increase of complex stress (for example by introduction of aryl substituents to the phosphorus atom in the chelating ligand). The idea of optical activity transfer from chiral ligand to substrate can be reduced to the following. The chiral information is passed from asymmetric centers in the chelate framework through PPh_2 groups to the catalytic metal center and then to the substrate.

1.11 Catalytic Hydrogenation by Heterometallic Polymers

Immobilization of several different metal complexes on the polymer support may result in a number of interesting features in a hydrogenation reaction. Similar catalytic systems, such as compact metals, catalysts of skeletal type and heterometallic complexes anchored to oxides, have already found their uses[18] in heterogeneous catalysis. One such approach consists of sequential support on the oxide surface of two layers of transition and non-transition ones metals. The latter serves as a trap for transition metal ions preventing their migration on the surface, thus retarding catalyst aggregation [262]. But more interesting is the case when each of the bound metal complexes take its part in hydrogenation giving rise to polyfunctional catalysis.

A comparison of hydrogenation by unsaturated substrates with both Pt/nylon and Pt–Sn/nylon (obtained by nylon impregnation with both H_2PtCl_6 and $SnCl_2$ and subsequent reduction with H_2 at 343 K) catalysts has been carried out [263,264]. In hydrogenation both of hydrocinnamic aldehyde and PhA (at 283 K) tin addition led to a sharp fall in

---

[18] Simulation of catalytic hydrogenation of alkynes in the presence of double metal centers in solutions has been carried out in [261].

the C=C hydrogenation rate and simultaneously to an increase of C=O reduction at low Sn/Pt ratios. The activated action of SnCl$_2$ is the result of π-acceptor complex properties promoting an reduction of electron density on the Pt atom, which is favorable for both H$_2$ and substrate coordination. Hydrogenation of both acrolein and cinnamic aldehyde (at 318 K) in the presence of Pt catalyst merely gives saturated aldehydes due to lower a C=C bond energy value in comparison with that of C=O. Addition of small Sn moieties led to a sharp selectivity rise in the direction of unsaturated alcohol formation due to preferably C=O hydrogenation, and a loss of catalyst activity when Sn/Pt>1.2. Tin introduction causes a double effect: firstly, due to their acid properties, Sn ions activate the C=O group making it more active then the C–C unit; secondly, an increase of Sn loading leads to an increase of the electronic effect on Pt ions provoking catalyst poisoning[19].

The poisoning action of Sn^{2+}, Pb^{2+}, Cd^{2+} and Fe^{3+} ions on both activity and selectivity of immobilized Pd catalyst in hydrogenation of heptene, PhA and benzaldehyde may be connected [265] with the generation of a layer of impurities on the surface of the catalyst. The removal of impurities with HClO$_4$ or H$_2$SO$_4$ solutions regenerates the catalyst. The formation of a suboxide titanium layer on the surface of Rh particles is also accompanied by strong metal-support interaction and causes the interlocking of chemisorption centers [266].

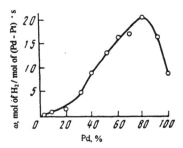

Figure 1.27 The dependence of catalytic activity of Pd–Pt immobilized clusters stabilized with PVPd on their composition (H$_2$ pressure 0.1 MPa, 303 K).

Heterometallic platinum–rhodium catalysts bound to polyamides show high activity and selectivity in the hydrogenation of nitrocompounds [267]. An active catalyst for acetylene hydrogenation to ethylene has been obtained by MoFeS-complex bound to PVP; the catalyst reveals high stability to the action of acids and bases and is easily regenerated. More often however each of the components of immobilized heterometallic system exhibits its individual features albeit partially modified. In particular this is a characteristic of immobilized heterometallic cluster complexes (see Section 1.8). At the same time, the presence of the peak in activity at the ratio Pd/Pt=4:1 (Figure 1.27) in selective hydrogenation of 1,3-COD to cyclooctene in the presence of colloid-dispersed Pt and Pd species stabilized with PVPd points to bimetallic cluster formation during cooperative reduction of PdCl$_2$ and H$_2$PtCl$_6$ in H$_2$O/C$_2$H$_5$OH in the presence of PVPd.

The role of second metal in immobilized metal complexes is ambiguous. Thus, the activation of Wilkinson's complex with the silver salt of poly(styrene)sulfonic acid is accompanied by free triphenylphosphyne bonding, favorable for rhodium coordinative unsaturation [268]. This procedure enables catalyst activity to increase by 15 times.

---

[19] Platinum and palladium polytin catalysts anchored to SiO$_2$ showed high stability (TN>9 000) in both carbonyls and nitrocompounds reduction.

The possibilities of bringing together bifunctional processes in the presence of immobilized heterometallic complexes, where the product of the first stage acts as substrate in the second, will be analyzed in detail in Chapters 3 and 5. One of the hydrogenation examples is bifunctional catalysis of Reppe's reaction [269] with macrocomplexes of $RhCl_3$ and $Fe(CO)_5$; hydroformylation of pentene-1 with subsequent hydrogenation of the aldehydes formed to alcohols. Similar behavior is shown in a number of other bifunctional processes such as isomerization, alkylation, di- and tri-, cyclomerization, etc., with subsequent hydrogenation.

1.12   H–H Activation Reactions and Hydrogen Transfer under the Action of Immobilized Metal Complexes

Immobilized metal complexes find their use in activation of H_2, hydrogen/deuterium exchange with water and alcohols, in *para-ortho*-conversion and in hydrogen transfer from one substrate to another.

1.12.1 Hydrogen/deuterium exchange reactions

Hydrogen/deuterium (H/D) exchange with water and alcohols is, from the one hand, a very convenient reaction for comparison of homogeneous and immobilized catalyst activities, and, from the other hand, a practically important technique for heavy water production

$$D_2(gas) + H_2O(liquid) \leftrightarrow HD(gas) + HDO(liquid).$$

In addition H/D exchange is widely used to identify association and dissociation stages in adsorption process as well to reveal the mechanism of proton transfer between reagents, exchange reactions, etc. Finally localization, concentration and separation of tritium is a very important ecological problem.

The kinetic features of H/D exchange in the presence of both homogeneous and immobilized rhodium complexes (the support is anionite containing amino groups) are similar [270]. The initial reaction rate is defined by an equation of the form

$$w = \frac{k_1[Rh]_0[D_2]}{1 + K[D_2]}$$

where $K = k_1/k_{-1}$ is equilibrium constant for hydrogen absorption (k_1 is hydrogen absorption constant rate, molmin^{-1}; k_{-1} is hydride complex decomposition constant rate, min^{-1}). Activation energies for D/H exchange are similar for the above systems. All the differences in catalyst activities are caused by activation entropy changes for the reaction control stage [225,271]. Heterogenized catalysts show higher stability than those in solution (Figure 1.28).

Examination of Figure 1.28 indicates that diverse factors can influence catalyst properties including the nature of the metal complex. Sometimes homogeneous catalysts are more active than heterogeneous (Table 1.20). However, heterogenized ones sometimes reveal a tendency to increase significantly the activity after several hours of working. This fact may have a reason that both composition and structure of operating metal complex fixed may be quite different from the starting one. Moreover, in the course of catalyst preparation, metal particles of different size can also be formed. Actually, rhodium particles of 50–100 nm incorporated into a polymer matrix have been detected by electron microscopy [272]. Small particles of Rh or Ru are supposed to be capable of H/D exchange. Addition of reducing agent (TiCl$_3$) causes the exchange rate to rise by several orders of magnitude. HCl doping acts as cocatalyst to activate the process.

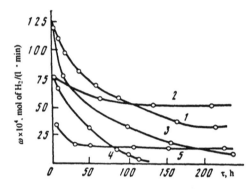

Figure 1.28   Changes in catalytic activity in D/H exchange. Systems: 1, Rh(alizarin red)Cl$_2$([Rh]$_0$=1×10^{-3} mol/g; 2, RhCl$_3$-vophatit-ES ([Rh]$_0$=2.5×10^{-5} mol/g); 3, Rh(alizarin red)Cl$_2$/vophatit-ES ([Rh]$_0$=1.25×10^{-5} mol/g); 4, Rh(indigo sulfonate)/vophatit-ES ([Rh]$_0$=2.5×10^{-5} mol/g). Conditions: 293 K, H$_2$ pressure 0.1 MPa.

Extreme instability of traditional heterogeneous catalysts for D/H exchange with water in the liquid phase (as a rule, metal complexes of the platinum group bound to different supports) is caused by capillary water condensation and consequently active site blocking, and is known to be a significant shortcoming. To avoid this effect different procedures of increasing surface hydrophobity have been derived including microcapsulation of catalyst particles into a polymer shell.

Table 1.20   Comparison of homogeneous and heterogenized (on phosphynated CSDVB) catalyst activities in D/H exchange[a] between D$_2$ and H$_2$O or C$_2$H$_5$OH [272].

Bound complex	Initial exchange rate (mol D_2/g M s)×10^6	
	H_2O	C_2H_5OH
[(C$_2$H$_4$)$_2$RhCl]$_2$	76	1.9
(C$_2$H$_4$)$_2$Rh(AcAc)	8.3	6.9
[(C$_{10}$H$_{12}$)RhCl]$_2$	105	25.0
Ru$_5$Cl$_{12}$$^{2-}$	–	0.4
RuCl$_2$(PPh$_3$)$_3$	–	1.8
RuCl$_2$[(CH$_3$)$_2$SO]$_4$	–	0.5
RhCl(PPh$_3$)$_3$ (unsupported)	–	63.0
RuCl(PPh$_3$)$_3$ (unsupported)	–	113
RuCl$_2$[(CH$_3$)$_2$SO]$_4$ (unsupported)	–	2.1
RuCl$_2$[(CH$_3$)$_2$SO]$_2$PPh$_3$ (unsupported)		11.9

[a]Reaction conditions: 298 K, D$_2$ pressure 88 kPa.

Thus, a hydrophobic catalyst has been obtained [273] by bonding a Pt compound to CSDVB and further reduction with hydrazine hydrate which yields more large particles than when H$_2$ is used as reductant. The greatest activity in tritium (T)/H exchange (303–343 K) is reached at a Pt content of 0.5–1.5%. However, PTFE (teflon) is more often used as the hydrophobic material [274–279]. Finely dispersed particles of platinum group metals as well Pt/C have been supported on polymers in the form of powder, net, paper, beads, textile, pellets and even Raschig's rings also made of teflon. These catalysts have been used for removal of tritium from reactors, as heat-evolved elements of atomic power stations and also for processing of exhausted nuclear fuel. The increase of catalyst hydrophobity causes a rise in activity; regeneration is reduced to simple heating for removal of trickle water from catalyst pores.

A packed column filled with catalyst is more often used for this process. Water mist and H_2 are blown through the column in a parallel current or opposing streams for the best contact of water with hydrogen on the surface of the catalyst. Independence of isotopic exchange rate on the water flow rate is reached under these conditions. D/H exchange between D_2 and H_2O is characterized [274] by two factors: the coefficient of mass transfer in the catalytic reaction HD(gas)+H_2O(vapor) and coefficient of mass transfer in the following reaction HDO(vapor)+H_2O(liquid) occurring within water droplets. The coefficient of total mass transfer becomes constant within some range above the reactor inlet.

Sometimes it is possible to make intermetallic compounds or their   hydrides hydrophobic which form active catalysts for isotopic exchange (for example $LaNi_5$) [280]. The powder of $LaNi_5$ is mixed with 60% water/PTFE emulsion. Granules of diameter 1–5 mm are prepared from this substance. Species are activated by heating at 520 K. Isotopic exchange rates ($\times 10^3$) in the presence of $LaNi_5$/PTFE, Pd/PTFE and Pt/C/PTFE are respectively equal to 2.5×10^{-2}, 1.8 and 4600 mol/(g min) (H_2 pressure 0.1 MPa, T=363 K). The order of the reaction in respect to H_2 or HT (catalyst $LaNi_5$/PTFE) is equal to 0.5 thus indicating that the reaction-controlled stage is surface dissociation of hydrogen isotopes. The latter is also confirmed by the value of the kinetic isotope effect $K_{HD}/K_{HT} \approx 11.6$ (353 K).

The catalyst composed of hydrophobic films PTFE–CSDVB of thickness 200 µm with sputtered Pt coverage (1–1.4%) is active in separation and concentration of tritium from tritium-containing waters (systems HTO–H_2O/H_2, HDO–H_2O/H_2) [185]. The efficiency of isotopic separation depends both on the activity of immobilized metal complex and on the characteristics (structure and thickness) of the cross-linked and hydrophilic layers. Such a catalyst enables isolation of 95–99.5% of tritium from radioactive liquid wastes of enrichment coefficient greater than 10^3. Catalysts derived from Pt, Rh, Pd supported on PTFE with a grafted PhAA layer (PTFE-gr-PhAA) also showed high efficiency in isotopic exchange [271]. Their specific activities are equal to 356, 22 and 2 mol of H_2/(mol of M min), respectively.

Enzymes and hydrogenaze are also capable of H–H bond activation (see for example [1]). Mainly HD has been obtained from the H_2–D_2O mixture by enzymatic H_2 cleavage

$$En + H_2 \xrightarrow{\ k_1\ } EnH^- + H^+$$

$$EnH^- + D^+ \xrightarrow{\ k_2\ } En + HD$$

As mentioned above (see Section 1.8) finely dispersed (0.8–2.5 nm) metal particles and colloids (1–100 nm) encapsulated in polymers also show activity in H_2/D_2 exchange.

Thus, Rh, Pd and Pt species stabilized with PVPd [178] and some other polymers [281] are effective in D_2/H_2 exchange as well as in both ethylene hydrogenation and D_2/H_2 exchange [282]. The decrease of Pd particle size in Pd/graphite systems has been shown to result in a reduction in C_2H_4 hydrogenation activation energy and a progressive increase of the activation energy for D_2/H_2 exchange. However, the decrease of particle size is accompanied by both geometrical changes in the surface of Pd particles and reorganization of their electronic state. Accordingly it is difficult to identify the influence of each individual factor on the catalytic properties. It should be noted in this connection that for small Pd particles ($d=1.0-1.8$ nm, at their surface density $(1-4)\times10^{15}$ per sm^2) the value of the $Pd_{3d5/2}$ bond energy decreases and tends to that for solid Pd with a decrease of particle numbers [283]. The rate of H_2/D_2 exchange is greatest for particles of $d=1.3$ nm, E_a decreasing with increasing particle sizes. These facts suggest that in small Pd particles the valency electrons are more strongly bound to the metal nuclei than in large aggregates. As a result the activation energy for the reaction with H atoms is greater.

Hydrogen/D_2 exchange in an equimolar mixture of H_2/D_2 in the presence of Au (0.3%) stabilized with PTFE (this catalyst has been obtained by support impregnation with a solution of hydroauric acid in acetone, with heating and activation with H_2) is characterized by $E_a=29$ kJ/mol [284]. The dependence of the reaction rate on the total pressure is described by the equation $w=k[KP/(1+KP)]^2$, where K is the adsorption rate constant.

It is most probable that H–D exchange proceeds via an intermediate H_2 D_2 complex adsorbed by active gold sites of the catalyst surface. The value of the adsorption heat (5.9 kJ/mol) indicates chemisorption-type bonding of both H_2 and D_2.

Information about H–D exchange with hydrocarbons in the presence of immobilized metal complexes is rather scarce. Besides the C_2H_4/D_2 exchange mentioned above, destructive hydrogenation accompanied by C_2H_6/D_2 exchange (353 K, $C_2H_6:D_2=10^{-1}\div10^{-3}$, catalyst on the base of sputtered rhodium) [285] may be mentioned. H–D exchange proceeds through a sequence of transformations in adsorbed intermediate complexes and requires the presence of surface hydrogen atoms, while whereas alkane desorption proceeds with participation of D_2 extracted from the gas phase.

Hydrogen/D_2 exchange can be used for deuteration of support surfaces, although there are detailed data only for deuteration of oxide surfaces (separation by cyclic heating of D_2O under vacuum) [286]; the value of $\alpha=99.6\%$ is reached over six cycles of Al_2O_3 surface treatment. Experimental evidence of H–D exchange in the polymer chain (CSDVB) in the presence of immobilized cobalt was obtained in the course of Fischer–Tropsch conversion [287]. Hydrogen/D exchange with the polymer chain was shown to proceed via the intermediate formation of a cobalt–hydride complex.

Hydrogen/D exchange with molecules other than H_2O or ROH protolytic reagents, such as NH_3 or amines, is also of great interest. Experimental data on the subject are however extremely scarce. For instance, membrane catalyst, H_2/Pd-membrane/organic compound, is active in H–D exchange with complex nitrogen-containing acids[20] in the solid state [288].

---

[20] Some other types of exchange such as that with participation of radiohalogen labels are of special interst.

1.12.2 Ortho-para- hydrogen conversion

Both H_2/D_2 exchange and o-H_2/p-H_2 conversion proceed through heterolytic dissotiative adsorption of hydrogen followed by chemical bond formation with the metal ion (as a rule in the lowest valent state), conversion rates being lower than exchange rates. Rhodium complexes are most effective in this conversion [1,289]. Immobilized analogues of these complexes are also used to find their use as ortho-para-H_2 conversion catalysts. Thus, polymers derived from pentafluorothiophenoxides show semiconductor properties that have been used for o-H_2 transformation into p-H_2 [290].

1.12.3 Hydrogen transfer in the presence of immobilized metal complexes

Catalytic reduction of organic compounds with the participation of chemically bound hydrogen molecules seems to be a promising process. These reactions can be catalyzed by immobilized metal complexes in a manner similar to activated hydrogen transfer under the action of enzymes. Selective hydrogenation of ketones, aldehydes and aromatics, as well dehalogenation, etc., can be carried out in this way. There are numerous examples of such reactions. We shall note only some characteristics of these processes.

$RuH_2(PPh_3)_4$ bound to different polycarboxylate matrices is active in transfer hydrogenation between alcohol and aldehyde [27].

$$normal\ C_6H_{11}CHO\ +\ \langle \bigcirc \rangle\text{—OH} \longrightarrow normal\ C_6H_{11}CHOH\ +\ \langle \bigcirc \rangle\text{=O}$$

Scheme 1.23

$Ru(CO)_2(OCOCH_3)_2(PPh_3)_2$ complexes catalyze hydrogen transfer through the initial stage of dissociation of the one carboxylate group with the second remaining in the vicinity of the metal ion coordination sphere. Vacant carboxylate groups of macrocomplexes also can take part in aldehyde transformation lowering the selectivity of transfer reaction. The effect is more extensive after catalyst re-use when additional carboxylic groups are released due to partial metal leaching. Metal polymers obtained by copolymerization of bis(acrylate)-bistriphenylphosphyne complexes such as $(CH_2=CHCOO)_2Ru(PPh_3)_2$ or $(CH_2=CHCOO)_2Ru(PPh_3)_2(CO)_2$ with MMA, are most appropriate for these purposes. There are no free carboxyl groups in such polymers so their selectivity reaches 100%.

Immobilized tetrasulfate ruthenium complexes bound to CSDVB, modified with 3(5)-methylpyrazole groups, are effective in hydrogen transfer from propanol-2 to cyclohexanone in an alkaline meium [291].

The catalytic system $Rh_2(OCOCH_3)_4$ doped with electron donor additives is effective in transfer hydrogenation between propyl-2 alcohol and ketones (cyclohexanone, acetophenone, mesityl oxide, etc.) [292]. The catalysts for hydrogen transfer between isopropyl alcohol and acetophenone or hexene-1 have been obtained by the $Rh(norbornadiene)_2ClO_4$ complex bonding to PS modified with imidazole or diphenylphosphyne groups [293]. The value of the conversion reached 94%, 1-phenyl-ethyl alcohol and hexane being the reaction products. Multiple catalyst re-use led to a slow complex transformation to metallic rhodium. The reductions of aldehydes, ketones, and esters by polymer-bound aluminum hydride are also interesting [294].

From other examples let us note hydrogen transfer from HCOOH to α,β-unsaturated ketones transforming into saturated compounds under the action of Rh, Ir and Ru complexes bound to PS [295], hydrodehalogenation of p-bromotoluene and *gem*-dihalogenidecyclopropanes by hydrogen transfer from propyl-2 alcohol in the presence of Rh and Ru complexes immobilized by polymers [296,297], etc.

Mechanisms of such transformations are usually similar to those proceeding under the action of homogeneous catalysts.

1.13 Hydrosilylation

Hydrosilylation is the process whereby alkenes, alkynes, ketones are reduced by hydrogen chemically bound to the silane molecule, $X^1X^2X^3SiH$ (X^1, X^2, X^3=H, Cl, Et, Pr, OEt, Ph, etc.)

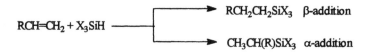

$$RCH{=}CH_2 + X_3SiH \longrightarrow \begin{cases} RCH_2CH_2SiX_3 & \beta\text{-addition} \\ CH_3CH(R)SiX_3 & \alpha\text{-addition} \end{cases}$$

Scheme 1.24

This is one of the most important reactions for the synthesis of different organosilicon compounds [298]. The principle catalysts are Pt, Pd, Rh, Ru, Ir, Co, Ni and Fe complexes, chloroplatinic acid being most widely used. The reaction mechanism involves oxidative addition of $HSiX_3$ to give the corresponding intermediate compounds proceeding under rather mild conditions.

The employment of immobilized metal complexes has promoted progress in this field [298]. A number of reviews devoted to hydrosilylation by immobilized metal complexes have appeared [4]. In this Section we will examine only the most interesting, recent results.

Immobilized catalysts not only encourage an increase of the total reaction rate but also promote both regioselectivity with respect to the products of α,β-addition and enantioselectivity with respect to compounds with an asymmetric carbon atom.

Thus, chloroplatinic acid, $PtCl_2(MeCN)_2$ and $[RhCl(CO)_2]_2$ immobilized on polyamides showed high activity in hexene-1 hydrosilylation [299]. It is important that activity decreases with increasing a distance between support amide groups and that it correlates with polymer crystallinity. The same features have been observed for platinum and rhodium complexes with polyphenylquinoxaline [300]. An extremely active catalyst for hydrosilylation of hexene-1 with triethylsilane has been obtained by coordination of $Rh(PPh_3)Cl$ with dicarbaunedecarborate-conaining polyamide [301]. Some hydrosilylation reactions for compounds with C=C bonds are given in Table 1.21.

Macromolecular Pt chelates with phosphyne and aminophosphyne nodes are of special interest as a new type of highly effective catalyst, their low molecular mass analogues being inactive under identical conditions.

Bonding of metal complexes to ion-exchange resins favor the hydrosilylation process and allows it to be controled [302–305]. $[PtCl_6]^{2-}$ bound to anionite AB17-8 is effective in hydrosilylation of heptene-1 and styrene with dimethylphenylsilane; the rise of TN leads to catalyst development. The order of increasing activity of immobilized metal complex is $Co^{2+}<Ni^{2+}<Pd^{2+}<Ir^{4+}<Ru^{3+}<Ir^{3+}<Pt^{4+}<Pt^{2+}$. At the same time a decrease of

trans-ligand influence has been observed for immobilized acidic platinum-row complexes. Ru, Os and Pd complexes are labile in both ionic exchange and hydrosilylation reactions. Both steric factors and substituent interactions with active sites are of great importance for hydrosilylation processes.

In addition to noble metal compounds, nickel complexes [306] are effective in hydrosilylation. Selectivity can be controlled by combinations of L and X in immobilized NiL_2X_2 (X=acidic ligand, L=phosphyne).

Effective catalysts for hydrosilylation have been produced by Pd^{2+} and Rh^{3+} immobilization on supports of mixed-type prepared by radical polymerization of acrylonitrile, vinylpyrrolidone and vinylpyridine in the presence of SiO_2 [49]. The rates of triethoxysilane addition to different unsaturated compounds under the action of these catalysts increases in the order of the substrates: hexene-1>heptene-1>styrene>dodecene>tetradecene>MMA>acrylic acid>vinylacetate>allyl alcohol>cyclohexene, acrylonitryle, trichloroethylene. As a rule, a rise of electronegativity of the silane substituent increases reaction rate. The order of activities of immobilized rhodium catalysts is [307] $HSi(OEt)_3$>$HSiEt_3$>$HSiCl_3$, whereas for platinum catalysts it is $HSiCl_3$>$HSi(OEt)_3$>$HSiEt_3$. The activities of the dihydrosilanes is higher then that of the monohydrosilanes.

Immobilization of metal centers by σ-bonds or polydentate bonds leads to an increase in catalyst stability not only for hydrosilylation but also for other catalytic processes. Supported Ir complexes are not only effective catalysts but also inhibit intramolecular reaction, such as alkene isomerization.

Immobilized complexes are also effective in acetylene hydrosilylation, for example in trichlorosilane addition to acetylene giving vinyltrichlorosilane (even in the gas phase). Polymer-sulfur-containing platinum complexes show high stability even after eight turnover cycles [308]. In the first stage $MeSiHCl_2$ addition to acetylene givee methylvinyldichlorosilane and then 1,2-bis(dichloromethylsilyl)ethane. Catalyst activity depends on the S/Pt ratio and shows its maximum value[21] at S/Pt=2.

Table 1.21 Hydrosilylation of C=C compounds by polymer complexes with aminophosphyne chelate ligands [311].

Catalyst	Substrate	Product yield (%)
]–(P)CNIrCl(CO)	CH_2=CH_2+$MePr_2SiH$	83
]–(P)CNIrCl(CO)	$CH_2CHSiMe_3$+Ph_3SiH	85
]–(PN)$_2$RuCl$_2$	$CH_3(CH_2)_3CH$=CH_2+$(EtO)_3SiH$	47
]–(PN)$_2$RuCl$_2$	$CH_3(CH_2)_3CH$=CH_2+$(MeO)_3SiH$	52
](PPh$_2$)$_2$PtCl$_2$	$CH_3(CH_2)_3CH$=CH_2+Me_2PhSiH	0
]–(NMe$_2$)$_2$PtCl$_2$	$CH_3(CH_2)_3CH$=CH_2+Me_2PhSiH	100
]–(P)CNPtCl$_2$	$CH_3(CH_2)_3CH$=CH_2+Et_3SiH	95

[21] Rhodium phosphyne catalysts bound to phosphynated SiO_2 are effective in acetylene hydrosilylation with R_3SiH yielding R_3SiCH=CH_2 and $R_3SiCH_2CH_2SiR_3$ [309] while phosphyne Rh^+ and Ru^{2+} complexes immobilized on modified SiO_2 effectively catalyze gas-phase hydrosilylation (353-443 K) with a selectivity with respect to vinyltrichlorosilane of up to 94% [310].

Phenylacetylene hydrosilylation has been observed in the presence of Pt, Pd, Ni and Co complexes with 1-vinylimidazoles [312]. Higher activity is showen by Pt complexes and it changes in the order of silanes: $Et_3SiH<(EtO)_3SiH<MeEt_2SiH<EtSiHCl_2<<CH_3SiHCl_2<<SiHCl_3$. The reaction specificity mainly depends on the nature of substituent at the silicon atom.

Clusters of platinum metals are more active in hydrosilylation than mononuclear species [313]. $Rh_4(CO)_{12}$ is more active than $Co_2(CO)_8$ but $Rh_6(CO)_{16}$ is less effective due to the high strength of cluster skeleton.

Metal polymer complexes (Rh-containing first of all) bound with chiral support ligands are active in asymmetric hydrosilylation of asymmetric ketones[22], especially of acetophenone, methyl benzyl ketone and isobutyroketone, cyclic ketones being less active than dialkyl aryl ketones. Hydrolysis of the products formed yields asymmetric alcohols.

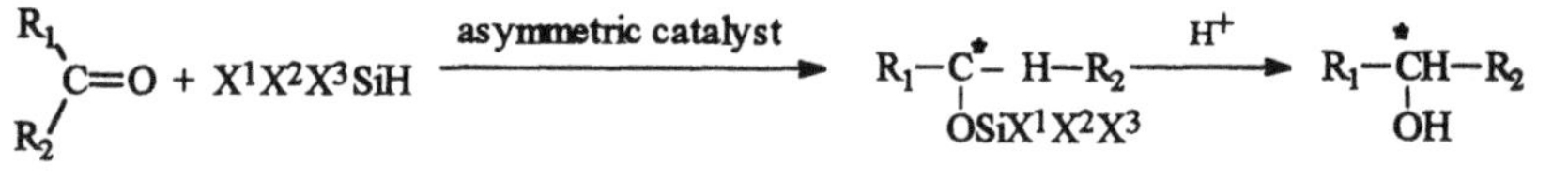

Scheme 1.25

Thus in hydrosilylation of acetophenone with Rh^+ complexes bound to PS modified with diphenylphosphyne (reagent Ph_2SiH_2, ketone/[Rh]=35–50; 24 h) the optical yield with respect to (S)(–)phenylmethylcarbinol was 29%, while for a homogeneous catalyst it was 25% [315]. However the optical yield for homogeneous catalysts can reach 36.5% [316]. As a rule, the macroligand surface upgrades all kinds of selectivity. Such factors as the nature of both reactants and metal complex (both metal ion and ligand surroundings, their dentatity and chirality) are the means of controlling hydrosilylation, including the asymmetry [317].

It is important that asymmetric reduction of polymer reagents can be carried out with the use of hydrosilylation agents. Thus, asymmetric poly(3-methyl-3,3-butene-2-ol), which cannot be obtained by direct polymerization of the respective monomer, has been produced by reduction of polyisopropenylmethylketone with Rh^+ complexes containing asymmetric ligands (THF, 323 K, Ph_2SiH_2) [318]. Optical rotatory power depends on reaction conditions, such as solvent nature and temperature, rising with amount of conversion. It is also interesting that asymmetric reduction of the monomeric ketone pinacolone and a copolymer of isopropenyl methyl ketone with styrene (1:1) under the action of Rh^+ in THF yields the respective alcohols with a similar rotatory power (–3°).

By use of hydrosilylation it is possible to estimate the reactivity of multiple bonds of different types. As has been shown with the example of oligo(butadiene) of molecular mass 15 000 [319], the silane addition (H_2PtCl_6, 303–383 K, toluene) proceeds mainly at the 1,2-double bonds, both 1,4-*cis*- and 1,4-*trans*-isomers showing low reactivity due to steric factors. Meanwhile, the opposite addition has been observed when a polymer with Si–H groups (polydimethylsiloxane) has been used as the hydrosilylating agent [320].

Grafted polysiloxane(polymethylmethacrylate) and polysiloxane(polystyrene) copolymers have been obtained by hydrosilylation. Polyalkylarylphosphyne(siloxanes)

---

[22] Asymmetric hydrosilylation of prochiral olefins has been studied only for homogeneous rhodium catalysts [314].

have been employed as the support for Rh and Ir complexes for preparing the catalysts for some other reactions [321].

There is no information in the literature about hydrogermylation, hydrostannylation, hydrocyanation, etc., catalyzed by immobilized metal complexes, although there is no restriction in principle to their existence.

1.14 Reduction of Macromolecules

In a reductive medium containing both mobile metal complexes and those bound to polymer matrix, hydrogenation of the polymer matrix unsaturated bonds or its functional groups may take place. Such a process can influence the properties of bound metal complexes. Metal leaching of fixed metal complexes is a model of polymer hydrogenation reactions under the action of homogeneous catalysts. However, information on this subject is mainly concentrated in patents.

The scale of polymers reactivities as substrates is rather broad. Thus, even such inert polymers as poly(α-olefins) have been hydrogenated in the presence of supported palladium complexes [20]. Polystyrene is not hydrogenated with rhutenium polymer catalysts even under rather forcing conditions (423 K, H_2 pressure 7 MPa) [218]. At the same time, the copolymer of styrene with cyclopentene (M_n=10 000) has been hydrogenated (with 20% benzene ring hydrogenation) to obtain a thermal shock-resistant polymer that shows low intrinsic viscosity in the melt and is compatible with other polymers [322]. Copolymers of styrene with amino groups have been obtained by hydrogenation of their nitroderivatives [323]. Hydrogenations of both polyethylene glycol and polypropylene glycol with high conversion degrees have been carried out in the presence of metal complex catalysts, including immobilized series [324].

However, the most fully explored is the process of unsaturated polymer hydrogenation, especially that of polymers derived from conjugated dienes. Thus it has been shown that during hydrogenation of polybutadiene in toluene (333 K) in the presence of RhCl(PPh₃)₃, polymer molecules of different levels of saturation are formed; the fraction of completely hydrogenated molecules [η] linear decreases with a rise in the extent of conversion.

It is interesting to note the behavior of polydienes obtained in the presence of neodymium catalysts in this reaction (see Chapter 3). The order of polymer activities in hydrogenation is [325]: 1,4-*cis*-polybutadiene>1,4-polyisoprene>>1,4-*trans*-poly-2,4hexadiene>1,4-*cis*-poly-2,3-dimethylbutadiene>1,4-*trans*-poly-2,4-hexadiene. The product of polybutadiene (PB) hydrogenation showed physico-chemical properties together with some other properties similar to linear high-density polyethylene while hydrogenated polyisoprene (PI) yielded an alternating copolymer of ethylene and propylene. In 1,4-*trans*-poly-2,4-hexadiene about 13% of the double bonds were not hydrogenated due to steric reasons. *cis*-Units of 1,4-*cis*-poly-2,3-dimethylbutane were hydrogenized more readily than *trans*-, the accompanying destruction of polymer backbone being observed for this polymer unlike to both PB and PI. Polybutadiene of number average molecular mass 10 000 (90% of 1,2-units) was selectively hydrogenated at 1,2-units in the presence of RhCl(PPh₃)₃ [326].

Butadiene-styrene and isoprene-styrene copolymers are readily reduced in both aliphatic and aromatic solvents (H_2 pressure 10 MPa, substrate concentration 1–25%) under the action of nickel [327] or Ziegler catalysts such as Ni(OCOR)₂–Et$_n$Al(OR′)$_{3-n}$ [328] as well as Cp₂TiCl₂–LiBu. This needs no LiBu addition in hydrogenation of living polymers obtained in the presence of organolithium compounds [329]. Diaryltitanocenes can be used for this purpose [330]. Weatherproof and heatproof hydrogenated diene polymers have been obtained [331] in the presence of the noble metal colloids mentioned above stabilized with polymers, Ni, Cu and Pd complexes anchored to SiO_2 [332], as well palladium salts of carbonic acids and Rh and Ni compounds [333]. High-transparency optical materials suitable for manufacture of perfect video- and compact disks have been produced [334] by reduction of a copolymer of tetracyclododecene with norbornadiene synthesized by ring-opening polymerization in the presence of immobilized Ni or Pd catalyst. Hydrogenation in the presence of immobilized metal complexes is also used [335] for divinyl compound and separation from monomers with close boiling points is necessary to avoid the impaired action of dienes. Unsaturated polymers including butadiene-styrene and nitrile rubbers can be hydrogenated both in solution and as latex in water in the presence of Ag, Cu, Fe, Ni, etc., salts [336].

Polyconjugated polymers with C=N= bonds are reduced under the action of trialkylborons [337]. Methods of selective hydrogenation of unsaturated polymers containig nitrile groups with conservation of nitrile units, for example of copolymers of dienes with acrylonitrile, in the presence of Wilkinson's catalysts [338,326–336,339–341], Ru complexes [342], palladium complexes including immobilized complexes [343] have been suggested. It was found that rate-controll stage of the reaction is the interaction with the double bond, because nitrile groups inhibit the reaction [344].

Immobilized on the surface of polycationic polymers MCl^{2-} complexes (M=Pd, Pt) are highly effective for the reduction of large biomolecules with H_2 under mild conditions [345]. Hydrogenolytic cleavage of polymer benzyl ester bond of enzyme unit- has been observed under the action of Pd black. This is an effective and mild way for the final stage of enzymes unblocking in the solid state synthesis of enzymes [346]. Colloidal Pd on the surface of PVPd forms a selective catalyst in the hydrogenation of biological membranes [347].

Information about the hydrogenolysis of metal polymer catalysts themselves is rather scant. That the reduction of nickel-poly-β-ketoester complex which is accompanied by metal-polymer bond hydrogenolysis and followed by poly-β-oxyester formation should be mentioned [348]. At the same time, hydrogenation of metal-containing monomers (and those with immobilized metal complexes) proceeds at multiple carbon-carbon bonds [349].

The examples presented demonstrate the possibility of polymer matrix self-hydrogenation in the course of intended catalytic reaction. One must take this into account for optimization and control of the catalyzed reaction.

1.15 The Connection between Homogeneous, Heterogeneous and Immobilized Hydrogenation Catalysts

As already mentioned above, immobilized catalysts are located in a category between homogeneous and heterogeneous catalysts and generally combine both their advantages. This is particularly well demonstrated by hydrogenation reaction. Immobilized catalysts show higher stability; in many cases they are as selective as homogeneous catalysts but as a rule they are less effective than homogeneous. However there numerous examples of the SCA of immobilized catalysts being a few orders of magnitude above that of homogeneous analogues (e.g., see [22,72,8,102,144,350–352]). Usually these are the cases where immobilization prevents internal deactivation processes, such as dimerization or aggregation of active sites. Examples are olefin hydrogenations in the presence of immobilized titanocenes (see Table 1.5) and hydrogenation of acetylenes under the action of molybdenum catalysts [112] (Figure 1.29). Sometimes metal-polymer catalysts compare favorably with homogeneous analogues in selectivity and especially regioselectivity. The process of metal complex immobilization often causes the activation of transition metal-to-ligand bonds excluding the necessity for further metal complex activation. For example, it has been shown that heterogenized analogues of $PdCl_2 \cdot (PPh_3)_3$ possess lower electron density than palladium and chlorine atoms (see Table 1.9) facilitating the coordination of such nucleophilic species as hydride-ions or double bonds of hydrogenated substrates.

Figure 1.29 A comparison of acetylene conversion (%) in the presence of $Mo[N(CH_3)_2]_4$ (1) and its immobilized analog (2).

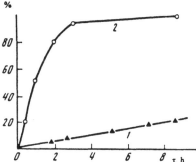

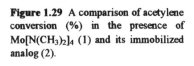
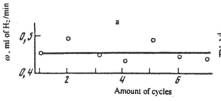
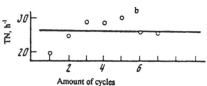

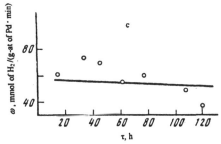

Figure 1.30 The dependence of metal-polymer catalyst activity on turnover cycles in hydrogenation of styrene (a) and pentene-1 (b); the changes of catalytic activity with time for butylvinyl ester and nitroanthraquinone hydrogenation (c). Catalysts: (a) rhuthenium immobilized by ionite; (b) rhodium on a copolymer of maleic acid; (c) palladium on anionite.

Thus, when active sites are hydride ions of transition metals immobilized on polymer supports and hydrogenation proceeds via reagent coordination with the active center, then the immobilized metal complexes act as homogeneous entities. However these show high stability and are easily separated from the reaction medium (see also Section 1.5) for multiple re-use without loss of activity and even without any induction period. The numerous examples of recycled styrene [162] and pentene-1 [103] hydrogenation may be noted in Figure 1.30.

There are also many common features in the behavior of immobilized and heterogeneous hydrogenation catalysts (Ir/SiO_2, Pd/Al_2O_3, Pd/C, Rh/C, Pt-black, etc.;

see [133,124]) although the activation conditions and their action may be significantly different.

Table 1.22 Comparison of hexene-1 and α-methylstyrene hydrogenation in the presence of different catalysts [215].

Catalyst	Initial rate of hydrogenation (mmol/(g-at of M s))	
	hexene-1	α-methylstyrene
Immobilized metal complex	680	940
5% Pd/C	230	700
5% Rh/C	2120	270

Table 1.23 Comparison of activities and selectivities of different catalysts in hydrogenation of *p*-nitrochlorobenzene

Catalyst	Initial rate (mmol of H_2/min)	Selectivity in respect to p-nitrochlorobenzene (%)
$PdCl_2 \cdot PE$-gr-P4VP (1% of Pd)	0.024	50
Second catalyst use	0.025	52
Third catalyst use	0.026	51
$PdCl_2(C_6H_5CN)_2$	0.091	35
Second catalyst use	Inactive	–
Pd/Al_2O_3 (5% Pd)	0.127	18
Second catalyst use	0.077	25
Third catalyst use	0.046	26

As shown in previous Sections (1.7 and 1.8), the hydrogenation reaction is often accompanied by the formation and further aggregation of free metal particles bound to the polymer support. The active species of conventional heterogeneous catalysts (excluding skeleton contacts, metal oxides and some others) are also metal particles, highly dispersed on the support surface. It follows that the differences between these catalysts are rather quantitative: in particle size, catalyst surface area, the character of the particle distribution, the strength of the bond with the support and stabilization degree. In the best metal polymer catalysts practically all metal atoms are stabilized with polymer matrices and all act as active sites, whereas in conventional heterogeneous catalysts the fraction of active species is not very high. Thus, in hydrogenation of ethylene in the presence of Ir/SiO_2 only 20% of metal atoms take part in catalysis [353]. The relative activities of heterogeneous and heterogenized catalysts depend also on the nature of the substrate (Table 1.22).

In the course of hydrogenation heterogeneous catalysts experience irreversible changes which cause aggregation of metal particles and a loss of activity in re-use. This effect is characteristic of *p*-nitrochlorobenzene hydrogenation (Table 1.23).

To demonstrate simultaneous increase of both catalyst activity and selectivity as the result of immobilization we shall mention quinoline reduction in the presence of Rh[+] bound to phosphynated CSDVB. The rate of the reaction is 22 times higher than that of homogeneous catalyst. The immobilized catalyst yields only 9,10-dihydroacridine, whereas up to 50% of 1,2,3,4-tetrahydroacridine has been formed in the homogeneous system [126].

The catalytic activity of Pd·P4VP complexes in allyl alcohol hydrogenation is greater by an order of magnitude than that of Pd-black [29.30] and by a few orders than that of conventional catalyst Rh/C in olefin hydrogenation [170–174].

To summarize all the analyzed data, we should note that immobilized hydrogenation catalysts show a number of specific features which identify them as a separate catalyst type. Optimization of catalyst preparation and hydrogenation reaction conditions ensured that the catalyst will work with the highest activity and selectivity. Matrix selectivity is the most important property of immobilized catalysts allowing specific hydrogenation of certain substrates from a mixture.

There are many examples that show that the polymer is not only an inert support but is also a reactive participant capable of electron density redistribution on fixed metal complexes. Even for supported, free metal particles, a strong interaction between metal atoms and support, depending on polymer redox properties, has been observed (e.g. Pt/nylon-66 [354]). The activity of supported Pt in benzene hydrogenation has also been shown to be dependent on these interactions.

As the number of homogeneous and heterogeneous catalysts for hydrogenation of aromatics and heterocyclic compounds is rather scant, development of commercial immobilized catalysts for these purposes seems to be promising [224]. This hope is connected with the reduced sensitivity of immobilized polymer catalysts to poisoning by heterocyclic compounds.

Immobilization of metal complexes is also a mean for studying the hydrogenation mechanism. Immobilization in many cases stabilizes catalyticaly active but normally unstable structures. Thus, immobilized particles of $[Ru_3O(OCCH_3)_6L_3]^+$ have been detected as active centers [161] in cyclooctene hydrogenation with rhuthenium complexes immobilized on carboxylate-polymer matrices. However, the activity of immobilized metal complexes is often lower than that of homogeneous catalysts due to reduced availability of active sites. To realize all catalyst possibilities, optimization of reaction conditions is required.

Nevertheless, there is still no clear idea of the nature of the active sites of immobilized metal complexes and even of the oxidation state of the transition metal ions during the course of hydrogenation. The majority of workers suppose that free metal particles stabilized by the polymer are the active species. This point of view is widely held[23]. At the same time there are other viewpoints . The presence of metal oxides has been proved by different methods. ESCA has showen [357] that rhodium in phosphynated polymer is in an oxidized state both before and after hydrogenation; Pt^{2+} species have been detected in platinum immobilized catalysts. There are many such examples .The presence of both Pd^0 and Pd^{2+} species in palladium phosphyne catalysts immobilized on PS in the course of hydrogenation has been shown [358] by both ESCA and X-ray diffraction methods. The Pd^0 fraction increases with a rise of the initial amount of Pd^{2+} in the catalyst. Interaction of preactivated catalysts with O_2 leads to palladium oxidation. It has been shown that there are two types of active site on the catalysts surface. These are free metal palladium, which promotes H_2 activation, and Pd^{2+}, catalyzing hydrogenation. However, such a mechanism is rather complicated because synchronous H_2 activation

---

[23] The criteria for catalytic reaction on polynuclear metal compounds (cluster catalysis) are stated in [355] and metods of cluster attachment to polymers in [356].

and substrate hydrogenation must proceed on an active site of specified geometrical configuration; this is hardly probable for immobilized catalysts. Activation of such catalysts is also of great importance [359–361]. The processes of immobilized palladium reduction is dependent on the nature of functional groups, reaction temperature, and other factors [362].

In principle, catalysts based on immobilized Pd^{2+} may reveal three types of active species: Pd^{2+}, Pd^{1+} and Pd^0, especially in the hydrogenation of aromatics and nitro compounds [363]. Rhodium species in all oxidative states have been revealed [364] in activated rhodium catalysts (Rh^{3+} immobilized on polyamines). The reversible transformation of all forms and correlation with their catalytic action have been observed. Moreover, active species in the same oxidative state can exhibit different activity due to the cooperative influence of the polymer backbone. Metal complexes may be located both at the end and within a macroligand, etc. [365]. The subject of cooperative interactions will be considered later in Chapter 5.

References

1. James, B.R. (1973) *Homogeneous Hydrogenation*. New York: Wiley.
2. Shmidt, F.K. (1986) *Kataliz Kompleksami Metallov Pervogo Perekhodnogo Ryada. Reaktsii Gidrirovaniya i Dimerizatsii* (Catalysis by First Transition Row Metal Complexes. Reactions of Hydrogenation and Dimerization), pp. 158-230. Irkutsk: University Press.
3. Sokol'skii, D.V. and Sokol'skaya, A.M. (1970) *Metally – Katalizatory Gidrogenizatsii* (Metals as Catalysts of Hydrogenation). Alma-Ata: Nauka.
4. Hartley, F.R. (1985) *Supported metal complexes. A new generation of catalysts*. Dordrecht: Reidel.
5. Klyuev, M.V., Pomogailo, A.D. (1988) *Gidrirovanie na metallosoderzhashchikh polimerah: osobennosti, problemy, perspektiy* (Hydrogenation by Metal-Containing Polymers: Specific Features, Problems, Outlook). Preprint. Chernogolovka: Institute of Chemical Physics, Russian Academy of Sciences.
6. Masters, K. (1983)) *Gomogennyi Kataliz Perekhodnymi Metallami* (Homogeneous Catalysis by Transition Metals). Moscow: Mir.
7. Harmon, R.E., Gupta, S.K. and Brown, D.J. (1973) *Chem. Rev.*, **73**, 21.
8. Bayer, E. and Schumann, W. (1986) *J. Chem. Soc., Chem. Commun.*, 949.
9. Albright, R.L. (1986) *React. Polym.*, **4**, 155.
10. Alami, S.W. and Caze, C. (1987), *Eur. Polym. J.*, **23**, 883.
11. Bergbreiter, D.E. (1985) *Am. Chem. Soc. Polym. Prepr.*, **26, 74**.
12. Bergbreiter, D.E. and Chandran, R. (1987) *J. Am. Chem. Soc.*, **109**, 174.
13. Bergbreiter, D.E., Blanton, J.R., Chandran, R., *et al.* (1989) *J. Polym. Sci., Polym. Chem. Ed.*, **27**, 4205.
14. Butler, J.S., Gordon, B. and Harrison, I.R. (1988) *J. Appl. Polym. Sci.*, **35**, 1183.
15. Pomogailo, A.D., Lisitskaya, A.P., Ponomarev, A.N., *et al.* (1973) *Abstracts of XVII Conference on high molecular compounds*. Moscow: Nauka.
16. Kritskaya, D.A., Pomogailo, A.D., Ponomarev, A.N. and Dyachkovskii, F.S. (1979) *Vysokomol. Soedin., Ser. A*, **21**, 1107.
17. Garnet, J.M., Levot, R. and Long, M.A. (1981) *J. Polym. Sci., Polym. Lett. Ed*, **19**, 23.
18. Hartley, F.R., Murray, S.G. and Nicholson, P.N. (1982) *J. Mol. Catal.*, **16**, 363.

19. Australian Patent 535,546, 1976.
20. Berenblyum, A.S., Karel'skii, V.V., Mund, S.L., *et al.* (1985) *Khim. Tekhnol. Topl. Masel*, 12.
21. (1986) *Huaxue Tongbao (Chemistry)*, N6, p.23.
22. McDonald, R.P., and Winterbottom, J. (1977) *J.Rev. Port. Quim.*, 19, 269.
23. Klyuev, M.V. (1986) *Proceedings of the V International Symposium on the Relation between Homogeneous and Heterogeneous Catalysis.* 2, p.123. Novosibirsk:
24. Chandrasekaran, E.S., Grubbs, R.H. and Brubaker, C.H. (1976) *J. Organomet. Chem.*, 120, 49; (1975) *J. Am. Chem. Soc.*, 97, 2128.
25. Pomogailo, A.D. (1988) *Polimernye Immobilizovannye Metallokomplrksnye Katalizatory* (Polymer-Immobilized Metal Complexes*)*. Moscow: Nauka.
26. Zurakowska-Orzagh, J., Skupinska, J. (1985) *Polimery*, 30, 133.
27. Valentini, D.Zh., Chekki, A., Bunio, S.D., *et al.* (1986) *Proceeding of the V International Symposium on the Relation between Homogeneous and Heterogeneous Catalysis*, 2, p.185. Novosibirsk: Institute of Catalysis Siberian Department USSR Academy of Sciences.
28. Rajca, I. (1984) *Chem. Stosow.*, 28, 253.
29. Sokol'skii, D.V., Zharmagambetova, A.K., Mukhamedzhanova, S.G., *et al.* (1985) *Dokl. Akad. Nauk SSSR*, 283, 678; (1986) *Izv. Akad. Nauk Kaz. SSR, Ser. Khim.*, 86.
30. Bekturov, E.A., Kudaibergenov, S.E., Saltybaeva, S.S., *et al.* (1985) *React. Polym.*, 4, 49; (1986) *Makromol. Chem,. Rapid. Commun.*, 7, 181; (1987) *React. Kinet. Catal. Lett.*, 33, 387.
31. Kopylova, V.D. and Astanina, A.N. (1987) *Ionitnye Kompleksy v Katalize* (Ionite Complexes in Catalysis), pp.82–83. Moscow: Khimiya.
32. Bandermann, F. and Joppich, G. (1985) *Chem.-Ing.-Tech.*, 57, 986.
33. Abubakirov, R.Sh., Sytov, G.A., Perchenko, V.N., *et al.* (1983) *Dokl. Akad.Nauk SSSR*, 282, 109; (1986) *Izv. Akad. Nauk Kaz. SSR, Ser. Khim.*, 20.
34. Suzuki, N., Ayaguchi, Y. and Izawa, Y. (1983) *Chem. Ind.*, 515.
35. Cum, G. (1984) *J. Chem. Technol. Biotechnol. A.*, 34, 416.
36. Wang, Y.-P. and Neckers, D.C. (1985) *React. Polym.*, 3, 191.
37. Jaffe, I., Segal, M. and Efraty, A. (1985) *J. Organomet. Chem.*, 94, C17.
38. Li, N.-M. and Frechet, J.M.J. (1987) *React. Polym.*, 6, 311.
39. Belyi, A.A., Chigladze, L.G., Rusanov, A.L., *et al.* (1983) *Izv. Akad. Nauk SSSR, Ser. Khim.*, 2678; (1987) *Izv. Akad. Nauk SSSR, Ser. Khim.*, 2153.
40. Belyi, A.A., Chigladze, L.G., Shapiro, E.S. and Vol'pin, M.E. (1984) *Abstracts of the IV International Symposium on Homogeneous Catalysis, Leningrad, 1984.*, vol.4, p.31. Chernogolovka: Institute of Chemical Physics, Russian Academy of Sciences.
41. Belyi, A.A., Chigladze, L.G., Rusanov, A.L.and Vol'pin, M.E.. (1989) *Izv. Akad. Nauk SSSR, Ser. Khim.*, 1961.
42. Gvinter, L.I., Ignatov, V.M., Suvorov, L.N. and Sharf, V.Z. (1987) *Neftekhimiya*, 27, 348.
43. Bhaduri, S., Khawaja, H. and Khanwalkar, V. (1981) *J. Chem. Soc,. Dalton. Trans.*, 447; (1982) *J. Chem. Soc., Dalton. Trans.*, 445.
44. Gogak, D.T. and Ram, R.N. (1989) *J. Mol. Catal.*, 49, 285.
45. Moberg, C. and Rakas, L. (1987) *J. Organometl. Chem.*, 335, 125.
46. He, B. and Wang, L. (1987) *Gaofenzi Xuebao –(Acta Polym. Sinica)*, 404.
47. Nakamura, Y. and Hirai, H. (1975) *Chem. Lett.*, 823.

48. Pomogailo, A.D. and Uflyand, I.E. (1991) *Makromolekulyarnye Metallokhelaty* (Macromolecular Metal Chelates). Moscow: Nauka.
49. Yanzhu, Z., Yongjum, L., Lingzhi, W. and Yingyan, J. (1980) *Fundam. Res. Organomet. Chem., Proceedings of the China–Japn.–USA Trilateral Seminar, p.77.* Peking:
50. Huang, H.-Y., Ren, C.-Y., Zhou, Y.S. *et al.* (1985) *Organosilicon and Bioorganosilicon Chem.: Proceedingd of the VII International Symposium on Organosilicon Chemistry,* (Kyoto, 1984), p.275. Chichester:Wiley.
51. Chen, Yu., Xiao, Ch., Luo, Ch. and Liu J., (1986) *J. Polym. Commun.,* 375.
52. Kalinin, V.N., Mel'nik, O.A., Sakharova, A.A., *et al.* (1985) *Izv. Akad. Nauk SSSR, Ser. Khim.,* 2242; (1984) *Izv. Akad. Nauk SSSR, Ser. Khim.,* 2151.
53. Kovredov, A.I., Dobroserdova, N.B., Ausheva, F.M., *et al.* (1988) *Neftekhimiya,* **28,** 33.
54. U.S. Patent 4,363,744, 1988.
55. Kats, G.A., Komarova, L.G. and Rusanov, A.A. (1991) *Izv. Akad. Nauk SSSR, Ser. Khim.,* 960.
56. Kaneda, K., Imanaka, T. and Tneranishi, S. (1985) *Polymer Materials Science and Engineering Proceeding of the Americal Chemical Society Division of Polymer Materials Science and Engineering,* vol. **53,** (Fall Meel., Chicago, 1985) Washington DC; vol.30, p.210.
57. Akanatsu, A.and Akabori, S. (1961) *Bull. Chem. Soc. Jpn.,* **34,** 1067; (1962) *Bull. Chem. Soc. Jpn.,* **35,** 1706.
58. Zhang, S.L., Xu, Y. and Ziao, S. (1986) *J. Catal.,* **7,** 364.
59. Jaske, G., Lauke, H., Mauermann, H., *et al.* (1985) *J. Am. Chem. Soc.,* **107,** 8111.
60. Fotcher, G., Le Marechal, J.F. and Marquet-Ellis (1982) *J. Chem. Soc., Chem. Commun.,* **224,** 251.
61. Sokol'skii, D.V. and Noskova, N.F. (1977) *Katalizatory Tipa Tsiglera-Natta v Reaktsii Gidrirovaniya* (Ziegler–Natta Catalysts in Hydrogenation Reactions). Alma–Ata: Nauka.
62. Noskova, N.F. and Sokol'skii, D.V. (1982) *Kinet. Katal.,* **23,** 1382.
63. Reguli, J. and Stasko, A. (1982) *Chem. Listy,* **76,** 1085.
64. Svoboda, P., Hetfleis, J., Schulz, W. and Proceiyus, H. (1986) *React. Kinet. Catal. Lett.,* **32,** 39.
65. Eisen, M.S. and Marks, T.J. (1992) *J. Am. Chem. Soc.,* **114,** 10358.
66. Sakai, M., Fujii, T., Kismito, T., Sazuki, K., *et al.* (1986) *Nippon kagaku kaishi (J. Chem. Soc. Jpn. Chem. Ind. Chem. Jpn.),* 126.
67. Biang-Hung, C., Grubbs, R.H. and Brubaker, C.H. (1985) *J.Organomet. Chem.,* **280,** 365.
68. Korolev, A.V., Brodskii, A.R., Noskova, N.F. and Sokol'skii, D.V. (1987) *Dokl. Akad. Nauk SSSR,* **296,** 379.
69. Yanagida, S., Hanazawa, M., Kabumoto, A., *et al.* (1987) *Synth. Met.,* **128,** 785.
70. Negishi, E. and Takahashi, T. (1994) *Acc. Chem. Res.,* **27,** 124.
71. Pomogailo, A.D. and Savostiyanov, V.S. (1988) *Metallosoderzhashchie Monomery i Polimery na ikh Osnove* (Metal-Containing Monomers and their Polymers). Moscow: Khimiya.
72. Neckers, D.C. (1987) *J. Macromol. Sci.,* **24,** 431.
73. Ishiyama, J., Shirakawa, T., Kurokawa, Y. and Imaizumi, S. (1988) *Angew. Makromol. Chem.,* **156,** 179.
74. Matheo, J.P. and Srinivasan, M. (1991) *Polym. Int.,* **24,** 249.

75. Chen, Z. and Jiang, Y. (1982) *Fundam. Res. Organomet. Chem., Proceedings of the China–Jap.–US Trilateral Seminar*, (Peking, 1980). p.629. New York: Wiley.
76. Rayer, G.P., Wen-Shiung, C. and Hattokimi, S. (1985) *J. Mol. Catal.*, **31**, 1.
77. Cherkashin, G.M., Shuikina, L.P., Parenago, O.P. and Frolov, V.M. (1985) *Kinet. Katal.*, **26**, 1110.
78. Xiao, C., Lin, Y., Ren, X. and Chen, Y. (1986) *J. Wuman Univ. Nat. Sci.*, 65.
79. Andersson, C., Larsson, R. and Yom-Tov, B. (1983) *Inorg. Chim. Acta* I, **76**, L185.
80. Tsubai, A., Urata, X., Miyadzaki, T. *et al.* (1985) *Shokubai (Catalyst)*, **27**, 84.
81. Grubbs, R.H. and Sweet, E.M. (1978) *J. Mol. Catal.*, **3**, 259.
82. Kincaid, B.M., Reed, J., Esenberg, P. and Boon-Keng, T. (1977) *J. Am. Chem. Soc.*, **99**, 2375.
83. Allum, K.G., Hancock, R.D., Howell, I.V. *et al.* (1975) *Organomet. Chem.*, **87**, 189.
84. Bernard, J.R., Chauvin, Y. and Commereuc, D. (1976) *Bull. Soc. Chim. Fr.*, Pt2, 1163; 1168.
85. Torronis, S., Innorta, G., Foffani, A. *et al.* (1985) *Abstracts of the. XII International Conference on Organometallic Chemistry*, (Vienna, 1985), vol. S1, p.473.
86. Grubbs, R.H. and Kroll, L.C. (1973) *J. Macromol. Sci. Chem.*, **7**, 1047.
87. Rollman, L.D. (1972) *Inorg. Chim. Acta*, **6**, 137.
88. Grubbs, R.H. and Sweet, E.H. (1975) *Macromolecules*, **8**, 241.
89. Bernard, J.R., Can, H.-V. and Teicher, S.J. (1975) *J. Chim. Phys. Phys.-Chem. Biol.*, **72**, 729; (1976) *J. Chim. Phys. Phys.-Chem. Biol.*, **73**, 799.
90. Holly, N.L. (1978) *J. Chem. Soc., Chem. Commun.*, 1074; (1979) In *Fundamental Research in Homogeneous Catalysis*, Edited by. M.Tsutsui, vol.3, p.461. New York: Plenum Press.
91. Patchornik, A., Ben-David, Y. and Milstein, D. (1990) *J.Chem. Soc. Chem. Commun.*, 1090.
92. Uematsu, T., Umino, M., Shinazu, S. *et al.* (1986) *Bull. Chem. Soc. Jpn.*, **59**, 3637.
93. Wang, Y.P. and Neckers, D.C. (1985) *React. Polym.*, **3**, 181.
94. Neckers, D.C. (1985) *Metal-Containing Polymeric Systems. Organometallic Polymers*, Eds. J.E.Sheats, C.E.Carraher, and C.U.Pitman, p.385–403, New York; Plenum Press.
95. Zhang, K. and Neckers, D.C. (1983) *J. Polym. Sci., Polym. Chem. Ed.*, **21**, 3115.
96. He, B. and Wang, L. (1989) *Chin. J. Appl. Chem.*, **6**, 52.
97. Pomogailo, A.D. and Klyuev, M.V (1985) *Izv. Akad. Nauk SSSR, Ser. Khim.*, 1716.
98. Braca, G., Sbrana, G., Valentini G., *et al.* (1980) *J. Mol. Catal.*, **7**, 457.
99. Brown, J.M. and Molinari, H. (1979) *Tetrahedron Lett.*, 2933.
100. Zhou, Z., Zhang, M.-Zh., Chen, C.-Zh. and Sun Y. (1987) *J. Catal.*, **8**, 69.
101. Bar-Sela, G. and Warchawsky, A. (1983) *React. Polym.*, **1**, 149.
102. Zudin, V.N., Chinakov, V.D., Nekipelov, V.M., *et al.* (1986) *Proceedings of the V International Symposium On the Relation between Homogeneous and Heterogeneous Catalysis* ,vol. **2**, p.97. Novosibirsk: Institute of Catalysis Siberian Department USSR Academy of Sciences.
103. Valentini, G., Sbrana, G. and Braca, G. (1981) *J. Mol. Catal.*, **11**, 383.
104. Okano, T., Tsukiajama, K. and Konisni, H. (1982) *J. Chem. Lett.*, 603.

105. Tyurenkova, O.A. and Chimarova, L.A. (1969) *Zh. Fiz. Khim.*, **43**, 2088; (1970) *Zh. Fiz. Khim.*, **44**, 88; 379; 2278.
106. Rasadkina, E.N., Kuznetsova, T.V., Teleshova, A.T. *et al.* (1974) *Kinet. Katal.*, **15**, 969.
107. Rasadkina, E.N., Rozhdestvenskaya, I.D. and Kalechits, I.V. (1973) *Kinet. Katal.*, **14**, 1214; (1975) **16**, 1465; (1976) **17**, 916; (1978) **19**, 793.
108. Drago, R.S., Nyberg, E.P. and El. Amma, A.G. (1981) *Inorg. Chem.*, **20**, 2461.
109. Seikaky, K., Nasaya, Y. and Takeo, T. (1972) *Shokubai (Catalyst)*, **14**, 154.
110. Braca, G., Valentini, G. and Ciardelli, F. (1978) *Chem. Ind.*, **60**, 530.
111. Koide, M., Sugamumma, N., Tsuchida, E. and Kurimura, Y. (1981) *Makromol. Chem.*, **182**, 1391.
112. Oguni, N., Shimazu, C., Iwamoto, Y. and Nakamura, A. (1981) *Polym. J.*, **13**, 845.
113. Jarell, M.S., Gates, B.C. and Nicholson, E.D. (1978) *J. Am. Chem. Soc.*, **100**, 5727.
114. Grubbs, R.H., Gibbons, C., Kroll, L.C. *et al.* (1973) *J. Am. Chem. Soc.*, **95**, 2373.
115. Bonds, W.D., Jr., Brubaker, C.H., Jr., Chandrasekaran, E.S. *et al.* (1975) *J. Am. Chem. Soc.*, **97**, 2128.
116. Grubbs, R.H., Lau, C.P. and Brubaker, C.H. Jr. (1977) *J. Am. Chem. Soc.*, **99**, 4517.
117. Bowden, C., McLellan, T.J., McDonald, R.P. and Winterbottom, J.M. (1982) *J. Chem. Technol. Biotechnol.*, **32**, 567.
118. Corain B., Basato M., Main P. *et al.* (1987) *J. Organomet. Chem.*, **326**, C43.
119. Keim, W., Mastrorilli, P., Nobile, C.F., *et al.* (1993) *J. Mol. Catal.*, **81**, 167.
120. Shah, J.N., Ram, R.N. (1992) *J. Mol. Catal.*, **77**, 235; (1993) **83**, 67.
121. Klyuev, M.V. (1972) Izv. Vyssh. Uchebn. Zaved., Khim. Khim. Tekhnol., **25**, 751.
122. Karakhanov, E.A., Neimerovets, E.V. and Dedov, A.G., (1985) Vestnik Mosk.Univ., Ser.2: Khim., **26**, 429.
123. Karakhanov, E.A., Pshezhetskii, V.S., Dedov, A.G. *et al.* (1984) *Dokl. Akad.Nauk*, **275**, 1098.
124. Kaneda, K., Terasawa, M., Imanaka, T. and Teranishi, S. (1978) *Fundamental Research on Homogeneous Catalysis: Proceedings of the I International. Conference Homogeneous Catalysis*, (Corpus Christi (Tx.),1978), vol.3,p.671. New York: Plenum Press.
125. Pittman C.U., Jr. and Ng, O. (1978) *J. Organomet. Chem.*, **153**, 85.
126. Fisch, R.H., Thormodsen, A.D. and Heinemann, H. (1985) *J. Mol. Catal.*, **31**, 191.
127. Pittman, C.U. (Jr.), Ng, O., Hirao, A. *et al.* (1977) *Relations entre catalyse homogene et catalyse heterogene: II International. Colloques*, (Lyon, 1970), p.50.
128. Cocco, G., Enzo, S., Pinna, F. and Strukul, G. (1983) *J. Catal.*, **82**, 160.
129. Pinna,F., Strukul, G., Fasserini, R. and Zingales,A. (1984) *J. Mol. Catal.*, **24**, 79.
130. Beringhelli, T., Gervasini, A., Morazzoni, F. *et al.* (1984) *J. Catal.*, **88**, 313.
131. Sgorlon, S., Pinna, F. and Strukul, G. (1987) *J. Mol. Catal.*, **40**, 211.
132. Karklin', L.N., Klyuev, M.V. and Pomogailo, A.D. (1980) *Proceeding of the IV All union Conference: Synthesis and Study of Inorganic Compounds in Non- Water Medium*, (Ivanovo, 1980), p.73. Ivanovo: Chemical Technological Institute.
133. Terasawa, M., Kaneda, K., Imanaki, T. and Teranishi, S. (1978) *J. Catal.*, **51**, 406.
134. Guyot, A., Graitall, Ch. and Bartolin, M. (1977) *Relations entre catalyse homogene et catalyse heterogene: II International Colloques.*,(Lyon,1977), p.277.
135. Cernia, E.M. and Graziani, M. (1974) *J. Appl. Polym. Sci.*, **18**, 2725.

136. Bronshtein, L.M. and Valetskii, P.M. (1993) *Vysokomol. Soedin.*, **35**, 1878.
137. Sidorov, A.I., Sul'man, E.M., Bronshtein, L.M., *et al.* (1993) *Kinet. Katal.*, **34**, 87.
138. Bergbreiter, D.E. and Bursten, M.S. (1981) *J. Macromol. Sci. A*, **16**, 369.
139. Strukul, G., Bonivento, M., Graziani, M., *et al.* (1975) *Inorg. Chim. Acta*, **12**, 15.
140. Innorta, G., Modelli, A., Scagnolari, F. and Follani, A. (1980) *J. Organomet. Chem.*, **185**, 403.
141. Nicolaides, C.P. and Coville, N.J. (1981) *J. Organomet. Chem.*, **222**, 285.
142. Sokol'skii, D.V. (1985) *Kataliticheskie Reaktsii v Zhidkoi Faze* (Catalytic Reactions in Liquid Phase): *Review and Report on VI All Union Conference.*(Alma-Ata, 1985), p.3–31. Alma-Ata: Nauka.
143. Regen, S.L. and Lee, D.P. (1977) *Makromolekules*, **10**, 1418.
144. Valentini, G., Braca, G., Sbrana, G. *et al.* (1980) *IUPAC. Macro Florence: International Symposium on Macromolecules*, (Pisa,1980), vol.4, p.143. Preprint. Pisa: Pisa University.
145. Newkome, G.R. and Yoneda, A. (1985) *Makromol. Chem., Rapid. Commun.*, **6**, 77.
146. Potapov, G.P. (1983) *React. Kinet. Catal. Lett.*, **23**, 281.
147. Sukhobok, L.N., Potapov, G.P., Luksha, V.G., *et al.* (1982) *Izv. Akad. Nauk SSSR, Ser. Khim.*, 2307; 2310.
148. Sokol'skii, D.V., Zharmagambetova, A.K., Bekturov, E.A. *et al.* (1988) *Zh. Fiz. Khim.*, **62**, 2056.
149. Hirai, H., Komatsuzaki, S. and Thosima, N. (1986) *J. Macromol. Sci., Chem.*, **23**, 933; (1984) *Bull. Chem. Soc. Jpn.*, **57**, 488.
150. Kobayashi, N. and Iwai, K. (1980) *J. Polym. Sci. Polym. Chem. Ed.*, **18**, 23.
151. Braca, G., Carlini, C., Ciardeli, F. and Sbrana, G. (1987) *Proceedings of the VI International Congress on Catalysis,*(London, 1986), vol.1, p.528.
152. Karklin, L.N., Klyuev, M.V. and Pomogailo, A.D. (1983) *Kinet. Katal.*, **24**, 408.
153. Karakhanov, E.A., Neimerovets, E.B., Pshezhetskii, V.S. and Dedov, A.G. (1986) *Khim. Geterotsikl. Soedin.*, 303; (1984) *Khim. Geterotsikl. Soedin.*, 1032.
154. Dedov, A.G., Loktev, A.S., Pshezhetskii, V.S. and Karakhanov, E.A. (1985) *Neftekhimiya*, **25**, 452.
155. Hirai, H., Chawanya, H. and Thoshima, N. (1981) *Makromol. Chem., Rapid. Commun.*, **2**, 99.
156. Perchenko, V.N., Mirskova, I.S. and Nametkin, N.S. (1980) *Neftekhimiya*, **20**, 518.
157. Klyuev, M.V. (1984) *Zh. Org. Khim.*, **23**, 581.
158. Uematsu, T., Saito, F., Miura, M. and Hashimoto, H (1977) *Chem. Lett.*, 113.
159. Pomogailo, A.D., Irzhak, V.I., Burikov, V.I. *et al.* (1982) *Dokl. Akad. Nauk SSSR*, **266**, 1160.
160. Bergbreiter, D.C.and Mariagnanam, V.M. (1993) *J. Am. Chem. Soc.*, **115**, 9295.
161. Nicolaides, C.P. and Coville, N.J. (1984) *J. Mol. Catal.*, **24**, 375.
162. Joo, F. and Beck, M.T. (1984) *J. Mol. Catal.*, **24**, 135.
163. Naahtgeboren, A.J., Nolte, R.J.M. and Drenth, W. (1981) *J. Mol. Catal.*, **11**, 343.
164. DeCroon, M.H.J., Colnen, J.W.E. (1981) *J. Mol. Catal.*, **11**, 301.
165. Guyot, A., Graitall, C. and Bartholin, M. (1977/1978) *J. Mol. Catal.*, **3**, 39.
166. Echmaev, S.B., Ivleva, I.N., Pomogailo, A.D. *et. al.* (1986) *Kinet. Katal.*, **27**, 394.
167. Bartholin, M., Graillat, C. and Guyot, A. (1981) *J. Mol. Catal.*, **10**, 361.
168. Bhaduri, S. and Khwaja, H. (1983) *J. Chem. Soc., Dalton. Trans.*1., 419.
169. Thornion, E.W., Knozinger, H., Tesche, B. *et al.* (1980) *J. Catal.*, **62**,117.

170. Hirai, H. (1979) *J. Macromol. Sci., Chem.*, **13**, 633.
171. Hirai, H., Nakao, Y. and Toshima, N. (1978) *Chem. Lett.*, 545; (1976) *Chem. Lett.*, 905; (1978) *J. Macromol. Sci., Chem.*, **12**, 1117; (1979) *J. Macromol. Sci., Chem.*, **13**, 727.
172. Hirai, H., Chawanya, H. and Toshima, N. (1984) *J. Chem. Soc. Jpn. Chem. Ind. Chem.*, 1027; (1986) *Kobunshi Ronbunshu (High Polym. Jpn.)*, **43**, 161.
173. Thiele, H. and Levern, H.S. (1965) *J. Colloid. Sci.*, **20**, 679.
174. Hirai, H., Ohtaki, M. and Komiyama, M. (1987) *Chem. Lett.*, 127.
175. Nakao, Y. and Kaeriama, K. (1983) *Kobunshi Ronbunshu (High Polym. Jpn.)*, **40**, 279.
176. Kurihara K., Fendler J.H., Ravet I. and Nady J.B. (1986) *J. Mol. Catal.*, **34**, 325.
177. Thosima, N. and Takahashi, T. (1988) *Chem. Lett.*, 573.
178. Thoshima, N. (1985) *Shokubai (Catalyst)*, **27**, 488; (1987) *Kobunshi Ronbunshu (High Polym. Jpn.)*, **36**, 670.
179. Sanchez-Delgado, R.A., Andriollo, A., Pupa J. and Martin, G. (1987) *Inorg. Chem.*, **26**, 11867.
180. Sokol'skii, D.V. and Murzagalieva, S.F. (1988) *Proceedings of the VII All Union Conference on Catalytic Reactions in Liquid Phase.* (Alma-Ata,1988), Part I., p.50. Alma-Ata: Nauka.
181. Otero-Schipper, Z., Lieto, J. and Gates, B.C. (1980) *J. Catal.*, **63**, 175.
182. Lieto, J., Rafalko, J. and Gates, B.C. (1979) *J. Catal.*, **62**, 149.
183. Pierantozzi, R., McQuade, K.J. and Gates, B.C. (1979) *J. Am. Chem. Soc.*, **101**, 5436.
184. Inoue, H., Nagao, Y. and Haruki, E. (1985) *J. Chem. Soc. Chem. Commun.*, 501.
185. Yeh, W.V. and Shapley, J.R., (1986) *J. Organomet. Chem.*, **315**, C29.
186. Inoue, H., Nagao, Y. and Haruki, E. (1986) *J. Chem. Soc. Chem. Commun.*, **1**, 165.
187. Lieto, J., Wolf, M., Matrana, B.A. *et al.* (1985) *J. Phys. Chem.*, **89**, 991.
188. Albanesi, G., Bernardi. R. and Moggi P. (1986) *Gazz. Chim. Ital.*, **116**, 385.
189. Attali, S. and Mathein, R. (1985) *J. Organomet. Chem.*, **291**, 205.
190. Costiglioni, M., Giordano, R. and Soppa, E. (1987) *J. Organomet. Chem.*, **319**, 167.
191. Sachtler, W.M.H. (1983) *Chemtech.*, **13**, 434.
192. Uematsu T., Kawakami T., Saitho F. *et al.* (1981) *J. Mol. Catal.*, **12**, 11.
193. Suzuki, N., Ayaguchi, Y., Tsukanaka, T. *et al.* (1983) *Bull. Chem. Soc. Jpn.*, **56**, 353.
194. Kacher, M.K., Mogilevich, G.E. and Pankov, A.G. (1987) *Osnovnoi Organicheskii Sintez i Neftekhimiya* (General Organic Synthesis in Petrochemistry), N23, pp. 89-92.
195. Perchenko, V.N., Kalashnikova, I.S., Kamneva, G.L. and Obydennova, I.V. (1987) *Neftekhimiya*, **27**, 219.
196. Frolov, V.M., Karakhanov, E.A., Parenago, O.P. *et al.* (1985) *React. Kinet. Catal. Lett.*, **27**, 375.
197. Parenago, O.P. (1988) *Selektivnoe gidrirovanie nenasyshchennykh soedinenii v prisutstvii palladiikompleksnykh katalizatorov s azot- i serosoderzhashchimi ligandami* (Selective Hydrogenation of Unsaturated Compounds in the Presence of Palladium Complexes with Nitrogen-containing Ligands). *Dissertation*, Moscow.
198. Karakhanov, E.A., Neimerovets, E.B., Grishina, I.A. and Dedov, A.G. (1985) *Voprosy Kinetiki i Kataliza: Kinet. Katal. s Uchastiem Moleculyarnogo Vodoroda*

(Some Aspects of Kinetics and Catalysis: Kinetics and Catalysis with Participation of Molecular Hydrogen). p.50. Ivanovo: Chemical Technological Institute.

199. Frechet, J.M.J. and Amaratunga, W. (1982) *Polym. Bull.*, **7**, 361.

200. Hirai, H., Chawanya, H. and Toshima N. (1984) *J. Chem. Soc. Japan. Chem and Ind. Chem.*, 1027.

201. Bayer, E., Schumann, W. and Geckeler, K.E. (1986) *Proceedings of the. II International Symposium on Polymer Renewable Resource Materials,* (Miami Beach, 1985), p.115, New York: American Chemical Society.

202. Zhang, Y. and Zhang, M. (1988) *Huaxue Shiji (Chem. Reag.),* **10**, 121.

203. Karakhanov, E.A., Neimerovets, E.B., Samoshin, V.V. *et al.* (1987) *Proceedings of the IV All Union Symposium on Heterogeneous Catalysis in Chemistry of Heterocyclic Compounds,* (Riga,1986), p.132. Riga: Zinatne.

204. Gol'dberg, Yu,.Sh., Iovel', I.G. and Shimanskaya, M.V. (1980) *Katalizatory soderzhashchie nanesennye kompleksy* (Catalysts Containing Immobilized Complexes). Part 2, p.26. Novosibirsk: Institute of Catalysis Siberian Department USSR Academy of Sciences.

205. Galvagno, S., Donato, A., Neri, G. *et al.* (1988) *React. Kinet. Catal. Lett.,* **37**, 443.

206. Carlini, C. and Sbrana, G.J. (1981) *J. Macromol. Sci. Chem.,* **16**, 323.

207. Sbrana, G.J., Braca, G., Valentini, G. *et al.* (1977/1978) *J. Mol. Catal.,* **3**, 111.

208. Spitsyn, V.I., Pirogova, G.N., Korosteleva, R.I. and Kalinina G.E. (1988) *Dokl. Akad. Nauk SSSR*, **298**, 149.

209. Grubbs, R.H. and Kroll, L.C. (1971) *J. Am. Chem. Soc.,* **93**, 3062.

210. Regen, S.L. and Lee, D.P. (1978) *Isr. J. Chem.,* **17**, 284.

211. Nakamura, H. and Hirai, H. (1976) *Chem. Lett.,* 165.

212. Kolot, V.N., Lyubimova, M.V., Litvin, E.F. *et al* (1982) *Izv. Akad. Nauk SSSR, Ser. Khim.,* 1420.

213. Sabadie, J. and Germain, J. (1974) *Bull.Soc. Chim. Fr.,* Pt.2, 1133; (1977) *Bull.Soc. Chim. Fr.,* Pt.1, 616.

214. Card, R.J. and Neckers, D.C. (1977) *J. Am. Chem. Soc.,* **99**, 7733.

215. Nakao, Y. and Fujishide, S. (1980) *Chem. Lett.,* 673; (1981) *J. Catal.,* **68**, 406.

216. Li, H., He, B. (1994) *Abstracts of the 6th International Conference. on Polymer Supported Reactions in Organic Chemistry,* (Venice,1994) p.80. Venice: Venice University.

217. Moreto, J.M., Albaiges, J., and Camps, F. (1974) *Proceedings of the International Symposium on Relations between Heterogeneous. and Homogeneous Phenomena,*(Brussels, 1973), S.4.1.

218. Ciardelli, F. and Pertici, P. (1985) *Z. Naturforsch.* **406**, 133.

219. Izakovich, E.N., Shupik, A.N. and Shul'ga, Yu.M. (1989) *Izv. Akad. Nauk SSSR, Ser. Khim.,* 1499.

220. Saha, Ch.R. and Bhattacharya, S. (1987) *J. Chem. Technol. Biotechnol.,* **37**, 233.

221. Pavlenko, N.V., Tripol'skii, A.I. and Golodets, G.I. (1988) *Kinet. Katal.,* **29**, 746.

222. Guo, X.Y., Zang, H.J., Li, Y.-J. and Jiang, Y.-Y. (1985) *Macromol. Chem., Rapid Commun.,* **5**, 507; (1986) *Chem. Reagent.,* **8**, 134.

223. Dedov, A.G., Neimerovets, E.B., Ryabov, A.B. *et al.* (1987) *Vestn. Mosk. Univ., Ser.2: Khim.,* **28**, 388.

224. Karakhanov, E.A., Dedov, A.G. and Loktev, A.S. (1985) *Usp. Khim.,* **54**, 289.

225. Khidekel', M.L., Bulatov, A.V., Lobach, A.S. and Chepaikin E.G. (1986) *Proceedings of the V International Symposium on the Relation between ·*

Homogeneous and Heterogeneous Catalysis, vol. 2, Pt.2, p.204. Novosibirsk: Institute of Catalysis Siberian Department USSR Academy of Sciences.

226. Ivanova, V.V., Rakovskii, S.K. and Shopov, D.M.(1982) *Proceedings of the II Scientific Conference on Complex Compound Chemistry*, (Burgas,1982), p.31. Academia.

227. Gryaznov, V.M. (1981) *Metally i Splavy kak Membrannye Katalizatory* (Metals and Alloys as Membrane Catalysts). Moscow: Nauka.

228. Gryaznov, V.M. (1986) *Z. Phys. Chem.*, **147**, 123.

229. Gudeleva, N.N., Nogerbekov, B.Yu. and Mustafina, R.G. (1988) *Kinet. Katal.*, **29**, 1488.

230. Nikolaev, N.I. (1980) *Diffuziya v Membranakh* (Diffusion in Membranes), Moscow. Khimiya.

231. Dubyaga, V.P., Perepechkin, L.P. and Katalevskii, E.E. (1981) *Polimernye Membrany* (Polymer Membranes). Moscow: Khimiya.

232. Whitney, D.H. and Wnek, G.E. (1985) *Polymer Material Science and Engineering. Proceedings of the American Chemical Society Division of Polymer Materials Science and Engineering.*(Fall. Meeting Chicago, 1985), Washington. D.C.,vol. **53**, p.659.

233. Ishiyama J. and Kukorawa Y. (1987) *Kagaky Koge (Chem. Ind.)*, **38**, 607.

234. Brunner, H. (1986) *J. Organomet. Chem.*, **300**, 39.

235. Kagan, H.B. (1988) *Bull. Soc. Chim. Fr.* Part I., 816.

236. Klabunovskii, E. (1992) *Kinet. Katal.*, **33**, 282.

237. Stille, J.K. and Paranello, G. (1983) *J. Mol. Catal.*, **21**, 203.

238. Stille, J.K. (1984) *J. Macromol. Sci,. Chem.*, **21**, 1689.

239. Stille, J.K. (1989) *React. Polym.*, **10**, 165.

240. Aglietto, M., Chellini, E., D'Antone, S. *et al.* (1988) *Pure Appl. Chem.*, **60**, 415.

241. Brunner, H., Bielmeir, E. and Wiehl, J. (1990) *J. Organomet. Chem.*, **384**, 233.

242. Krentsel', L.B., Litmanovich, A.D., Plate, N.A., Karpeiskaya, E.I., Godunova, L.F. and Klabunovskii, E.I. (1992) *Vysokomol. Soedin*, **34**, 10.

243. Karpeiskaya, E.I., Godunova, L.F., Levitina, E.S. (1992) *Izv. Akad. Nauk SSSR, Ser. Khim.*, 2368.

244. Deschenaux, R. and Stille, J.K. (1985) *J. Org. Chem.*, **50**, 2299.

245. Selke, R. (1986) *J. Mol. Catal.*, **37**, 227.

246. Selke, R. (1987) *Progress of the Prague Meeting on Macromolecules XXX Microsymposium: Polymer Supported Organic Reagents and Catalysts*, (Prague, 1986), p.34. Prague: Institute of Macromolecular Chemistry Czechoslovak Academy of Sciences.

247. Brunner, H.K.A., Kunz M. and Halhammer E. (1986) *J. Organomet. Chem.*, **308**, 55.

248. Yoshinaa, K., Kito, T. and Onikubo, K. (1983) *Nippon Kagaku Kaishi (J. Chem. Soc. Jpn. Ind. Chem)*., 345.

249. Chan, A.S.C. and Landis, C.R. (1989) *J. Mol. Catal.*, **49**, 165.

250. Guo W., Chen, Y. and Ma Y.(1985) *Huaxue Tongbao (Chemistry)*, 16.

251. Tai, A., Tsukioka, K., Imachi, Y *et al.* (1984) *VIII International Congress on Catalysis.*(Weinheim,1983), vol.5, p.531.

252. Kinting, A., Krause, H. and Capka, M. (1985) *J. Mol. Catal.*, **33**, 215.

253. Tungler, A., Acs, M., Mathe, T. *et al.* (1985) *Appl. Catal.*, **17**, 127.

254. Cesarotti, E., Ugo, R. and Vitiello, R. (1981) *J. Mol. Catal.*, **12**, 63.

255. Mutin, R., Abboud, W., Basset, J.M. and Sinou, D. (1985) *J. Mol. Catal.*, **33**, 47.

256. Klabunovskii, E.I. (1985) *Catalytic Reactions in Liquid Phase: Review and Report on the VI All Union Conference.*(Alma-Ata, 1984), p.147, Alma-Ata:

257. Klabunovskii, E.I., Pavlov, V.A. and Fridman Ya.L. (1984) *Izv. Akad. Nauk SSSR, Ser. Khim.*, 1005.

258. Baker, G.L., Fritschel, S.J. and Stille, J.K. (1983) *Initiat Polymer: 183 American Chemical Society National Meeting* (Washington D.C., 1983) p.137.

259. Bakos, J.F., Toth, I. and Heil, B. (1984) *Tetrahedron Lett.*, **25**, 4965.

260. Halpern, J. (1983) *Pure Appl. Chem.*, **50**, 99.

261. Cowie, M., Sutherland, B.R. and Vaartstra, B.A. (1985) *Abstracts of the XII Conference on Organometallic Chemistry.*(Vienna, 1984) vol.2, p.325.

262. Ermakov, Yu.I., Zakharov, B.A. and Kuznetsov, B.N. (1980) *Zakreplennye Kompleksy na Okisnykh Nositelyakh v Katalize* (Complexes Immobilized on Oxides in Catalysis). Novosibirsk: Institute of Catalysis Siberian Department USSR Academy of Sciences.

263. Galvagno, S., Poltarzewski, Z., Donato, A. *et al.* (1986) *J. Mol. Catal.*, **35**, 365.

264. Poltarzewski, Z., Galvagno, S., Pietropaolo, R. and Staiti, P. (1986) *J. Catal.*, **102**, 190.

265. Mallat, T., Szabo, S. and Petro, J. (1985) *Acta Chim. Hung.*, **119**, 127.

266. Sadeghi, H.R. and Henrich, V.E. (1984) *Appl. Surface Sci.*, **19**, 330.

267. Nurgozhaeva, Sh.Kh. (1981) *Fiziko-Khimicheskie Issledovaniya Slozhnykh System* (Physico-Chemical Studies of Complex Systems). Alma-Ata: Nauka.

268. Bergbreiter, D.E., Bursten, M.S., Lynch T.S. *et al.* (1982) *Fundamental Research on Organometallic Chemistry: Proceedings of the China–Jpn.–US. Trilateral Seminar* (Peking, 1980), New York: 1982, p.373.

269. Kawabata, Y., Pittman, C.U., Jr. and Kabayashi, R. (1981) *J. Mol. Catal.*, **12**, 113.

270. Sakharovsky, Yu,A., Rosenkevich, M.B., Lobach, A.S., *et al.* (1978) *React. Kinet. Catal. Lett.*, **8**, 195.

271. Lobach, A.S., Filatova, M.P., Savostiyanov, V.S. and Khidekel', M.L. (1984) *Abstracts of the IV International Symposium on Homogeneous Catalysis,* (Leningrad 1984), vol.4, p.269. Chernogolovka: Institute of Chemical Physics, Russian Academy of Sciences.

272. Strathdell, G. and Given, R. (1974) *Can. J. Chem.*, **52**, 3000.

273. Mao, S., Miao, Z.Z., Gao W. *et al.* (1987) *J. Nucl. Radiochem.*, **9**, 34.

274. Kitamoto, A., Takashima, Y. and Shimizu, M. (1983) *J. Chem. Eng. Jpn.*, **16**, 495.

275. Matsuda, S. *Chemen* (1983), **21**, 48.

276. Asakura, Y. and Uchida, S. (1982) *J. Nucl. Sci. Technol.*, **19**, 953.

277. Asakura, Y., Tsuchiya, H., Yusa, H. and Matsuda, S. (1980) *Trans. Am. Nucl. Soc.*, **34**, 436.

278. Andreev B.M., Magomedbekov E.P., Firer A.A. and Naaev, I.Yu. (1986) *Kinet. Katal.*, **27**, 224.

279. Jpn. Patent 54-82492, 1982; 54-85966, 1981; 55-110252, 1982; 57-13387, 1983; 54-169446, 1981; 54-169447, 1981.

280. Stepanov, Yu.P., Gorshkov, A.I., Baikov, Yu.M. and Dunaev, T.Yu. (1983) *Kinet. Katal.*, **24**, 494.

281. Shao, R., Zhang, Z., Ding, S. *et al.* (1982) *Huaxue Tongbao.(Chemistry)*, 9.

282. Takasu, V., Sakuma, T. and Matsuda, Y. (1985) *Chem. Lett.*, 1179.

283. Takasu, V., Akimaru, T., Kasahara, K. *et al.* (1982) *Proceedings of the VII International Conference on Vacuum Metall.* (Tokyo 1981), p.719.

284. Iida, I. (1979) *Bull. Chem. Soc, Jpn.*, **52**, 2858.

285. Frennet, A., Gruco, A., Degols, L. and Lienard, G. (1987) *Acta Chim. Hung.*, **124**, 5.
286. Gaetano, K. and Pagni, R.M. (1985) *J. Org. Chem.*, **50**, 499.
287. Ferkins, P. and Vollhardi, K.P.C. (1979) *J. Am. Chem. Soc.*, **101**, 3985.
288. Filikov, A.V. and Miyasoedov, N.F. (1985) *J. Radioanal. Nucl. Chem. Lett.*, **93**, 355.
289. Kriss, R.U. and Eisenberg, R. (1989) *J. Organomet. Chem.*, **359**, C22.
290. Jpn.Patent 49-21983, 1981.
291. Smirnova, Ya.B., Isaeva, V.I. and Sharf, V.Z. (1994) *Izv. Akad. Nauk SSSR, Ser. Khim.*, 1497.
292. Oro, L.A. and Sariego, R. (1982) *React Kinet. Catal. Lett.*, **21**, 445.
293. Marces, R. (1986) *React Kinet. Catal. Lett.*, **31**, 337.
294. Boga, C., Contento, M., Manescalchi, F. (1989) *Makromol. Chem., Rapid. Commun.*, **10**, 303.
295. Azran, J., Buchman, O. and Blum, J. (1981) *Tetrahedron Lett.*, **22**, 1925.
296. Dovganyuk, V.F., Sharf, V.Z., Saginova, L.G. *et al.* (1989) *Izv. Akad. Nauk SSSR, Ser. Khim..*, *777*; (1990), 268.
297. Isaeva, V.I., Smirnova, V.V. and Sharf, V.Z. (1992) *Izv. Akad .Nauk SSSR, Ser. Khim.*, 65; (1993) 613.
298. Yur'ev, V.P. and Salimgareeva, I.M. (1982) *Reaktsii Gidrosililirovaniya Olefinov* (Reactions of Olefins Hydrosililation). Moscow: Nauka.
299. Michalska, Z.M. and Ostaszewski, B. (1986) *J. Organomet. Chem.*, **299**, 259.
300. Lu, F., Wang, D. and Yu, Y. (1985) *Gaofenzi Xuebao (Polym. Commun.)*, 154.
301. Sun, C.-T., Nin, S.-Y., Zhang, S. *et al.* (1984) *Acta Chim. Sinica*, **42**, 549.
302. Reihsfel'd, V.O. and Skvortsov, N.K. (1985) In: *Khimiya Geterogenizirovannykh Soedinenii* (Chemistry of Heterogenized Compounds), edited by *A.Ya.Yuffa*, p.111, Tyumen: University Press.
303. Reihsfel'd, V.O., Skvortsov, N.K. and Brovko, V.S. (1986) *Proceedings of the V International Symposium on the Relation between Homogeneous and Heterogeneous Catalysis*, vol.2, p.215, Novosibirsk: Institute of Catalysis Siberian Department USSR Academy of Sciences.
304. Zaslavskaya, T.N., Filippov, N.A. and Reikhsfel'd, V.O. (1981) *Zh. Obshch. Khim.*, **51**, 107.
305. Nikitin, A.V. and Reikhsfel'd, V.O. (1985) *Zh. Obshch. Khim.*, **55**, 2079.
306. Skvortsov, N.K., Voloshina, N.F., Moldovskaya, N.A. and Reikhsfel'd, V.O. (1988) *Zh. Obshch. Khim.*, **58**, 1831; (1985), **55**, 2323.
307. Capka, M. and Hetflejs, J. (1974) *Collect. Czech. Chem. Commun.*, **39**, 154.
308. Hu, C.-Y., Han, X.-M. and Jiang, Y.-Y. (1986) *J. Mol. Catal.*, **35**, 329.
309. Pukhnarevich, V.B., Burnashova, T.D., Omel'chenko, G.P. *et al.* (1986) *Zh. Obshch. Khim.*, **56**, 2092.
310. Marcinies, B., Foltynowiez, Z., Urbaniak, W. and Perkowski, J. (1987) *J. Appl. Organomet. Chem.*, **1**, 267.
311. Michalska, Z.M. (1986) *Proceedings of the V International Symposium on the Relation between Homogeneous and Heterogeneous Catalysis*, vol.2, p.203. Novosibirsk: Institute of Catalysis Siberian Department USSR Academy of Sciences.
312. Kopylova, L.I., Ivanova, N.D., Voropaev, V.N., *et al.* (1985) *Zh. Obshch. Khim.*, **55**, 1036.

313. Magomedov, G.N.-I., Druzhkova, G.V., Syrkin, V.G. and Shkol'nik, O.V. (1980) *Koord. Khim.*, **6**, 767.
314. Brunner, H. (1983) *Angew. Chem.*, **95**, 921.
315. Dumont, W., Poulin, J.-C., Dang, T.-P. and Kagan, H.B. (1973) *J. Amer. Chem. Soc.*, **95**, 8295.
316. Kinting, A. (1986) *Z. Chem.*, **26**, 180.
317. Reikhsfel'd, V.O., (1986) *Abstracts of the VI All U'nion Conference. on Chemistry and Application of Organosilica Compounds*, (Riga, 1986), p.6. Riga: Zinantne.
318. Kawai, M. and Masuda, T. (1981) *Polym. J.*, **13**, 573.
319. Khananashvili, L.M., Kopylov V.M., Shkol'nik M.I., *et al.* (1985) *Vysokomolek. Soedin.*, *B*, **27**, 752.
320. Chujo, Y., Murai, K. and Yamashita, Y. (1985) *Makromol. Chem.*, **186**, 1203.
321. Ejike, E.N., Parish, R.V. and Jideonwo, A. (1989) *J. Appl. Polym. Sci.*, **38**, 271.
322. Jpn.Patent. 59-147674, 1979.
323. Appl. 3608975 BRD, 1981.
324. Patent 4366333, 1982; Rum.Patent 87828, 1982.
325. Hsieh, H.L. and Yeh, H.C. (1985) *Polym. Prepr. Am. Chem. Soc. Div. Polym.Chem.* **26**, 25.
326. Mohammadi, N.A. and Rempel, G.L. (1989) *J. Mol. Catal.*, **50**, 259.
327. Czech.Patent 227,105,1979.
328. Czech. Patent 226,896, 1979.
329. Patent 4,501,857, 1979; Jpn. Patent 60-79005, 1979.
330. US. Patent. 4,673,714. 1980; 4,656,230, 1980.
331. Jpn. Patent 58-5303 1977, 58-5304 1977.
332. US Patent 4629767 1978.
333. Jpn. Patent. 59-1170, 1978;60-292120, 1987; 60-292121, 1987; 61-61687, 1987; 60-272547, 1987.
334. Jpn. Patent 48-114132, 1987; 60-26024, 1987.
335. Jpn. Patent 60-185733,1987.
336. Jpn. Patent 44-52950, 1987.
337. Kolontarov, I. Ya., Polyakov, Yu.N. and Shagov, V.S. (1985) *Dokl. Akad. Nauk Tadzh. SSR, Ser. Khim.*, **28**, 284.
338. Mohammadi, N.A. and Rempel, G.L. (1987) *Macromolecules*, **20**, 2362.
339. Rosedale, J.H. and Bates, F.S. (1988) *J. Am. Chem. Soc.*, **110**, 3542.
340. Zuchowska, D., Burczyk, L., Meissner, W. and Tukasik, Z. (1987) *Przem. Chem.*, **66**, 380.
341. Appl. 3329974 BRD, 1982.
342. Appl. 3541689 BRD, 1982.
343. Appl. 35414403 BRD, 1984.
344. Rempel, G.L. (1987) *Abstract Paper 31 IUPAC Macromolecule Symposium.* Meseburg: Meseburg University.
345. Chao, S., Simon, R.A., Mallouk, T.E. and Wringhton, M.S. (1988) *J. Am. Chem. Soc.*, **110**, 2270.
346. Colombo R. (1980) *Chem. Lett.*, 1119.
347. Joo, F. and Benko, S.. (1992) *React. Kinet. Catal. Lett.*, **48**, 619.
348. Davydova, S.L., Alieva, E.D. and Plate, N.A. (1968) *Izv. Akad. Nauk SSSR, Ser. Khim.*, 2844.

349. Klyuev, M.V., Tereshko, L.V., Dzhardimalieva, G.I. and Pomogailo, A.D. (1986) *Izv. Akad. Nauk SSSR, Ser. Khim.*, 2531.
350. Kalk P., Roilbeanc R., Gaset A. *et al.* (1980) *Tetrahedron Lett.*, **21**, 459.
351. Card, R.J. and Ziefuer C.E. (1979) *J. Org.Chem.*, **44**, 1096.
352. Bartholin, M., Canan, J. and Gyuot, A. (1977) *J. Mol. Catal.*, **2**, 307.
353. Norval, S.V., Thomson, S.S. and Webb G. (1980) *Appl. Surface Sci.*, **44**, 51.
354. Staite, P., Galvagno, S., Antonucci P. *et al.* (1984) *React. Kinet. Catal. Lett.*, **26**, 111.
355. Laine, R.M. (1982) *J. Mol. Catal.*, **14**, 137.
356. Gates, B.C. and Lamb, H.H. (1989) *J. Mol. Catal.*, **52**, 1.
357. Jones, R.A. (1985) *J. Chem. Soc., Chem. Commun.*, 373.
358. Andersson, C. and Larsson, R. (1983) *J. Catal.*, **81**, 194.
359. Modelli, A., Scagnolari, F., Innorta, G. *et al.* (1984) *J. Mol. Catal.*, **24**, 361.
360. Bar-Sela, G. and Warshawsky, A. (1990) *J. Polym. Sci.,: Polym. Chem. Ed.*, **28**, 1303.
361. Mani, R., Mahadevan, V. and Srinivasan, M. (1990) *Br. Polym. J.*, **22**, 177.
362. Belyakova, Z.V., Pomerantseva, M.G. and Chernyshev, E.A. (1982)) *Sovremmennye Problemy Khimii i Khimicheskoi Promyshlennostu* (Advanced Problems of Chemistry and Chemical Industry: Review of Technical and Economical Investigations of Scientific Research Institutes), N16/137, p.1. Moscow: Scientific Institute of Technological Applications.
363. Echert, H., Fabry, G., Raudaschl, G. and Seidel, C. (1983) *Angew. Chem.*, **95**, 894.
364. Fridman, A.I., Patsevich, I.V. and Galushko, T.V. (1990) *Kinet. Katal.*, **31**, 979.
365. Lawrence, S.A., Sermon, P.A. and Feinstein-Jaffe, I. (1989) *J. Mol. Catal.*, **51**, 117.

CHAPTER 2

OXIDATION REACTIONS CATALYZED BY
IMMOBILIZED METAL COMPLEXES

Electrophilic oxidation by dioxygen and hydroperoxides is an important reaction in organic synthesis. This reaction permits the introduction of oxygen-containing groups into nearly any compound and the preparation of valuable substances. Oxidative processing (as well as hydrogenation and polymerization) of organic raw materials with the use of the oxygen in air as a cheap oxidant is highly effective and a very important process used in the chemical industry. Such reactions with participation of dioxygen are a main part of the complex enzymatic processes occurring in nature.

Oxidation also plays an important role in catalysis because it permits the optimization of methods of directed transformations of various compounds into substances of homogeneous functional content, to suppress unwanted reactions, and to prevent further destructive oxidation of the desired products. Stabilization during storage is an important problem because all products of the chemical and the food industries contain microadmixtures of metal ions (predominantly ions of iron and copper) which are often fixed as various complexes and facilitate decomposition and degradation. Therefore, it is important to study the effect of various homogeneous or heterogeneous catalysts that may be capable of inhibiting, accelerating or making the oxidation more selective[1].

These approaches are often used also for solving ecological problems, for instance for extraction of toxic chemicals, such as ketones, aldehydes, organic acids, sulfur-containing compounds etc., from industrial waste waters. Undoubtedly, oxygen from the air or dilute solutions of H_2O_2 are ecologically suitable oxidants for large-scale processes, and metal ions can be used as catalysts. It should be noted that as early as 1960 the Wacker process of acetaldehyde production was proposed in Germany in which ethylene and oxygen are passed through a solution of $PdCl_2$ and $CuCl_2$ in dilute hydrochloric acid.

The search for and use of new catalysts including immobilized species, as well as the increase of our understanding of mechanisms and kinetics of oxidation, may help to solve many of these problems.

The very possibility of using immobilized systems in oxidation processes is based on the achievements of homogeneous catalysis. Transition metal ligands which are retained in coordination sphere after multiple catalytic acts have been found. The use of polymer groups as ligands permits the ligand surroundings to be varied and regulation of the catalytic properties of the complexes because of the flexibility of the macromolecular chains, their ability to adopt various conformations, and the possibility of creating

---

[1] Moreover, the possibility of non-catalytic oxidation is argued [1] because direct reactions of O_2 with molecules of organic compounds are spin-forbidden and should have high values of E_a. Traces of ions of transition metals are always present in reaction media.

various spatial distributions of metal centers immobilized on the polymer chains. Depending on the chemical nature of initial components, immobilized complexes can be soluble or insoluble in the reaction mixture; therefore, it is possible to transform homogeneous into heterogeneous catalysts and vice versa.

In this chapter data on oxidation of paraffins, alkanes, aromatic hydrocarbons and function-containing compounds are analyzed. Special attention is paid to O_2 transfer (likewise in oxygen-binding membranes), addition of oxygen to multiple bonds (epoxidation), and oxidation of optically active substrates (in this case immobilized complexes serve as enzyme models). It is supposed that this analysis enables us to determine features of immobilized catalysts in comparison with homogeneous analogues and to compare the features of these systems with those of other catalytic systems. Note that these reactions have not actually been analyzed anywhere except in review [2].

2.1 Basic Laws of Oxidation Processes

Oxidation processes consist of a large number of consecutive and parallel reactions with participation of initial substances, intermediates and free radicals contributing to chain reactions [3]. Hydroperoxides formed in the first step of a reaction of hydrocarbon (RH) oxidation decompose in two ways: with formation of stable products (alcohols, ketones, aldehydes, acids etc.) and with generation of new chains. Radical-chain oxidation can be catalytic or non-catalytic.

A mechanism of non-catalytic oxidation in the early stages of RH oxidation can be represented by the following simplified scheme.

(a) Regular chain generation:

$$RH + O_2 \xrightarrow{k_i} R^\cdot + HO_2^\cdot$$

$$2RH + O_2 \xrightarrow{k_i'} 2R^\cdot + H_2O_2$$

Chain generation in the case of initiated oxidation (In_2 is an initiator):

$$In_2 \xrightarrow{k_i''} 2 In^\cdot$$

$$In^\cdot + RH \longrightarrow R^\cdot + InH$$

(b) Growth of the chain:

$$R^\cdot + O_2 \longrightarrow RO_2^\cdot$$

$$RO_2^\cdot + RH \xrightarrow{k_2'} R^\cdot + ROOH$$

(c) Degenerated branching of chains:

$$ROOH \longrightarrow RO^\cdot + OH^\cdot$$

$$ROOH + RH \longrightarrow RO^\cdot + R^\cdot + H_2O$$

(d) Chain decomposition of hydroperoxide:

$$R^\cdot + ROOH \diagup \begin{matrix} ROH + RO^\cdot \\ \\ RCOR + RH + HO^\cdot \end{matrix}$$

(e) Molecular decomposition of hydroperoxide:

$$ROOH \longrightarrow RCOR + H_2O$$

(f) Termination of chains:

$$
\left.
\begin{array}{l}
R^{\cdot} + R^{\cdot} \\
R^{\cdot} + RO_2^{\cdot} \\
RO_2^{\cdot} + RO_2^{\cdot}
\end{array}
\right\}
\xrightarrow{k_6}
\left\{
\begin{array}{l}
R - R \\
ROOR \\
RCOR + O_2 + ROH \\
\text{molecular products}
\end{array}
\right.
$$

Scheme 2.1

The rate of chain generation w_i in the absence of an initiator or catalyst is very low $(10^{-9}$–10^{-7} mol/l s$)$, and if the concentration of hydroperoxide does not exceed ~0.01%, the chain process proceeds mainly due to free radicals produced in reactions of degenerated branching (see reactions in [4]). At $[ROOH]$~10^{-6} mol/l the rates of chain generation and branching become commensurable; the main reaction for chain generation is bimolecular in the case of oxidation of paraffins and alkylaromatic hydrocarbons and trimolecular in the case of oxidation of olefin hydrocarbons and tetralin. Due to relatively high strength (120–150 kJ/mol) of the –O–O– bond in hydroperoxides (intermediate products and concurrently the initiators of chain reactions) the degenerated branching proceeds with low rates that results in the appearance of a marked induction period for low-temperature non catalytic oxidation. In these reactions free radicals attack only hydrogen atoms (the energy of rupturing the C–H bond, for instance, in the methylene group of n-alkane is 370 kJ/mol). In addition, the chain process for liquid-phase oxidation of hydrocarbons is characterized by substantially longer chains ($v = w/w_i \geq 10$) and higher selectivity than in the gaseous phase.

The general formula for the rate of a developed process of oxidation is

$$w = -\frac{d\,[RH]}{dt} = -\frac{d\,[O_2]}{dt} = k_2 k_6^{-1/2}\,[RH] w_i^{1/2}$$

the ratio of rate constants for chain growth and termination being a characteristic of the ability of hydrocarbons to be oxidized [5]. For instance, at 303 K, the values of $k_2 k_6^{-1/2}$ are 0.014, 0.049, 0.21, 1.5 and 2.3 $l^{1/2}$ mol$^{-1/2}$ s$^{-1/2}$ for toluene, n-xylol, ethylbenzene, cumene and cyclohexene, respectively. Relative rates of autoxidation of some unsaturated and alkylaromatic compounds are 10.15 for cyclohexene, 9.34 for cyclopentene, 8.50 for 2-methylbutene-2, 8.21 for dicyclopentadiene, 7.82 for norbornene, 1.30 for n-ethyltoluene, 1.00 for isopropylbenzene, 0.83 for ethylbenzene, 0.73 for n-xylol and 0.28 for toluene [6]. On bimolecular termination the chain length decreases with increasing w_i: $v = w/w_i = k_2 k_6^{-1/2}[RH]\,w_i^{-1/2}$, and at $w_i = k_2^2 k_6^{-1}[RH]^2$ the chain reaction is transformed to a non-chain radical reaction.

In the presence of homogeneous (molecular-dispersed) metal–complex catalysts the rates of some reactions and even the direction of the process can be substantially changed. Ions of the transition metals, M^{n+}, can, in principle, affect the direction of any of the above steps (a)–(f) of the oxidation process. The role of metals is, however, unclear in most cases because of, on the one hand, the variety of redox states of metals and their ability to promote different steps, and, on the other hand, reactions of metal ions with radicals with formation of products which do not continue the chain reaction. Under these conditions primary products can produce free radicals, therefore the reaction scheme for the catalyzed, homogeneous oxidation becomes very complex [7–11].

The key role of M^{n+} ions in liquid-phase oxidation of hydrocarbons is in the acceleration of initiation, as a rule a limiting step for a free radical chain process. This results in a one-half-order reaction with respect to catalyst concentration and a first-order reaction with respect to hydrocarbon concentration. Reaction acceleration (most often revealed by the disappearance of the induction period) can be achieved in the following three ways.

1. The first is due to catalytic decomposition of hydroperoxides that are present at very low concentrations and stable under mild reaction conditions [8,12,13]. This process, leading to formation of alkoxy and hydroperoxy radicals, is described by the Haber–Weiss scheme [14]:

$$\text{ROOH} + M^{n+} \longrightarrow \text{RO}^{\cdot} + {}^{-}\text{OH} + M^{(n+1)+}$$

$$\text{ROOH} + M^{(n+1)+} \longrightarrow \text{ROO}^{\cdot} + H^{+} + M^{n+}$$

It was shown in a number of studies that formation of a complex of ROOH with M^{n+} or $M^{(n+1)+}$ precedes these reactions. Decomposition of coordinated hydroperoxides can proceed by two mechanisms: (i) molecular (heterolytic), leading to the formation of non radical products, and (ii) radical, the main route to decomposition in developed radical-chain processes. Catalytic decay of hydroperoxides is faster by several orders of magnitude than the thermal decomposition. At the same time intensive molecular decay of hydroperoxides can lead to deceleration of oxidation.

2. Another route to acceleration of initiation is through activation of dioxygen by a metal ion and the subsequent interaction of the complex formed and the oxidizable substrate [8]:

$$M^{(n+1)+} + O_2 \longrightarrow {}^{\delta}M^{n+} \cdots O_2^{\delta} \xrightarrow{\text{RH}} \begin{cases} \text{MOOH} + R^{\cdot} \\ M^{(n+1)+}\text{(OH)} + \text{RO}^{\cdot} \end{cases}$$

Oxygen activation is substantially disguised by interaction of metal complexes with hydroperoxides; therefore the rate of reaction of "activated" oxygen is, as a rule, lower than the rates of free radical formation in this reaction. Nevertheless, initiation by activation of molecular oxygen is the most promising route from the viewpoint of selectivity: it is possible to introduce O_2 directly into the C–H bond of oxidizable substrates.

Peroxides of transition metals are themselves active intermediates in heterolytic and homolytic liquid-phase reactions of catalytic oxidation of alkenes, aromatic hydrocarbons and alkanes [15]. Heterolytic oxidation is characterized by (i) a requirement for a free coordination volume near the transition metal atom (Ti, V, Mo, W, Co, Rh, Pd, Cu) of a peroxo complex, (ii) formation of a strained metal ring as an intermediate, and (iii) a high degree of selectivity of the reaction. Homolytic oxidation proceeds via M–O bond cleavage in peroxo complexes together with the formation of a biradical, and is characterized by lower selectivity.

3. Numerous experimental findings confirm that activation of a coordinated substrate involves electron transfer from substrate to metal ion [7]. Most often this transfer proceeds via formation of a cation-radical from the substrate with subsequent loss of a proton:

$$RH + M^{n+} \longrightarrow [RH]^{+} + M^{n-1}$$

$$RH^{+} \longrightarrow R^{\cdot} + H^{+}$$

$$R^{\cdot} + O_2 \longrightarrow ROO^{\cdot} \xrightarrow{RH} products$$

A high degree of selectivity of autoxidation of organic compounds catalyzed by M^{n+} is caused also by the interaction of the substrate with a peroxide derivatives of the metal [16]. Autoxidation of organic compounds can proceed via a metal ion:

$$RH + M^{n+} \longrightarrow M^{(n-1)+} + R^{\cdot} + H^{+}$$

However, this reaction, in contrast to decomposition of intermediate peroxides, is slow; therefore the route of autoxidation via a metal ion is ineffective.

Detailed studies of oxidations controlled by homogeneous catalysts reveal the dual function of transition metal compounds: initiating and inhibiting processes. For instance, M^{n+} ions can, especially at a low degree of oxidation, inhibit oxidation by interacting with peroxides:

$$M^{n+} + RO_2^{\cdot} \longrightarrow M^{(n+1)+}(OH)^{-} + RCOR$$

and competing with chain growth reactions. The dual function of the M^{n+} ions accounts for a critical phenomenon [17]: the oxidation rate increases to a certain level with increase of the catalyst concentration, and then drastically decreases. This phenomenon is caused by the initiating function of catalyst at its low concentration (in this case chains are terminated by bimolecular recombination without participation of catalyst) and by the inhibiting function of catalyst at its high concentration (in this case chains are terminated by interaction of radicals with the catalyst).

Heterogeneous catalysts of various types are used for the oxidation of organic compounds. For instance, solid catalysts immobilized on Al_2O_3, SiO_2, alumosilicates, zeolites, MgO, Cr_2O_3, TiO_2, SnO_2, diatomite, ZrO_2, $CaSO_4$, carbides, nitrides, etc. are used in gaseous-phase processes. To prepare a hydrophobic support, the material is treated (as in hydrogenation processes) by polymers: silicon oils, poly(tetrafluoro-ethylene) (PTFE), polyethylene (PE), or polystyrene (PS) [18]. As a rule, the main role of metal oxides (including immobilized species) in liquid-phase oxidation of hydrocarbons is to decrease the activation energy of the initiation step.

Supported catalysts obtained by the method of impregnation, though widespread in heterogeneous oxidation, have substantial shortcomings: easy washing out of the active component in the course of reaction and decreasing effectiveness due to formation of polymolecular structures on their surface during their preparation. For instance, the impregnation of silica gel by $KMnO_4$ results in formation of a discreet monolayer of physically sorbed salt, in many cases in the course of the process colloid particles of

MnO_2 and a mixture of physically adsorbed molecules of $KMnO_4$ and Mn^{4+} are formed [19,20].

Most salts, including transition metal acetates, used in liquid-phase oxidation of hydrocarbons are incompatible with the medium. Incompatibility diminishes their effectiveness and puts tough restrictions on concentration ratios. To overcome this obstacle, attempts have been made to include MX_n into inverted micelles of ionogenic non–ionic surface-active substances (SAS) [21]. In micelles the rate of these oxidation reactions increases by 3–10 times. Using metal salts as a one-electron oxidant in the interphase and micelle catalysis makes it possible to regenerate reduced metals in the aqueous phase and to switch from stoichiometric oxidation to catalytic processes [22].

Heterogeneous and homogeneous catalysts increase their activity mainly due to their influence on hydroxide decay. The decay can proceed, depending on the nature of MX_n and reaction conditions, by the following mechanisms: (i) a hetero-homogeneous process (formation of radicals on the catalyst surface with subsequent chain reactions of these radicals in the aqueous phase), (ii) an exclusively heterogeneous reaction (all steps of the oxidation proceed on the catalyst surface without escape of radicals into the aqueous phase), (iii) a mixed process (simultaneous development of the process both on the surface and in the aqueous phase) [23–25].

The hetero-homogeneous mechanism is realized most often but, in the presence of catalysts facilitating only heterogeneous oxidation, only one reacting particle may be activated and, thus, the problem of oxidation selectivity can be solved. The scheme of heterogeneous radical (non-chain) hydrocarbon oxidation carried out in a "surface cell" can be represented in the following way:

$$]\cdots\overset{\delta^-}{O}=\overset{\delta^+}{O} \longrightarrow]\cdots OOH \longrightarrow]-ROOH$$

$$\nearrow \qquad \swarrow$$

$$RH \qquad \dot{R}$$

In recent years immobilized catalysts have taken an important place in the oxidation of organic compounds. However, researchers sometimes believe that the small number of metal polymer systems found for oxidation reactions (in contrast to hydrogenation reactions) is caused by their sensitivity to oxidative transformations [26]. Indeed, this feature has restricted the use of metal polymers in these processes. Maybe, this explains why oxidative catalysis was not correctly analyzed (see, for example, [27]). Meanwhile, immobilization of MX_n on polymers permits (i) the use of supported structures that have no soluble analogues, (ii) restrictions to be overcome on the solubility of metal complexes in the reaction medium, (iii) the effectiveness of using a transition metal to increase due to its more efficient use in a single process and for its repeated use, (iv) the prevention of contamination of the product required by transition metals, (v) the activity and selectivity of metal complexes to be affected in specified steps of the process by varying the nature of macromolecular ligand, etc.

Oxidants immobilized on polymers [2,28,29] can be used in these reactions. Perbenzoic and other peracids (produced, for instance, by reaction of carboxyl-substituted PS with hydrogen peroxide in methanesulfonic acid [30,31]), diarylselenoxide or telluride on PS [32], phenyl selenoacid and its arsenic analogue grafted to PS [33], cross-linked PS modified by $As(O)(OH)_2$ groups (arsonic acid) at the para-positions of benzene rings (7–50%) [34], iodobenzene, hypochloride anion, periodites on a polymeric support [35], a polymer with iodoxyl groups [36], polymeric N-oxides etc., are used as immobilized oxidants.

Advantages of oxidants bound to polymeric supports are obvious: simplicity and ease of process control, selectivity of catalyzed reactions and their ability to discriminate properties of macroporous and gel resins on the basis of reactivity of the supported oxidants.

2.2 Activation and Binding of Dioxygen by Metal Polymers

As mentioned above, one way to accelerate catalytic oxidation is the activation of the O_2 molecule by a metal complex with subsequent interaction of the activated oxidant with a substrate. The term "oxygen activation" implies the direct interaction of the O_2 molecule with transition metal ions and the formation of intermediates participating directly in oxidation of organic compounds. Investigation of these reactions is interesting from the viewpoint of searching for new and effective catalysts and modeling complex processes in biosystems.

The main results from studies of these reactions in homogeneous systems with free, unbound metal complexes are the following ([7], pp. 71–119; [37,38]).

Interaction of a metal complex with O_2 proceeds via consecutive one-electron transfer steps from the metal center: the formation of a superoxo complex (MO_2) containing a superoxide ion (monohaptocoordinated O_2), and the formation of a μ–peroxo complex (MOOM) which binds O_2^{2-} (dihaptocoordinated O_2). The latter is an intermediate in the reduction of molecular oxygen to H_2O_2:

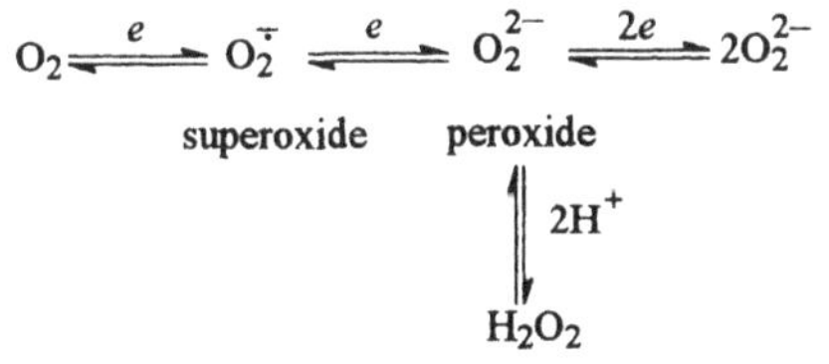

$$O_2 \underset{\longleftarrow}{\overset{e}{\longrightarrow}} O_2^{-} \underset{\longleftarrow}{\overset{e}{\longrightarrow}} O_2^{2-} \underset{\longleftarrow}{\overset{2e}{\longrightarrow}} 2O_2^{2-}$$

superoxide   peroxide

$$\Big\Updownarrow 2H^{+}$$

$$H_2O_2$$

For example, superoxo cobalt(3+) is formed as an intermediate and is rapidly transformed into μ-peroxo cobalt(3+):

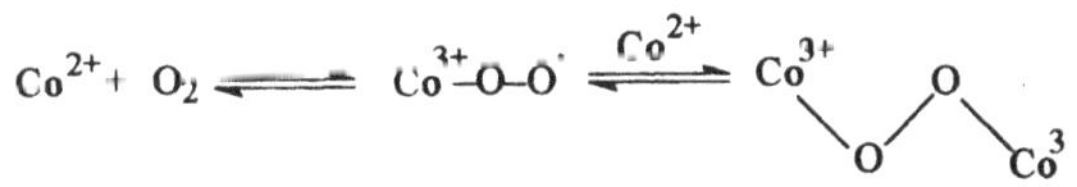

Low-valence transition metal ions (Co^{1+}, Ir^{1+}, Pd^0, Pt^0 and others) are able to form monomolecular peroxo complexes:

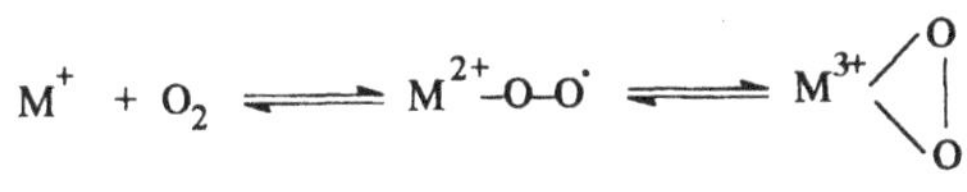

In addition, O_2 can be incorporated into M–H or M–R bonds with formation of hydroperoxo and alkylperoxo metal complexes, respectively. A generalized scheme for the main reactions of O_2 is given in Scheme 2.2.

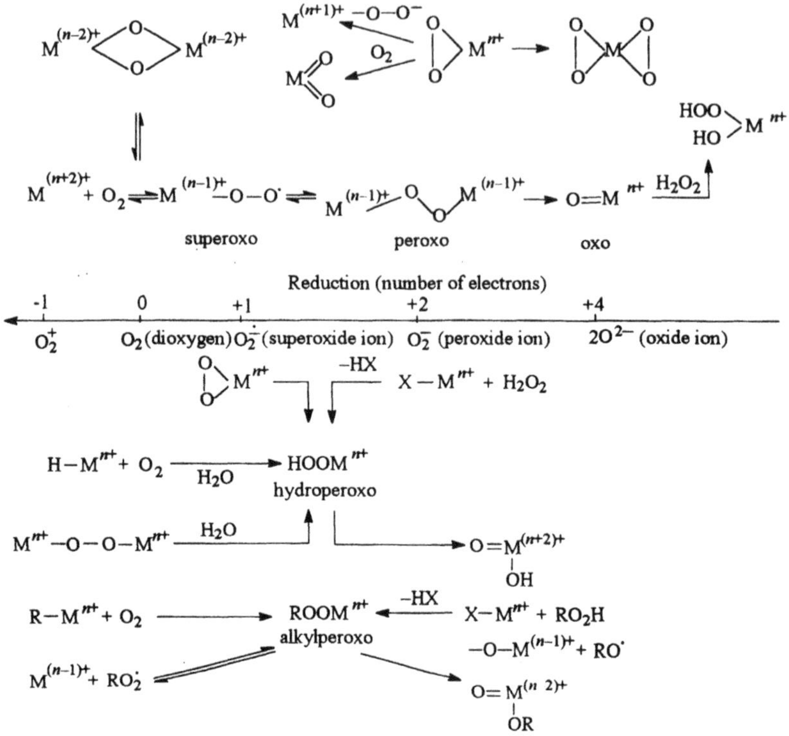

Scheme 2.2

Features of the polymeric ligand affect the kinetics and even the directions of these reactions. For instance, macrocomplex Co^{2+}–polyethyleneimine (PEI) can reversibly bind O_2 with formation of a μ-peroxo adduct stable in aqueous solution at room temperature. This is characterized by a high dissociation constant k_d because of a distortion of the μ-peroxo structure caused by steric hindrances and electrostatic repulsion in the polycation:

$$2Co^{2+}L_6 + O_2 \underset{k_d}{\overset{k_0}{\rightleftharpoons}} L_5\text{–}Co^{3+}\text{–}O\text{–}O\text{–}Co^{3+}\text{–}L_5$$

where L is the monomer link $-NH-CH_2-CH_2$.

The efficiency of these polymeric complexes at pH 7.4 approaches that of hemoglobin and myoglobin [39]. The comparison of reversible O_2 binding by the polymeric complex and its low molecular mass analogue ethylenediaminetetraacetate (EDTA) (Figure 2.1) [40] shows that the value of k_d for reaction

$$L_5\text{–}Co^{3+}\text{–}O\text{–}O\text{–}Co^{3+}\text{–}L_5^+ \ \ EDTA \ \ \xrightarrow{k_d} \ \ Co^{2+}\text{–}EDTA + O_2 + \text{polymer}$$

is higher by 3–5 times for the polymeric complex than for monomeric. Interaction of $RhCl_3$ with PEI in alcohol leads to a partial reduction of Rh^{3+} to Rh^{2+}, polymeric complexes of which are able to react further with oxygen[2] in air with formation of $Rh^{3+}-O_2^-$ superoxide complex [43].

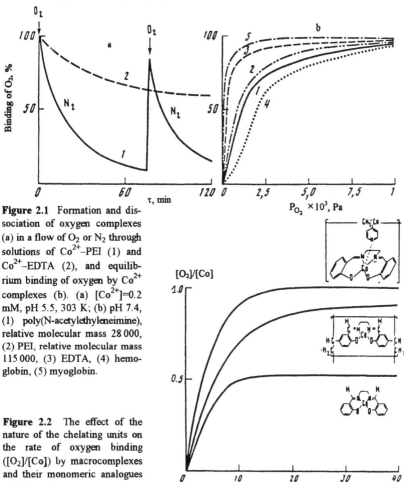

Figure 2.1 Formation and dissociation of oxygen complexes (a) in a flow of O_2 or N_2 through solutions of Co^{2+}–PEI (1) and Co^{2+}–EDTA (2), and equilibrium binding of oxygen by Co^{2+} complexes (b). (a) $[Co^{2+}]=0.2$ mM, pH 5.5, 303 K; (b) pH 7.4, (1) poly(N-acetylethyleneimine), relative molecular mass 28 000, (2) PEI, relative molecular mass 115 000, (3) EDTA, (4) hemoglobin, (5) myoglobin.

Figure 2.2 The effect of the nature of the chelating units on the rate of oxygen binding ($[O_2]/[Co]$) by macrocomplexes and their monomeric analogues (293 K, solid state).

Macromolecular metal chelators are more often used for these purposes. For instance, the Co^{2+} ion immobilized on polychelators (obtained by copolymerization of 2-hydroxy-5-vinylbenzaldehyde or divinylsalen with styrene, as well as by reaction of polymer analogue transformations) combines with O_2 to form a mononuclear superoxo complex.

---

[2] It was found [41] that VCl_4/SiO_2 systems can generate O_2^- radicals as the result of partial hydrolysis of a surface vanadium complex caused by silanol groups of the support. In some cases O_2 interacts chemically with a supported complex, for example in the reaction of amine complexes of Co^{2+} on poly-4-vinylpyridine (P4VP) with O_2 [42].

This can be regenerated because μ-peroxo complexes are not formed in these systems. In most cases the affinity of polymeric complexes with O_2 is higher than that of monomeric complexes due to differences in the strength of the metal–oxygen bond and in the solvation of O_2 (gas)–O_2 (polymer) adducts [44–46]. The increase in the affinity of polymeric complexes to O_2 was also observed for bis(dimethylglyoxymato)Co^{2+} supported on a copolymer with imidazole groups [47]. The high rate of O_2 binding (90% of O_2 reacts after several minutes) was found for Co^{2+} polymeric chelates with N_2O_2 blocks [44,48–51]. Formation of binuclear ($O_2/Co=0.5$) and mononuclear ($O_2/Co=1$) complexes in the case of low molecular mass and polymeric chelators, respectively, was corroborated. Their structure can be represented by Scheme 2.3 [49].

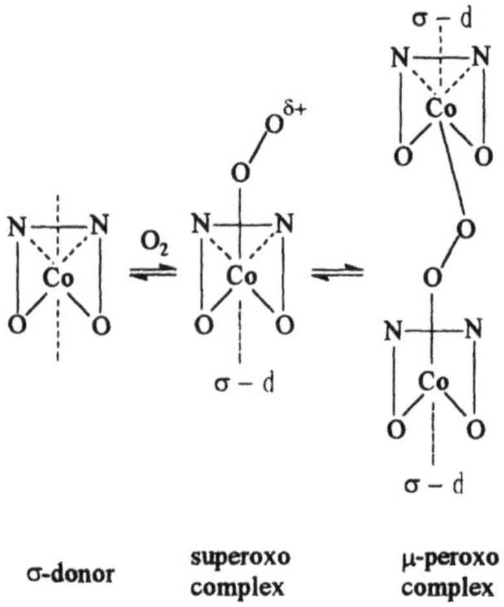

σ-donor   superoxo complex   μ-peroxo complex

Scheme 2.3

The O_2 binding depends on the nature of the chelating units of macrocomplexes [50] (Figure 2.2). The formation constant for the oxygenated Co^{2+}–salen complex is $K=500$ l/mol in the case of a polymeric ligand and 13 l/mol in the case of the low molecular mass analogue, Py–Co^{2+}–salen. In some cases complexes of O_2 with polymeric cobalt-containing Schiff bases can be isolated and characterized [52]. The structure of chelating units is distorted in the course of O_2 fixation and oxidation of Co^{2+} to Co^{3+}. The distortion was experimentally proven for Co^{2+} on poly(styrenesulfonate)-2 (2-pyridylazol)-naphthol. Macrochelates retain their activity up to 353 K and are stable catalysts that can be regenerated. The main parameters for O_2 binding on synthetic and natural carriers are given in Table 2.1.

In addition to cobalt macrocomplexes, immobilized Cu^{2+} compounds are most often used for O_2 activation. For instance, the rate of O_2 binding by Cu^{2+} poly(acroleinoxime) complexes decreases with increasing molecular mass of the macroligand and concentration of immobilized Cu^{2+} [53]. The process proceeds by a cyclic ("ping-pong") redox mechanism:

$$Cu^{2+} \underset{O_2}{\overset{\sim CH=NOH}{\rightleftharpoons}} Cu^{I}$$

The relation between amounts of adsorbed $(Q^a_{O_2})$ and desorbed $(Q^d_{O_2})$ oxygen and concentration of oxime groups (Q_L) is given by the formula:

$$Q=(Q^a_{O_2}-Q^d_{O_2})/Q_L=0.26-0.34$$

It is plausible that the same mechanism is valid for the reversible adsorption of oxygen (and carbon oxide) by Cu^{1+}, Cu^{2+} and Pt^{2+} polymeric chelates on 2,3,6,7-octanetetraon-tetraoximes [54].

Table 2.1 Thermodynamic parameters for the reversible binding of O_2 by polymeric Co^{2+}-containing complexes and low molecular mass and natural analogues [55].

Co^{2+} complex	K_{O_2} (mM^{-1})	ΔH (kJ/mol)	ΔS $(J/mol\ deg)$
Py–Co–PP [a]	1.3	−34.3	−21
CS4VP$_{23}$–Co–PP [b]	2.3	−30.9	−189
CS4VP$_{42}$–Co–PP	7.6	−30.9	−180
Py–Co–salen [c]	0	−	−
CS4VP$_{23}$–Co–salen	8.4	−96.6	−256
CS4VP$_{34}$–Co–salen	19.1	−96.6	−248
Coboglobin	50	−55.9	−223
Myoglobin	2000	−76.0	−252

[a] PP=Protoporphyrin-IX, Py=pyridine.
[b] Copolymer of styrene (CS) with 4-vinylpyridine (the subscript is the content of links of 4-vinylpyridine).
[c] μ-dioxo dimer is formed.

Vanadium dichloride and $MnCl_2$ polymeric complexes (with chelating units of the Schiff base-type) bind reversibly 0.4–0.5 mol of O_2 per mol of M^{2+} at room temperature [56]. High values of magnetic moments (5.5 μ_B for Mn^{2+} complexes and 4.5 μ_B for V^{2+}) testify to the formation of bridge peroxo complexes with an O_2 molecule between two metal atoms (as in cobalt complexes, see Scheme 2.3).

The most important success in activation of molecular oxygen was achieved for immobilized complexes of Fe^{3+}, Mn^{3+} and Co^{3+} with porphyrins and phthalocyanines. These are studied as model systems of metal enzymes (metal-containing electron carriers), for example cytochrome P450, myoglobin and hemoglobin [57–59]. Molecules of metal porphyrins and their derivatives have, on the one hand, a central metal ion in various redox states which has uncoupled electrons and, on the other hand, a planar porphyrin ring with a developed system of π-conjugated bonds. These features open up interesting possibilities for using compounds of this type in oxidation reactions of organic compounds (see Section 2.4). The catalytic reactions with participation of O_2 proceed by two mechanisms: donor (electron transfer from catalyst to substrate) and acceptor processes (electron transfer from oxidizable substrate to catalyst).

Many such complexes are known as carriers of oxygen to molecules of oxidizable substrate [55,60]. In these systems it is probable that the polymeric matrix prevents irreversible interactions of O_2 with active centers (the phenomenon of oxygen protection) and impedes oxidation of easily oxidizable substrates. The period of half-transformation of oxide complexes is several hours under optimal conditions. Polymer ligands, like protein globins, prevent dimerization of active complexes holding them at a distance (the matrix isolation). For instance, in contrast to reactions of O_2 with $Co^{2+}-$ tetraphenylporphyrin and Fe^{2+}–tetraphenylporphyrins where μ-dioxo dimer and μ-oxo dimer, respectively, are formed, in polymeric complexes such reactions are suppressed. In addition, polymer ligands, isolating metal centers from the protic medium, suppress monomolecular oxidation. The polymer effect is accounted for by conformational features of macrochains; oxidative complexes are stabilized due to flexible motions and a specific microenvironment. Metal porphyrins immobilized in hydrophobic regions of water-soluble polymers are used in attempts at making synthetic O_2 carriers for reversible oxidation in water and other protic media. For instance, O_2-binding centers, five-coordinated high-spin Fe^{2+}–porphyrin complexes, are oxidized in solution to low-spin Fe^{3+} complexes:

$$Fe^{2+} + O_2 \rightleftharpoons \left[\overset{\overset{\displaystyle \ulcorner O=O}{|}}{Fe^{2+}} \longleftrightarrow \overset{\overset{\displaystyle \ulcorner O-O^{\cdot}}{|}}{Fe^{3+}} \right] \xrightarrow{k_d} Fe^{3+} + O_2^{\cdot}$$

To protect O_2-binding centers (even in aqueous solutions), polymers and copolymers derived from 1-vinylimidazole, 1-vinyl-2-methylimidazole, water-soluble copolymers with 4-vinylpyridin (partially quaternized), 1-vinyl pyrrolidone and others are used [61]. Creating steric hindrances and using low temperatures are other approaches.

The rate of O_2 absorption by polymer heme is higher than that by free heme (rate constants are 10^{-1} and 10^{-4} s^{-1} in H_2O, respectively). The value of k_0 is varied by one order of magnitude depending on the conformation of the polymer ligand. The four planar coordination sites of heme are occupied by porphyrin nitrogen, the fifth axial coordination site is occupied by the nitrogen of the polymer ligand, and the sixth site is occupied by Cl, OH, solvents (dimethyl formamide (DMFA), H_2O) or polymer. The number of axial coordination ligands is about 2.5 for polyvinylimidazole and decreases with decreasing 1-vinylimidazole content of the copolymer.

This structure resembles the structure of the active center of hemocyanine [62]. In deoxyhemocyanine each Cu^{1+} ion is held by three imidazole ligands in a bimetal center, localized in a hydrophobic region made of triptophan and phenylalanine residues. In oxyhemocyanine a peroxo dicopper complex is formed:

deoxy-He

Cu···Cu 0.34–0.37 nm

oxy-He

Cu···Cu 0.36 nm

Authors of [63] think that the behavior of polymeric hemocomplexes resembles that of oxyhemoglobin: the CO flow through solutions results in the formation of hemocarboxy complexes, which testifies to Fe^{2+} formation in the course of oxidation.

Decomposition of oxygenil complexes proceeds by a first-order reaction and the time of half-transformation (8 min for copolymers at 243 K) depends on pH; these observations testify to irreversible oxidation of the complexes due to addition of a proton to the oxygen-containing adduct. The rate constant depends markedly on the environment of the active center: the value of k_d decreases after adding salts, polyelectrolytes and SAS to the reaction mixture.

Triple amphiphilic block copolymers (of PEG and styrene) including substituted Fe^{2+}–phenylporphyrin form oxygenated complexes in dry toluene, toluene saturated with H_2O, and even in aqueous solutions at pH 7 and room temperature [64] due to a hydrophobic protective region (a micelle) surrounding the complex.

An approach based on the addition of Fe^{2+}–heme to coacervate, a polymeric composition consisting of polyelectrolytes (mixture of polycations, N,N-dimethyliminoethylene-N,N-dimethyliminomethylene-1,4-phenylenedichloride and anions of acrylic and methacrylic acids) [63], is also interesting. The heme derivative is immobilized in the coacervate and acquires the ability to form oxygen adducts even in buffer solution at pH 10.

The ability of polymeric complexes to bind dioxygen is used in selectively permeable membranes applied for oxygen enrichment [65–70]. For instance, [67] details of a study of the passing of O_2 through a membrane obtained by interaction of Co^{2+}–N,N'-bis(salicylidene)ethylenediamine complex with poly(octyl methacrylate-co-4-vinylpyridine). The solution of the polymer and the complex was held on a teflon plate in an inert atmosphere, and a transparent flexible film (40–60 µm) of metal polymer was formed after drying under vacuum. The color of the membrane changed reversibly from brown to dark violet after exposure to oxygen or nitrogen, and after several minutes (Figure 2.3) an equilibrium complex was formed:

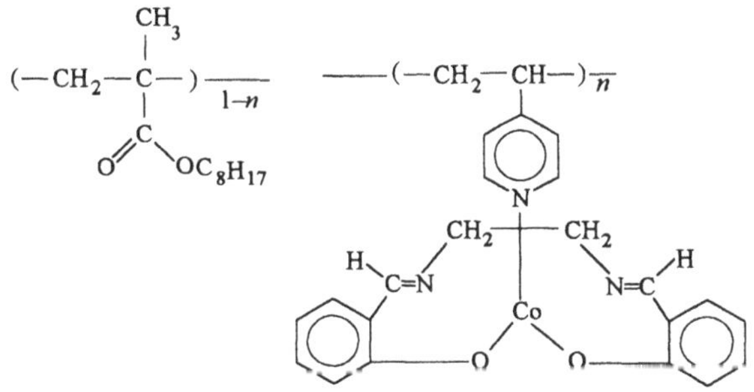

The O_2 binding constant determined by a photometric method is $K=1.1\times10^4$ l/mol (303 K). The Langmuir adsorption type and reversibility of O_2 binding by this membrane were confirmed experimentally. The photodissociation of bound oxygen (by flash photolysis) results in $k_0=3.85\times10^5$ l/(mol s) and $k_d=39$ s^{-1}. The coefficients of permeability to O_2 and N_2 for the membrane were measured with a low-pressure apparatus: the permeability coefficient P_{O_2} decreases with increasing O_2 pressure to 100 torr, then remains constant, whereas P_{N_2} does not depend on pressure (Figure 2.3b).

The reversible formation of Co^{3+} peroxo complexes (membrane derived from $CoCl_2$ and poly(N-acetylethyleneimine-co-oxy-1-chlormethylethylene)) diminishes with increasing content of amino groups and the degree of membrane cross-linking. This is explained [65] by suppression of formation of µ-dioxo complex adducts due to steric hindrances

(Figure 2.4). Under optimal conditions (content of nitrogen 54 mmol/g, cross-linking degree 6 %, [Co]=0.73 mmol/g), $P_{O_2}=3\times10^{10}$ ml·cm/(cm s cm Hg) and the activation energy for O_2 transfer is 17 kJ/mol. These values are equal to 8.1×10^{-10} and 29 for initial macroligand and 2.6×10^{-10} and 20 kJ/mol for teflon, respectively.

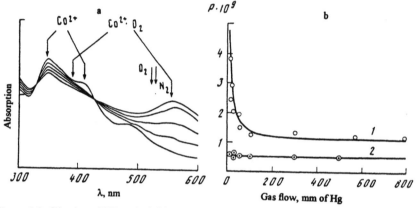

Figure 2.3 Kinetics of UV and visible spectra at binding of O_2 by a membrane obtained by interaction of $Co^{2+}-N,N'$-bis(salicylidene)ethylenediamine complex with poly(octyl methacrylate-co-4-vinylpyridine) (a) and the effect of the pressure of the gas flow on the permeability coefficient of the membrane (b). (a) 303 K, [Co]/[N]=1:10, [Co]=1.2 mass%, the membrane thickness is 40 μm; (b) membrane thickness 62 μm, (1) O_2, (2) N_2.

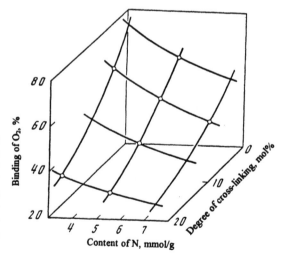

Figure 2.4   Effect of the amino group content and the degree of membrane cross-linking on O_2 binding by membranes formed from $CoCl_2$ and poly(N-acetylethylene-imine-co-oxy-1-chlormethyl-ethylene). [Co]=0.05 mmol/g, pH 5.8, acetate buffer, 298 K.

The permeability and selectivity (P_{O_2}/P_{N_2} and P_{CO_2}/P_{O_2}) of different gases depend also on the nature of the transition metal introduced into membrane (Table 2.2). The transition metal compounds decrease the free volume in the membranes, and the gas permeability but increase the selectivity P_{O_2}/P_{N_2}. The basic problems of gas separation were analyzed in a recent review [71].

Table 2.2 Permeability[a] and selectivity of a membrane derived from poly(butylmethacrylate-co-4-vinylpyridine) (PBMVP) and low molecular analogues for different gases.

Membrane	Permeability, $P \times 10^{10}$						
	O_2	N_2	CO_2	CO	H_2	O_2/N_2	CO_2/O_2
PBMVP	8.5	2.6	23	0.92	18	3.3	2.7
Triethylenetetramine	7.4	2.3	–	–	18	3.2	–
PBMVP–Cu^{1+}	5.4	1.3	16	–	15	4.3	3.0
PBMVP–Fe^{2+}	5.8	1.3	11	–	–	4.4	3.0
PBMVP–Cr^{3+}	6.5	1.6	10	1.2	–	4.0	1.6
PBMVP–Co^{2+}	3.8	1.0	–	1.4	1.5	3.7	–
PBMVP–Mn^{2+}	4.4	1.4	3.9	0.58	13	3.2	0.73
TETA–Mn^{2+}	6.8	2.0	8.8	0.88	–	3.5	1.3
TETA–Cu^{2+}	11.8	4.1	21.9	–	–	2.9	1.9

[a]The ratio of links 4-vinylpyridine/butylmethacrylate=26/74, [M^{n+}]/[4-vinyl-pyridine]=1/5.

The search and study of macromolecular oxygen-fixing systems is interesting also from the point of view of making effective and selective oxidation catalysts.

2.3 Hydrocarbon Oxidation by Immobilized Complexes

In principle, the mechanism of oxidation in the presence of immobilized complexes resembles that for catalysis by water-soluble metal complexes, but, as a rule, the polymeric nature of the reagents leads to distinct and characteristic features of the oxidation. Hydrocarbon oxidation is a model reaction which enables special features of these catalytic processes to by analyzed. In addition, this resembles, to a considerable extent, enzymatic catalysis; it also proceeds at low temperatures with high selectivity and requires small quantities of catalyst. This resemblance enables us to use the Michaelis–Menten equation for the oxidation reaction (as well as for hydrogenation, see Chapter 1):

$$E + S \underset{k_{-1}}{\overset{k_1}{\rightleftharpoons}} ES \xrightarrow{k_2} E' + P$$

where E is a catalyst, S is a substrate, P are products, E' is rapidly oxidized to E, $K_M = (k_{-1}+k_2)/k_1$ is the Michaelis constant, ES is a Michaelis–Menten complex. The initial (w_0) and maximum (w_M) rates are related by the equation:

$$\frac{1}{w_0} = \frac{K_M}{w_M [S]} + \frac{1}{w_M}$$

2.3.1 Oxidation of paraffins and cycloparaffins

Researchers have begun to study these reactions relatively recently; nevertheless, they have already accumulated substantial experimental material concerned with oxidation of the highest hydrocarbons.

Methane oxidation (in the gas phase) was studied only for metal complexes on inorganic supports. For instance, the process of extensive oxidation of CH_4 on a chromium–cupric catalyst is characterized by a substantial share of hetero-homogeneous oxidation even at a low degree of conversion of methane in the gas phase [72]. It was found [73] that the promotion of a vanadium-containing supported catalyst by palladium increases the initial rate of complete oxidation of methane, but does not change the mechanism of this reaction[3].

Methane oxidation by heteropolyacids (including those immobilized on mineral supports [75]), for instance by $H_3PMo_{12}O_{40}$, is a stable process with major products CH_3OH, CO, CO_2 and CH_2O. The sequence of catalytic activity of transition metal polyphthalocyanins for propane oxidation is similar to that for oxidation of phenol and cumene ($Co^{2+} \geq Mn^{2+} \geq Fe^{2+} >> Cu^{2+}$) [76].

It is interesting that the gas-phase oxidation of n-butane by lamellar graphite compounds (LGC) is catalyzed only by the graphite support itself [77], its activity being comparable to those for oxides of some metals.

In the presence of macrocomplexes of Fe^{3+} with polyethylene glycol (PEG) (relative molecular mass $(10–35) \times 10^3$), hexane is oxidized by H_2O_2 in the mixture H_2O/CH_3CN at 298–313 K to hexanone-2 and hexanone-3 [78]. It was found that complexes of Cr, V, Mn, Co and Mo, supported on poly(vinyl alcohol) (PVAl) and poly(furfurolidine) resin, are effective catalysts for the liquid-phase oxidation of n-pentadecane [79]. During the onset of the reaction catalyzed by both free and immobilized complexes the kinetic curves of O_2 binding have strongly pronounced autocatalytic features and are described by semiempirical equation $d[O_2]/dt=k_1+k_2[O_2]$ where k_1 depends on macrocomplex concentration, and k_2 is determined by the properties of the macrocomplex. In the initiated process the activity of oligomeric Ti–O–Sn catalysts approach those of cobalt stearate ($CoSt_2$), the most active catalyst for oxidation of paraffins, and $MnSt_2$, an industrial catalyst (Table 2.3).

Table 2.3 Rate constants of O_2 binding for the reaction of n-pentadecane oxidation [79].

Catalyst	k_1 (ml/min)	k_2 (min^{-1})
$CoSt_2$	1.227	0.010
$MnSt_2$	0.551	0.0022
Soluble Ti–O–Sn oligomer (relative molecular mass 1250)	0.312	0.0092
Solid Ti–O–Sn oligomer	0.268	0.0047

Products of $n-C_{15}H_{32}$ oxidation by oligomeric catalysts contain appreciable amounts of ketoacids. Note that stearates of Cr, Ni and Ce affect substantially the rate of free radical

---

[3] It was supposed [74] that the high activity and selectivity of silica powder, containing phosphorous and vanadium in the oxidation of n-butane is explained by formation of oxidized groups containing vanadium and phosphorous with optimal redox properties.

formation at the initial stage of pentadecane oxidation (chlorobenzene, 363–393 K) [80]. Non-catalytic reaction of chain generation (403–433 K) is a first-order reaction with respect to substrate and dissolved O_2 concentrations and is described by the rate constant $K_0=3.97\times10^{14}\exp[-(176.2\pm 16)/RT]$ l/(mol s).

In the presence of metal complexes the formation of free radicals proceeds via a catalyst–O_2 complex; in the case of $CoSt_2$, chains are generated in a triple complex catalyst–O_2–substrate.

Liquid-phase oxidations of solid paraffins in the presence of heterogeneous and colloidal forms of manganese are accompanied by a substantial (as compared with homogeneous catalysts) increase of acid yield [81].

The effectiveness of n-paraffins oxidations by Co^{3+} ions supported on activated coal is high, but the selectivity is low: the ratio between fatty acids, esters, ketones and alcohols is 3:3:3:1 [82]. The mechanisms of oxidation with the participation of homogeneous and heterogenized catalysts are similar; in both cases the Co^{3+} ion is the active center. Products of oxidation of liquid paraffins (a fraction with relative molecular mass 321 at 403 K) in the presence of metal carbonates contain hydroperoxides and carbonic acids (up to 63.5%); the metal activity decreases in the row Co>Mn>Zn>Ni>Cr [83].

There are only scarce data on catalytic oxidation of cyclohexane. Oxidation by O_2 in the presence of cluster and polynuclear iron carboxylates produces a mixture of alcohols and aldehydes or ketones [84]. Substituted cobalt tetraazoporphyrins enable the mild oxidation of cyclohexane to be carried out by cumene hydroperoxide with high yield using a catalyst (up to 1200 mol of oxidized substrate per mol of Co) [85].

I think that the success of activation of alkanes in solution by transition metal complexes under mild conditions [86], and the recent data on oxidation of methane and its homologues by O_2 in the presence of immobilized iron complexes [87], promises intensive development of the oxidation of hydrocarbons by immobilized complexes in immediate future.

2.3.2 Oxidation of linear olefins

Oxidations of linear olefins proceed by a free-radical mechanism; the accumulation of epoxides, ROOH, RCHO, ketones and RCOOH in the course of the reaction testifies to the chain character of these reactions. The main requirement for these processes is the selectivity: non-catalytic oxidation of propylene (at 423 K) results in the formation of more than 20 products.

Oxidation of ethylene by oxygen at 383 K in acetic acid in the presence of metal–polymer catalysts (complexes of platinum group metals supported[4] on PE, PS, PVC, PVAl, PAN, PVP or teflon), results in formation of methyl methacrylate and a small quantity of ethylene oxide [89]. Acrylic acid is obtained by oxidation of propylene (in water at 338 K) in the presence of palladium on coal (Pd/C) by two steps: at first to acroelin, then to acid with a selectivity of up to 91% (in the presence of radical inhibitors) [90]. The catalyst is regenerated easily by removal of adsorbed acrylic acid. Complexes of LGC with transition metal salts are also effective catalysts for the gas-phase oxidation of propylene [77]. Solutions of giant clusters such as

---

[4] The selective homogeneous oxidations of olefins with terminal double bonds to methyl ketones by means of O_2, ROOH and H_2O in the presence of platinum group metals have been analyzed in review [88].

$Pd_{561}Phen_{60}(OAc)_{180}$ oxidize propylene to allyl acetate [91]. Low-temperature oxidation of isobutylene to methacrolein in the presence of photoimmobilized copper-containing catalysts is carried out with 100% selectivity. In the case of supported tungsten complexes the highest yield is produced at 360 K [92]; active centers of these catalysts include the oxocomplexes WO_3^+ coordinated with an organic ligand.

Oxidation of olefins (C_6) by O_2 at 313–383 K in aqueous solutions containing Pd^{2+} and Cu^{2+} salts in the presence of water-soluble polyoxyalkylethylene glycol [93] results in increases of both rates and selectivities of oxidation. This also facilitates isolation of the reaction products and improves the solubilities of catalyst components. Obtaining carbonyl compounds by paraphase oxidation of olefins, catalyzed by the same salts in the presence of oligomeric SAS, exhibits the same features [94].

The rate of oxidation of hexene-1 by dioxygen in the presence of Co^{2+}–bis(salicylidene-γ-iminopropyl)methylamine complex does not depend on the addition of inhibitors, therefore this is not a free-radical reaction [51]. Oxidation of olefins by O_2 and H_2O_2 at 333 K using this catalyst leads to selective formation of $CH_3C(O)R$ and the corresponding secondary alcohols; products of allyl oxidation, which are characteristic of olefin autoxidation, were not found. This reaction proceeds via oxidation of primary alcohol by a Co^{2+} chelate with the formation of the corresponding aldehyde and H_2O_2. The latter forms cobalt hydroperoxide which is attached to the olefinic double bond with formation of alkyl peroxide, and products are produced as result of decomposition of this alkyl peroxide. The catalyst is deactivated due to both oxidation of the ligand of the chelator and irreversible formation of the μ-peroxo dicobalt complex. Photocatalytic oxidation of hexene-1 by homogeneous and heterogenized Ti^{4+} complexes is accompanied by reduction of Ti^{4+} to Ti^{3+} and the formation of acetic, propionic and butyric aldehydes, together with the peroxide [95,96].

Catalytic oxidation of octene-1 by O_2 in the presence of Rh^{3+} or Pd^{2+} complexes on phosphorus–molybdenum acids $H_{3+n}PMo_{12-n}V_nO_{40}$ in alcohol at 333 K is accompanied by the formation mainly of octane-2-one; in the presence of immobilized Pd^{2+} the share of isomerization products increases substantially [97].

The products of decene-1 oxidation in the presence of Wacker-type catalysts (Pt^{2+}) immobilized on a number of polymeric supports carrying cyanomethyl polybenz-imidazole were regarded simply as "ketone products" [98]. If $CuCl_2$ is used as a cocatalyst, primarily methyl ketone is formed. The use of cross-linked supports increases catalytic activity because in this case coordinationally unsaturated centers of Pd^{2+} are formed. It is important that oxidation proceeds with a high rate even at 393 K in 2-methoxyethanol; formation of Pd^0 was not found in this case. Under such conditions homogeneous systems are not active because of the reduction of Pd^{2+}.

Interesting features were found [99] for liquid-phase oxidation of decene-1 and mixtures of aliphatic α-olefins C_{11}–C_{14} by means of macromolecular Co^{2+} complexes supported on oxygen-containing (PVAl, PAA, PMAc) and nitrogen-containing (polyacrylamide, PAAm, PAN, copolymers of acrylic acid and acrylamide (CAAA), PAN oxime (PANO)) polymers. The oxidation rate depends substantially on the nature of the polymer ligand: oxygen-containing polymeric supports are the most active. Oxygen binding by Co^{2+} complexes increases in the ligand sequence (Figure 2.5): PVAl>PMAc>CAAA>PAN>PAAm>PANO.

The dependence of the oxidation rate on concentration of peroxides added in the start of the reaction has a maximum: peroxides at first accelerate oxidation in the

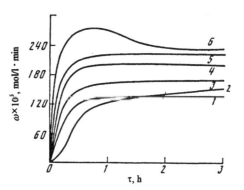

Figure 2.5 Kinetics of oxidation of mixtures of aliphatic α-olefins C_{11}–C_{14} by means of Co^{2+} complexes supported on PANO (1), PAAm (2), PAN (3), CAAA (4), PMAc (5) and PVAl (6) under comparable reaction conditions.

presence of immobilized complexes, and later inhibit it. Ions with configuration d^7 (Co^{2+}) and d^5 (Mn^{2+}) are especially active, therefore it was concluded that ions with odd number of d-electrons more easily form σ-bonds with coupling of valent electrons between metal and ligand. In many cases the contribution of the σ-bond to the total energy of the coordination bond forms a major part. Participation of uncoupled d-electrons in ligand binding facilitates the formation of the activated olefin–catalyst complex and intermediate radicals characterized by a high degree of delocalization. The properties of the macroligand affects the dynamics of redox processes: ligands in reaction complexes should be mobile because a part of them are ousted by the olefin or oxygen during the course of the oxidation. Strong binding of ligands, for instance amines, with metal complexes leads to inhibition of oxidation. This concept also explains the effect of macrochain length on the catalyst activity observed in many cases.

Polymers with conjugated bonds often catalyze liquid-phase oxidation of alkanes, for instance of propylene [100].

It is important that supported clusters can be used for olefin oxidation. For example, a mixture of clusters $Co_4(CO)_{12}$ and $Rh_4(CO)_{12}$ at a molar ratio 1–3/1, immobilized on amine-containing ion-exchange resin [101], catalyzes the oxidation of 2-olefins to the corresponding alcohols.

2.3.3 Oxidation of cycloolefins

Both uncatalyzed and catalyzed (by soluble complexes of transition metals) oxidation of cyclohexane has been used as a model reaction for the investigation of the mechanism of hydrocarbon oxidation. Immobilized complexes are also used in model systems. Cyclohexane oxidation proceeds in two ways, with participation of C=C or C–H bonds:

$$RO_2^. + \langle\text{cyclohexene}\rangle \longrightarrow ROOH + \langle\text{cyclohexyl}\rangle^.$$
$$\text{ROO}^.$$

$$\langle\text{cyclohexyl}\rangle^. + O_2 \longrightarrow \langle\text{peroxo}\rangle$$
$$\text{ROO}^.$$

and

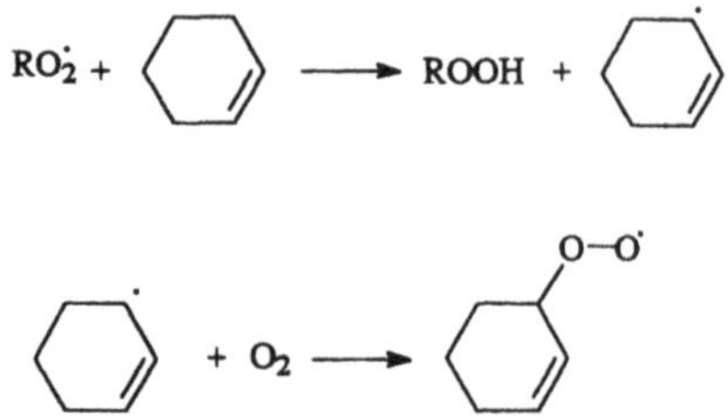

The rate of liquid-phase oxidation of cyclohexane by dioxygen in CCl_4 at 333 K in the presence of $MoCl_5$ on LGC is higher than that of thermal oxidation [102]; 2,3-epoxy-cyclohexane, cyclo1-hexene-ol-2 and cyclohexyl hydroperoxide (CHHP) are major products. In the case of homogeneous systems a critical behavior at variations of concentration was observed. This testifies to an inhibiting role of $MoCl_5$ in this reaction: so-called molybdenum "blues" Mo^{5+}–Mo^{6+} are formed when the catalyst concentration exceeds a critical value. It is plausible that the oxidation is catalyzed by Mo^{6+} and inhibited by Mo^{5+}. Incorporation of Mo^{5+} into graphite does not change substantially its catalytic properties or critical phenomena but prevents its dissolution in the reaction mixture. The catalyst participates in degenerated branching, molecular decomposition of CHHP and epoxidation of cyclohexene[103]:

$$\mathrm{ROOH} + \mathrm{Mo}^{5+} \xrightarrow{\text{fast}} \mathrm{Mo}^{6+} + \mathrm{RO}^{\cdot} + {}^{-}\mathrm{OH}$$

$$\mathrm{ROOH} + \mathrm{Mo}^{6+} \underset{k_{-1}}{\overset{k_1}{\rightleftharpoons}} [\mathrm{Mo}^{6+} \cdots \mathrm{ROOH}]$$

$$[\mathrm{Mo}^{6+} \cdots \mathrm{ROOH}] \xrightarrow{k_2} \mathrm{Mo}^{5+} + \mathrm{RO}_2^{\cdot} + \mathrm{H}^+$$

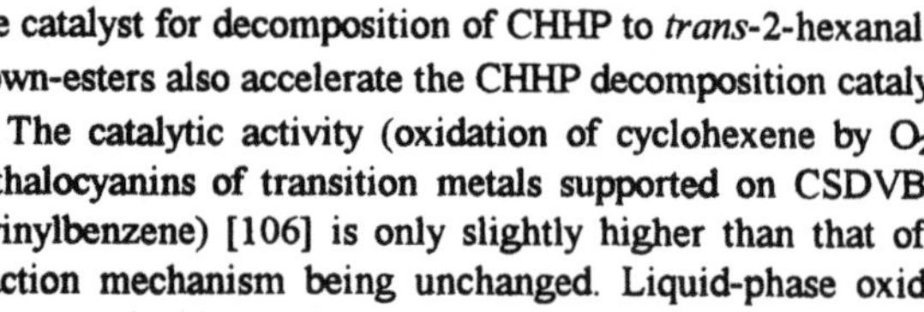

Impregnation of various polymers by K_2PdCl_4 solution results in the formation of an active catalyst for decomposition of CHHP to *trans*-2-hexanal [104]; additions of PEG and crown-esters also accelerate the CHHP decomposition catalyzed by $VO(AcAc)_2$ [105].

The catalytic activity (oxidation of cyclohexene by O_2 at 351 K) of chloro-sulfur phthalocyanins of transition metals supported on CSDVB (copolymer of styrene with divinylbenzene) [106] is only slightly higher than that of unsupported analogues, the reaction mechanism being unchanged. Liquid-phase oxidation of cyclohexene in the presence of a bimetallic Co–Mo catalyst supported on PVAl or poly(furfurolidine) resin [79] has the same features: the chain radical mechanism is retained, and the sum of concentrations of the alcohol and ketone formed always exceeds the concentration of cyclohexene oxide[5].

[5] Iminodiacetate complexes of Co^{2+}, Mn^{2+} and Cr^{3+} supported on silica are effective in liquid-phase cyclohexene oxidation which proceeds by a hetero-homogeneous radical chain mechanism [107–109].

Oxidation of cyclohexene by polymeric cobalt-containing catalysts [110–114] has many significant features. In the presence of Co(AcAc)$_2$ (AcAc=acetylacetonate) the process is characterized by an induction period, after which the rate rapidly reaches its highest value and decreases thereafter (Figure 2.6(1)). In the case of heterogenized catalysts[6] there is no induction period, the oxidation rate is constant (PE-gr-poly(cobalt acrylate) (PE-gr- =polyethylene-grafted), copolymer of styrene with cobalt acrylate, or PE-gr-PAA-Co^{2+} as catalysts) or increases (PE-gr-poly(acrylamide–Co^{2+} chloride complex), poly(acrylamide–Co^{2+} chloride complex), poly(acrylamide–Co^{2+} nitrate complex)). Catalysts prepared from PE-gr-(poly(Co^{2+} acrylate) and a copolymer of cobalt diacryl with styrene are the most active; in their presence the oxidation commences and proceeds thereafter at a constant rate.

O$_2$, mol/l

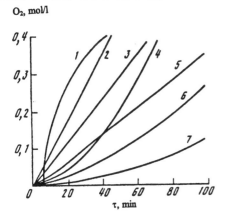

τ, min

Figure 2.6   Oxygen binding for cyclohexene oxidation in the presence of Co-containing catalysts (333 K, without solvent). (1) Co(AcAc)$_2$, (2) PE-gr-poly(cobalt acrylate), (3) copolymer of styrene with cobalt acrylate, (4) PE-gr-poly(acrylamide-Co^{2+} chloride complex), (5) PE-gr-PAA–Co^{2+}, (6) poly(acrylamide–Co^{2+} chloride complex), (7) poly(acrylamide–Co^{2+} nitrate complex).

It is important that metal polymer catalysts retain their activity at high degrees of oxidation and can be reused, after their isolation from reaction mixture, many times. The products formed in the presence of most such catalysts are approximately the same (Figure 2.7): major products are CHHP, cyclohexenone, cyclohexenol and cyclohexene oxide.

Figure 2.6 also shows that an increase of Co^{2+} content in the polymers (see data for homo- and copolymers obtained from metal-containing monomers) does not increase the catalytic activity. Therefore, in heterogeneous catalysis it is expedient to use metal polymers in which Co^{2+} is localized only on the surface or in a thin grafted layer near the surface as was mentioned above (Chapter 1) for the hydrogenation processes.

Oxidation of cyclohexene in the presence of metal polymers consists of elementary processes occurring on the surface (decomposition of ROOH, formation of molecular products) and chain radical transformations as a consequence of escape of radicals into the volume. The chain reaction was proved by adding free radical acceptors (F). It was checked in preliminary studies that the acceptors did not interact with the surface of PE-gr-PAA–Co^{2+}; inhibitors immobilized on supports may be used for such confirmation tests.

---

[6] Polymerization, copolymerization and grafted polymerization, leading to formation of such catalysts (metal polymers), are described in a book [115].

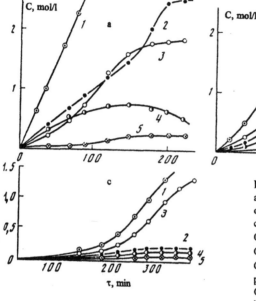

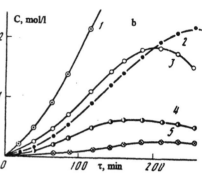

Figure 2.7 Kinetics of product accumulation during cyclohexene oxidation in the presence of various catalysts at 333 K. (a) PE-gr-poly Co(acrylate)₂, (b) PE-gr-poly CoCl₂·4acrylamide, (c) CoCl₂·4acrylamide. (1) sum of the products, (2) cyclohexenone, (3) CHHP, (4) cyclohexenol, (5) cyclohexenoxide.

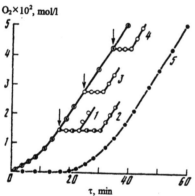

Figure 2.8 Inhibition of cyclohexene oxidation by dimer F–F. Concentration of cyclohexene 5 mol/l, catalyst PE-gr-PAA-Co^{2+}, 328 K. Concentration of F–F $1.7×10^{-4}$ (1, 3, 4), $3.4×10^{-4}$ (2), $1.7×10^{-5}$ mol/l (5). Arrows indicate the addition of inhibitors.

The addition of inhibitor in the stationary phase of the process results in lag periods which do not depend on the extent of oxidation (Figure 2.8). The rate of initiation at the beginning of the process is 30 times lower than that in the stationary phase. Thus, the processes on the surface must determine the rate of chain generation, and free radicals are not formed in the bulk of solution.

At the beginning of the process (in the absence of ROOH) Co^{2+} compounds can interact with O₂ with formation of free radicals [116]:

$$Co^{2+} + O_2 \rightleftharpoons [Co^{3+} \cdots ^\cdot O{-}O^-]^{2+} \xrightarrow{\text{RH}} Co^{2+} + R^\cdot + HO_2^\cdot$$

In developed process the main contribution to the chain initiation results in the decay of coordinated ROOH:

$$Co^{2+} + n\,ROOH \underset{}{\overset{K}{\rightleftharpoons}} [Co^{2+} \cdots n ROOH] \xrightarrow{k_i} RO^{\bullet} + [Co(OH)]^{2+} + (n-1)\,ROOH$$

The rate of the initiation is given by the equation:

$$w_i = \frac{k_i K [ROOH]_0^n [Co^{2+}]_0}{1 + K[RCOOH]_0^n}$$

For heterogenized catalysts the values of k_i (a measure of the catalyst efficiency for the radical decay of CHHP) and K are determined by diffusion, sorption and desorption processes, as well as by the activity of Co^{2+} ions on the polymer surface [24]. In the stationary phase, the value of w_i does not depend on ROOH (at $[ROOH] > 1 \times 10^{-2}$ mol/l the oxidation rate is constant from the outset, the acceleration period is absent). This means that $K[ROOH]_0 \gg 1$ and $w_i = k_i [Co^{2+}]_0$. It was found that $k_i = 2.5 \times 10^{-4}$ s^{-1}, $K > 10^2$ l/mol (for reactions with participation of $Co(AcAc)_2$ $n = 2$, $K = 5.6$ (l/mol)^2). The main kinetic parameters for oxidation promoted by Co^{2+}–polymer complexes are given in Table 2.4.

Table 2.4 Kinetic parameters of cyclohexene oxidation in the presence of immobilized catalysts [114] (333 K, without solvent).

Catalyst	$[Co^{2+}] \times 10^3$	$w_i \times 10^6$	$w \times 10^4$	$w/[Co^{2+}] \times 10^2$	$k_2 k_6^{-1/2}$
	(g at/l)	(mol/l s)	(mol/l s)	(s^{-1})	(l (mol s)$^{-1/2}$)
PE-gr-poly(acrylate Co^{2+})	0.88	6.5	1.4	160	5.4
Copolymer of styrene and Co^{2+}–diacrylate	10	2.8	0.97	9.7	5.8
PE-gr-poly(acrylamide–CoCl$_2$)	0.71	7.3	1.3	180	4.8
Homopolymer derived from CoCl$_2$·4AAm	7.1	1.8	0.75	11	5.5
Homopolymer derived from Co(NO$_3$)$_2$·4AAm	6.4	0.21	0.22	3.4	4.9
PE-gr-poly(Co^{3+}·acrylate)	1.0	1.8	0.62	6.2	4.6
Without catalyst (AIBN as an initiator)	–	–	–	–	3.7

The value of K and the independence of w_i on [ROOH] indicate that near the surface of the heterogeneous metal–polymer catalyst the local concentration of hydroperoxide is high. It was shown that in the vicinity of catalytic centers (Co^{2+} and Co^{3+} ions on PE–gr–PAA), there are always vacant carboxyl groups able to bind CHHP molecules by hydrogen bonds. It is known [117] that the association constant of organic acids with hydroperoxides in the liquid phase reaches $\sim 10^4$ l/mol, in agreement with the equilibrium constant of CHHP with the surface of PE–gr–PAA–Co^{2+} ($K > 10^2$ l/mol).

Figure 2.9 Kinetics of O_2 absorption and peroxide accumulation for oxidation of α-pinene in the presence of PVAl–Co^{2+}. Concentration of the catalyst is 5×10^{-2} mol/l, 323 K.

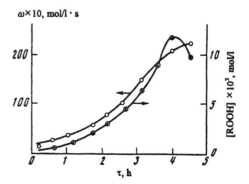

The small increase of $k_2 k_6^{-1/2}$ (see Table 2.4) cannot lead to formation of substantial quantities of ketone, alcohol and epoxide, which is characteristic of catalytic oxidations, therefore it is plausible that the oxidation products are formed also as a result of the termination reaction of the first order with participation of immobilized Co^{2+} ions:

$$Co^{2+} + RO_2^{\cdot} \longrightarrow Co^{3+} + \text{molecular products}$$

Thus, a high local concentration of CHHP molecules is created at free COOH groups in the vicinity of active centers during cyclohexene oxidation on metal polymer catalysts.
In the stationary phase the local concentration is constant and does not depend on the concentration of ROOH in the solution. Probably, the transfer of ROOH molecules to the active center is carried out initially via their migration across the two-dimensional catalyst surface, and only after their decay the radicals formed escape into the volume where the chain radical oxidation of hexene is developed.

We can control the process by changing the ratio between vacant functional groups of polymeric support and immobilized metals ("the loading degree"); thus, in such systems there is an additional control lever that is absent from homogeneous catalysis.

It is important that after 2–5 oxidation cycles and washing the catalyst with benzene, the structure and activity of the catalyst and the ratio between carboxylate ions and COOH groups do not differ from those of the initial catalyst.

In these metal–polymer catalysts Co^{2+} ions exist in both octahedral (in copolymers of styrene with cobalt diacrylate, PE–gr–poly(cobalt acrylate) and others) and tetrahedral (homopolymer of $CoCl_2 \cdot 4AAm$ and a product of its grafting to PE) coordination. The transition $Co^{2+} \rightarrow Co^{3+}$ during catalysis is accompanied by a distortion of coordination. However, there is no direct correlation between catalytic activity and the coordination type of the cobalt ions in the polymer. It is interesting that for cyclohexene oxidation in the presence of mono- and binuclear Cu^{2+} chelates (333 K, O_2, the increase of activity by 8–30 times compared with the uncatalyzed reaction) the oxidation rate increases after changing the configuration of the chelating unit from the square-planar to distorted-tetrahedral, whereas the nature of the ligands of the chelating unit (the Schiff base) does not affect the content and ratio of oxidation products [118]. The activity of the metal polymers is influenced substantially by the nature of the solvent ([119] Ru^{2+} complexes as catalysts) and polymeric support (poly(siloxane)-supported tetra(4-pyridyl)-porphyrinatomanganese(III) system [120]).

Data on the oxidations of other cycloolefins are very scant. For instance, liquid-phase oxidation of vinylcyclohexene and its substituted derivatives by O_2 in the presence of metal complexes gives vinylcyclohexan-1,2-diol and the corresponding substituted products [121].

Concerning bicyclic olefins, oxidation of α-pinane catalyzed by polychelates and Co^{2+}–polymer complexes [99] should be mentioned[7]. The rates of O_2 absorption and hydroperoxide accumulation are nearly constant (Figure 2.9). The process proceeds by a radical chain mechanism, the main products being α-pinane oxides, verbenol and verbenone. This means that the main direction of oxygen attack, besides the double bond, is the secondary C–H bond in the allyl position relative to the double bond.

2.3.4 Oxidation of alkylaromatic hydrocarbons

Substrates of this type are oxidized by metal complexes only due to alkyl substituents. However, benzene can be involved in these processes. For instance, under the action of complexes of transition metals (Fe^{2+}, Fe^{3+}, Cu^{2+}, Cu^{1+}, Pd^{2+}, Ni^{2+}, Pt^{2+}, Co^{2+} and Rh^{3+}) immobilized on PVS, PEG and other polymers, both homo- and heterogeneous hydroxylation of benzene occur (H_2O_2, aqueous-organic media, 323 K) [123]. The selectivity with respect to phenol formation reaches 83–91%.

Kinetic features of toluene oxidation (353 K, O_2) in the presence of homogeneous cobalt acetate systems and their heterogenized analogues (on the surface of microporous CSDVB) were compared in [124]. Homogeneous oxidation in acetic acid proceeds effectively to benzoic acid and is inhibited by dioxane, whereas heterogenized systems are inactive without the additions. The reaction rate is proportional to the concentrations of peroxides formed as a result of conversion of dioxane under the action of immobilized cobalt complexes, therefore it was concluded that such immobilized catalysts are inactive in the direct oxidation of toluene by dioxygen. Perhaps in heterogenized systems complexes [R–C(O)OH...Co^{3+}...RC(O)OH], more stable than in homogeneous systems, are formed; these complexes decompose slowly. The activities of heterogenized catalysts are lower than of homogeneous species and depend on the degree of cross-linking of the polymer.

Oxidation of ethylbenzene (EB) in the presence of immobilized complexes has been studied in more detail[8]. Co^{2+} and Mn^{2+} supported on polypropargylacrylate, PAN and polyoxyphenylbenzeneterephthalamine, were used as catalysts [126–128]. Complexes of Co^{2+} with a copolymer of acrylonitrile and propargylacrylate possess the highest activity and selectivity (up to 96% with respect to acetophenone (AP)). Kinetics of the accumulation of α-phenylethyl hydroperoxide (HP), methylphenylcarbinol (MPC) and AP confirm that the oxidation proceeds by the consecutive-parallel scheme:

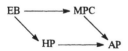

At the beginning of the reaction initiation results from O_2 activation:

---

[7] An analysis of catalytic oxidation of mono-, bi- and tricyclic olefins by hydroperoxides in the presence of homogeneous catalysts (Rh, Ir, Mo and V complexes) is given in review [122].

[8] This reaction is considered to be a model for the non-catalytic oxidation of organic compounds [125].

$$M^{2+} + O_2 \longrightarrow [\, M^{3+} \cdots O_2^- \,] \xrightarrow{\text{RH}} [\, M^{3+} \cdots OOH] + C_6H_5\dot{C}HCH_3 \longrightarrow$$

$$\longrightarrow M^{2+} + \underset{\underset{OOH}{|}}{C_6H_5C\,HCH_3} \longrightarrow \underset{\underset{O}{\parallel}}{C_6H_5C\,CH_3}$$

Most probably, AP and MPC are formed in a single act of recombination of peroxide radicals:

$$RO_2^{\cdot} + RO_2^{\cdot} \longrightarrow ROH + R^ICOR^I$$

A new phenomenon, "the structural tuning" of active centers of metal polymer catalysts to a hydrocarbon substrate, was found in liquid-phase oxidation of alkylaromatic hydrocarbons, xylol and EB, by molecular oxygen [129–133]. Initially, a water-soluble Co^{2+} complex is formed, for instance with a copolymer of diethyl ether of vinylphosphonic acid and acrylic acid in the presence of EB or α-phenylethyl hydroxide. A structural rearrangement, specific to the reacting compounds, is further fixed in the process of cross-linking of the polymeric support by N,N'-methylenediacrylamide with subsequent removal of pattern substrates from cross-linked formations. This tuning leads to a substantial increase of activity and selectivity of the catalyst in comparison with a catalyst obtained without directed generation of its structure (Figure 2.10), and also to a substantial increase of the rate of formation of all three major products of oxidation. The initial rate of HP formation increases more than the rate of HP decomposition to MPC and AP (Figure 2.11).

When the concentration of HP reaches a certain value, the formation of AP and MPC is stopped, and the selectivity of transformation of EB to HP is close to 100%. This is explained by catalyst modification due to HP adsorption. Other products of EB oxidation do not possess such properties. The tuning of the catalyst to HP demonstrates the possibility of increasing the selectivity of metal polymer catalysts with respect to selected substrates [134].

The temperature dependence of the reaction rate (Figure 2.12) resembles, to a certain extent, those for the hydrogenation processes (see Section 1.5) in the presence of immobilized catalysts: the reaction rate increases with increasing temperature (E_a=42 kJ/mol) and reaches a maximum at 398 K, then sharply decreases. The rising temperature increases the rate of HP decay but selectivity with respect to HP decreases.

The above-mentioned complexes of Co^{2+} and Cu^{2+} with PAN, propargyl methacrylate and others are active only in oxidation of monoethylbenzene but not diethylbenzenes [135,136]. Note that polymeric catalysts of this type are stable in the oxygen-containing medium even at 393 K.

The specially tuned polymer–cobalt catalyst (as described above) is effective in liquid-phase oxidation of styrene [137]. It is known that styrene can be oxidized in two ways: to olefin oxide and RCHO (or ketone), or to the copolymer of the olefin and O_2 (polyperoxide); to improve the selectivity, we must suppress the second alternative, an oxidative polymerization. The rate of formation of the monomeric products (benzaldehyde, styrene oxide) increases and the rate of polymerization decreases in the presence of the metal polymer: the ratio between the rates of accumulation of benzaldehyde, styrene oxide and polymer is 1:0.9:0.6 for the non-catalytic process and 1:0.6:0.8 for the catalytic process. Thus, metal polymers effectively suppress (after regeneration of the catalyst this ratio is 1:0.46:0.03) the unwanted process, the copolymerization of styrene with oxygen.

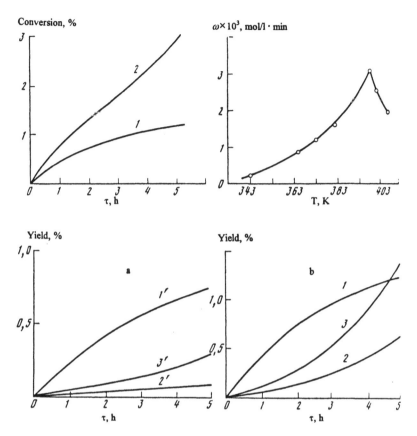

Figure 2.10 Kinetics of liquid-phase oxidation of EB in the presence of "untuned" (*1*) and "tuned" (*2*) macrocomplexes with respect to EB. Without solvent, 363 K.

Figure. 2.11 Kinetics of formation of α-phenylethyl HP (1, 1′), MPC (2, 2′) and AP (3, 3′) in the presence of "untuned" (a) and "tuned" (b) macrocomplexes. Reaction conditions are the same as in Figure 2.10.

Figure 2.12 The temperature dependence of O_2 binding for the oxidation of EB in the presence of a "tuned" complex of Co^{2+} with a copolymer of diethyl ether of vinylphosphonic and acrylic acids cross-linked by N,N'-methylene diacrylamide. The catalyst concentration is 20 mg of Co/l, without solvent.

A polymeric film, deposited on the catalyst surface, plays an important role in the heterogeneous catalytic oxidation of styrene[8]. It is known, for oxidation of styrene by dioxygen, that the addition of polymeric products formed in the reaction, as well

---

[8] A thin (~0.25 mm) polymer film, for instance copolymer of fluoroethylene and perfluorovinyl ether, can simultaneously serve as a reactor and heat exchanger in which a circulating liquid abstracts the heat of isothermal decomposition of hydroperoxides (for instance, cumene) during the formation of ketones and alcohols [138].

as other polymers, for instance poly(propylene glycol), results in an increase of selectivity of styrene epoxidation: the oxide yield reaches 98% of the styrene involved in the reaction [139]. Styrene can be selectively oxidized to phenylacetaldehyde, for instance in an aqueous solution with Pt^{2+} and Cu^{2+} in the presence of PEG (n=10–100) [140]. Styrene can also be oxidized to polyperoxystyrene [141].

Liquid-phase oxidation of cumene by transition metal complexes (Co, Mn, Cu, Fe) immobilized on ion-exchange resins has been studied extensively [142–148]. The process proceeds both by chain radical and molecular mechanisms, the catalyst participating both in initiation and termination of chains. The main results of investigations of these systems are the following. Complexes with amine and pyridine groups have the highest catalytic activity (major products are cumene hydroperoxide (CHP) and dimethyl phenyl carbinol (DMPC)), and ionites with carboxyl and sulfur-containing groups have the lowest. Ionites with phosphate groups occupy an intermediate place. Amine ionite complexes are characterized by high selectivity in the oxidation of cumene to DMPC; the selectivity of phosphor-containing ionites for CHP increases in the sequence $Co^{2+}<Mn^{2+}<Cu^{2+}<Ni^{2+}$. Iron ionite complexes are inactive in these reactions.

High selectivity with respect to CHP at relatively low catalyst activity is explained [148] by participation of the catalyst in chain initiation due to hydroperoxide decomposition by a radical mechanism. In the presence of PVP complexes of Cu^{2+} the cumene oxidation proceeds simultaneously by a radical chain hetero-homogeneous and molecular mechanisms [142]. Incomplete inhibition even on adding substantial amounts of ionol (an ion-exchange resin) testifies to the initiation of the reaction not only by ROOH decay, but also by O_2 activation on the catalyst with subsequent development of the process both by radical and molecular mechanisms:

$$Cu^{2+} + ROOH \longrightarrow Cu^+ + RO_2^. + H^+$$

$$Cu^+ + O_2 \rightleftharpoons [Cu^{2+}\cdots O_2^-]$$

$$[Cu^{2+}\cdots O_2^-] + RH \begin{cases} Cu^{2+} + HO_2^- + R^. \\ Cu^+ + ROOH \end{cases}$$

Cu^{2+} is immobilized on these polymers as isolated mononuclear tetracoordinated complexes, $Cu(N)_4$, and as clusters, in regions with high local concentrations of Cu^{2+} [144]. The proportion of isolated complexes increases in the course of the reaction; this leads to a sharp decrease of catalyst activity and to partial washing of copper from ionite granules. Moreover, the polymeric matrix (for example, CH_3 groups of the copolymer of 2-methyl-5-vinylpyridine with DVB [142]) is partially oxidized. This leads to a rearrangement of the Cu^{2+} macrocomplex with three and four monodentate pyridine ligands to a complex with two bidentate chelating ligands:

The effect of the properties of the support's functional groups on liquid-phase oxidation of cumene was also found for catalytic oxidation in the presence of CHP (353 K, O_2, 2.8% CHP, ionic-immobilized Co^{2+}) [149]. Without metal polymers the oxidation is autocatalytic, the reaction rate increasing in the presence of cobalt supported on weakly or moderately acidic ionites. The oxidative activity of the Co^{2+} ion bound to carboxylate groups increases with increasing concentration (compare with oxidation of cyclohexene by carboxyl metal polymers, Section 2.3.3). It was found for catalysts containing residues of phosphorous and phosphoric acids[9] that ionites with $P(OH)_2$ groups are more active than systems with sulfo groups having high acidity (these cause decomposition of CHP to phenol and acetone).

Metal chelators, including polymeric species, are also effective in oxidation of cumene. The activity of polymeric Schiff bases in the autoxidation of cumene decreases in the sequence of M^{2+}: Mn>Co>Ni>Cu>Zn [151]. The selectivity also depends on the nature of chelating units, for instance the azomethine type. Polymeric Schiff complexes of V^{2+} and Mn^{2+} oxidize cumene (303–423 K, $[S]/[K_T]=10^3$, reaction time 15 h) [56]. At the optimal temperature (373 K) the degree of cumene transformation is above 50% and the major product is DMPC. Cobalt, Ni, Cu and Zn chelates with N and S atoms as donors (for example, benzylidene-2-mercaptoaniline [152]) inhibit cumene oxidation because they are acceptors of peroxide radicals[10].

Metal complexes immobilized on supports of mixed type (inorganic materials, the surface of which is modified by polymer grafting; methods of obtaining these materials are summarized in [156]) take an important place in reactions of liquid-phase oxidation. For instance, thermal treatment of SiO_2–PAN (obtained by radical polymerization of

---

[9] Polymers derived from urea and phosphorus-containing compounds with coordinated salts of Ni, Co, Fe and Cu are effective and selective catalysts for the oxidation of cumene to CHP (O_2, 373 K) [150].

[10] Inhibition or activation in the presence of electron–donor compounds is also often observed in homogeneous catalytic oxidation of cumene. For instance, electron–donor additions accelerate oxidation in the presence of $Ni(AcAc)_2$; this is explained [153] by the increasing ability of complexes of Ni^{2+} with these additives to form active adducts with O_2. The effective charge of the nitrogen atom in heterocyclic bases (pyridine>quinoline>acridine>2,2-dipyridyl, 1,10-phenanthroline) correlates with their activity in the oxidation of cumene at 383 K in the presence of $M(AcAc)_2$ [154]. A flow of nitrogen oxides through reaction mixture leads to a complete cessation of both the initial and the developed process because of the formation of nitrosyl complexes of $M(AcAc)_2$ (where M=Co, Cu, Ni), coordinationally saturated polynuclear compounds, and because of oxidation of the metal chelate ligand [155].

acrylonitrile in butanol in the presence of SiO$_2$) in air at 463–523 K leads to formation of a cyclized polymer bound to SiO$_2$. Its interaction with ethanol solutions of MX$_n$ produces metal complex catalysts for oxidation of EB or cumene [157]. Their activity in oxidation of cumene decreases in the sequence Cu^{1+}>Cu^{2+}>Co^{2+}>Mn^{2+} and depends on the conditions of PAN cyclization, the molar ratio N/M and the temperature (see curves with a maximum in Figure 2.13).

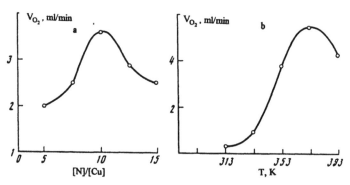

Figure 2.13 Dependence of the rate of cumene oxidation on the molar ratio N/Cu in Cu^{1+}–PAN (a) and temperature (b). (a) [Cu]=0.01 mmol, [cumene]=0.05 mol; P_{O_2}=0.1 MPa, T=355 K, (b) [N]/[Cu]=10.

If the complexes are unstable, they can be assembled from components on a support (ion-exchange resins, chitosane, activated coal and others). This method was used in [158] for complexes with trioxyanthraquinone. Their activity in cumene oxidation decreases in the sequence Mn^{2+}>Cu^{2+}>Co^{2+}>Fe^{3+}>VO^{2+}>AIBN.

It is interesting that donor–acceptor immobilization of hemin on the surface of silica with chemically bound donor *trans*-ligands (imidazole, aminopropanol) permits the creation of a catalyst for the mild oxidation of hydrocarbons, for instance cumene, at 293 K [159].

Metal complexes prepared from nitrogen-containing coal possess high selectivity with respect to hydroperoxide in the oxidation of EB and isopropylbenzene [160] probably because of formation of a longer chain of corresponding hydroxides: a correlation between oxidation rate, radical generation, chain length and strength of oxygen bond is observed.

Heterogeneous catalysts for the liquid-phase oxidation of hydrocarbons are obtained also from humic acids extracted from brown coals, for instance coals of the Kansko-Achinskii coal basin [161].

Thus, the oxidations of alkylaromatic hydrocarbons in the presence of immobilized complexes have been studied in some detail.

2.3.5 Oxidation of tetralin

The catalytic oxidation of tetrahydronaphthalene, a product of the partial hydrogenation of naphthalene, is considered in a separate section because of numerous interesting features found for this reaction in the presence of immobilized catalysts.

Tetralin is oxidized mainly to tetralin hydroperoxide, tetralon-1 and tetralol-1. The initial rate of liquid phase catalytic oxidation of tetralin (O$_2$, 393 K, Co^{2+} complex) is higher by nearly 4 orders of magnitude than that of the non-catalytic oxidation [162]. The hydroperoxide accumulation decreases in the presence of Co^{2+}. This testifies to catalytic decomposition of hydroperoxide and free radical generation by a scheme of secondary catalysis. It is important that the Co^{2+} complex itself is able to produce free radicals by its interaction with tetralin. Additions of AIBN or tetralin hydroperoxide accelerate oxidation of tetralin in the presence of Co^{3+} 2,4-pentadionates (333 K), whereas Co^{2+} chelates markedly inhibit the reaction [163]. The traditional scheme of catalytic autoxidation of tetralin, in which the role of cobalt complexes is only in radical decomposition of the peroxide formed, does not describe the kinetics of the process.

Complexes of Co^{2+}, Co^{3+}, Cu^{2+} and Ni^{2+} with bioligands (S-cystine, 5-oxy-1,4-naphthoquinones and others), as well as binuclear clusters of copper which model some natural bioenergetic systems[11], catalyze liquid-phase oxidation of tetralin by dioxygen with high (up to 98%) selectivity with respect to tetralon production [166]. Binuclear copper complexes, also special structures, for instance oxohemocyanine complexes (see Section 2.2), are the most effective.

Copper ions supported on thermostable polyazopolyarenes, are highly labile and reactive [167]. A plausible model of copper-containing centers of oxidation is a square-planar complex (complex I). After adding tetralin hydroperoxide, the phenylazo group of the polymer is oxidized to an azooxy group (complex II) with opening of one coordination site:

I   II

The combined action of tetralin and its hydroperoxide on the polymeric complex results in transformation of complex I to complex II; the addition of Ph$_3$P results in the reverse transformation.

Tetralin is oxidized on ionite complexes of transition metals (Mn^{2+}, Cu^{2+}, Co^{2+}, VO^{2+}, 363 K, 3 h, liquid-phase oxidation) with a low (8–12%) degree of oxidation [168]; major products are tetralin hydroperoxide (21–32%), 1-tetralon and 1-tetralol (the ratio between these reaches 23). At the same time the autoxidation of tetralin catalyzed by complexes of Co^{2+} with pyridine [169] or alkylphosphonium [170], grafted onto

---

[11] The study of tetralin oxidation by heterogenized complexes of Co^{2+} on silochrome (S_{sp}=140 m^2/g) shows that methods of obtaining the complexes substantially affect the activity and selectivity, and that chemical immobilization results in more effective catalysts than the impregnation method [164,165].

colloidal copolymers of styrene, acrylic or methacrylic acid and DVB, is nearly 10 times more rapid than when using homogeneous catalysts. The major product is 1-tetralon.

The complex of $CoCl_2$ and polyurethane (PU) $(OCH_2NHCO)_n$, where $n=4-8$, is an effective catalyst for the liquid-phase oxidation of tetralin (air, bubbling, 363 K, 4 h) [171–173]. Under optimal conditions 1-tetralon is formed selectively; the selectivity is observed also at a high degree of transformation, the ratio ketone/alcohol being 20–25; tetralin hydroperoxide, an intermediate, is not found. The reaction is of first order with respect to substrate and one-half order with respect to catalyst; these observations testify to an initiating function of the catalyst. The initiation rate in the presence of $CoCl_2 \cdot PU$ is $w_i = 1.2 \times 10^{-3}$ mol/(l s), and $E_a = 53.2$ kJ/mol.

These catalysts can be recycled, and their activity and selectivity can be varied by varying the structure and length of chain. With increasing relative molecular mass of PEG from 4000 to 40000 the rate of tetralin oxidation in the presence of supported MCl_2 (M=Co, Mn and Cu) increases by 10 times (Table 2.5).

Table 2.5 Dependence of initial rates of hydroperoxide decomposition w_{i0} and tetralin oxidation w on molecular mass of PEG (355 K, $[ROOH]_0 = 0.4$ mol/l, O_2, catalyst content 0.25–1.0% of tetralin) [174].

Molecular mass	w_{i0} (mol O_2/(g ion Co·min))	w_i (mol ROOH/(g ion Co·min))	w_{i0}/w_i
4 000	180	35	5
20 000	900	40	22.5
40 000	1800	33	54.5

The data show that complexes $MCl_2 \cdot PEG$ are initiators, and molecular mass affects only the initiation. The rate of HP decay, in contrast to oxidation, does not depend on the molecular mass of the polymer.

Activation energy E_a (46 kJ/mol) does not depend on molecular mass of PEG, and additional cross-linking of chains inhibits the process. Therefore the variations of the rate constants with varying molecular mass are explained by variation of the preexponential factor in the equation for reaction rate, i.e. by entropy factors [175,176].

Here we encounter a new effect in catalysis, the effect of the energy of the non-equilibrium conformations of the polymer which can arise during the course of reaction through interaction of substrates with the immobilized metal ions. This energy increases with increasing chain length and under certain conditions (for instance, if the rate of the catalytic step is higher than that of the relaxation rate of the non-equilibrium molecule of the polymer) it can be transferred to reacting components, accelerating their transformation.

The following example demonstrates the effect of non-equilibrium conformations. Macrocomplexes of $MCl_2 \cdot PEG$ in tetralin in the absence of O_2 exist as globules the sizes of which are less than the sizes of the free PEG molecules. The activation of O_2 on a metal ion can be accompanied by a temporary breaking of one of the M–O bonds (like the ousting of a ligand from a complex). This results in a non-equilibrium conformation which tends to unfold:

$$\underset{\underset{Cl}{\overset{\displaystyle M}{\diagdown}}Cl}{\overset{\displaystyle \smile O \diagup \diagdown O \smile}{}} + O_2 \underset{K_r}{\rightleftharpoons} \underset{\underset{Cl}{\overset{\displaystyle M}{\diagdown}}Cl}{\overset{\displaystyle \smile O \diagup \diagdown O \smile}{}} \underset{O_2}{\vphantom{}} \xrightarrow{RH} \underset{\underset{Cl}{\overset{\displaystyle M-Cl}{\uparrow}}}{\overset{\displaystyle \smile O \diagup \diagdown O \smile}{}} \underset{OOH}{\vphantom{}} \xrightarrow{RH} \underset{\underset{Cl}{\overset{\displaystyle M}{\diagdown}}Cl}{\overset{\displaystyle \smile O \diagup \diagdown O \smile}{}} + \dot{R} + OH_2^{\bullet}$$

This process is reversible and characterized by the value of K_r: $-RT\ln K_r = \Delta G$, where ΔG is the change in the free energy of macrocomplex formation. ΔG can be divided into two parts: $\Delta G = \Delta G_x + \Delta G_c$, where ΔG_x is the contribution from rupture of old and formation of new chemical bonds, and ΔG_c is the work of chain stretching that was "stored" during production of the macrocomplex, and is determined mainly by the entropy share $\Delta G_c = \Delta H_c - T\Delta S_c$ (just as the value of $T\Delta S_c$ depends on the molecular mass of PEG, it increases with increasing mass). This follows from the formula

$$K_r = \exp(\Delta S_x/R)\exp(\Delta S_c/R)\exp(-\Delta H_x/RT) = K_0\exp(-\Delta H/RT),$$

the values of K_r for polymeric complexes and their low molecular mass analogues differing by the factor $\exp(\Delta S_c/R)$, which is a part of the preexponential factor and depends on the molecular mass of the macroligand. In other words, the proportion of complexes bound with O_2 increases with increasing macrochain length; the value of this proportion is contained in the expression for the effective constant of the initiation step.

Thus, if the rate of a catalytic reaction is higher than the relaxation rate of non-equilibrium conformations of the macrochain, the activity of immobilized complexes in the oxidation reaction increases; the additional cross-linking (for instance, in complexes with Co/PEG= 2 or 3) results in a decrease of the effective chain length and diminishes this effect [175].

It is important that the polymer length affects substantially the distribution of products of the liquid phase oxidation of tetralin: a major product (up to 98%) is tetralin hydroperoxide. Probably, the decay of HP, in contrast with the reactions of generation of free radicals in which polymer chains are involved, proceeds via formation of HP complexes with Co^{2+}, and in this case the conformational transformations of chains do not play a significant role. The values of the ratio $w_{i0}/w=5-50$ (see Table 2.5) account for the selective accumulation of HP in the oxidation process. Then HP molecules decompose to molecular (95%) and radical (~5%) products [174]. Radicals generated on the complex escape into the liquid phase and carry out the oxidation with a chain length of about 60.

It is plausible that the effect of non-equilibrium conformations accelerates other processes catalyzed by soluble immobilized macrocomplexes. The effect resembles the energy recuperation in enzymatic reactions: the energy evolved in the course of reaction is not dissipated into heat, but is used for activation of catalyzed reaction.

2.3.6 Oxidation of aromatic hydrocarbons with condensed benzene rings

Data on the oxidation of naphthalene, the simplest compound of this class, are scant. Oxidation of naphthalene, 2-methyl-, 2,3- and 2,6-dimethylnaphthalene in AcOH by a 60% aqueous solution of H_2O_2 in the presence of Pd^{2+} immobilized on polystyrene modified by sulfo groups was studied in [177,178]. The selectivity of this reaction with respect to methylnaphthalenes (for instance, oxidation of 2-methylnaphthalene to 2-methyl-1,4-naphthaquinone) is higher than that with respect to methylbenzene. The conversion and selectivity increase with increasing temperature (to 373 K).

Table 2.6 Oxidation of condensed aromatic compounds in the presence of free and immobilized manganese acetate tetraporphynate [181].

| Substrate | Catalyst | K_{eff} | Product | Yield (%) | |
| | | | | per sub-strate | per por-phyrin |
		(l/mol s)			
Anthracene	free	4	9,10-Antraquinone	80	200
	immob.	12	Endepoxide of anthracene	50	125
9,10-Dimethyl-anthracene	free	0.06	9,10-Dimethylanthra-quinone	10	25
	immob.	0.02	Endepoxide of 9,10-anthracene	10	25
9,10-Diphenyl-anthracene	free	0.08	9,10-Diphenylanthra-quinone	20	50
	immob.	0.02	Endepoxide of 9,10-diphenylanthracene	10	25
Bianthryl	free	0.07	Bianthryl quinone	25	63
	immob.	0.02	Endeperoxide of bianthryl	10	25
Phenanthrene	free	1.50	Phenanhrenequinone	30	70
	immob.	0.60	Endeperoxide of phenanthrene	20	50

Oxidation of anthracene is interesting from the viewpoint of obtaining anthraquinone, a valuable product. The oxidation can be carried out by a non-catalytic method using actually any oxidant, but the non-catalytic process requires rigid conditions and is accompanied by formation of a number of side products, especially anthrone and bianthrone:

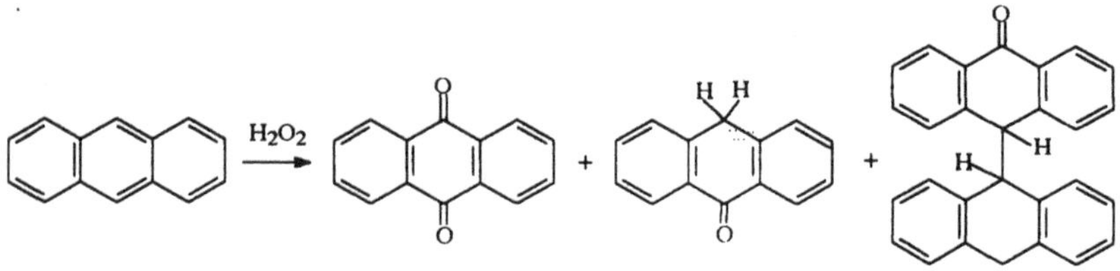

Non-catalytic oxidation of anthracene by hydrogen peroxide to 9,10-anthraquinone in AcOH solution (353 K) proceeds at a low rate and low selectivity due to interaction of anthracene with the peroxide formed. In the presence of NH_4VO_3 or $VO(AcAc)_2$ anthracene is oxidized even at 302–323 K with a yield of 60–95% [179]. At the same time, the H_2O_2 disproportionation catalyzed by vanadium complexes is inhibited in the presence of anthracene. Therefore, the process is determined by the rate of formation of singlet oxygen and its complexes with vanadium.

The use of VO^{2+} as catalyst and heterogenized Cu^{2+}, immobilized on the same CSDVB granules with azomethine blocks as promoters permits to rise the conversion of anthracene to 98% and the yield of anthraquinone to 43% [180]. If the most active and most often used oxide catalyst $Cu_{0.6}V_2O_5$ is used, these values are equal to 46 and 27%, respectively.

The non-chain radical anaerobic oxidation of anthracene and its derivatives and phenanthrene in the presence of monomeric (soluble) and polymeric (insoluble) manganese porphyrates results in the formation of different products. These are the corresponding quinones in the presence of monomeric catalysts and exclusively endeperoxides in the presence of polymeric catalysts (Table 2.6) [181]. In this system dialkyl- and diarylsubstituted alcohols are oxidized to stereoisomeric epoxy alcohols in the presence of monomeric metalporphyrins and are not oxidized in the presence of polymeric metalporphyrins.

2.4 Polymeric Metalporphyrins as Catalysts for Olefin Oxidation

Polymeric metalporphyrins, obtained by polymer analogous transformations and copolymerization of porphyrin-containing monomers, and by the method of condensational "assembling" [182], are synthetic models for natural cytochromes P-450 carrying out detoxication of living organisms by hydroxylation of saturated and aromatic hydrocarbons and epoxidation of olefins by dioxygen [183]. The high efficiency of cytochromes P-450, in comparison with model systems, is explained in the following way. The protein part of the enzymes (oxygenases, peroxidases, etc.) facilitates a proper spatial orientation of oxidizable substrate relative to the oxygen–heme complex (see Section 2.2) and creates optimal microsurroundings of the active center [184].

The active site including a central metal atom is served by the large conjugated π-electron system of a macrocyclic ligand. The active center can be modified by electron transfer to a support via an axial coordination bond. The activity depends, to a certain extent, on the nature of the polymer, the bond type and the hydrophilic character of the metalporphyrin [185, 186]

Alkanes, olefins and oxygen-containing compounds can be oxidized by three different mechanisms: (1) activation of dioxygen (see Section 2.2), (2) initiation of free radical oxidation, (3) non-chain radical oxidation [187, 188]. The chain radical oxidation proceeds by activation of coordinated dioxygen due to formation of a charge-transfer complex (CTC). In the case of tetraphenylporphyrin (TPP) as ligand, CTCs are superoxide complex $TPP–Co^{3+}\cdots O_2^{-\cdot}$ and peroxide complex $TPP–Mn^{4+}\cdots O_2^{-}$.

Chain radical autoxidation (such processes are characterized by the formation of a variety of products) can proceed both by abstraction of an active proton from a substrate and by addition of the superoxide (or peroxide) complex to a double bond in the substrate:

$$TPP-Mn^{4+}\cdots O_2^- \begin{array}{c} \nearrow TPP-Mn^{4+}\ ^-O_2H\ +\ R^{\cdot} \\ \\ \searrow \underset{\underset{\underset{O-O-Mn^{4+}-TPP}{|}}{|}}{-C-\overset{\cdot}{C}-} \end{array}$$

$$\searrow C=C \diagup$$

The catalytic transformation occurs via an intermediate triple complex with singlet dioxygen.

Metalporphyrins in the presence of a reductant (sodium ascorbate, $NaBH_4$, H_2–colloidal Pt, etc.) demonstrate high catalytic activity also in reactions of non-chain radical aerobic oxidation of olefins. In these reactions a superoxide (or peroxide) complex serves as a carrier of oxygen to the substrate, the metalporphyrin being oxidized to an inactive form. The role of the reductant is in the regeneration of the active form of metalporphyrin, i.e. in the closing of the catalytic cycle:

$$TPP-Mn^{4+} + BH_4^- + C_2H_5OH \longrightarrow TPP-Mn^{2+} + B(OC_2H_5)H_3^- + H_2O$$

The properties of the metalporphyrins affect substantially their activity: the immobilization of complexes of phthalocyanine with Ni, VO, Co, Fe and Mn on polymeric support CSDVB results only in an insignificant increase in their activity in oxidation of cyclohexene at 351 K [106]. At the same time complexes of Cu and Fe with chlorine-containing polyphthalocyanines show high selectivity in the oxidation of cyclohexene and cumene to hydroperoxide [189,190]; these catalysts can serve, depending on their concentration, as initiators or inhibitors of the process. The semiconductor properties of the catalyst, stipulated by the system of conjugated bonds in the polymer, play an important role [189]. This leads to the formation of ion-radicals on the catalyst surface which are active in the oxidation of hydrocarbons to hydroperoxides. The strong electric fields near the active center, caused by separated charges (metal ion–electron density of the porphyrin ring), facilitate the formation of a cyclic triple intermediate complex during the catalytic oxidation of olefins [191]. The spatial stabilization of the active centers also plays an important role. The hydrophobic evironment near the immobilized molecules of metalporphyrins inhibits (and even prevents) formation of oxodimers in the presence of O_2 and subsequent irreversible oxidation of M^{n+}, as it has been mentioned for other metal–polymer catalysts.

The immobilization of tetra-p-aminophenylporphyrinate of manganese acetate, one of the most active catalysts for oxidation, leads to formation of soluble porphyrin-containing metal-copolymers [192, 193] and increases by 20–40 times their specific catalytic activity (SCA) in oxidation of cholesterol by dioxygen. This can be explained by the participation of hydrophobic fragments of macrochain in effective sorption of substrate. Due to this sorption, the local concentration of cholesterol in the vicinity of the active center increases (compare with the increase of the local concentration of CHHP at oxidation of cyclohexene on immobilized Co^{2+}-containing catalysts (Section 2.3.3). The concentration increase depends on the ratio between hydrophilic and hydrophobic properties of fragments of immobilized catalyst and on the content of metalporphyrin in the catalyst.

An unusual temperature dependence (Figure 2.14) was found for the effective rate constant K_{eff} for the cholesterol oxidation with a maximum at room temperature [193]. This temperature dependence differs from those observed with the use of oligomeric and monomeric catalysts and testifies to phase transitions in these systems (as mentioned for hydrogenation, Section 1.5). One of these phase transitions may be caused by disordering of polymer fragments with simultaneous breaking of hydrogen or other (maybe coordinational) bonds between polymer fragments. The characteristic viscosity changes as $\eta = \eta_0 \exp(-E/RT)$ in the same temperature region that testifies to the possibility of conformational transitions in metalporphyrin-containing macromolecules. For immobilized systems, especially for catalysis by insoluble macrocomplexes, the maximum of the dependence of SCA on temperature is found rather often. The conformation of metal polymers in solutions fluctuates substantially at temperatures near the phase transition; therefore, the rate constant also fluctuates (at least due to fluctuations of the preexponential factor which depends on steric factors, or due to fluctuations of interactions between active centers). Therefore, at phase transition temperatures there is a set of different rate constants corresponding to metastable conformational states of polymer. This results in non-Arrhenius dependence, i.e. a maximum in the dependence of SCA on temperature.

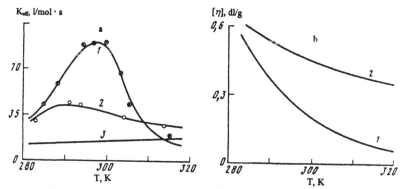

Figure 2.14 Temperature dependences of (a) the effective rate constant of cholesterol oxidation in the presence of NaBH$_4$ and copolymers of N-Mn-porphynateacrylamide and (b) the characteristic viscosity of copolymers. (a) 1 with methylmethacrylate, 2 with butyl methacrylate, 3 monomeric tetra-n-aminophenylporphyrin–Mn^{2+} acetate. (b). 1 with methyl methacrylate, 2 with polymethyl methacrylate.

Insoluble polymeric catalysts, porphynates of manganese (MnP) and iron (FeP) immobilized on ion-exchange polymeric films [181,194] are inactive in oxidation of cholesterol and androstane due to rigid ionic fixation. The best catalysts are MnP immobilized by the morpholine ion, a bifunctional bridge, on a hydrolyzed copolymer of tetrafluoroethylene and perfluoro-3,6-dioxo-5-methyl-8-sulfonylfluorooctene-1 containing terminal sulfogroups:

$$-(\!-CF_2-CF_2-)_{\overline{n}} \quad -(\!-CF_2-CF-)_m$$

where X is a resin anion and ∇ is the plane of porphyrin ring.

In the presence of both monomeric and polymeric catalysts with NaBH$_4$ as a reductant, a single product 3β,5α-cholestenediol is formed with 5α-hydroperoxocholestan-3β-ol as an intermediate. This shows that the immobilization of MnP does not change the oxidation mechanism, but changes the kinetic parameters of the process: K_{eff} increases from 9 to 28 l/(mol s), TN increases from 100–200 to 4000. The immobilization of MnP molecules on insoluble polymers completely excludes their association which is typical of their homogeneous analogues: the value of K_{eff} remains unchanged at varying catalyst concentrations and is close to the value for diluted solutions of MnP. The inactivation of such systems is explained [186] by a variety of side processes resulting in the washing of morpholine and, in its absence, decomposition of peroxide compound the formed Mn–O–O.

Note that polyphthalocyanines of various metals can also act as the above-considered immobilized complexes, namely, they can accelerate the decay of hydroperoxides, for instance cumene [195], and, thus, accelerate the oxidation of unsaturated compounds, compared with the non-catalytic reaction.

Finally, it was found [189,190] that the presence of metal in some polyphthalocyanines is not necessary for olefin oxidation, but the activity of these "non-metallic" polymers is low. The same finding was made for heterocyclic polymers containing nitrogen atoms in the main chain[12], and for graphite slowly catalyzing cumene oxidation [197].

It seems that polymeric metalporphyrins are a very promising type of heterogeneous catalyst for liquid-phase oxidation and have advantages in comparison with other metal polymers: high thermal and chemical stability under the conditions of an intensive chain radical process or in the presence of reductants of various nature. The advantages of these catalysts are realized in oxidation of various substrates and especially thiols (see Section 2.8).

2.5 Catalase-Type Activity of Immobilized Complexes

Metal polymers are interesting as models of catalase, peroxidase, pyrocatechase, proteolytic and other enzymes. Liquid-phase decomposition of hydrogen peroxide is a convenient model for various redox processes including enzymatic ones.

Homogeneous decomposition of H_2O_2 in the presence of ions of transition metals (mainly Fe^{2+} and Cu^{2+}) proceeds by the Haber–Weiss chain reaction scheme according to which hydrogen peroxide serves alternatively as an oxidant and a reductant in reactions with reduced and oxidized forms of metals:

$$HOOH + M^{n+} \longrightarrow HO^{\cdot} + {}^{-}OH + M^{(n-1)+}$$

$$HOOH + M^{(n-1)+} \longrightarrow HOO^{\cdot} + H^{+} + M^{n+}$$

The alternation of these two reactions, caused by the catalyst, results in constant concentrations of different valent forms of the transition metal; the values of the concentrations depend on the activity of reduced and oxidized forms of metal ion in reaction with H_2O_2 and other products of the reaction mixture [8]. To possess the high redox activity, the metal ion or complex should be coordinationally unsaturated. One of the valent forms of the metal can be stabilized due to complexing[13]. Such complexing can affect electronic structure and, hence, the redox potential of the metal and activity and selectivity of the newly formed catalyst. From this viewpoint immobilized metal complexes are promising catalysts [198].

However, the activities of Fe^{2+}, Cu^{2+}, Co^{2+}, Mn^{2+} and other metal ions immobilized on sulfocationites and anionites are relatively low, especially in the case of ionites in H^+

---

[12] For instance, polyphenylcyaneamide oxidises cumene with the HP yield 30–45% (373 K, 7 h) [196]. A correlation between the catalytic activity of the polymers and their electric properties (length of effective conjugation, specific electroconductivity, concentration and mobility of charge carriers) was found.

[13] However, there are other opinions. For instance, for homogeneous H_2O_2 activation catalyzed by V^{5+} complexes, complexes of vanadium with one, two and even three dioxygen anions O_2^{2-} were found to be involved in the key step of the process [179]. Two O_2^{2-} ligands in the complex of V^{5+} with three peroxoligands disproportionate with formation of free or coordinated singlet dioxygen.

form [199–201]. An increase in the extent of polymer's cross-linking leads to a further decrease of the activity. The catalase-type activity of Cu^{2+} complexes immobilized on vinylpyridine ionites (AN-40, AN-25, AN-42 and AN-23) depends on the nature of ligands; the features of these systems were explained rather well using electronic ideas and taking into account steric hindrances that arise on complex formation and affect both the effective charge of metal ions and the saturation degree of their coordination sphere [200,202]. The stability and activity of complexes decrease with increasing rigidity of the polymer matrix. Figure 2.15 shows that the ability of transition metal complexes, immobilized on ionite AN-40, to decompose H_2O_2 varies by more than 500 times on varying the nature of the transition metal [168]; the sequence of activity of the complexes coincides with the order of their stability: $Cu^{2+}>Ni^{2+}>Co^{2+}>Mn^{2+}>Zn^{2+}$. Probably, this sequence is determined by the lability and stability of the coordination sphere and by the ability of the ions to change their valency. The differences in these factors for different ions can even lead to different mechanisms of catalysis. A relation between the rate of H_2O_2 decomposition and energies of sorption and hydration was found for catalysis by ionite complexes [203]. Manganese oxides immobilized on nitrone are very active [204].

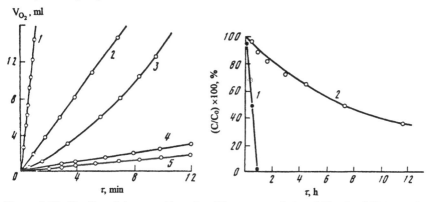

Figure 2.15 Kinetics of decomposition of a 1% aqueous solution (100 ml) of H_2O_2 in the presence of 0.5 g of ionite AN-40 containing ions of Cu^{2+} (1), Ni^{2+} (2), Co^{2+} (3), Mn^{2+} (4) and Zn^{2+} (5). $[M^{2+}]=0.25$ mg of ion/g.

Figure 2.16 Comparison of the stability of free $Fe^{3+}PC$ (1) and $Fe^{3+}PC$ immobilized on PS (2). $[H_2O_2]=5\times10^{-2}$ mol/l.

The kinetics of the heterogeneous decomposition of H_2O_2 in the presence of Ni^{2+} amine complexes immobilized on ion-exchange resin Dowex 50W was explained in [205] by formation of an intermediate that serves simultaneously as an inhibitor of the decomposition. It is interesting that the value of E_a decreases from 33.5 to 25.5 kJ/mol with increasing degree of cross-linking.

Polyphthalocyanines have relatively high peroxidase activity decreasing in the order Fe>Co>Mg>Ni>Cu [206]. The reaction sequence for H_2O_2 decomposition depends on phthalocyanine (PC) concentration: the order is one, two or intermediate at low, high and intermediate PC concentrations, respectively. This is explained by a varying of the degree of coating of the catalyst by the substrate [207].

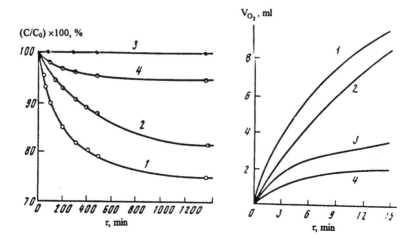

Figure 2.17 Decomposition of H_2O_2 in the presence of metal chelates under comparable conditions.

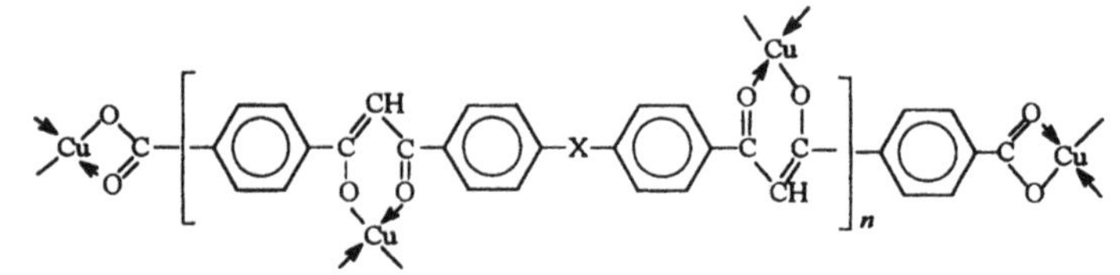

where X=O (1) or CH₂ (2);

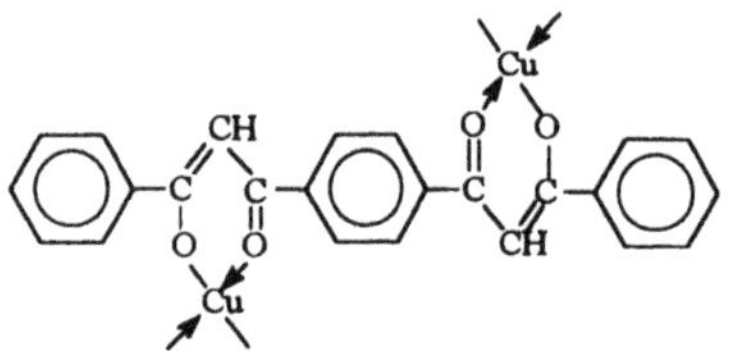

Figure 2.18 Evolution of O_2 from a 1% solution of H_2O_2 at 313 K in the presence of polyacrylates of Cu^{1+} (1), Co^{2+} (2), Fe^{3+} (3) and Ni^{2+} (4).

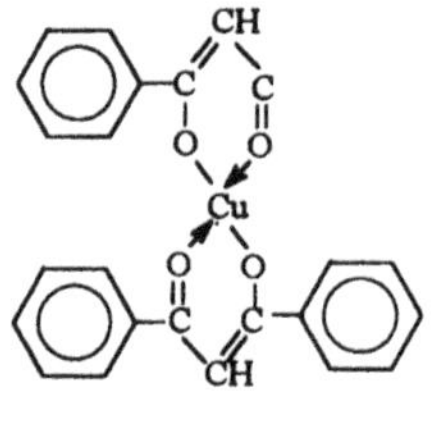

Iron phthalocyanine immobilized on PS [208] is an active catalyst for H_2O_2 decomposition (K_m=2.8×10⁻² mol/l at pH 7 and 298 K). Its characteristic feature is high stability: after 12 h its activity decreases by only 60%, whereas a homogeneous catalyst deactivates after 0.5 h (Figure 2.16). The activation energy is only 7.6 kJ/mol for the reaction with participation of the immobilized catalyst and 18.5 kJ/mol for the homogeneous catalyst.

The same features were also found for Fe^{3+} and Co^{2+} phthalocyanines covalently bonded with poly(2-vinylpyridine-co-styrene) [209].

It is important that even small changes in the structure of the polychelate chain lead to a substantial change in activity of Cu^{2+} complexes, monomeric analogues being less active (Figure 2.17). The catalase-type activity of Cu^{2+} complexes immobilized on CSDVB modified by Schiff bases was studied comprehensively [180,210,211].

The catalase-type activity of Cu^{2+} complexes of similar structure immobilized on PS depends on the nature of the chelating unit (Table 2.7). In all cases the reaction is first order with respect to substrate and catalyst, but parallel non-catalytic processes also occur. The decomposition rate is described by the equation $w=w_0 + K[Cu^{2+}][H_2O_2]$ where w_0 is the sum of rates of H_2O_2 decomposition in homogeneous media (0.013×10^{-4} mol/l), on vessel walls (0.248×10^{-4} mol/l) and on the surface of the support (without Cu^{2+}) (1.38×10^{-4} mol/l). Both immobilized and free Cu^{2+} complexes catalyze H_2O_2 decomposition by a chain radical mechanism. The effect of the polymeric support results in an increase of catalase-type activity of complexes of Cu^{2+} with Schiff bases (as compared with low molecular mass analogues) and in the unusual decrease of the activation energy (from 84 to 62–81 kJ/mol, depending on the nature of support). These facts testify to a variety of factors affecting catalase-type activity of immobilized metal complexes.

Table 2.7 Kinetic parameters of H_2O_2 decomposition in water in the presence of macromolecular chelates of Cu^{2+} (6×10^{-4} mol/l, 303 K, polymeric support CSDVB) [180].

Polymeric support	$w\times10^4$ (mol/l min)	$K\times10^2$ (l/mol min)	E_a (kJ/mol)	$A_0\times10^{-12}$
(P)—CH_2—NH_2	1.20	11.5	76.4	3.01
(P)—$CH_2N{=}CH$—⟨○⟩N	1.56	16.7	70.3	0.30
(P)—$CH_2N{=}CH$—⟨○⟩ÓH	1.35	1.39	81.5	18.80
(P)—$CH_2N{=}CH$—⟨○⟩	1.70	16.1	62.3	0.01

At the same time a correlation between the preexponential factor and E_a indicates a similarity of mechanisms of H_2O_2 decomposition in these systems. The absence of correlation between E_a and the rate constants in the sequence of ligands with similar structures probably testifies to the important role of entropy factors that depend also on the nature of the polymeric support. It is interesting that a substantial decrease of E_a, in comparison with soluble analogues, was also found for H_2O_2 decomposition in the presence of Fe^{3+} complexes immobilized on the surface of oxidized coal [199].

Polymers obtained by homopolymerization of di- and triacrylates of transition metals (especially $Cu(acrylate)_2$ and $Co(acrylate)_2$), systems with a high degree of cross-linking, are active catalysts that can be recycled [212] (Figure 2.18).

In heterophase processes the proportion of the active component that is accessible to reagents should be taken into account. This was done for H_2O_2 decomposition in the presence of the Wilkinson complex $RhCl(PPh_3)_3$ and its heterogenized analogues (CSDVB, 2% and 5% of DVB, modified by chelating ethylenediamine groups) [213,214].

Kinetic data (the reaction orders with respect to H_2O_2 and Rh are 0.4 and 0.6, 0.3 and 0.4, and 0.6 and 0.7, and E_a=21.5, 22.6 and 29.0 kJ/mol for free Wilkinson complex and for Wilkinson complex immobilized on modified CSDVB with 2% and 5% of DVB, respectively) show that only a part of the Rh participates in a process characterized by a constant K^*. This constant is determined in the following way.

The following scheme of H_2O_2 decomposition is considered [215]:

$$H_2O_2 \rightleftharpoons HO_2^- + H^+$$

$$]{-}Rh_a + HO_2^- \xrightarrow{k_1}]{-}Rh_a \cdot HO_2^-$$

The second molecule of H_2O_2 interacts with this intermediate:

$$]{-}Rh_a \cdot HO_2^- + H_2O_2 \xrightarrow{k_2}]{-}Rh_a + O_2 + H_2O + HO^-$$

Scheme 4

However, $[]{-}Rh]=[]{-}Rh_a]-[]{-}Rh_i]$, where $[]{-}Rh]$, $[]{-}Rh_a]$ and $[]{-}Rh_i]$ are concentrations of total, accessible and inaccessible rhodium, respectively. Then

$$d[O_2]/dt=K^*([]{-}Rh]-[]{-}Rh_i])[H_2O_2]$$

$$K^*=k_1k_2/[H^+]=const$$

$$w=d[O_2]/dt=K^*[Rh]^m[H_2O_2]^n .$$

It was found that K^* is proportional to $[Rh]^m$ l^m/min^{-m} where m=1, 0.7 or 1.3 for catalysis by free Wilkinson complex and Wilkinson complex immobilized on modified CSDVB with 2% and 5% DVB, respectively. The value of K^* varies with varying temperature and, in principle, may depend on H_2O_2 concentration.

Nevertheless, only transition metals localized on the surface or in the very thin layer near the surface of the polymer have the highest activity (as well as in other catalytic processes). For instance, Cu^{2+} chelate on the open surface of CSDVB (grafting by UV irradiation of acrylonitrile, reduction by $LiAlH_4$, formation of bis(carboxylmethyl)amino groups under the action of chloroacetic acid) is by 2–5 times more effective in H_2O_2 decomposition at 298 K than regular macrocomplexes [216].

The kinetics of H_2O_2 decomposition in the presence of a complex of Cu^{2+} with a polyampholyte (copolymer of 2-methyl-5-vinylpyridine and acrylic acid) resembles that in the presence of enzymes where double and triple enzyme–substrate complexes are taken into account [217]. The rate constant depends strongly on pH (Figure 2.19) because of the formation of complexes of different content, K_m=2.67×10^{-3} mol/l at pH 8.5. There is a correlation between activity and content of diamagnetic Cu^{2+} complexes. Probably, these diamagnetic species are binuclear complexes with strong antiferromagnetic interaction between copper ions, like the active centers of hemocyanine (see Section 2.2).

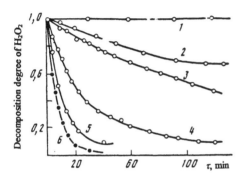

Figure 2.19 Peroxidase-type activity of a copolymer of 2Me5VP with acrylic acid (1) and Cu^{2+} complex supported on the copolymer, at different pH (2–6). (2) pH 7.0, (3) 8.0, (4) 9.0, (5) 10.0, (6) 11.0. [copolymer]=2×10^{-3} mol/l, [Cu^{2+}]=5.0×10^{-4} mol/l, T=298 K, μ=0.1.

Cu^{2+} complexes on poly(acrylhydroxamic acid) are well known catalysts for H_2O_2 decomposition [218]. Inhibition of these processes by ionol testifies to heterogeneous stages.

Bis(ethylenediamine)–Fe^{3+} complexes immobilized on PAAc have activity comparable to catalase activity[14] [222–225]. The first step in H_2O_2 decomposition is H_2O_2 activation by the polymeric chelate with participation of amino groups (Scheme 2.5):

---

[14] Enzyme catalase is an effective catalyst for H_2O_2 decomposition. A molecule of the iron protein with relative molecular mass approx. 250 000 contains four heme groups, each containing protoporphyrin IX (a macrocycle) with a high-spin Fe^{3+} ion [219]. There are many heterogenized enzymatic systems, obtained by immobilization of catalase on synthetic polymers (biomimetic polymers), which decompose H_2O_2 to H_2O and O_2, for instance derived from PEG [220]. Some anticancer compounds, coordinational complexes of Pt, Rh and Pd with components of DNA and RNA, reveal a catalytic activity in H_2O_2 decomposition [221], the catalytic activity of the compounds correlating with their therapeutic effect. On the basis of these observations, a molecular mechanism for the effect of macrocomplexes with bioligands on cancer cells was proposed, and the molecular consideration of cancer phenomena was unified.

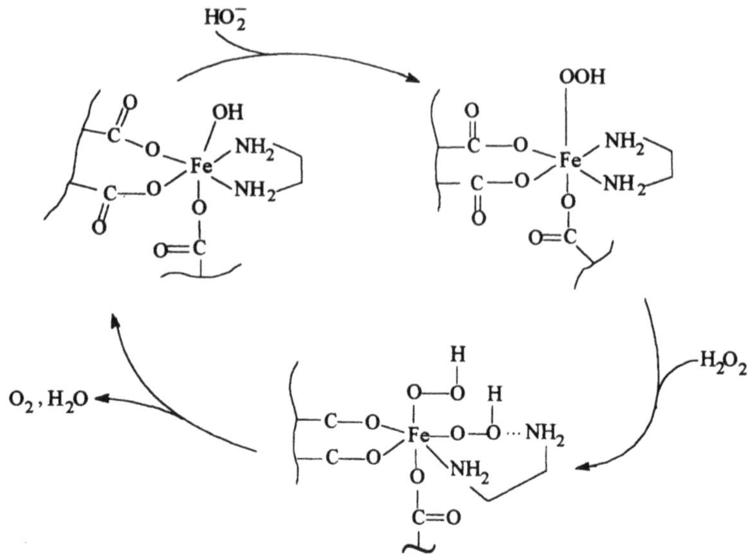

Scheme 2.5

For H_2O_2 decomposition catalyzed by liver catalase the E_a is higher, 105 kJ/mol, whereas for H_2O_2 decomposition in the presence of these macromolecular complexes the E_a is lower, 96 kJ/mol at pH 10.7.

Finally, it should be noted that many organic polymers, for instance poly(acrylonitrile) and poly(aminoquinones) [218], themselves possess activity for the disproportionation of hydrogen peroxide.

2.6 Metal Polymers as Catalysts for Olefin Epoxidation by Alkylhydroperoxides

Olefin epoxidation

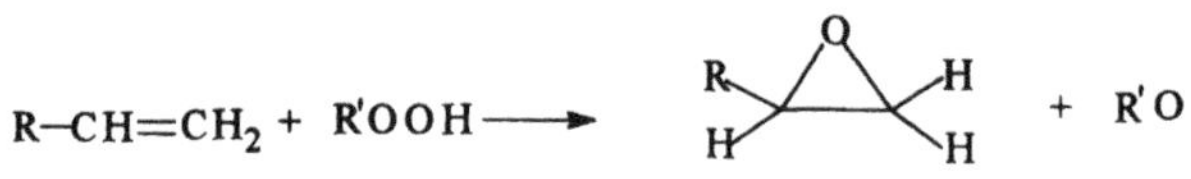

$$R-CH=CH_2 + R'OOH \longrightarrow \quad + \quad R'O$$

has many features in common with biological monooxygenation. The chemistry of oxiranes is developing in the direction of industrial production of epoxide compounds. This reaction requires heterolysis of the peroxide bond. The presence of carbonyl groups facilitates the heterolysis in peracids: the Polezhaev reaction is a molecular pathway for olefin epoxidation with participation of RC(O)OOH. Epoxidation occurs due to electrophilic attack, therefore di- and tri-substituted double bonds are more easily epoxidized. In diens it is exclusively the double bonds that are epoxidized. Epoxidation often competes with spontaneous or olefin-induced decomposition of hydroperoxide (HP) or peracids. At least five mechanisms of olefin epoxidation are known, radical and molecular mechanisms are the most common [226].

In the presence of metal complex catalysts epoxidation proceeds under mild conditions [226–229] because in the transition state

$$\left[\text{Catalyst} \cdots O_2^- \cdots \overset{\displaystyle C}{\underset{\displaystyle C}{\|}} \right] \longrightarrow O\underset{C}{\overset{C}{\triangleleft}}$$

the bond O–O is more easily heterolyzed. An active catalyst for epoxidation should be, on the one hand, a weak oxidant and, on the other hand, a strong Lewis acid [230]. The catalytic activity of compounds depends on their ability to catalyze the transfer of oxygen from HP to olefin. The best catalysts are compounds of Mo, V and W in a high valence state; these compounds have a large positive charge and are able to accept electron couples on vacant orbitals and to form complexes, of various stability, with olefins and HP [231]. Compounds of Fe, Co, Mn, Ni and Cu possess diametrically opposite properties; their single function is homolysis of the peroxide bond by a one-electron mechanism with formation of free radicals.

One of the indications of the catalytic activity of Mo^{6+} complex is its ability to form mixed-ligand mono- and di-peroxo complexes, active agents of direct epoxidation [232]:

$$Mo^{6+}\!\!\overset{O}{\underset{O}{\triangleleft}} + \;>\!C\!=\!C\!<\; \longrightarrow Mo^{6+} + \;\overset{O}{\triangle}$$

$$\overset{O}{\underset{O}{>}}\!Mo\!\overset{O}{\underset{O}{<}} + \;>\!C\!=\!C\!<\; \longrightarrow Mo^{6+}\!\!\overset{O}{\underset{O}{\triangleleft}} + \;\overset{O}{\triangle}$$

Formation of mixed-ligand complexes in reactions with organic peroxides is a limiting step of the whole process. Formation of diolate Mo^{6+} complexes is a first step for homogeneous catalytic oxidation of hydroperoxide:

$$R\!\!\overset{O}{\diagdown}\!\!\underset{R}{\diagup}\!O\!\!\overset{\overset{\displaystyle O}{\|}}{\underset{\underset{\displaystyle O}{\|}}{Mo}}\!\!O\!\!\overset{\diagup R}{\underset{\diagdown R}{}}$$

monoperoxo complexes being by two orders of magnitude more active than diperoxo complexes. The selectivity of direct epoxidation (the ratio of the yield of olefin oxide (moles) to equivalents of oxygen consumed) increases with increasing temperature because of the higher value of E_a for this reaction in comparison with non-selective oxidation.

It could be expected that immobilization prevents elution of M^{n+} because of the formation of diolate complexes[15] or oxidation of ligand which binds the catalytically active complex to the support. However, in recent reviews concerned with the epoxidation mentioned above, data on this subject are absent.

Catalysts formed from various ion-exchange and epoxide resins and artificial fibers rapidly lose their activity because of elution of molybdenum. The complex $MoO_2(AcAc)_2$ immobilized on anionites [234] is also rather unstable and transforms in the course of the reaction to the diolate complex. The oxalate complex $[Mo_2O_4(C_2O_4)_2]^{2+}$ on anionite is more stable with respect to substitution of oxalate ions in the coordination sphere of Mo^{5+}, but it is oxidized slowly by the peroxo ligand. Nevertheless, by proper immobilization of molybdenum dioxyacetylacetonate or vanadium oxyacetone [235,236] effective catalysts for epoxidation of cyclopentane, cyclohexene, styrene etc. by tert-butyl hydroperoxide (TBHP) or 1-phenylethyl, at 323–353 K, were created.

The same comments are valid for chelate complexes of $MoO_2(AcAc)_2$ and $VO(AcAc)_2$ with polymers derived from chloromethylated PS [237]. After activation of the catalyst by TBHP, epoxidation of TBHP in dichloroethane at 353 K proceeds with high rate and is described by the equation:

$$W = k_{eff}[TBHP]^{\alpha}[CH]^{\beta}[Ct]^{\gamma},$$

where $\gamma = 0.3$ and α and β for different supports vary from 0.7 to 1.6 and from 0.2 to 0.4, respectively. It is interesting that macrocomplexes with a ratio of ligand:metal=2:1 are most stable and resistant to leaching (during 10 cycles). The metals of the active center are in the highest valence state.

It was found [238] that Mo^{6+} complexes supported on polybenzimidazole are very active and stable in epoxidation of propylene by TBHP (the Halcon (or Arco) process). The yield of propylene oxide reaches 99.8%. It is obvious that these catalytic systems will be used in industry.

The basicity of pristine anionites is important for catalytic activity of immobilized molybdate ions [239].

Catalysts obtained by immobilization of molybdenyl groups on PVAl, poly(propargyl alcohol), thermally treated PVAl or furfurolidine resin, are active and stable during 500 h of continuous process [79,240–242]. The initial rate of cyclohexene epoxidation by EB hydroperoxide (EBHP) or TBHP is given by the formula:

$$W_0 = (k_1 + k_2[ROOH])[ROOH]_0$$

This formula takes into account the polymer swelling in the course of the reaction under the action of both reaction products and HP itself: the first and second terms correspond to reactions on unswelled and swelled catalyst, respectively. The differences in activity of various regions of catalyst are explained by different accessibility of active centers to substrate. This is proved by a linear dependence of activity on the swelling

---

[15] As a rule, they are obtained by interaction of corresponding polymers with 0.1–0.01 K_2MnO_4 or $NaVO_3$ at pH 3–4 and by subsequent treatment of the obtained products by traditional methods [233].

degree, which can be controlled in various ways (including variation of the cross-linking degree of the polymer chains).

Kinetic data show that epoxidation occurs on the interaction of olefins with a HP adsorbed onto a catalyst. The rigid structure of the heterogenized catalyst impedes formation of a bond between Mo^{6+} and ROOH; it is proven by low values of the sorption coefficient K_s (Table 2.8) in comparison with the formation constants of complexes of molybdenyl diolates with hydroperoxides in homogeneous media (3–5 l/mol). At the same time such impediments are insignificant for sorption of reaction products, and values of K_{in} are close to those for homogeneous catalyst in the presence of molybdenyl dithiolates. The properties of HP also affect substantially the kinetic parameters. Constants for the formation and deactivation of the active complex in homogeneous and heterogenized systems are close, but, in spite of low values of SCA of immobilized complexes, the efficiency of heterogenized systems is higher due to the limited solubility of molybdenum complexes.

Table 2.8 Kinetic parameters for cyclohexene epoxidation (353 K).

Hydroperoxide	K_s, l/mol	K_{in}, l/mol	E_a, kJ/mol
EBHP	0.25±0.10	6.0±0.6	100±2
HPC[a]	0.40±0.10	15.0±2.0	85±4
TBHP	0.10±0.05	18.0±2.0	60±2

[a]Hydroperoxide of cumene.

The rate of epoxidation of cyclohexene in the presence of heterogenized complexes of molybdenum and their deactivation are very considerable as a function of the density and acid strength of a pendant group linked to the coordinating molybdenum center [243]. In this case multifunctional resins, based on stepwise polyaddition polymers such as bis-amine and bis-thiol, were used as supports.

The nature of the olefin also affects the reaction rate, for instance cyclohexene is a highly reactive substrate, whereas the reactivity of 1-nonene is substantially lower. The topochemistry of formation of reaction products was studied in the system with 1-nonene as a substrate and polypropargyl alcohol as a support [242]. ROOH decomposes only to ketones in "unswelled" regions of catalysts, whereas in "swelled" regions the ratio alcohol/ketone is 4:1, i.e. active centers of "swelled" polymer resembles those of homogeneous systems. The selectivity of the process is determined mainly by the ability of initial and required products to form complexes. In addition, free radical decomposition of HP occurs only due to low-valence forms of molybdenum, Mo^{5+}, its proportion being small.

Thus, the change in the degree of cross-linking of the macroligand may increase the ketone yield and the ratio between epoxidation and HP decomposition, and may approach the selectivity with respect to oxides of that for homogeneous catalysts.

An interesting comparison of homogeneous and heterogenized catalysts of epoxidation was carried out in [244] for Mo^{5+} dithiocarbomate complexes immobilized on chloromethylated PS:

In the presence of TBHP Mo^{5+} is oxidized to Mo^{6+}, which carries out the reaction. An analogous homogeneous system, $Mo_2O_3[(S_2CN(C_2H_5)_2]_4$ and TBHP, is degraded rapidly due to a side process, ligand oxidation. This leads to a low yield of cyclohexene oxide (20% vs. 70% for the heterogenized system). In the immobilized systems there are diffusional barriers restricting ligand oxidation, although an excess of TBHP leads to abstraction of 10% of Mo from the catalyst. At optimal ratios of reacting components (TBHP/Mo=1:1, cyclohexene/TBHP=25:1) these systems can be used many times.

Investigation of the epoxidating properties of Mo^{6+} immobilized on cellulose phosphate (CH as a substrate, 333 K, TBHP) led to interesting conclusions [245]. A protonated form of the grafted complex

has the highest activity. Catalytic activity and selectivity (the cyclohexene oxide yield is 80–90% of consumed TBHP, for epoxidation of allyl chloride the epichlorhydrin yield is 75%) is not changed during repeated use of the same weighing of catalyst.

The reaction mechanism includes interaction of immobilized catalyst with HP and olefin and formation of a bisubstrate mixed-ligand complex (Scheme 2.6).

The acid form of molybdenum phosphate cellulose is active because protonation of the grafted complex increases the electron-accepting ability of molybdenum due to the relatively stronger polarization of the peroxo ligand[16].

Kinetic curves (Figure 2.20) are not described by simple equations of first and second order. The first order for CH and TBHP is observed at relatively low concentrations of reagents due to intermediate binding of a substantial proportion of the reagents with the catalyst. The inhibiting effect of TBHP at high concentrations due to formation of a complex of the catalyst with two molecules of TBHP was mentioned above. This impedes the interaction of Mo^{6+} with CH; the effect of reaction products, tert-BuOH and CHO, is the same. The epoxidation rate is described satisfactorily by kinetic equations that take account of the chemosorption of reagents and products on Mo-containing active centers and the rate constant for the decomposition of a triple complex of catalyst with both reagents (as considered above).

---

[16] Cellulose phosphate forms at pH 1.8 active and selective (95% with respect to olefin oxides) catalysts based on V^{5+} [246]; the scheme for this process is as follows:

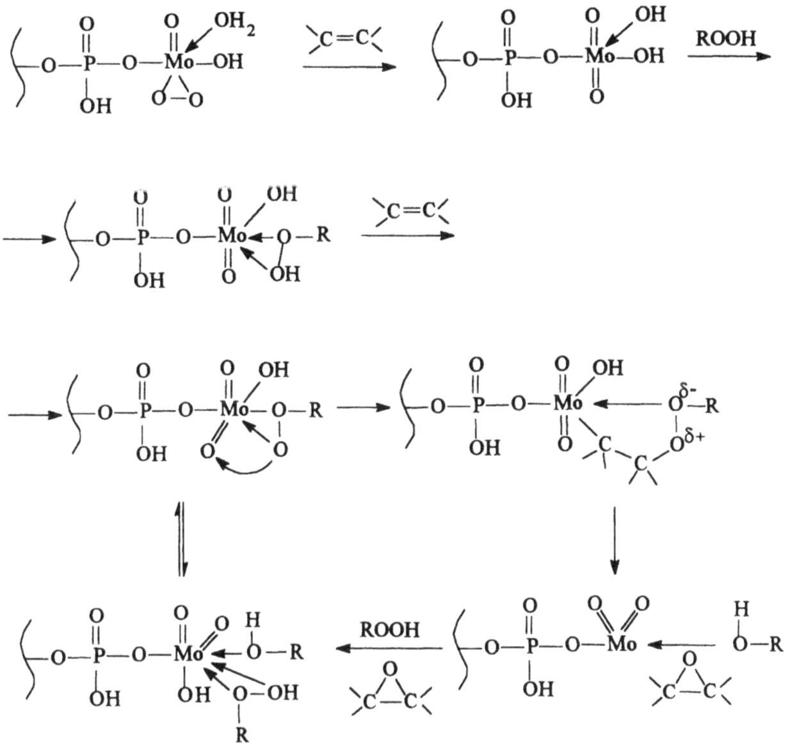

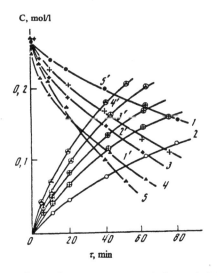

Scheme 2.6

C, mol/l

Figure 2.20 Kinetic curves for consumption of TBHP $(1–5)$ and accumulation of CHO $(1'–5')$ in the course of epoxidation of CH in the presence of molybdenum phosphate cellulose at different temperatures. $(1,1')$ 333 K, $(2,2')$ 383 K, $(3,3')$ 343 K, $(4,4')$ 348 K, $(5,5')$ 351 K. [Mo]=5.6×10^{-4} mol/l, [HP]$_0$=0.25 mol/l, [CH]$_0$=4.7 mol/l, benzene as solvent.

τ, min

A comparison of rate constants and the enthalpies of formation of the intermediate complexes of the catalyst with reagents leads to an important conclusion: the decomposition of triple complex [TBHP¨Mo¨CH] on the surface of the support is characterized by substantially higher activation energy and a higher positive activation entropy than decomposition in solution: E_a and $\Delta S^\neq$ are equal to 136 kJ/mol, 290 J/(mol deg) and 72 kJ/mol, 166 J/(mol deg), respectively. This difference is caused [245] by the heterogeneous character of the system: the triple biligand complex immobilized on a polymeric support is stabilized in an ordered state by an additional interaction with the polymer surface (this is also observed for polymerization processes in the presence of immobilized complexes, see Section 3.1.10). On the one hand, a higher activation energy is required for decomposition of the ordered complex, on the other hand, release of reaction product into solution is accompanied by abrupt disordering of the system. This phenomenon is common for catalysis by immobilized complexes and was quantitatively analyzed for the first time for this reaction.

Tetrabromoxomolybdate immobilized by the method of ion exchange on CSDVB modified by quaternized amino groups is a selective catalyst of olefin epoxidation in the presence of TBHP [247]. It is important that the selectivity can be controlled by varying the nature of the amino groups of the support due to the changing stability of the Mo^{5+} complexes. The immobilized molybdenum complexes are most stable in epoxidation of α-methylstyrene in the presence of TBHP: selectivity increases with increasing number of catalytic cycles and reaches 96–98% [248]. Probably, the active surface centers are formed in the course of catalytic epoxidation.

A prospective process is the epoxidation of olefins (for instance, epoxidation of cyclooctene by H_2O_2) catalyzed by colloidal catalysts prepared by emulsion copolymerization as cationic or anionic ion exchange latexes in the presence of $(NH_4)_6Mo_7O_{24}\cdot4H_2O$ [228]. The anionic transition metal complex catalysts bind to cationic particles, which serve as sites of locally high concentration of reactant and catalyst in aqueous dispersions.

Cyclohexene is effectively epoxidized by H_2O_2 in the presence of polymers with organophosphoryl units as macroligands for peroxotungstic acid $H_2W_2O_{11}$ [249].

The active complex obtained in the presence of H_2O_2 by the interaction of MoO_3 with CSDVB containing triethylenetetraamine groups is a promising catalyst for this reaction [247].

Mn^{3+} porphyrin complexes immobilized on cation polymeric latexes (obtained by emulsion copolymerization of vinylbenzyl chloride, divinylbenzene (DVB) and octadecyl ester as the third monomer with subsequent quaternization by trimethylamine) epoxidize, in combination with NaOCl, linear and cyclic olefins at a high rate [250].

A comparison of catalytic properties of MoO_2^{2+} and VO^{3+} ions immobilized on macro- and microporous PS with chelating functional residues of iminodiacetic acid, bis(phosphonomethyl)amine and bis(2-hydroxyethyl)amine, leads to interesting conclusions [251]. In the presence of TBHP, the polymeric complexes are active catalysts of cyclohexene, E-geraniol and linalool, an isomer of geraniol:

$$CH_3-\underset{\underset{CH_3}{|}}{C}=CH-CH_2-CH_2-\underset{\underset{CH_3}{|}}{\overset{\overset{OH}{|}}{C}}-CH{=}CH_2$$

Molybdenum compounds have the highest activity for CH epoxidation; V^{5+} immobilized on macroporous support by residues of iminodiacetic acid or bis(phosphonomethyl) amine has the highest activity for E-geraniol oxidation (100% conversion and 98% regioselectivity are achieved after 1 h at 355 K). However, catalysts derived from Mo^{6+}

are more stable; after five cycles the release of the metal is actually not observed, whereas under the same conditions the vanadium content decreases by 15–25% and selectivity diminishes to 93–95%. Analysis shows [233] that for E-geraniol epoxidation, the efficiency and regioselectivity of polymeric vanadium-containing catalysts and regular catalysts derived from VO(AcAc)$_2$ are close. Nevertheless, the macrocomplexes should be used for production of fragrant compounds (derivatives of E-geraniol and linalool are fragrant compounds) because in this case the full extraction of the residual catalyst from oxidates is more important than activity.

The mechanisms of epoxidation by soluble and immobilized complexes derived from VO(AcAc)$_2$ and poly(vinylbenzoylacetone) are similar in the case of epoxidation of allyl alcohol and geraniol with TBHP as an epoxiding agent [252]. Under the action of soluble and immobilized catalysts products of the same content are formed: up to 98% of 2,3-epoxygeranyl acetate and 2% of its isomer 6,7-epoxygeranyl acetate. In this case the heterogenized catalyst can be regenerated many times.

Ti^{4+} ions on methylated CSDVB, modified by cyclopentadienyl rings, epoxidize cyclohexene (the yield is as much as 88%) and cyclooctene (in yields of up to 37%) (353 K, 18–20 h, TBHP) [253]. Such catalysts can be regenerated in THF by water-free HCl. Ti(O-iso-Pr)$_4$ complexes on polystyrene with chiral groups are active in asymmetric epoxidation of geraniol by TBHP to (S,S)-2,3-epoxygeraniol [254]. The enantiomeric excess reaches 66%. Recycling of the catalyst leads to a small decrease of asymmetric induction (loss of one-third of activity after five regeneration cycles).

There are sparse data on epoxidation of cyclohexene and cyclooctene in the presence of immobilized zirconocene and hafnocene [255].

The use of metals (for instance, Au and Pt) in an ultrafine colloidal state [256] is a new route for the catalytic epoxidation of unsaturated compounds; stable colloids are obtained by reduction of corresponding metal ions in the presence of protective reagents: synthetic polymers, artificial liposomes and others (this subject was considered in detail in Section 1.8).

For olefin oxidation in [34] aqueous H$_2$O$_2$ was used as an oxidant together with cross-linked PS, benzene rings of which were modified at p-positions by residues of arsonic acid, As(O)(OH)$_2$. TiCl$_4$ on cross-linked polystyrenetellurium acid is an effective catalyst for epoxidation (with H$_2$O$_2$ as an oxidant) of a variety of unsaturated compounds: aromatic and non-aromatic hydrocarbons, alcohols and their derivatives [257,258]. For instance, CH conversion reaches 97.2% and selectivity with respect to epoxide reaches 95.6%. This catalyst has no low-molecular mass analogue. Epoxidation can be carried out also by means of immobilized peracids (peroxo acids) [30]; dioxythane [259], iodosobenzoic and other residues on a polymer matrix can also serve as epoxiding agents. For instance, oxidation of stilbene, styrene and chalcone in the presence of Zn^{2+}–biomycin proceeds via formation of a complex of iodosobenzene with the catalyst, subsequent interaction of the complex formed with alkane, and then formation of epoxide and C$_6$H$_6$ elimination [260].

The analysis given above shows that to create effective immobilized catalysts of epoxidation, a number of problems must be solved, the most important of which are the following.

The binding of active ligands should be strong, to prevent the ousting of ligands by reagents. Ligands should be stable to oxidation by HP. The immobilized complex should possess active sites accessible to the substrate. The molecule of HP is activated in a mixed-ligand complex and interacts with olefin epoxides or alcohols, depending on the nature of bond being formed (of π- or σ-type). Therefore, the electronic environment of the catalyst should prevent homolytic decomposition of the hydroperoxide or its one-electron oxidation with formation of radicals, and should facilitate heterolytic cleavage of the O–O bond. Finally, the transfer of one oxygen atom to a double bond of olefin, a main step, should be of synchronous type:

Epoxidation of alkylhydroperoxides in the presence of immobilized complexes, as well as epoxidation in analogous homogeneous systems, are characterized by phenomena of autoacceleration and inhibition. These phenomena draw attention to the processes facilitating or inhibiting the incorporation of reagents into the coordination sphere of catalyst.

Olefin oxides are reactive compounds and undergo, to a lesser extent on immobilized catalysts, various chemical transformations (opening and isomerization of the epoxide ring, oxidation of methylene groups, etc.) with formation of glycols, esters and other oxygen-containing products. Therefore, the content of the reaction mixture becomes more complex with increasing transformation; and as clear from the analysis given above, immobilized catalysts are more suitable for solving the problem of selectivity.

It is plausible that immobilization of two different transition metals on the same polymeric support (within the same reaction center or statistically on the same granule of catalyst) will result in synergistic increases of activity and selectivity[18].

Immobilized complexes are not used for obtaining olefin oxides by means of co-oxidation of unsaturated and organic compounds, for example olefins and RCHO, whereas co-oxidation is widespread in homogeneous catalysis [229]. Probably, it is explained by unsatisfactory knowledge of the oxidations of oxygen-containing substrates in the presence of immobilized macrocomplexes.

2.7 Macromolecular Complexes for the Oxidation of Oxygen-Containing Substrates

Oxidation of oxygen-containing products of partial oxidation of hydrocarbons is a promising method of obtaining synthetic fats, ionogenic SAS, herbicides, drugs, monomers for obtaining polymers, etc. Two problems are most important for production of these compounds. Selectivity is the first. The oxidation rates of alcohols, aldehydes and ketones are, as a rule, higher than those of the initial hydrocarbons. Therefore, it is difficult to stop the oxidation before formation of the thermodynamically stable final products, CO_2 and H_2O, and kinetic control of the transformation degree is very difficult. Extraction of the remaining catalyst from reaction products is the second problem. Immobilization of metal complexes can solve this problem.

---

[18] For example, the synergistic effect was found for conjugated epoxidation and oxidation of cyclohexene in the presence of chromium- and molybdenum-containing complexes [260] or iminodiacetate complexes of Cr^{3+}–V_2O_5 [261] immobilized on silica gel. The synergistic effect is caused by autoacceleration of the process by one product of oxidation (cyclohexene) and/or by formation of mixed-metal complexes.

Mechanisms of liquid-phase oxidation (including homogeneous oxidation in the presence of transition metal compounds) were analyzed in detail in [7,262]. This makes substantially easier the task of comparative analysis of the behavior of free and immobilized complexes in these processes.

2.7.1 Oxidation of aliphatic alcohols

The principal scheme of oxidation of primary and secondary alcohols with formation of carbonyl compounds and H_2O_2 is given below:

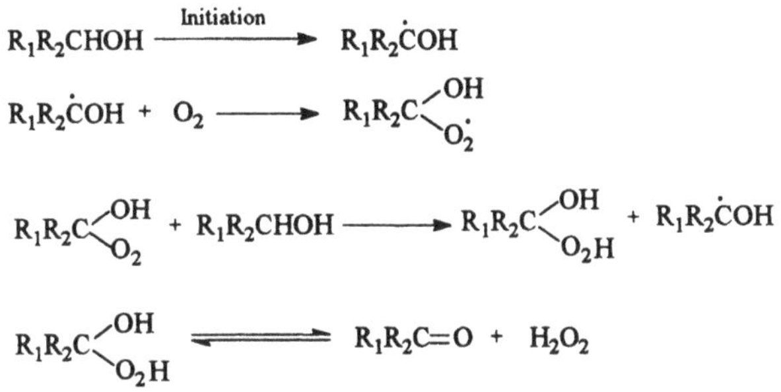

$$R_1R_2CHOH \xrightarrow{\text{Initiation}} R_1R_2\dot{C}OH$$

$$R_1R_2\dot{C}OH + O_2 \longrightarrow R_1R_2C\begin{smallmatrix}OH\\O_2^{\cdot}\end{smallmatrix}$$

$$R_1R_2C\begin{smallmatrix}OH\\O_2\end{smallmatrix} + R_1R_2CHOH \longrightarrow R_1R_2C\begin{smallmatrix}OH\\O_2H\end{smallmatrix} + R_1R_2\dot{C}OH$$

$$R_1R_2C\begin{smallmatrix}OH\\O_2H\end{smallmatrix} \rightleftharpoons R_1R_2C{=}O + H_2O_2$$

Scheme 2.7

Initiation in the presence of Co and Mn salts includes, as a rule, homolytic cleavage of metal alkoxides, a limiting step of the process. For instance, stoichiometric oxidation of *tert*-amyl alcohol in acid media results in formation of acetone according to the following scheme:

$$EtMe_2COH + XCo^{3+} \longrightarrow EtMe_2CO{-}Co^{3+} + HX$$

$$EtMe_2CO{-}Co^{3+} \longrightarrow EtMe_2CO^{\cdot} + Co^{2+}$$

$$EtMe_2CO^{\cdot} \longrightarrow Me_2CO + \dot{E}t$$

Selectivity in cleavage of the C–C bond of the alkoxy radical depends on the stability of the radical produced.

Formation of acids follows a similar scheme with oxidation of Co^{2+} to Co^{3+}:

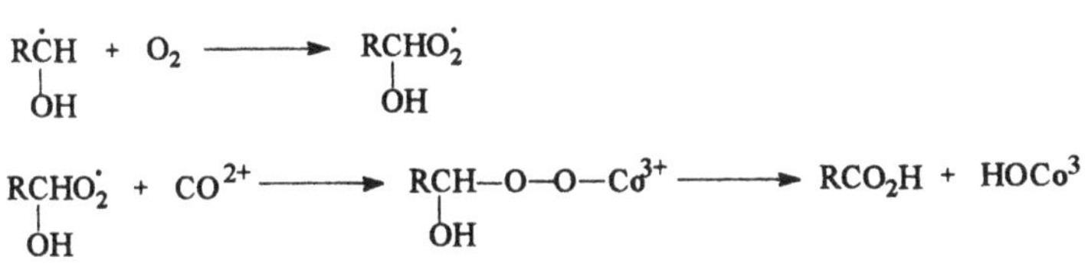

$$R\underset{\underset{OH}{|}}{\dot{C}}H + O_2 \longrightarrow R\underset{\underset{OH}{|}}{CHO_2^{\cdot}}$$

$$R\underset{\underset{OH}{|}}{CHO_2^{\cdot}} + Co^{2+} \longrightarrow R\underset{\underset{OH}{|}}{CH}{-}O{-}O{-}Co^{3+} \longrightarrow RCO_2H + HOCo^3$$

Cobalt complexes are dimerized to inactive compounds, and additiives, for instance PPh_3, are required to prevent this. Accordingly realization of homogeneous catalytic oxidation of such substrates is a difficult task even without taking into account the factors of selectivity[19]. For instance, liquid-phase oxidation of isobutyric aldehyde to isobutyric acid in the presence of $Mn(AcAc)_2$ at 288–348 K and catalyst concentration higher than 10^{-5} mol/l is accompanied by a decrease of selectivity, increase in the yield of CO, CO_2 and propane, i.e. increase in the proportions of the reactions of decarboxylation and decarbonylation [267].

Two systems of oxidation of isopropyl alcohol to acetone and isopropyl acetate (323 K, O_2, glacial acetic acid), in the presence of $Co(OCOCH_3)$ or Co^{2+} immobilized on sulfonated CSDVB, were compared in [268]. In both systems the catalyst initiates decomposition of peroxides to radicals and oxidation occurs only in the presence of peroxide. These facts testify to the chain character of the process. It is interesting that the metal–polymer complex also catalyzes oxidation of isopropanol, the mechanisms of oxidation in liquid and gaseous phase being the same.

Chromium complexes immobilized on a polymeric support are used most often. For instance, chromate ions immobilized on quaternized ammonium resin were used for the oxidation of primary and secondary alcohols [269]; bichromate on polyvinylpyridine is active in the oxidation of various alcohols (1-butane-, 3-pentane-, 1-hexane- and 1-phenylethanol, and benzyl alcohol) to the corresponding aldehydes (hexane, 343 K) [270,271]. The highest oxidative activity of metal–polymer catalysts was found in the case of benzyl, 1-phenyl and cinnamyl alcohols; conversion reaches 99% after 24 h, the activity of polyvinylpyridine bichromate being by 3–4 times higher than that of polyvinylpyridinechlorchromate. It is important that these catalysts are easily regenerated (by washing with NaOH and HCl) without losing activity in subsequent cycles. The same is valid for catalytic oxidation of octanol-2 to octanone-2 (343 K, cyclohexane) in the presence of ionic (solution of CrO_3 in excess of HCl) or neutral (CrO_3) chromium complexes immobilized on a macroporous (r>590 nm) copolymer of styrene with 4-vinylpyridine and DVB or polymer obtained by postgrafting of styrene and 4-vinylpyridine to CSDVB with S_{sp}=640 m^2/g [271].

Complexes immobilized on the macroporous copolymer are the best catalysts (the yield of octanone-2 reaches 37% for ionic and 80% for neutral complexes) due to better accessibility of reaction centers to substrate. Polyvinylpyridine chromate was used in [272] for selective oxidation of 1,4-cyclohexanediol; the process proceeds predominantly with participation of only one OH group. Chromate ions on polybenzimidazole are active catalysts for obtaining acetals and ketals from carbonyl compounds even at 293 K [273].

Immobilized copper–chromium catalysts have undoubtedly an ecological future: for instance, purification of exhaust gases (by oxidation of butyl acetate) was optimized in [274].

There are only scarce data on the activity of immobilized manganese-containing complexes in the oxidation of alcohols. For instance, saturated and unsaturated alcohols are oxidized in the presence of heterogenized $KMnO_4$ [275]; in this case unsaturated alcohols are oxidized without cleavage of the C–C bond the yield of corresponding ketones being 82–100%.

Immobilized metals of the platinum group are effective catalysts for oxidation of oxygen-containing compounds. For instance, polysilazane supported on SiO_2 (a support of mixed type) forms a complex with H_2PtCl_6, with a coordination bond N–Pt [276,277]. The complex is active in the oxidation of ethanol to acetaldehyde and isopropanol to

---

[19] Alcohol oxidations in the presence of soluble complexes of V^{5+} [263] and Ru^{3+} [264,265] are also complex. A kinetic model of liquid-phase radical chain catalytic oxidation of aldehydes by Co complexes heterogenized on rubber support was proposed in [266]. This takes into account the distribution of catalyst both in solution and in the porous support.

acetone (yield of products can be as much as up to 90% at room temperature and atmospheric pressure of O_2) and is very stable. Its activity depends only on the ratio N/Pt and reaches a maximum at N/Pt=4. Electrochemical oxidation of MeOH in a solution of $HClO_4$ on Pt deposited on the surface of solid polymeric electrolyte (SPE), an ion-exchange membrane of "naphthol" type, was considered in [278]. The efficiency of this process is caused by stabilization of Pt ions in different oxidation states on the surface of the SPE.

The limiting step for oxidation of butane by oxygen to butyric acid in the presence of Pt complexes [279] is dehydrogenation of butanol to aldehyde which is oxidized further to acid. As mentioned in Section 2.3.2 (see also [90]), oxidation of propylene proceeds by two steps: first to acrolein, then to final product with high rate and selectivity (up to 91%) in the presence of a radical inhibitor to suppress side reactions. Isopropanol is oxidized relatively easily by O_2 to acetone in the presence of $PdCl_2$ on heteropolyacid [280]. It is interesting that heteropolyacids $H_6[Co^{2+}W_{12}O_{40}]$, $H_4[SiMo_{12}O_{40}]$ and $H_3[PMo_{12}O_4]$ immobilized on P4VP are stable heterogenized catalysts for oxidation of butanediol-1,4 to γ-butyrolactone [281].

Other mixed-metal immobilized catalysts are rarely used for oxidation of oxygen-containing substrates, although homogeneous analogues are used for this purpose. For instance, regioselective oxidation of "internal olefins" carrying neighbor oxygen functions, $RCH–CHCH_2COR^1$ in $RC(O)CH_2CH_2COR^1$, in the presence of $PdCl_2–CuCl_2$ in an H_2O/dioxane mixture was studied in [282].

2.7.2 Oxidation of phenols by dioxygen

Systematic studies of the oxidations of other oxygen-containing substrates have been less comprehensive than those of alcohols. Only those identified as giving the most substantial achievements are considered here.

Dihydroxybenzenes (pyrocatechol, hydroquinone and others) are oxidized in the presence of complexes immobilized on PEI at room temperature in aqueous solution saturated with O_2 at pH 7–10 [283]. The rate of phenol oxidation catalyzed by metal polymers is higher than that of autoxidation, the sequence of activity being Cu>Co>Fe>Mn. The process starts by transfer of two electrons from substrate to oxygen via a system of molecular orbitals of a metal atom with suitable ionization potential and electronegativity. Perhaps, the transfer proceeds most easily in immobilized binuclear Cu^{2+} complexes[20].

Isolation of the catalytically active binuclear Cu^{2+} complex from H_2O by a polymeric support results in the higher activity of metal polymer catalysts. This isolation diminishes also the inhibition by substrate [286]. For instance, in a planar complex of Cu^{2+} with triple block-copolymer styrene-1-vinylimidazole-PEG (a plastoquinone model) ions of Cu^{2+} are surrounded by hydrophobic domains in an aqueous medium [287]. This facilitates the unusual outersphere reactions of electron transfer: reduction of Cu^{2+} to Cu^+ by ascorbic acid and oxidation of Cu^+ to Cu^{2+} by $Fe^{2+}(Phen)_3$. The rate of the first reaction is lower and the rate of the second reaction is higher than that for complexes with low molecular mass analogous ligands.

---

[20] Note for comparison that planar, binuclear Cu^{2+} chelates with low molecular mass ligands are more active than mononuclear complexes in the oxidation of hydroquinone by dioxygen [284], as well as binuclear complexes of Cu^{2+}, Ni^{2+}, Co^{2+}, Mn^{2+} and Ti^{4+} in the oxidation of 3,5-di-tert-butylpyrocatechol to quinone [285].

Catalytic properties depend also on the composition of the supporting copolymer. For instance, a complex of Cu^{2+} with a hydrophilic copolymer of 4-vinylpyridine and N-vinyl-pyrrolidone with the molecular ratio of links 23:77 is most active in oxidation of hydroquinol to p-benzoquinone (298 K, $H_2O/MeOH$, 1:1 v/v): values of $K_{eff} \times 10^5$ s^{-1} are equal to 11.7, 190.6, 4.9 and 627 for complexes of Cu^{2+} with pyridine, P4VP, poly(N-vinylpyrrolidone) and this copolymer, respectively [288]. The value of K_{eff} depends on the molecular ratio of monomer links of the copolymer (Figure 2.21). Probably, main causes of this phenomenon are the change in macrochain conformation due to chain lengthening after introduction of N-vinylpyrrolidone and subsequent formation of less stable and more active Cu^{2+} complexes, and the change in pH of the medium. Oxidation of hydroquinone in the presence of Cu^{2+} complexes with partially quaternized P4VP obeys the Michaelis–Menten law [289]. The viscosity of the system decreases with increasing pH of solution, concentration of electrolyte, temperature and degree of the macroligand quaternization; a decrease of viscosity favors the formation of intramolecular chelates, and subsequent polymer chain contraction facilitates charge transfer between neutral hydroquinone or a semihydroquinone anion and a Cu^{2+} ion in the polymeric complex. This results in an increase of catalytic activity, whereas opposite factors facilitate expansion of polymer chain and inhibit formation of chelate, and, as a result, the catalytic activity decreases.

An interesting method of obtaining immobilized complexes based on the important role of binuclear Cu^{2+} complexes was used in [290,291]. Preformed binuclear complexes

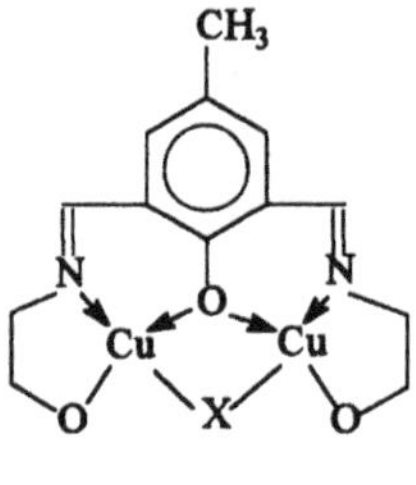

X=Cl, OH

bound by reactive polymer groups are active in the reaction:

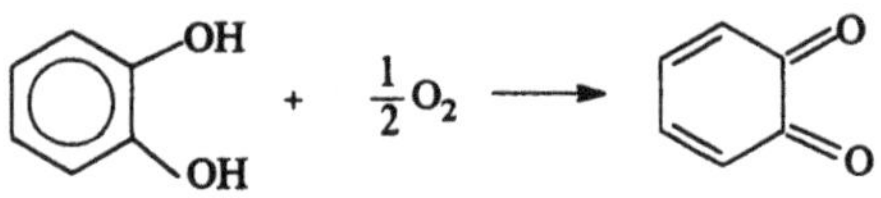

The activities of polymeric catalysts derived from complexes of Mn^{2+} with PEI are substantially higher than the activities of low molecular mass analogues, and approach the activity of pyrocatechase: 5.6×10^2, 4×10^{-3} and 1.6×10^3 cycles/s, respectively [292]. Metal polymer chelates, for instance derived from modified CSDVB, are active and stable catalysts for the oxidation of phenols as well as for alcohols. The rate of oxidation of hydroquinone or 2,6-dehydroxyphenylacetic acid at 323 and 333 K decreases in the sequence of metals $Au^{3+} > Ag^{1+}$, $Cu^{2+} > Hg^{2+}$ or $Cu^{2+} > Au^{3+} > Ag^{1+} > Hg^{2+} > Fe^{3+}$, respectively. Oxidation of 2,6-dehydroxyphenylacetic acid proceeds by the Michaelis–Menten mechanism. Cobalt–porphyrin complexes bound to a polymer are effective catalysts of oxidation of hydroquinone to quinone [293]. Complexes with Schiff bases catalyze an interesting reaction [294]: oxidation of a trans-2 (1-propenyl)-4,5-methylenedioxyphenol to kapron (a type of nylon), at room temperature in CH_2Cl_2 (yield 78–94%). Co^{2+} chelates are the most active; it is supposed that the reaction proceeds with participation of free radicals.

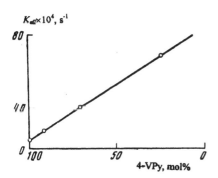

$K_{ef} \times 10^4$, s^{-1}

80

40

0

100   50   0

4-VPy, mol%

Figure 2.21 Correlation of the rate of hydroquinone oxidation with the content of 4VP links in a copolymer of 4VP with N-vinylpirrolidone. Molecular ratio hydroquinone/4VP links/$CuSO_4$=10:10:1, ratio H_2O/CH_3OH = 1·1 (v/v), 298 K

2.7.3 Features of oxidation of sterically hindered phenols

Immobilized catalysts are used most often for the selective oxidation of alkylphenols for the following reasons. The first step of their oxidation is formation of phenoxy radicals. The radicals are then transformed to dimeric and polymeric products linked by C–C and C–O bonds. Two radicals are dimerized with formation of cyclohexanedione (fast oxidation) or dienes and diphenols (slow oxidation, phenolate ions or phenol molecules undergo electrophilic substitution by the radical formed). o,o^1- and p,p^1-bound dimers are formed predominantly if two out of three sites in o- and p-positions are occupied by substituents. This route was used for the synthesis of 2,3,5-trimethylhydroquinone, a half-product of vitamin E:

$$H_3C \quad \overset{OH}{\underset{CH_3}{\bigcirc}} \quad CH_3 \quad + O_2 \quad \xrightarrow{M^{n+}} \quad H_3C \quad \overset{O}{\underset{O}{\bigcirc}} \quad CH_3 \quad CH_3$$

However, realization of this reaction in homogeneous conditions leads to wastage of the catalyst and contamination of the products formed by the catalyst.

Cationic latexes mentioned above do not affect substantially the oxidation of 3,4-dimethoxybenzyl alcohol, 4-hydroxy-3-methoxytoluene and 3,4-dimethoxytoluene catalyzed by cobalt phthalocyanine tetra(sodium sulfate) [295].

A simplified scheme for the oxidation of sterically hindered phenols taking into account only formation of major products is given in Scheme 2.8.

Thus, although selective oxidation of phenols is important for practical applications, this is a complex problem in organic synthesis.

In the presence of metal–polymer catalysts (most often complexes of Cu^{2+} and Co^{2+}) the major reaction products are quinones, diphenoquinones, diphenohydroxyquinones and phenoxyl polymers [21] ([7], p. 368–382). The complex of Cu^{2+} with P4VP was the first polymeric catalyst to be used for the oxidation of sterically hindered phenols [296]. Phenols were oxidized by O_2 in CH_3OH, DMSO or DMFA, then a soluble polymeric macrocomplex was precipitated by ether. Reaction products were the corresponding p-benzoquinones and dioxydiphenyls (Table 2.9).

---

[21] Oxidative polycondensation of phenols in the presence of immobilized catalysts will be considered in Section 3.6.1.

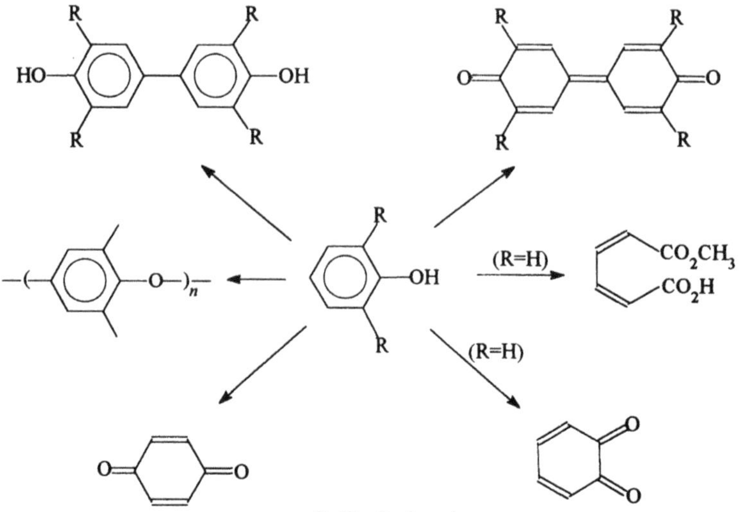

R=H, alkyl, aryl

Scheme 2.8

Table 2.9 Oxidation of substituted phenols in the presence of Cu^{2+}–P4VP [296].

Substrate	Product	Yield (%)
2,4,6-Tri-*tert*-butylphenol	2,6-Di-*tert*-butylbenzoquinone	95
2,4-Di-*tert*-butylphenol	2,2^I-Dioxy-3,3^I,5,5^I-tetra-*tert*-butyldiphenol	70
2,6-Dimethoxyphenol	1,1^I,2,2^I-Tetramethoxy-*n*-benzoquinone	95
2,6-Di-*tert*-butyl-4-methylphenol	2,2^I,5,5^I-Tetra-*tert*-butylbenzoquinone	25

The following scheme shows a mechanism for the oxidative dimerization of dialkylphenols in the presence of immobilized Cu^{2+} complexes [297–299]:

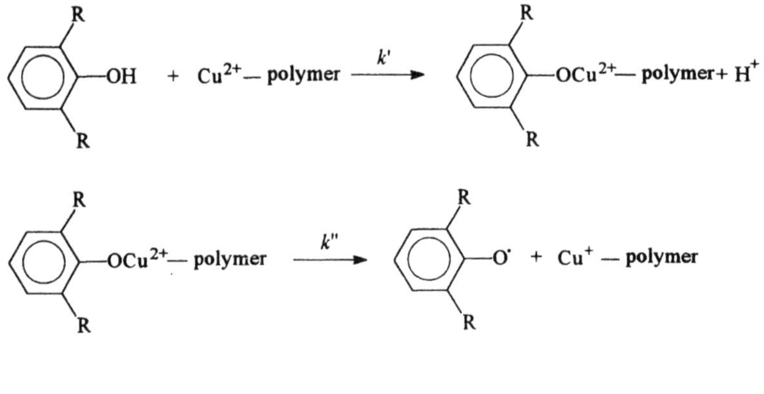

One-electron reduction of the Cu^{2+}–phenolate formed results in the production of a phenoxy radical which is oxidized via the intermediate dihydroxybiphenyl to diphenoquinone:

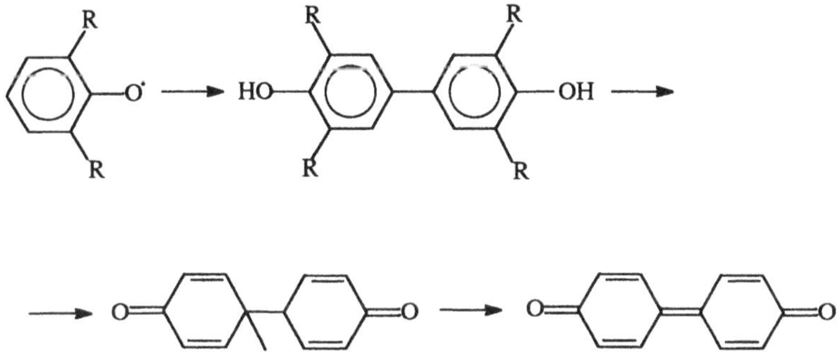

At the same time the active centers are regenerated by oxidation of the Cu^{+1} formed to Cu^{2+} (the "ping-pong" mechanism):

$$Cu^{1+} - polymer + 2H^+ \xrightarrow[k_0]{O_2} Cu^{2+} - polymer + H_2O$$

Oxidation of macrocomplexes has a variety of features in comparison with their low molecular mass analogues. For instance, it was found for complexes of Cu^{1+} with polyvinylazole and imidazole that oxidation of metal polymers is characterized by higher values of enthalpy (29–61.7 and 13.5 kJ/mol, respectively) and lower values of the entropy factor (14.3–92 and 154.6 kJ/(mol deg), respectively) [300,301]. The differences in kinetic and activation parameters are caused by the changing conformation of the polymeric ligand in the course of oxidation of Cu^{1+} to Cu^{2+}:

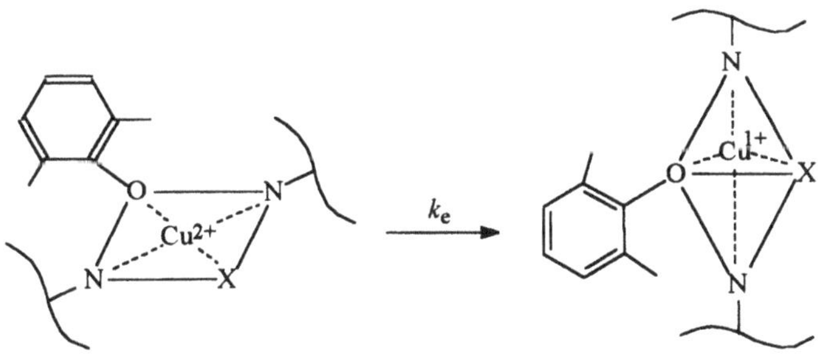

Scheme 2.9

More detailed consideration of these processes is given in Chapter 5. Here we only note the deeply rooted idea that the acceleration of reactions in the presence of metal complexes is caused by increasing rates of redox transformations in hydrophobic regions of a polymer (it seems that this idea was proposed for the first time for oxidation of xylenol by complexes of Cu^{2+} with copolymers of styrene and 4-vinylpyridine [288]) is not always valid.

Thus, the transformation scheme can be represented as follows:

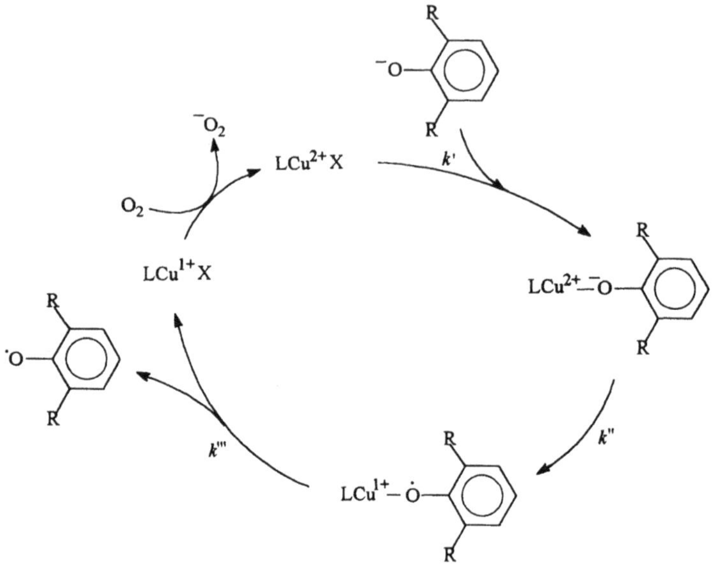

Scheme 2.10

According to [302], active centers in copper-containing systems are bridged binuclear dihydrocomplexes of Cu^{2+}

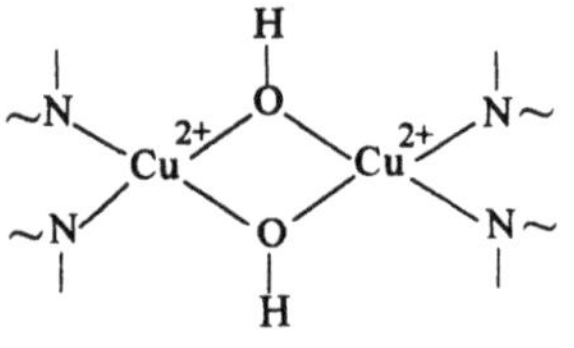

For instance, in the case of the copolymers of styrene and 4-vinylpyridine or dimethylaminomethylstyrene with different contents of nitrogen-containing fragments [N] and a constant ratio [N]/[Cu^{2+}] the reaction rate increases with an increasing proportion of [N] to 0.4. This is explained [303] by the change in activation parameters caused by the increase in conformational mobility of intermediates due to a distortion of an intermediate catalyst–substrate complex before the electron transfer from substrate to Cu^{2+} (the rate limiting step of the reaction). In addition, at low content of [N] there are steric restrictions to the formation of binuclear Cu^{2+} complexes and intermediate complex Cu^{2+}–polymer ligand–substrate, because at least six styrene links are required for the formation of bridged binuclear oxocomplexes Cu^{2+} with copolymer of styrene and N-vinylimidazole [304] (Figure 2.22). These ideas are not trivial because usually macroligands are used to hold metal ions separately.

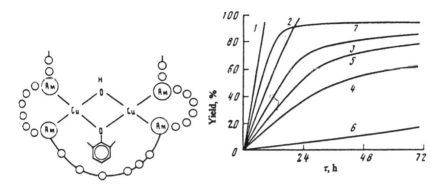

Figure 2.22 The supposed structure of the binuclear Cu^{2+} complex. Circles are styrene links.

Figure 2.23 The dependence of the oxidation rate of 2,6-di-*tert*-butylphenol on the nature of chelates bound to Cu^{2+}. (1) N_2O_2 (a low molecular mass chelate), (2) N_3O_2 (chelate of salcomine type), (3) a product of copolymerization of 2-hydroxy-5-vinylbenzaldehyde with styrene, (4) a copolymer of divinylsalene with styrene, (5–7) ligands obtained by the method of polymer-analogous transformations: (5) PS (microporous)+5-chloromethyl-2-hydroxybenzaldehyde, (6) chlormethylated PS+N_3O_2 , (7) P4VP+N_2O_2. [Phenol]=0.25 mol/ l, 293 K.

Chelate complexes of Cu^{2+} with tetradentate Schiff bases derived from salicylaldehyde and diamines (salcomines) are widely used for homogeneous oxidation of sterically hindered phenols by dioxygen. These are able bind O_2 reversibly (see Section 2.2) and activate oxygen in oxidation reactions with full conversion of phenols to quinones and diphenoquinones [305], the ratio between them depending on the structure of complexes and conditions of the reaction. The main features of homogeneous oxidation of sterically hindered phenols in the presence of the chelates are as follows. Oxygenated complexes of cobalt, $Co^{2+\cdots}O_2$, are plausible intermediates in the process; oxidation proceeds via the formation of phenoxyl radicals whereas the formation of a peroxo bridge Co–O–O–Co inhibits the process.

It was supposed that the immobilization of salcomine (N_3O_2 chelate) on a polymer may prevent the formation of the peroxo bridge because the rigid backbone of the polymer facilitates formation of monomeric oxygenated Co^{2+} complexes. To prove this supposition, a heterogeneous analogue of salcomine with monomeric Co^{2+} was synthesized from chloromethylated CSDVB modified by the multidentate ligand bis(3-aminopropyl)amine [306,307]:

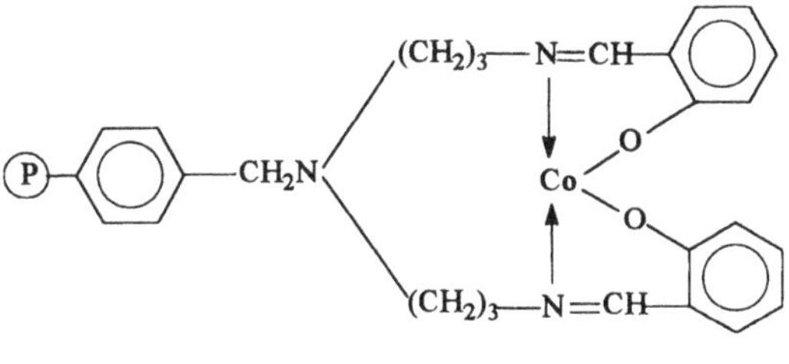

It turned out that the catalytic activity of the heterogenized salcomine (both on macroporous and gel-type polymer) in oxidation of 2,6-dimethylphenol and 2,6-di-*tert*-butylphenol is lower than that of homogeneous salcomine, and depends on the method of preparation (Figure 2.23). However, the heterogenized catalysts can be recycled many times without loss of catalytic activity, and in some cases 2,6-di-*tert*-butyl-1,4-benzoquinone is a major product. Using complexes of Cu^{2+} with various mono-, bi- and multidentate azomethine ligands for oxidation by oxygen and hydroxides, we can suppress unwanted free radical reactions by varying the structure of the coordinated center.

For instance, in the presence of cobalt phthalocyanine tetrasulfonate grafted to CSDVB (H_2O/MeOH, 297–343 K, O_2) 3,5,3',5'-tetra-*tert*-butyl-4,4-diphenoquinone is a single product of oxidation of 2,6-di-*tert*-butylphenol [308]. For the ungrafted catalyst the rate decreases by 10 times and additionally a small quantity of 2,6-di-*tert*-butyl-1,4-benzoquionone is formed [170]. A distortion of the structure of Co^{2+} chelates as a result of supporting on microporous ionite AN-20 with bidentate salicylidene groups leads to the formation of adducts inactive in the oxidation of 2,3,6-trimethylphenol [180,307]. Co^{2+} complexes undergo complex transformations in the course of the oxidation process. For instance, it was found by the stop flow method that Co^{2+} is oxidized to Co^{3+} by dioxygen in the system Co^{2+}–polystyrene sulfonate-2-(2-pyridylazo)-1-naphthol [309]. This system is important for understanding biological processes because the polyelectrolyte is an analogue of the protein constituent of enzymes. The reaction proceeds via formation of an intermediate complex; the higher rate of this reaction in comparison with that for low molecular mass analogues is explained by a stabilization of this intermediate by two mechanisms: electrostatic (the negative charge on polystyrene sulfonate is stabilized by the positive charge of the intermediate complex) and coordination (the oxygen atom is coordinated with Co^{2+}).

An important role of the stereostructure of macroligands was revealed for oxidative coupling of 2,6-xylenol with Co^{2+} chelates immobilized on atactic or isotactic polyacrylamide oximes (PAAO) obtained by the interaction of PAN with NH_2OH in DMFA [310]. For both supports the kinetics of the reactions obey the Michaelis–Menten equation; however, the value of k_2 (the rate constant of the second, rate-limiting reaction) is higher for the isotactic polymer (e.g., at 303 K these values are 0.304 and 0.284 min^{-1}, respectively), whereas the magnitudes of the Michaelis constants are close (0.274 and 0.282 mmol/l). It is supposed that these results are accounted for by the entropy increase: the value of $\Delta S^{\neq}$ is lower for the isotactic macrocomplex (Table 2.10) because its chain has a spiral, ordered structure.

Table 2.10 Thermodynamic and activation parameters for the oxidative coupling of 2,6-xylenol at 300 K [310].

$-\Delta G$ (kJ/mol)	ΔH (kJ/mol)	ΔS (kJ/(mol deg))	$\Delta G^{\neq}$ (kJ/mol)	$\Delta H^{\neq}$ (kJ/mol)	$\Delta S^{\neq}$ (kJ/(mol deg))
Isotactic PAAO–Co^{2+}					
20.6	58.4	260.4	88.2	35.7	−172.2
Atactic PAAO–Co^{2+}					
20.7	45.4	218.4	89.0	40.0	−163.8

In addition to dioxygen, H_2O_2 is an effective oxidant for 2,3,6-trimethylphenol (TMP) in the presence of immobilized Cu^{2+} chelates [311,312]. However, although the activity of these catalysts is high, their selectivity (with respect to 2,3,5-trimethyl-1,4-benzoquinone) is low (Figure 2.24). This may be explained by vigorous oxidative coupling of the substrate, oxidative cleavage of the quinone ring, etc. ESR spectra of the immobilized Cu^{2+} complexes change substantially in the course of the reaction because of gradual substitution of nitrogen atoms in the internal coordination sphere of copper ions by oxygen (compare with cumene oxidation, Section 2.3.4). For various chelating units in the initial macrocomplex, the same copper complexes are formed in the course of the catalytic reaction because of oxidation of the polymer matrix and elimination of part of the chelate cycles.

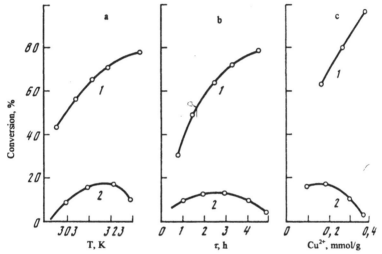

Figure 2.24 Dependences of the conversion of 2,3,6-trimethylphenol (1) and the selectivity of 2,3,5-trimethyl-1,4-benzoquinone formation (2) in the presence of a Cu^{2+} macrochelate with Schiff polybases on temperature (a), reaction time (b) and content of Cu^{2+} in the catalyst (c). (a) Molar ratio TMP/acetone/H_2O_2/Cu^{2+}=1:690:104:0.05, time of reaction 3h; (b) molar ratios are the same as for (a), T=313 K; (c) molar ratio TMP/acetonitrile/H_2O_2=1:350:104, T=323 K, time 3 h, sample weight 0.05 g.

Experimental data show that in the presence of immobilized Cu^{2+} complexes the oxidation of sterically hindered phenols by hydroperoxide, in contrast to the oxidation by dioxygen, proceeds by a radical mechanism.

The immobilization of soluble metal–polymer Cu^{2+} complexes (with P4VP, copolymers of styrene and 4-vinylpyridine and others) on SiO_2 (a support of mixed type) increases the activity of the system substantially (to 70%) in the oxidative coupling of 2,6-di-*tert*-butylphenol [183,313,314]. For both soluble and immobilized metal–polymer complexes the reaction is described by the Michaelis–Menten mechanism, the rate constant of the rate-limiting step being k_2=0.02 s^{-1} for both systems and the Michaelis constant being 0.5 and 1.1 mol/l for the free and supported macrocomplexes, respectively. The relative activity of the catalyst is almost independent of the molecular mass of the polymeric ligand. The more flexible, grafted polymeric complex has the highest activity among the immobilized catalysts; nevertheless, its activity is 6 times lower that of the free linear macrocomplex. The less flexible conformation of adsorbed metal–polymer is characterized by lower activity. Therefore, the metal–polymer immobilization on the surface of the mineral support affects the concentration of the Michaelis–Menten complex (ES) and does not affect the nature of the active centers. In the course of the process the maximum rate decreases by no more than one-half.

For oxidative dimerization of 2,6-dimethoxyphenol it was found [315] that Cu^{2+} complexes with oligomeric ligands are 2–3 times more active than with monomeric or polymeric ligands. The same "oligomeric" effect was found also for dimerization of hydroxyl-containing acetylenes [316] and was accounted for by the following reasons. Oligomers, as more strongly bonding ligands, carry out more effective coordination, have a high accessibility of each Cu^{2+} ions to the substrate, are more labile, and in this case active centers have better hydrophobic protection against admixtures.

2.7.4 Catalytic oxidation of CO

Catalytic oxidation of CO has been studied intensively over the last years, mainly with the purpose of removing CO from exhaust gases. Porous mineral supports (soaked in solutions of metal complexes of the platinum group and sometimes in solutions of Cu, Fe, Ni, Co and Cr compounds) are used most often for this purpose, as well as silicon-containing polymers [317]. Reversible binding of CO with polymeric Cu^{1+} and Pd^{2+} chelates derived from octane tetraontetraoxime was found in [318]. A Pd–Cu system able to oxidize CO was obtained in [319] by sorption of Pd^{2+} and Cu^{2+} ions on polyacrylamide gel swelled in aqueous solution. Unbound hexaaqua complexes of copper regenerate Pd^0 to Pd^{2+} in the course of the catalytic process. At Cu/Pd>100 the system is stable for 2–4 h at 25–55°C and E_a is equal to 27 kJ/mol. Rhodium, Ru and Pt immobilized on fully fluorinated polymeric supports naftion are highly effective in the oxidation of CO. These catalysts (including the membrane-type) are obtained by ion exchange with corresponding salts and by subsequent retention in a flow of H_2 at 473 K [320]. The size of the clusters formed is 2.8, 3.3 and 3.4 nm, respectively, the activity increases in the order Ru<Rh<Pt. Thin films of catalyst immobilized on a support (perfluoroethylenesulfoacid or partially sulfonated PS containing Ag, Cu, Rh, Ru, Pt or Ir cations) oxidize CO, NO, NH_3, N_2H_4, C_2H_5 and other gases at 373–473 K [321]. The process is accompanied by the formation of metal particles of 2.5–4.0 nm. For oxidation of CO to CO_2 in the presence of immobilized Rh, Ru or Pt, the activation energies are 66, 12 and 56 kJ/mol, respectively; probably, this reaction is diffusion-controlled. Non-woven carbon cloth formed from polyacrylonitrile is used to remove technological toxic gases from a flow: the cloth is obtained by oxidation of polyacrylonitrile by air at 573 K, with carbonization, activation and subsequent impregnation in ammonia solutions of Cu, Cr or Ag salts [322]. Transition metal salts supported on LGC are also highly active in the oxidation of CO [77], the catalytic activity of graphite itself being comparable to that of oxides of some metals[22].

---

[22] Rhodium supported on activated coal catalyzes at $T<373$ K formation of 3-pentanone from C_2H_4, CO and H_2 [323].

2.7.5 Polymeric oxidants as reagents

Finally, let us consider the most typical examples of the above-mentioned reactions where polymeric oxidants are used [319].

Phenylselenic acid grafted onto PS is an effective catalyst for the oxidation of olefins to *trans*-diols by aqueous H_2O_2 (yields are 71–98%) and it retains its activity after recycling [33]; the selenic catalyst promotes the process under substantially milder conditions than the arsenic analogue. Periodates and iodates (IO_4 and IO_3) immobilized on a polymeric support are used as oxidizing agents for oxidation of quinols, pyrocatechols, glycols and others [324]. The treatment of I^{3+} covalently bound to CSDVB by sulforyl chloride or peracetic acid results in the formation of dichloro- and diacetoxy-iodo copolymers that were used, with high yield, for oxidation of secondary alcohols to ketones, 1,2-diols to aldehydes and 1,4- and 1,2-benzenediols to quinones [325].

Gelatinous and network CSDVBs containing pyridodipyrimidine groups are active in autorecycling oxidation of cyclopentanol to cyclopentanone (393 K, 25 h) [326]. The activity of the polymeric catalyst is higher than that of the monomeric analogue; the yield of cyclopentanone decreases with increasing degree of cross-linking of the polymer. The content of functional groups on the polymer surface plays an important role in alcohol oxidation, as already mentioned for hydrocarbon oxidation.

For oxidation of ethanol, 2-propanol, 1- and 2-butanol, cyclohexanol and butyric, isobutyric, isovaleric and benzaldehydes by O_2, the catalytic properties of CSDVB containing sulfate and quinone functional groups are comparable to those of Pt, Rh and Ir complexes [327]. In the presence of Fe^{3+} polymers containing nitroxyl free radical groups oxidize benzyl alcohols [328].

Phenylselenic acid grafted onto polystyrene oxidizes alcohols (in the presence of TBHP and dichloromethane) to carbonyl compounds [33]. Ketones are oxidized by aqueous H_2O_2 in the presence of copolymers of styrene with DVB containing AsPh(OR), AsPh(OR)$_2$, AsPh(O)(OH) and As(O)(OH)$_2$ groups, where R is a hydrocarbon residue [329]. The process is characterized by high effectiveness: in the case of oxidation of 2-methylcyclohexanol the yield of lactone is 68% after 6 h, the oxidate being free from oxyacids (see also [11,322]).

Polymeric reagents are also used for oxidation of substrates of other types. Polymeric oxidants are used for oxidation of sulfur-containing compounds (for more details see the next section). For instance, tetrahydrothiophen is oxidized to sulfoxide and sulfone by perbenzoic acid on cross-linked (5–40%) CSDVB in tetrahydrofuran at 298 K [31]. The activity decreases with increasing degree of cross-linking and usually does not depend on whether the peracid is bound to the matrix directly or via an ethylene bridge. Diaryl selenoxide supported on PS oxidizes mildly and selectively thiols, sulfides and thiamides [32]. Oxidations of sulfides (most often n-Bu$_2$S, n-C$_8$H$_{17}$SMe, PhSMe and tetrahydrothiophen) in the presence of polymeric reagents proceed at high rates. For instance, sulfides are oxidized to sulfoxides selectively and with yields of up to 100% on a polymeric reagent obtained electrochemically from cross-linked 4-vinylpyridine hydrobromide [330]. The process is accompanied by quantitative evolution of H_2; this observation points to water molecules as the primary oxidant. The reagent can by effectively regenerated by continuous electrochemical oxidation.

Problems of oxidative stability of macroligands are of special interest. Cross-linked polymers considered in this section as supports are usually stable to oxidation at ambient temperatures. For instance, the oxidation of CSDVB in the presence of combustion catalysts (silver permanganate, mixtures of di- and trivalent cobalt with silver) starts and completes only at 528 K [331], whereas in the absence of such active catalysts the oxidation proceeds only to CO_2 at 535–773 K. This subject is considered in more detail in Section 2.10.

2.8 Oxidation of Sulfur-Containing Compounds in the Presence of Immobilized Metal Complexes

Organic and inorganic sulfur-containing compounds are, together with those containing oxygen, the most extensively studied substrates for oxidation in the presence of immobilized catalysts. Oxidation of sulfur-containing compounds is used not only for production of valuable products, but also for removal of mercaptan from oil cracking products, natural gas and gas condensate, for neutralization of thio salts and waste waters and for resolution of other ecological problems.

In this section catalytic oxidation of thiols, sulfides and thio salts is considered systematically; features of these reactions caused by immobilization of metal complexes on macromolecular supports are also analyzed.

The most investigated process is the liquid-phase autoxidation of thiols by molecular oxygen:

$$2HSH + O_2 \xrightarrow{\text{catalyst}} RSSR + H_2O_2$$

$$2RSH + H_2O_2 \longrightarrow RSSR + H_2O$$

$$\overline{4RSH + O_2 \longrightarrow 2RSSR + H_2O}$$

Scheme 2.11

H_2O_2 is formed in the reaction of RSH with O_2 catalyzed, for instance, by TPP–Co^{2+}, and decomposes in the reaction of RSH with H_2O_2 under the action of a base, for instance a polymeric one. The relative rates of these reactions, in principle controllable, determine the quantity of H_2O_2 accumulated in the reaction system [332]. The catalyst is deactivated mainly due to intoxication of active centers by sulfur-containing oxyacids formed in small quantities in the reaction of RSH with H_2O_2.

The role of base in RSH oxidation is rather complex. On the one hand, bases accelerate transformation of RSH into RSSR via intermediate reactive particles RS⁻, and on the other hand, bases neutralize sulfur-containing oxyacids formed in a side reaction. In the case of a polymeric base, the unstable reactive mononuclear adduct $RS^- - Co^{\delta+} - O - O^{\delta-}$ is formed, the share of the mononuclear adducts depend on the ratio q between Co^{2+} and macroligand links.

The effectiveness of metal polymeric catalysts is due to at least two factors: (i) more effective interaction between oxidation centers and basic centers, (ii) suppression of unwanted dimerization of Co^{2+}–phthalocyanine complexes due to their binding to the polymeric globule. In the homogeneous system TPP–Co^{2+}/NaOH in the presence of O_2 and a substrate the dimerization of TPP–Co^{2+} is unavoidable, and the binuclear dioxygen adducts formed RS–Co–O–O–Co–SR are rather stable and ineffective in this reaction. The main steps of the process are given in Scheme 2.12 [333].

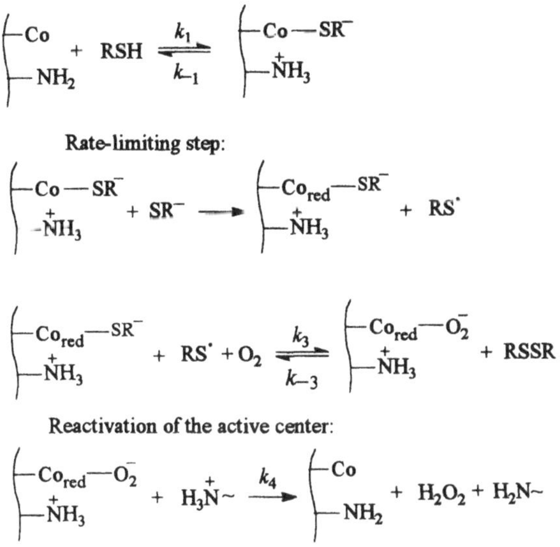

$$\left|\begin{array}{l}-Co \\ -NH_2\end{array}\right. + RSH \underset{k_{-1}}{\overset{k_1}{\rightleftharpoons}} \left|\begin{array}{l}-Co-SR^- \\ -\overset{+}{N}H_3\end{array}\right.$$

Rate-limiting step:

$$\left|\begin{array}{l}-Co-SR^- \\ -\overset{+}{N}H_3\end{array}\right. + SR^- \longrightarrow \left|\begin{array}{l}-Co_{red}-SR^- \\ -\overset{+}{N}H_3\end{array}\right. + RS^{\cdot}$$

$$\left|\begin{array}{l}-Co_{red}-SR^- \\ -\overset{+}{N}H_3\end{array}\right. + RS^{\cdot} + O_2 \underset{k_{-3}}{\overset{k_3}{\rightleftharpoons}} \left|\begin{array}{l}-Co_{red}-O_2^- \\ -\overset{+}{N}H_3\end{array}\right. + RSSR$$

Reactivation of the active center:

$$\left|\begin{array}{l}-Co_{red}-O_2^- \\ -\overset{+}{N}H_3\end{array}\right. + H_3\overset{+}{N}\sim \overset{k_4}{\longrightarrow} \left|\begin{array}{l}-Co \\ -NH_2\end{array}\right. + H_2O_2 + H_2N\sim$$

Scheme 2.12

The scheme includes a cooperative interaction between oxidation centers and basic centers (bifunctional catalysis). It is interesting that the immobilization of TPP–Co^{2+} and TPP–Fe^{3+} on the same polymeric matrix polyvinylamine (PVAm) does not increase its effectiveness in the reaction because here the degree of interaction between oxidation centers and basic centers does not increase.

The catalytic properties of TPP–Co^{2+} immobilized on PVAm are determined by a number of parameters: pH and ionic strength of the medium, temperature and molecular mass of the macroligand [332–336]. The latter parameter is most important because it determines the local density of active centers in the macromolecular globule; the influence of other parameters is indirect, via concentration of thiol anions in the vicinity of the active centers, as well as through the effect of electrostatic interactions on the chemical microsurroundings in the domains of the polymeric globule.

For instance, the share of separated centers[23] $\chi_{Co,1}$ depends on the polymerization degree (PD) of PVAm and the ratio $q=$[TPP–Co^{2+}]/[NH_2] (Table 2.11). Indeed, in samples with lower molecular mass the proportion of isolated Co^{2+} complexes is smaller. Binuclear complexes cannot be formed when each polymeric globule (the globule consists of a single polymer molecule because of very high dilution, <0.05% with respect to polymer mass) contains only one Co^{2+} ion, i.e. these conditions model infinite dilution.

The value of K_m and the activation parameters also depend on the PD of PVAm because of various conformational transformations of the macroligand globule. The

---

[23] A statistical analysis of the distribution of TPP–Co^{2+} on PVAm chains with different polymerization degrees (PD) shows [334] that the molar share of chains with PD$=i$ containing k molecules of Co^{2+} on each chain is given by the formula $\chi_{Co,\,k}=k(1-p)^2(pq)^{k-1}/[(1-p+pq)^{k+1}]$ where $q=$[TPP–Co^{2+}]/[NH_2].

charge density of polymer chains depends on the basicity of amino groups of the polymeric support. We can control the activity of metal polymers by varying the basicity. For instance, for TPP–Co^{2+} immobilized on quaternized 2,4-ionen (a polyammonium resin of a condensational type) with M_n 1740 and 6 600 K_m=11.3 and 6.8 mol/l, turnover number (TN)=3 700 and 3 300 s^{-1}, respectively, whereas for PVAm with M_n=3×10^4 and K_m=90 mol/l and TN=2800 s^{-1} [337]. However, at low ionen concentrations cobalt centers aggregate with formation of complexes with molar ratio [N$^+$]/[Co]=4:1 [338].

Table 2.11 Calculated values of $\chi_{Co,1}$ and experimental values of spesific catalytic activity (SCA) for oxidation of 2-mercaptoethanol by molecular oxygen in the presence of Co^{2+} phthalocyanine tetrasulfonate Na immobilized on PVAm of different PD[334].

$q\times10^2$	PD=50		PD=1680	
	$\chi_{Co,1}$ (%)	$w\times10^{-2}$ (ml of O$_2$/ /μmol of Co min)	$\chi_{Co,1}$ (%)	$w\times10^{-2}$ (ml of O$_2$/ /μmol of Co min)
1	45.04	5.8	0.32	2.0
0.2	82.95	6.6	5.27	3.6
0.1	90.88	6.6	13.93	3.9
0.02	98.07	8.4	56.04	6.8
0.01	99.03	9.2	73.31	8.4

Table 2.12 Comparative analysis of catalytic activity of polymeric and monomeric forms of Co^{2+} phthalocyanines [333].

Polymer	$K\times10^4$ (mol/l s)	Monomeric analogue	$K'\times10^4$ (mol/l s)	K/K'
PBAm	5.9	1,3-Propylenediamine	0.25	23
Poly(L-lysine)	3.5	L-lysine	0.27	13
PEI	1.9	Dimethylamine	0.17	11
2,4-Ionen	6.7	Tetramethylammonium hydroxide	0.17	40

For oxidation of 2-mercaptoethanol by molecular oxygen the activity of the catalyst with a polymeric ligand is 11–40 times higher than that of the monomeric analogue (Table 2.12). The activity of immobilized TPP–Co^{2+} complexes in oxidation of 2-mercaptoethanol depends on the nature of the support and decreases in the order: cationic polyelectrolytes>>cationic polymer colloids>cationic colloidal silica≈aqueous solution>ion exchange resins [339].

Addition of p-benzquinone inhibits the process; this observation proves radical steps in Scheme 2.12.

Oxidation of thiols by immobilized complexes kinetically resembles that by vitamin B_{12}[24] : in the presence of O_2 thiol anions transform into disulfides, whereas oxygen binds to the catalyst [340].

Heterogeneous immobilized catalysts are used for autoxidation of thiols, for instance TPP–Co^{2+} supported on cross-linked PVAm or grafted to the surface of CSDVB (obtained by radical graft polymerization of N-vinyl $tert$-butylcarbomate, hydrolysis of grafted chains, and interaction with Co^{2+} 4,4',4'',4'''-tetracarboxyphthalocyanine [334]).

Cationic latexes (copolymers of styrene, DVB and ionogenic monomers) with immobilized Co^{2+}–TPP are effective catalysts for the autoxidation of 1-decanethiols [343]. A comparative analysis of the effects of mononuclear and binuclear supported complexes of cobalt on autoxidation of thiols, thioglycols and thiolactic acid was carried out in [344,345]. Mononuclear TPP–Co^2 and binuclear Co^{2+}–dodecarboxyphthalo-cyanine were supported on ion-exchange resins; these chelates aggregate in aqueous solutions, but their immobilization (addition of DMFA, high pH of the medium) leads to the formation of monomeric species active in water and toluene at 298 K. In the course of oxidation of thio compounds the divalent cobalt chelates transform into monovalent, the oxidized form of cobalt being easily regenerated by air or by an aqueous solution of Fe^{3+}. Chelate complexes of Co^{2+}–phthalocyanine supported on porous CSDVB containing amino groups were used in [346] for demercaptation of acid oil products by oxidation by air in the presence of mobile or immobile catalyst layers.

In the case of n-butylmercaptane oxidation, catalytic activity of metal complexes immobilized on anionite AN-221 is determined by the nature of the transition metal and decreases in the order $Cu^{2+}>Fe^{2+}>Mn^{2+}>Co^{2+}>Cr^{3+}>Pd^{2+}$ (Figure 2.25) [168,347]. Dependences of SCA on concentration of Cu^{2+} in an ionite matrix often have a maximum (Figure 2.25b). The maximum can be caused by a number of factors, for instance by a changing content of coordination centers formed in the ionite matrix at a changing concentration of metal ions. The maximum is observed at low concentrations of copper ions; at these concentrations a maximum number of polymer ligands participate in copper coordination[25].

Catalysts on supports of mixed type seem to be promising for use in the oxidation of thiols. Note in this connection that the nature of the support affects substantially aerobic oxidation of ethylmercaptane at 293 K by immobilized cobalt–phthalocyanine [350]. The catalysts were obtained by deposition on SiO_2 or activated coal, impregnation of ZnO, or chemical immobilization with participation of $tert$-amino groups of amberlyst-A21; the sequence of SCA of these catalysts is amberlyst-A21>activated coal>

---

[24] Coenzyme PQQ accelerates by 30–50 times anaerobic oxidation of thiophenol, benzthiophenol, propyl- and $tert$-butylthiol to the corresponding disulfides (aqueous solution, Na_2CO_3, pH 6.6–7.4) [340]. Under anaerobic conditions $PQQH_2$ is rapidly formed. Vitamin B_{12} accelerates by 10^2–10^5 times anaerobic oxidation of thiophenol, 2-mercaptoethanol, butane-1,4-dithiol by flavin in the presence of polymer micelles with quaternized poly(1-vinyl-2-ethylimidazole) [341]. The effect of vitamin B_{12} is explained by the binding of thiol ions to the polymer micelle with subsequent formation of hydrophobic ionic pairs. Oxidation of 2-mercaptoethanol by Co and Fe phthalocyanines encapsulated into a polypeptide kinetically resembles catalysis by enzymes [342].

[25] The high activity of copper-containing aerosils (silica powder) in the oxidation of sulfur-containing compounds by dioxygen is caused by isolated copper ions [348]. Spin–spin interactions are negligibly small at distances between ions $r>1.5$ nm, and the share of these isolated ions increases with increasing surface concentration of Cu^{2+} ions [349].

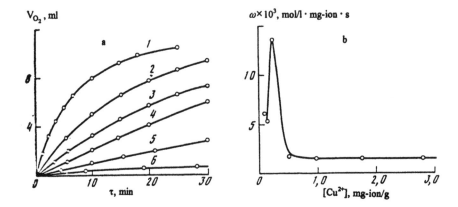

Figure 2.25 Kinetics of liquid-phase oxidation of n-butylmercaptane (a) and dependence of SCA on concentration of Cu^{2+} in ionite matrix (b). (a). 1, Cu^{2+}, 2, Fe^{2+}, 3, Mn^{2+}, 4, Co^{2+}, 5, Cr^{3+}, 6, Pd^{2+}. Solvent is o-xylol+3% NaOH, catalyst is anionite AN-22 containing 0.5 mg-ion of metal per g.

$>ZnO>SiO_2$, the values of SCA correlating with electron-donor properties of the support.

Data on oxidation of sulfoxides by immobilized complexes are scant. Mo^{5+} ions supported on chloromethylated CSDVB modified by dithiocarbomate groups, oxidize Me_2SO to Me_2SO_2 in the presence of a small quantity of TBHP[244], whereas in a homogeneous catalytic system $Mo_2O_3(S_2CNEt_2)_4$ TBHP decomposes rapidly due to side reactions. Note that water-soluble macrocomplexes $MoO_2(cis$-O-PEG$)_2$ (a model of molibdooxidase, hydrophobic surroundings of active center protects Mo^{6+}/Mo^{5+} against water) carry out oxidation of PPh_3 to Ph_3PO by dioxygen under mild conditions.

Oxidation of inorganic thio salts by dioxygen in the presence of complexes of transition metals immobilized on polymers is especially important for solving ecological problems. The most extensively investigated system of this kind is Cu^{2+} complexes immobilized on cross-linked (by means of 1,2-dibromethane) P4VP with various degrees of quaternization [351]. The increase of the quaternization degree leads to a substantial increase of the rate of oxidation of the thiosulfate ion SO_3^{2-} to tetra- and trithionates $S_4O_6^{2-}$ and $S_3O_6^{2-}$. The maximum rate is at pH 2.16 and the process obeys the Langmuir–Hinshelwood model which takes into account the coordination of O_2 and thio anion on Cu^{2+}. Complexes of Cu^{2+} with P4VP (relative molecular mass 3.4×10^5, low degree of cross-linking) catalyze oxidation of all three thio salts to sulfate. Among the polythionate ions the lowest initial oxidation rates were found for $S_4O_6^{2-}$, $S_3O_6^{2-}$ and $S_2O_3^{2-}$, whereas $S_2O_3^{2-}$ and SO_3^{2-} are oxidized substantially more rapidly. The rate limiting step of these oxidative processes is the delivery of O_2. It is interesting that $S_2O_6^{2-}$ is not oxidized.

Between 298 and 313 K the effective activation energies for oxidations of $S_4O_6^{2-}$, $S_3O_6^{2-}$ and $S_2O_3^{2-}$ are 44.1, 62.6 and 48.7 kJ/mol, respectively. The coefficient of O_2 transfer from liquid to catalyst surface calculated from kinetic data is equal to 8.7×10^{-4} m^2/s in all cases. H_2SO_4 formed as a result of oxidation of polythionate ions at pH 2 diminishes sharply the catalyst effectiveness because of destruction of the active complex.

The catalysts consisting of complexes of Cu^{2+} with P4VP are highly active and stable in oxidation of thio salts in waste water[26] if the concentration of the complexes does not exceed 1000 ppm [353]. The process can be carried out continuously in a reactor or in a system of two reactors at 293–318 K.

Macrocomplexes of Cu^{2+} with polythiosemicarbazide (obtained by condensation of N,N'-diaminopyridine with methylene-bis(4-phenylisocyanate), and $Cu(CH_3COO)_2 \cdot H_2O$, oxidize thiosulfate ions by means of dioxygen by a three-step mechanism: first of all $S_2O_3^{2-}$ to $S_4O_6^{2-}$, then $S_2O_3^{2-}$ to SO_4^{2-} and finally $S_4O_6^{2-}$ to SO_4^{2-} [354]. Thus, by varying the functional groups of a macroligand in the Cu^{2+} complex, we can change the routes of thiosalt oxidation:

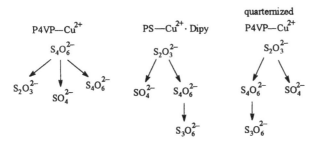

Scheme 2.13

Among other metals, complexes derived from $Cr_2O_4^{2-}$ are active in the oxidation of thiosalts [355]; a number of "non-metal" polymers based on viologen, quinones and others are able to oxidize mainly $S_2O_4^{2-}$.

Oxidation of sulfide ions (H_2S, Na_2S) by dioxygen in aqueous solutions in the presence of metal polymer catalysts is applied widely (perfluorinated polymer films, polyacrylamide hydrogel, polymer fibers, various ionites are used as supports) [356–358].

The rate of oxidation of Na_2S by dioxygen depends on the nature of the transition metal ions included in hydrogels and increases in the order $Fe^{3+} < Cu^{2+} < Ni^{2+} < Co^{2+}$ (Figure 2.26). It has been shown that Fe^{3+} or Cu^{2+} complexes of different content, mono and binuclear, exist in the hydrogel matrix. For instance, at the start of the process Cu^{2+} ions are distributed randomly in perfluorinated sulfocationite bilayer films of thickness 20–450 μm [359] (these polymeric membrane catalysts act as catalysts as considered in Section 1.9.3). After completion of the catalytic process all the copper is localized as sulfide complexes in a thin (5–10 μm) surface layer. A correlation was found between microwave conductivity and catalytic properties of binuclear and polynuclear copper complexes:

[26] Copper ions on polymeric supports are also promising catalysts for detoxication of cyanine-containing waste waters by means of H_2O_2 [352].

formed in carboxyl-containing fibers [360,361]. In this way extended electron-transferring chains are formed; these chains facilitate the electron transfer from sulfide ions to oxygen:

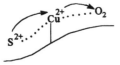

Synthetic Fe_2S_2 clusters, the structure of which resembles the structure of ferredoxin active centers, can be considered as convenient models for electron carriers in the respiration chain from the viewpoint of content of active centers, polymeric structure and enzyme function.

Figure 2.26 Effect of the nature of the transition metal in the hydrogel matrix on the rate of oxidation of Na_2S by dioxygen. 313 K, $[M^{a+}]=$ 1.8×10^{-6} g-ions, $[Na_2S]=1$ mmol/l; 1, Fe^{3+}, 2, Cu^{2+}, 3, Ni^{2+}, 4, Co^{2+}.

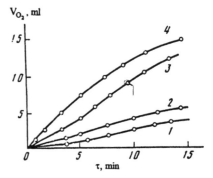

The same channels of electron transfer were found also for a bimetal[27] $Cr^{3+}-Cu^2$-containing catalyst for sulfate oxidation supported on a carboxyl-containing cationite, at a certain ratio of Cr^{3+}/Cu^2 [363].

The nature of the binding ligands is also important. For instance, the catalytic activity of ionite complexes in this reaction can be varied by 20 times by varying the nature of anionite functional groups. In this way we can also vary selectivity for H_2S oxidation, obtaining only elementary sulfur or only SO_4^{2-} anions [364].

Relatively simple and convenient methods for obtaining immobilized catalysts for oxidation of sulfites are known. The catalysts are obtained by mixing a metal complex with a solution or melt of a polymer [365]. For instance, metal phthalocyanines are mixed with PE or PP, the mixture is heated with stirring to the melting temperature of polymer, and then the catalytic mass is granulated or formed. Cobalt phthalocyanine (1% mass in benzene) is mixed with PE (99% mass), then the mixture is formed by removal of solvent. To increase the catalytic activity of the thermoplastic polymer, PE is grafted by 5–10% mass of acrylic acid, and 3–5 mg/g of $Co(OCOCH_3) \cdot 2H_2O$ is added. A catalyst for oxidation of sulfites may contain as a main component active silt (30–50% mass) supported on PE, PP or PS.

---

[27] Homogeneous synergistic bimetal systems are known. For instance, Cu^{2+} and Fe^{3+} chlorides formed in aqueous acid solution bimetal complexes with ratio Fe/Cu=1:1 the activity of which in oxidation of SO_2 to H_2SO_4 is higher than the sum of the activities of individual components [362].

Immobilization of Cu^{2+} or Pt^{2+} complexes on TiO_2 or Al_2O_3 impregnated with a teflon suspension results in a hydrophobic catalyst of mixed type [366]. The catalyst oxidizes effectively NaSH to $S_2O_3^{2-}$ by O_2 in aqueous solution. Due to a hydrophobic layer, reactions proceed on the surface between three phases: solid catalyst reacts with gaseous component and substrate in solution. This topographic feature is typical of catalysis by immobilized complexes.

Complexes supported on polymers may also serve as effective heterogeneous catalysts for reactions with participation of sulfur-containing compounds, for instance, reaction of SO_2 or 3-sulfolene with butadiene. This reaction leads to valuable monomers, cis, trans-2,5-divinylsulfolanes, and is catalyzed by Pd(AcAc)$_2$ supported on a copolymer of butylvinyl sulfoxide with vinylpyridine cross-linked by vinylbenzene [367].

2.9 Asymmetric Induction in Oxidation Reactions in the Presence of Immobilized Complexes

There are three different methods of synthesis of optically active compounds from optically inactive racemic mixtures: spontaneous, biochemical and chemical, the chemical method being the most common. Immobilized metal complexes are the best models of asymmetric induction by enzymes: they produce large quantities of enantiomeric products from small quantities of chiral compounds (this is already demonstrated for hydrogenation transformations in Section 1.10).

Two types of catalysts can be used for asymmetric induction: (i) catalysts supported on macroligands with optically active groups, (ii) metal complexes immobilized on optically inactive polymers which have their own optically active ligands. The first type of catalyst is more widespread. Products of polymerization and copolymerization of optically active monomers, or products of polymer-analogous transformations with participation of optically active reagents, are used as polymeric supports. The structure of macrochains is important for these processes (see below), therefore macrocomplexes (mainly with optically active polyamino acids) soluble in the reaction medium are used most often [368].

Complexes of Cu^{2+} with poly(L-histidine), Cu^{2+}-pLH, are more active in oxidation of ascorbic [28] or 2,5-dehydroxyphenylacetic acid at pH 4.3–5.0 than solutions of the low-molecular mass copper analogue [369]. The higher activity of Cu^{2+}-pLH is caused not by the structure of the macrocomplex, but by its polyelectrolytic nature: the concentration of reagents in the polyelectrolyte is higher than in solution. The active center of the Cu^{2+} complex at pH 5 is distorted square planar in which the metal ion is coordinated by three nitrogen atoms of imidazole rings of pLH and by at least one deprotonated peptide nitrogen atom. At pH 14, however, four peptide nitrogen atoms distort the squared coordination, and one imidazole group is bound to the axial site. The oxidation kinetics obey the Michaelis–Menten equation [370], and the "ping-pong" mechanism is realized according to which copper is reduced in the reaction with substrate and substrate is deoxidized by dioxygen, as in the oxidation of phenols (see Section 2.7).

---

[28] Ascorbate oxidase is a copper-containing enzyme catalyzing aerobic oxidation of vitamin C. Its polypeptide chain has a relative molecular mass of about 140 000 and contains 8–10 Cu^{2+} ions [312].

Oxidation of 3,4-dehydroxyphenylalanine by dioxygen in the presence of macrocomplex of a Cu^{2+} with poly(L-lysine) (Cu^{2+}–pLL) proceeds according to the scheme:

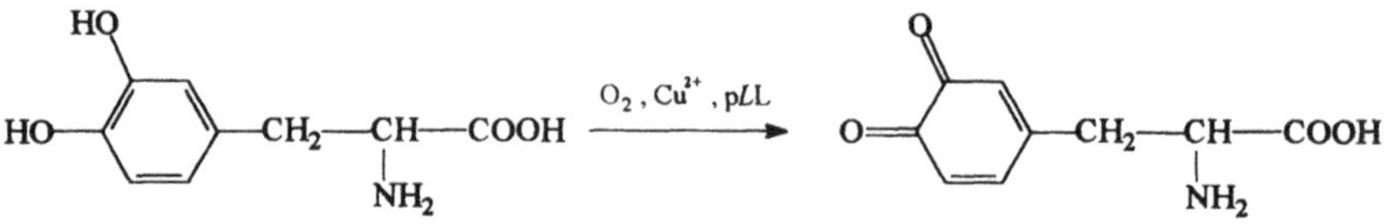

At pH 6.9 and 10.5 the rate of the reaction is higher than in the presence of Cu^{2+} with ethylenediamide or butylamine [371,372]. However, only at pH 10.5 does this macrocomplex act as a selective asymmetric catalyst, because at pH 10.5 the macrocomplex is soluble, whereas complexes with low molecular mass ligands are converted to insoluble $Cu(OH)_2$. At pH 6.9 electrostatic factors play an important role.

The asymmetric activity of the macrocomplex is determined by its spatial structure which varies with varying pH. At pH 8.5 copper ion coordinates with four nitrogen atoms of side chain amino groups, whereas at pH 10.5 copper ions coordinate peptide nitrogen atoms with substitution of a proton. In addition, at pH 10.5 the backbone chain of the polymeric ligand has a spiral conformation, the asymmetric induction on oxidation of 3,4-dihydroxyphenylalanine being proportional to the content of protein α-helices. For oxidation of D- or L-3,4-dehydroxyphenylalanine at pH 10.5 and 6.9 the activity (relative to the activity of a Cu^{2+} complex with a low-molecular mass analogue) and the stereospecific index (ratio between the initial rate constants of pseudo-first order reactions of oxidation of D and L-isomers, K_D/K_L) are 7 and 2.9 and 1.53 and 1.00, respectively.

However, to understand asymmetric induction we should take into account, in addition to the spatial structure of macroligands, the distance between supported Cu^{2+} complexes along the backbone chain. The active center of ascorbate oxidase consists of two identical subunits with relative molecular mass 30 000–40 000, each subunit consisting of two polypeptide chains [373]. It is plausible that at least two supported Cu^{2+} ions should also cooperate in the synthetic substrate-containing complex to ensure stereoselectivity. Due to bifunctional properties of 3,4-dihydroxyphenylalanine, this is coordinated in a transition complex with Cu^{2+} ions of the supported catalyst almost simultaneously both by its own amino acids and by the catechol groups [371] (Figure 2.27). The appearance of the stereoselectivity shows that these bifunctional properties are more important for binding of the D-isomer than its enantiomer.

Thus, the influence of chain conformation on asymmetric induction can be caused, on the one hand, by formation of a proper chain structure and, on the other hand, by differences in affinity of substrate enantiomers or the stability of intermediates[29].

These ideas were developed further in a study of the kinetics and mechanism of stereoselective peroxidase-type activity of immobilized complexes in oxidation of

---

[29] The same factors were found to be important in enzymatic catalysis. In mammal organisms the metabolism of S-nicotin (tobacco alkaloid) to S-cotinine lactame includes two-electron oxidation near the chiral C5 atom catalyzed by cytochrome P-450 [374]. The structure of the complex of S-nicotin with an active site of the cytochrome determines the stereochemical pathway of the reaction.

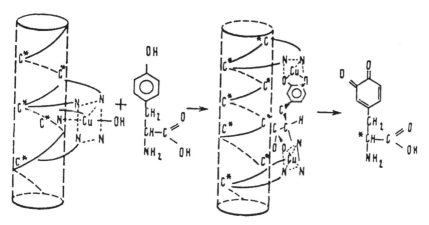

Figure 2.27   Mechanism of oxidation and bifunctional coordination of 3,4-dehydroxy-phenylalanine on Cu^{2+}–pLL (see text).

L-adrenalin or L-ascorbic acid [375–377]. Ions of a heme-like Fe^{3+} complex, *trans*-Fe(2,2',2",2'''-tetrapyridyl)-(OH)$_2^+$ supported on poly(L-glutamate) (FeL), or on poly(D-glutamate) (FeD), are active catalysts at pH 7. The peroxidase-type activity of FeD and FeL is higher than that of Cu^{2+}-pLH. The reaction proceeds simultaneously by catalytic and non-catalytic (non-catalytic electron transfer between ascorbate ion and H_2O_2) mechanisms. The reaction includes three steps: formation of an intermediate substrate-catalyst complex, intramolecular electron transfer and oxidation of both the low-valent metal ion and the substrate radical by H_2O_2. The step including the attack of the central metal ion is responsible for the stereoselectivity which arises only on participation of the polypeptide support in formation of the intermediate complex. The conformational asymmetry of the polypeptide matrix substantially affects the catalytic properties and selectivity of enantiomeric macrosystems FeD and FeL (Table 2.13).

Table 2.13   Characteristics of stereoselective oxidation of L-ascorbic acid by H_2O_2 in the presence of FeL and FeD (pH 7, 289.9 K)

f	α	K_{FeD}	K_{FeL}	K_{FeD}/K_{FeL}	$G_{LL}^{\neq}$ $G_{DL}^{\neq}$	E^a_{FeD}	E^a_{FeL}
		(l mol/s)			(J/mol)	(kJ/mol)	
0.01	0.05	3649	3665	1.0	0	13.0	13.0
0.04	0.18	485.5	317.7	1.5	962	–	–
0.06	0.30	409.3	208.7	2.0	1680	21.0	22.7
0.10	0.47	442.3	173.0	2.6	2430	52.5	47.9
0.20	0.68	414	106.4	4.0	3430	68.5	71.4

Values of E_a have been measured at 289–299 K.

The proportion of α-helices in the polypeptide chain depends on the degree f of its "loading" by metal complexes (the molar ratio between bound Fe^{3+} and macrochain links). The stereoselectivity factor K_{FeD}/K_{FeL} is equal to 1 at $f < 0.02$ (i.e. catalysis is non-stereospecific): the "renaturation" caused by the appearance of conformational asymmetry in bound molecules arises and increases with increasing f in the range f=0.02–0.20, i.e. with increasing proportion of α-helices in the polypeptide backbone.

Thus, the stereoselectivity of oxidation is determined by an entropy factor and especially by conformational rigidity of the initial Michaelis complex formed on the interaction of optically active ascorbate anion with chiral fragments of the polymeric catalyst. The conformational asymmetry of macroligands causes different steric strains in the LL- and DL-adducts, and these strains determine mutual orientation of centers and distances between the centers.

Changes in free energy of diastereomeric transition states

$$K_{DL}/K_{LL}= \exp[(G_{LL}^{*}-G_{DL}^{*})/RT]$$

reflect differences in steric factors, steric strains being less in diastereomer DL than in stereomer LL (Table 2.13). This results in more effective electron transfer due to more favorable activation entropy. At high degrees of ordering of spatial structure the polypeptide chain of catalyst the enantiomeric excess reaches 65% [378].

A model of a diastereomeric non-covalent complex with a mobile electron which takes into account different "freezing" of internal rotations[30] of the intermediate around the substrate-catalyst bond for different diastereomers was proposed on the basis of calculations of conformation energy [380].

Besides poly(aminoacids), other macromolecules, most often P4VP, are used for asymmetric oxidation. The structure of partially quaternized P4VP, the nature of alkylating agents, and conformation of macromolecules affect substantially the binding of Cu^{2+} and the formation of catalytic centers for oxidation of ascorbic acid by dioxygen [381, 382]. The activity of immobilized complexes is 200–1500 times higher than that of low molecular mass analogues and 300 times lower than the activity of ascorbate oxidase. The reaction proceeds via the formation of the intermediate complex P4VP–Cu^{2+}–ascorbate ($K\approx10^4$ l/mol). At pH 4.5 this step is rate-limiting, followed by intramolecular decay of the triple complex ($k\sim0.4$ s^{-1}). It is important that the interaction of P4VP with Cu^{2+} is not accompanied by formation of peroxide ions O_2^{2-} and their escape into the bulk of the liquid. A cyclic mechanism (of the "ping-pong" type) includes oxidation of Cu^{1+} ions; at pH 3.5 two-thirds of the total quantity of copper ions exist as polynuclear complexes.

Cu^{2+} macrochelates exhibit high catalytic activity in the oxidation of L-ascorbic acid, the activity being determined by the nature of the chelating unit [218] (Figure 2.28). It is interesting that the activity of the system Cu^{2+}–polyaminoorganosiloxane in the oxidation of ascorbates or hydroquinones by dioxygen at 298 K is close to the activity of Cu^{2+}-pLH [383]. In contrast to catalysis by complexes of Cu^{2+} with low molecular mass ligands, reactions involving catalysis by metal polymers are of zero order with respect to substrate at low concentrations. The influence of the polymer chain is also explained by a change in conformation in aqueous solution; the conformation depends on the mobility and hydrophobicity of the backbone chain and is controlled easily by the pH of the medium and the quantity of coordinated Cu^{2+} ions ("loading" of the chain).

---

[30] Probably, the nature of the solvent also affects these processes, as shown for the oxidation of ascorbic acid by Cu^{2+}, Ni^{2+} and Co^{2+} complexes [379].

Figure 2.28 Effect of the nature of the chelating unit on the rate of oxidation of *L*-ascorbic acid. (1) chelate 1, first use; (2) chelate 1, second use, (3) chelate 2, (4) chelate 4, (5) copper benzoate, (6) chelate 3, (7) without catalyst. The structures of chelates are given in Figure 2.17. $[Cu]=1\times10^{-4}$ mol, 290 K.

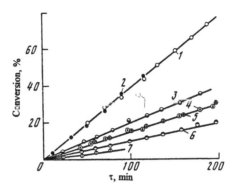

The complex latex–histamine–Cu^{2+} is catalytically active and stable over long periods in the oxidation of ascorbic acid at 288–309 K [384].

Monosugars, for instance glucose, are oxidized by dioxygen in the presence of a supported palladium catalyst [385]. *D*(+)-xylose and *D*(+)-glucose are oxidized by dioxygen to *D*-xylic and *D*-gluconic acid with a yield of up to 90% in the presence of a molybdenum complex and vanadyl immobilized in a matrix of polyacrylamide gel [386].

Oxidation of *L*-thiols by dioxygen or aqueous H_2O_2 catalyzed by protoheme chloride [387] or Fe*L* or Fe*D* [388] is described by the scheme:

$$H-S-CH_2-C\overset{COOH}{\underset{NH_2}{\cdots H}} \quad \xrightarrow[\text{catalyst}]{O_2 \text{ or } H_2O_2,} \quad \begin{array}{l} S-CH_2-C\overset{COOH}{\underset{NH_2}{\cdots H}} \\ \\ S-CH_2-C\overset{COOH}{\underset{NH_2}{\cdots H}} \end{array}$$

S-cysteine $\hspace{5cm}$ *L*-cystine

As a rule, the rate of this reaction is 2–3 orders of magnitude higher than the rate of Na_2S oxidation. At pH 7 the reaction is of third order ($k=6.1\times10^4$ $l^2/(mol^2$ s) at 299 K both for Fe*L* and Fe*D* , a catalyst–substrate complex being formed with $k\approx10$ $l/(mol$ s). The absence of stereoselectivity is caused by many factors including ligand exchange.

Macromolecular reagents can cause inversion and even disappearance of optical activity. For instance, a polymeric catalyst causes inversion of saccharose at 333 K [389]; racemization of α-amino acids is caused at 298–313 K by Cu^{2+} supported on a strong-base ainion-exchange resin partially modified by 5-sulfosalicylaldehyde [390].

Thus, the features of asymmetric oxidation of organic substrates resemble those of other processes of asymmetric induction (hydrogenation and hydroxylation of chiral amino acids and ketones and other processes). Asymmetric induction in hydroformylation will be considered in Chapter 4.

2.10 Catalytic Oxidation of Polymers

In Section 1.14 the effect of catalyzed reactions on polymeric supports was considered for the case of hydrogenation processes; in the present section oxidation of macromolecular supports is considered.

As already mentioned in this chapter, oxidation of a polymeric matrix, a reaction that parallels the main processes of oxidation of various substrates in the presence of immobilized catalysts, sometimes plays an important role. The oxidation of macroligands soluble under the conditions of oxidative processes is common; in this case, due to washing out of metal complexes, homogeneous metal–complex oxidation of polymers is realized. For insoluble supports, only a thin surface layer is oxidized. Inhibition and acceleration of oxidation and destruction of polymers in the presence of metal-containing compounds, as well as achievements of polymer stabilization, were analyzed recently in [391].

Some oxidation reactions of polymeric matrices proceed rather vigorously. For instance, the oxidation rate of a solution of PE in hexadecane in the presence of $MnSt_2$ is higher than the oxidation rate of hexadecane itself [392] (Figure 2.29). A kinetic scheme for the mixed initiation of these reactions was proposed. Deep oxidation and destruction of rather stable PE and even its crystalline regions was found in oxidative processes (333–403 K, O_2, cobalt bromide complexes) (Table 2.14). Unsaturated bonds of PE participate easily in various reactions including epoxidation under mild conditions (peracetic acid, 318 K, 2.5 h) with subsequent chemical transformations of oxirane groups and formation of various derivatives of PE [393]. It is important that the solubility of metal complexes in the oxidized polymer decreases by several orders of magnitude [394].

Figure 2.29 Dependence of the rate of oxidation of hexadecane and PE in hexadecane on concentration of consumed O_2 at 403 K. (1,2) $[CoSt_2]=$ 1×10^{-3} mol/l, (3,4) [Mn]/[Co]= 0.01, (1,3) hexadecane, (2,4) PE in hexadecane.

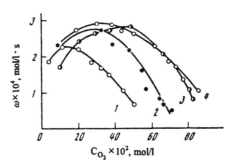

Table 2.14 Catalytic oxidation of PE crystals[a] [395].

T (K)	Time (h)	Number of groups[b] per 1000 C atoms		M_n	M_w
		Carbonyl groups	Carboxyl groups		
388	1	12	17	–	–
388	3	10	17	1300	2300
388	5	10	17	–	–
393	1	15	24	–	–
393	3	21.5	33	980	1000

[a]Pristine PE, M_n=15 500, M_w=130 000.
[b]After treatment of oxidized PE by concentrated HNO_3 at 373 K for 2 h.

Polydienes are especially labile to oxidation. Direct catalytic oxidation of 0.5–5% solutions of high molecular mass polydienes in decalin in the presence of silver catalysts results in the formation of oligomeric (PD=100–300) dienes with terminal OH groups and a small quantity of COOH groups [396]. Some Ni^{2+} chelates photostabilize cis-1,4-polybutadiene [397]. The kinetics and mechanism of the catalyzed oxidation of polybutadiene by molecular oxygen were studied in [398]. A controlled oxidation and destruction of PVS occurs under the action of aqueous solutions of H_2O_2 or NaClO [399]. The products formed, polyhydroxyketone and especially polyenolketone, are more labile to biodegradation than the pristine polymer. In this way, the problem of degradation of polymers after use could be solved.

Polyethylene glycol is often used as a macroligand. Melts of PEG can be oxidized by oxygen to hydroperoxides and aldehydes [400]. Terminal hydroxy groups of oligomeric PEG (relative molecular mass 1000–3000) are oxidized to carboxyl groups in acid media at 293–333 K under the action of $K_2Cr_2O_7$ [401]; PEG carbonic acids are obtained by oxidation of PEG in the presence of the catalyst $Cu(OCOCH_3)_2 \cdot H_2O$–NH_4VO_3 at 328 K, terminal aldehyde groups being introduced in the presence of $(NH_4)_2Ce(NO_3)_6$ or pyridine chlorchromate [402]. Salts of dicarboxylic acids are produced with high activity and selectivity in the presence of a base and aldehydes and salts of monocarbonic acids are not formed [403].

In principle, the kinetics of PEG solution oxidation fit the scheme of liquid-phase oxidation of oxygen-containing compounds by dioxygen considered in this chapter: various peroxide radicals, both with and without OH groups at α-carbon atom, are formed.

Intracomplex Cu^{2+} compounds, depending on content and structure of chelating unit, catalyze or inhibit degenerated, branched autoxidation of PEG [404].

It was found that tris(triethylsiloxy) vanadate acts as an inhibitor at low concentrations and as a catalyst of thermoxidation of polydiethylsiloxane oligomers at high concentrations [405].

Oxidation of poly-N-vinylcarbozole under the action of $Cu(BF_4)_2 \cdot 6H_2O$ at 298 K in CH_2Cl_2 solution is initiated by dissociation of an equimolar amount of CTC of Cu^{2+} with a carbozole group[31] [406]. One-electron oxidation results in formation of a polyvinylcarbazole cation radical, the first active intermediate. Initiators and catalysts retained in polymers after their synthesis can also serve as initiators of polymer oxidation; therefore the content of these active species should be strictly controlled. For instance, chains of industrially obtained polyvinylpyrrolidine can contain HPs, carboxyl and hydroxy groups and double bonds. Autoxidation of this polymer occurs with participation of Ce^{4+} ions used as the initiator for synthesis of the polymer [407]. The first step of the autoxidation is formation of HP groups. Formation of HP in PS was found at addition of O_2 to a solution of "live" polystyryl lithium [408].

Grafted polymers are sometimes used as macroligands in oxidation processes. The behavior of these macroligands can be different from that of homo- and copolymers. For instance, grafted copolymer cellulose–PAN is oxidized more rapidly than ungrafted cellulose, the oxidation rate increasing with increasing extent of grafting [409].

---

[31] Oxidation with participation of charge transfer complexes is a basic stage of chemical development of photoimages on carbazole-containing polymers.

Probably, the PAN grafting prevents cleavage of cellulose chains and accelerates oxidation of OH groups because of the more open structure of the grafted polymer.

Reactions of polymers with metal surfaces, for instance surfaces of reactors, are useful for the identification of side processes. For instance, during isothermal oxidation of PE, PP, PS, PETP and other polymers in air in the presence of copper, Cu^{2+} ions are transferred both in liquid (melted) and solid polymers independent of contact type (adhesional or frictional) [410].

Oxidation of a boundary layer of PE at 423–453 K by O_2 increases the adhesion of PE to steel; oxygen-containing groups formed in poly(methylmethacrylate) by photooxidation also increase adhesion between polymer and metal [411]. Diffusional escape of Cu^{2+} from interphase surfaces of the metal catalyst–polymer (isotactic PP) depends on the degree of preoxidation of PP and the temperature, and is independent of other factors [412]. On contact oxidation of rubber in the presence of catalytically active metals (Pd, Cu, Zn, Fe) metal ions are also transferred into the rubber matrix (a kind of metal dissolution). Moreover, rubber films oxidized on brass dissolve the elements of the alloy selectively [413].

Metal oligomers formed from oxidated dienes (oligopiperylenes) with Mn^{2+}, Co^{2+} and Pb^{2+}, are oxidation catalysts [414]. Attrition metals (Cu, Fe, Ni, Cr, Al) are active centers of catalytic oxidation and affect oxidative stability [415]. Nevertheless, metals and their compounds often serve as antioxidants of polymers (see, for example, [416–419]). Problems of autoxidation and stabilization are most important in the case of polymers with conjugated bonds, for instance polymers of acetylene-type, including doped polymers. These polymers are relatively rarely used as macroligands, therefore I refer readers to [420] where this problem is considered in detail.

2.11 Conclusions

Catalytic oxidation of hydrocarbons and function-containing substrates under the action of immobilized complexes is an effective way of optimizing the oxidation processes. Active and selective heterogenized analogues of homogeneous systems are known for most reactions. Variation of the nature of metal complexes and polymers enables catalysts with higher activity and selectivity than those of homogeneous and heterogeneous analogues to be obtained. Sometimes immobilized complexes have no homogeneous or heterogeneous analogues. Moreover, immobilized catalysts can be recycled relatively easily.

Polymeric matrices separate products from the catalyst, prevent irreversible interactions between active centers and dioxygen, and prevent the oxidation of easily oxidizable substrates or ligands. A hydrophilic or hydrophobic environment for the active center can be easily constructed.

A triple bisubstrate complex stabilized by functional groups of macroligand is formed on the polymer surface. Effective binding of substrate with the functional groups increases the concentration of substrate near the active center. The main function of polymeric supports is the matrix isolation of active centers (however, the polymer prevents monomerization of binuclear complexes in the case of sterically hindered phenols). Cycles of alternative reduction and oxidation of immobilized centers are more easily realized because of the conformational mobility of polymeric ligands. Functional groups can participate in these reactions; bifunctional catalysis with participation of oxidation centers and basic centers is often realized (see oxidation of thiols by Cu^{2+}

complexes supported on PVAm). Selection of functional groups of macroligands enables the route of reaction to be controlled (see oxidation of thio salts) and extended electron-transferring chains to be formed playing the role of electron carriers in oxidation reactions.

The structure of the macrochain is most important in stereoselective oxidation of organic substrates. Asymmetric induction by immobilized systems is caused by appropriate macrochain structure and by preferential binding of one enantiomer to the catalyst.

The catalysis of oxidation by immobilized complexes resembles catalysis by biocatalysts. Probably, this resemblance is explained by the molecular mobility both of proteins and macroligands in synthetic protein analogues: structural-dynamic reorganization of active centers, as well as redistribution of chemical bonds in an oxidizable substrate, are very important for catalytic reactions.

Recently new data on oxidation of various types of substrates by dioxygen in the presence of polymer-supported catalysts have been published: oxidation of catechol amines [421], hydrocarbons [422–424] and sodium sulfate [425], selective epoxidation [426, 427], hydroformylation of olefins [428], etherification [429], etc.

The main laws of oxidation by immobilized complexes are similar to those of other catalytic processes, hydrogenation, polymerization and hydroformylation.

References

1. Veprek-Siska, J. (1985/1986) *Oxid. Commun.*, **8**, 301.
2. Sherrington, D.C. (1988) *Pure and Appl. Chem.*, **60**, 401.
3. Emanuel, N.M., Denisov, E.T. and Maisus, Z.K. (1965) *Tsepnye Reaktsii Okisleniya Uglevodorodov v Zhydkoi Faze* (Chain Reactions of Hydrocarbon Oxidation in Liquid Phase). Moscow: Nauka.
4. Semenov, N.N. (1986) *Tsepnyye Reaktsii* (Chain Reactions). Moscow: Nauka.
5. Howard, J.A. (1972) *Free-Radical Chem.*, **49**, 49.
6. Pritzcow, W. and Rosner, H. (1975) *J. Prakt. Chem.*, **137**, 990.
7. Sheldon, R.A. and Kochi J.K. (1981) *Metal Catalyzed Oxidations of Organic Compounds*, pp. 275–288. New York: Academic Press.
8. Skibida, I.P. (1975) *Usp. Khim.*, **44**, 1729.
9. Haines, A.H. (1985) *Methods for Oxidation of Organic Compounds. Alkanes, Alkenes, Alkynes and Arenes*. London: Academic Press.
10. Almazov, T.G. and Margolis, L.Ya. (1988) *Vysocoselektivnye Katalizatory Okisleniya Uglevodorodov* (High-Selective Catalysts of Hydrocarbon Oxidation). Moscow: Khimiya.
11. Rubailo, V.L. and Maslov, C.A. (1988) *Zhydkofaznoye Okisleniye Nepredel'nykh Soedinenii* (Liquid Phase Oxidation of Unsaturated Compounds). Moscow: Khimia.
12. Stern, E.B. (1973) *Usp. Khim.*, **42**, 232.
13. Denisov, E.T. (1981) *Itogi Nauki Tekh. Kinet. Katal.*, **9**, 155.
14. Haber, F. and Weiss J. (1934) *Proc. R. Soc.* London, Ser. A, **147**, 332.
15. Minoun, H. (1984) *Proceedings of the III Boc Priestley Conference on Oxygen and Conversion of Future Feedstocks*, London, 1983), pp. 120–131.
16. Spirina, I.V., Maslennikov, V.P. and Alexandrov, Yu.A. (1987) *Usp. Khim.*, **56**, 1167.
17. Evmenenko, N.P., Gorokhavanskii, Ya.B. and Pylenko, Yu.I. (1972) *Dokl. Akad. Nauk SSSR*, **202**, 1117.

18. Jpn Pat. 56–108524, 1982.
19. Jazzaa, Al., Clark, J.H. and Robertson, M. (1982) *Chem. Lett.*, N 3, 405.
20. Climent, M.S., Marinas, J.M. and Senisterra, G.G. (1986) *React. Kinet. Catal. Lett.*, **32**, 177.
21. Shibaeva, L.V. (1986) *Proceedings of the VI All-Union Scientific Conference on Oxidation of Organic Compounds in the Liquid Phase* (in Russian), Vol. 2, p. 24.
22. Skarzewski, J. (1986) *Pr. Nauk. Inst. Chem. Org. Fiz.Politech.*, **30**, 1.
23. Blumberg, E.A., Norikov, Yu.D. (1984) *Itogi Nauki Tekh. Kinet. Katal.*, **12**, 3.
24. Norikov, Yu.D., Blumberg, E.A., Salukvadze L.V. (1975) *Problemy Kinetiki I Kataliza. Poverkhnostnye Yavleniya v Geterogennom Katalize* (Problems of Kinetics and Catalysis. Surface Phenomena in Heterogeneous Catalysis), Vol. 16, p. 150. Moscow: Nauka.
25. Norikov, Yu.D., Blumberg, E.A., Irmatov, M.D., Bochorishvili, Sh.K. (1975) *Dokl. Akad. Nauk*, **233**, 1187.
26. Holy, N.L. (1980) *CHEMTECH*, **10**, 366.
27. Hartley, F.R. (1985) *Supported Metal Complexes. A New Generation of Catalysts*, Dordrecht: D. Reidel.
28. Bugloss, A., Waterkouse, J.S. (1987) *J. Chem. Educ.*, **64**, 371.
29. *Polymers as Aids in Organic Chemistry* (1980) London: Academic Press.
30. Harrison, C.R., Hodge, P. (1974) *Chem. Commun.*, **24**, 1009; (1976) *J. Chem. Soc. Perkin Trans.*, pt. 1, No. 6, 605.
31. Greig, S.A., Hancock, R.D., Sherrington, D.C. (1980) *Eur. Polym. J.*, **16**, 293.
32. Hu, N.X., Aso, Y., Otsubo, T., Ogura, F. (1985) *Chem. Lett.*, **5**, 603.
33. Taylor, R.T., Flood, L.A. (1983) *J. Org. Chem.*, **48**, 5160.
34. Pittman, C.U., Jr. (1980) *Polym. News*, **6**, 223.
35. Salunkhe, D.G., Jagdate, M.H., Swami, S.S., Salunkhe, M.M. (1986) *Curr. Sci. (India)*, **55**, 922.
36. Stevenson, T.A., Taylor, R.T. (1988) *React. Polym.*, **8**, 7.
37. (1980) *Metal Ion Activation of Dioxygen*, New York: Wiley.
38. (1988) *Oxygen Complexes and Oxygen Activation by Transition Metals*, London:, Plenum Press.
39. Antonini, E., Brunori, M. (1971) *Hemoglobin and Myoglobin in Their Reaction with Ligands*, Amsterdam, North–Holland.
40. Nishide, H., Yoshioka, H., Wang, S., Tsuchida, E. (1985) *Makromol. Chem.*, **186**, 1523.
41. Fricke, R., Jerschkewitz, H.G., Kazanskii, V.B., *et al.* (1981) *React. Kinet. and Catal. Lett.*, **18**, 473.
42. Koda, G., Uchida, Y., Misono, A. (1970) *Inorg. and Nucl. Chem. Lett.*, **6**, 567.
43. Zamaraev, K.I., Parmon, V.A. (1983) *Usp. Khimii*, **52**, 1433.
44. Wohrle, D., Bohlen, H., Aringer, C., Pohl, D. (1984) *Makromol. Chem.*, **185**, 669.
45. Leal, O., Anderson, D.L., Bowman, R.G., *et al.* (1975) *J. Am. Chem. Soc.*, **97**, 5125.
46. Collman, J.P., Gagne, R.R., Kouba, J., Ljusberg–Wahren, H. (1974) *J. Am. Chem. Soc.*, **96**, 6800.
47. Drago, R.S., Caul, J.H. (1979) *Chem. Commun.*, **17**, 746; (1979) *Inorg. Chem.*, **18**, 2019.
48. Aeissen, H., Wohrle, D. (1981) *Makromol. Chem.*, **182**, 2961.
49. Wohrle, D., Bohlen, H., Meyer, G. (1984) *Polym. Bull.*, **11**, 151.
50. Wohrle, D., Bohlen, H. (1986) *Makromol. Chem.*, **187**, 2081.
51. Wohrle, D., Bohlen, H., Aringer, C., *et al.* (1983) *Abstracts of IUPAC Macro 83*,

(Bucharest, 1982) sect. 1, p. 354.

52. Bied-Charreton, C., Frostin-Rio, M., Pujol, D., *et al.* (1982) *J. Mol. Catal.*, **16**, 335.
53. Masuda, S., Nakabayashi, I., Ota, T., Takemoto, K. (1979) *Polym. J.*, **11**, 641.
54. Yukimasa, H., Sawai, H., Takizawa, T. (1979) *Makromol. Chem.*, **180**, 1681.
55. Tsuchida, E. (1979) *Macromol. Sci. Chem.*, **13**, 545.
56. Sawodny, W., Grunes, R., Reitzle, H. (1982) *Angew. Chem.*, **94**, 803,
57. Nolte, R.J.M., Razenberg, A.S.J., Schuurmann, R. (1986) *J. Amer. Chem. Soc.*, **108**, 2751.
58. DePorter, B., Meunier, B. (1985) *J. Chem. Soc., Perkin Trans.* 2, **11**, 1735.
59. Tyeklar, Z., Karlin, K.D. (1989) *Acc. Chem. Res.*, **22**, 241.
60. Savitskii, A.V., Nelyubin, V.I. (1975) *Usp. Khim.*, **44**, 214.
61. Ohyanagi, M., Nishide, H., Suenaga, K., Tsuchida, E. (1988) *Macromolecules*, **21**, 1590.
62. Robb, D.A. (1984) *Copper Proteins and Copper Enzymes*, Vol. 2, p. 207.New York: Pergamon Press.
63. Nishide, H., Ohno, H., Tsuchida, E. (1981) *Macromol. Chem. Rapid Commun.*, **2**, 55; (1982) **3**, 161.
64. Shigehara, K., Sato, Y., Tsuchida, E. (1981) *Macromolecules*, **14**, 1153.
65. Tsuchida, E., Nishide, H., Yoshioka, H. (1982) *Macromol. Chem. Rapid Commun.*, **3**, 693.
66. Tsuchida, E., Hasegawa, E., Kanayama, T. (1978) *Macromolecules*, **11**, 947.
67. Nishide, N., Kuwahara, M., Ohyanagi, M., *et al.* (1986) *Chem. Lett.*, **1**, 43; Idem. (1987) *Macromolecules*, **20**, 417.
68. Jarr, L.A., Way, J.D., Noble, R.D., *et al.* (1987) *AIChE J.*, **33**, 480.
69. Yoshikawa,M., Ezaki, T., Sanui, K., Ogata, N. (1988) *J. Appl. Polym. Sci.*, **35**, 145.
70. Tsuchida, E., Nishide, H., Yuasa, M., *et al.* (1985) *J. Chem. Soc., Dalton Trans.*, **2**, 275.
71. Yampolskii, Yu. P. (1989) *Zh. Vses.Khim. Ova.*, **30**, 679.
72. Zaitsev, A.M., Arnautova, G.M., Kuzmina, G.V., Dolmatova, O.A. (1983) *Kataliz I Katalizatory (Catalysis and Catalysts)*, Leningrad: pp. 57–64.
73. Chashechnikova, I.T., Golodets, G.I. (1982) *React. Kinet. and Catal. Lett.*, **21**, 59.
74. Zazhigalov, V.A., Zaitsev, Yu.R., Belousov, V.M. (1986) *React. Kinet. Catal. Lett.*, **32**, 209.
75. Moffat, J.B., Kasztelan, S. (1987) *J. Catal.*, **106**, 512; (1988) **109**, 206.
76. Mokshina, R.I., Poluectov, O.G., Polyak, S.S., *et al* (1990) *Kinet. Katal.*, **31**, 381.
77. Belousov, V.M., Pal'chevskaya, G.A., Zaitsev, Yu. R., *et al.*(1980) *Katal. Katal.*, No. 18, 43–47.
78. Inventor's Certificate 114707, USSR.
79. Litvintsev, Yu. I., Riko, E.A., Manakov, M.N., Lebedev, N.N. (1980) *Proceedings of the Symposium on Catalysts Containing Supported Complexes* (in Russian), Vol. 2, p. 151, Novosibirsk.
80. Krylova, S.V., Agabekov, V.E., Fedorishcheva, M.N., *et al.* (1987) *Neftekhimiya*, **27**, 125.
81. Galimova, R.A., Kuztetsova, I.M., Kuznetsova, N.N. (1981) *Khimiya i Tekhnologiya Pererabotki Nefti i Gasa (Chemistry and Technology of Gas and Oil Processing)*, pp. 5–7, Kazan.
82. Hashimoto, K., Inoue, K., Kinoshita, T., *et al.* (1984) *J. Chem. Soc. Jpn. Chem. Ind. Chem.*, **5**, 668.

83. Yadav, O.P., Kala, Ch., Haryana, Agr. (1986) *Univ. J. Res.*, **16**, 330.
84. Fr. Patent 2543542, 1985.
85. Barkanova, S.V., Dernacheva, V.M., Zheltukhin, I.A., *et al.* (1985) *Zh. Org. Khim.*, **21**, 2018.
86. Shilov, A.E. (1984) *Activation of Saturated Hydrocarbons by Transition Metal Complexes*, Dordrecht: Reidel.
87. Belova, V.S., Gimanova, I.M., Stepanova M.L., *et al.* (1991) *Dokl. Akad. Nauk*, **316**, 653.
88. Minoun, H. (1981) *Pure and Appl. Chem.*, **53**, 2389.
89. U.S. Patent 4211880, 1981.
90. (1984) *Chem. Eng. News*, **62**, 32.
91. Stolyarov, I.P., Vargaftik, M.N, Moiseev, I.I. (1987) *Kinet. Katal.*, **28**, 1359.
92. Belousov, V.M., Koshuda, E.V., Lyaschenko, L.V. (1986) *React. Kinet. Catal. Lett.*, **32**, 33; (1987) **33**, 99.
93. Jpn Patent 55–35061, 1981.
94. U.S. Patent 4237071, 1981.
95. Zombeck, A., Hamilton, D.E., Drago, R.S. (1982) *J. Am. Chem. Soc.*, **104**, 6782; (1987) **109**, 374.
96. Lyashchenko, L.V., Belousov, V.M., Stepanenko, V.I. (1981) *Abstracts of the V Conference on Oxidation of Heterogeneous Catalysis* (in Russian), Vol 2, p. 374, Baku.
97. Ali Bassam, El., Bregeault, J.-M., Martin, J. (1987) *J. Organometal. Chem.*, **327**, C9.
98. Tang, H.G., Sherrington, D.C. (1993) *Polymer*, **34**, 2821; (1993) J. Catal., **142**, 540.
99. Popova, T.K., Malikov, B.F., Popova, N.I. (1978) *Abstracts of the Conference on Mechanisms of Catalytic Reactions* (in Russian), Moscow, Nauka, Vol. 2, p. 43; Popova T.K., Stadnik L.N., Popova N.I. (1982) *Katalizatory i Protsessy Konversii Uglevodorodov (Catalysts and Processes of Hydrocarbon Conversion)*, p. 53, Kiev: Nauk. Dumka.
100. Makarov, G.G., Bobolev, A.V., Dunin, A.F. (1975) *Abstracts of the III Conference on Liquid Phase Oxidation of Organic Compounds* (in Russian), p. 34, Minsk.
101. Fin. Patent 71927, 1987.
102. Kovtyukhova, N.I., Belousov, V.M., Konishevskaya, G.A. (1982) *Neftekhimiya*, **22**, 612.
103. Kovtyukhova, N.I., Belousov, V.M., Novikov, Yu.N., Volpin, M.E. (1986) *Kinet. Katal.*, **27**, 1335; (1987) **28**, 353.
104. Acodedji, E., Lieto, J., Aune, J.P. (1983) *J. Mol. Catal.*, **22**, 73.
105. Artemov, A.V., Litvintsev, I.Yu., Vaishtein, E.F., *et al.* (1985) *Kinet. Katal.*, **26**, 621.
106. Gebler, M. (1981) *J. Inorg. Nucl. Chem.*, **43**, 2759.
107. Berentsveig, V.V., Chan, B. N., Barinova, T.V., Lisichkin, G.V. (1984) *Kinet. Katal.*, **25**, 1477.
108. Berentsveig, V.V., Chan, B. N., Rozentuller, B.V. (1984) *Vestn. Mosk. Univ., Ser. 2: Khim.*, **25**, 375; (1988) **29**, 390.
109. Berentsveig, V.V., Barinova, G.V., Staroverov, S.N. (1988) *Kinet. Katal.*, **29**, 978.
110. Rubailo, V.L., Nikitin, A.V., Kasparov, V.V., Pomogailo, A.D. (1983) *Abstracts of the IV All-Union Conference on Catalytic Reactions in the Liquid Phase* (in Russian), Vol. 2, p. 110, Alma-Ata: Nauka.

111. Nikitin, A.V., Maslov, S.A., Rubailo, V.L. (1986) *Abstracts of the VI All-Union Conference on Oxidation of Organic Compounds in the Liquid Phase*, Vol. 2, p. 26, L'vov.
112. Nikitin, A.V., Pomogailo, A.D., Rubailo, V.L. (1987) *Izv. Akad. Nauk SSSR, Ser. Khim.*, **1**, 36.
113. Nikitin, A.V., Maslov, V.S., Rubailo, V.L. (1987) *International Microsymposium on Oxidation of Organic Compounds*, p. 4, Tallinn.
114. Nikitin, A.V., Pomogailo, A.D., Maslov, S.A., Rubailo, V.L. (1987) *Neftekhimiya*, 27, 234.
115. Pomogailo, A.D., Savost'yanov, V.S. (1994) *Synthesis and Polymerization of Metal-Containing Monomers*, Boca Raton, Fl: CRC Press.
116. Simon, J., Le Moigue, J. (1980) *J. Mol. Catal.*, **7**, 137.
117. Emanuel,' N.M., Zaikov, G.E., Maizus, Z.K. (1973) *Rol' Sredy v Radikal'no-Tsepnykh Reaktsiyakh Organicheskikh Soedinenii* (Role of the Medium in Radical-Chain Reactions of Organic Compounds), Moscow: Nauka.
118. Solozhenko, E.G., Panova, G.V., Potapov, V.M., Garbar, A.V. (1986) *Zh. Obshch. Khim.*, **56**, 405.
119. Leising, R.A., Takeuchi, K.J. (1987) *Inorg. Chem.*, **26**, 4391.
120. Hilal, H.S., Kim, C., Schreiner, (1993) *J. Mol. Catal.*, **81**, 157.
121. Jpn Patent 57–24332.
122. Scnurpfeil, D. (1984) *Wiss. Z. Techn. Hochsch. (Chem. Cart. Chem. Cat.)* Leuna–Merseburg, **26**, 411.
123. Karakhanov, E.A., Narin, S.Yu., Filippova, T.Yu., *et al.* (1987) *Neftekhimiya*, **27**, 791.
124. Setenek, K. (1983) *Proceedings of the V International Symposium on Heterogenous Catalysis*, Vol. 1, p. 27, Sofia.
125. Emanuel', N.M., Gal, D. (1984) *Okislenie Etilbenzola. Model'naya Reaktsiya* (Oxidation of Ethylbenzene. Model Reaction), Moscow: Nauka.
126. Rashidova, S.Sh., Usmanova, M.M., Azizov, U.M., *et al.* (1980) *Proceedings of the Symposium on Catalysts Containing Supported Complexes* (in Russian), Novosibirsk, Vol 2, pp. 167, 171.
127. Rashidova, S.Sh., Atanasova, S.D., Ilyasova, D.D. (1979) *React. Kinet. Catal. Lett.*, **10**, 263.
128. Kayumova, Sh.A., Azizov, U.M., Rashidova, S.Sh., *et al.* (1984) *Kinet. Katal.*, **25**, 1356.
129. Efendiev, A.A., Shik, G.L., Vekilova, L.F., Shakhtinskii, T.N. (1980) *Proceedings of the Symposium on Catalysts Containing Supported Complexes* (in Russian), Vol. 2, p. 179, Novosibirsk.
130. Shakhtakhtinskii, T.N., Blumberg, E.A., Vekilova, L.F., *et al.* (1981) *Proceedings of the VI Soviet–France Seminar on Catalysis*, p. 144, Moscow.
131. Efendiev, A.A., Orudzhev, D.D., Shkhtakhinskii T.N., Kabanov, V.A. (1986) *Proceedings of the V International Symposium on the Relation Between Homogeneous and Heterogeneous Catalysis*, Vol. 2, p. 138, Novosibirsk.
132. Efendiev, A.A., Orudzhev, D.D., Shakhtinskii, T.N., Kabanov, V.A. (1987) *Kinet. Katal.*, **27**, 520.
133. Efendiev, A.A. (1994) *Macromol. Symp.*, **80**, 289.
134. Vekilova, L.F., Blumberg, E.A., Efendiev, A.A., *et al.* (1987) *Azerb. Khim. Zh.*, **3**, 79.
135. Kayumova, Sh.A., Azizov, U.N., Iskanderov, S.I. (1986) *Kinet. Katal.*, **27**, 1141.

136. Kayumova, Sh.A., Sulukvadze, L.V., Azizov, U.M., *et al.* (1987) *International Symposium on Oxidation of Organic Compounds*, p. 36, Tallinn.
137. Vekilova, L.F., Zasedatelev, S.Yu, Blumberg, E.A., *et al.* (1980) *Proceedings of the Symposium on Catalysts Containing Supported Complexes* (in Russian), Vol. 2, p. 141, Novosibirsk.
138. U.S. Patent 4322560, 1982.
139. Shipunova, N.A., Zasedatelev, S.Yu., Filippova, T.B., Blumberg, E.A. (1981) *Abstracts of the XII Mendeleev Meeting on Pure and Applied Chemistry* (in Russian), Vol. 3, p. 106, Moscow.
140. Jpn Patent 55–35063, 1981.
141. Mayo, F.R., Cais, R.E. (1981) *Macromolecules*, **14**, 885.
142. Berentsveig, V.V., Dotsenko, O.E., Kokorin, A.I., *et al.* (1982) *Izv. Akad. Nauk SSSR, Ser. Khim.*, **10**, 2211.
143. Berentsveig, V.V., Dotsenko, O.E., Kopylova, V.D., Boiko, E.T. (1982) *Kinet. Katal.*, **23**, 64.
144. Kokorin, A.I., Berentsveig, V.V., Kopylova ,V.D., Frumkina, E.L. (1983) *Kinet. Katal.*, **24**, 181.
145. Pokrovskaya, I.E., Bashurina, T.P. (1982) *Neftekhimiya*, **22**, 643.
146. Berentsveig, V.V., Syshchikova, I.G., Dotsenko, O.E., Evseev, A.M. (1984) *Zh. Fiz. Khim.*, **58**, 2722.
147. Berentsveig, V.V., Kopylova, V.D., Frumkina, E.L., Boiko, E.T. (1980) *Proceedings of the Symposium on Catalysts Containing Supported Complexes* (in Russian), Vol. 2, p. 121, Novosibirsk.
148. Berentsveig, V.V., Dotsenko, O.E., Kopylova, V.D., *et al.* (1983) *Neftekhimiya*, **23**, 238.
149. Setinek, K., Drapakova, S., Prokop, Z. (1986) *Collect. Czech. Chem. Commun.*, **51**, 1958.
150. U.S. Patent 4182909, 1981.
151. Kuruzu, Y., Storck, W., Manecke, G. (1975) *Makromol. Chem.*, **176**, 3185.
152. Petrova, J., Ivanov, S.K., Kirilov, M., Zdravkova, Z. (1984) *Oxid. Commun.*, **7**, 211.
153. Mosolova, L.A., Ovchinnikov, V.I., Dobrotvorskii, A.M., *et al.* (1980) *Izv. Akad. Nauk SSSR*, Ser. Khim., **4**, 731.
154. Kozlov, S.K., Ovchinnikov, V.I., Dobrotvorskii, A.M., *et al.* (1980) *Zh. Prikl. Khim.*, **53**, 1406.
155. Sirota, T.V., Kasaikina, O.T., Gagarina, A.B. (1986) *Abstracts of the VI All-Union Conference on Oxidation of Organic Compounds in the Liquid Phase* (in Russian), Vol. 2, p. 10, L'vov.
156. Pomogailo, A.D. (1986) *Synthesis and Properties of Polymer-Filled Polyolefins* (in Russian), Vol. 10, p. 63, Chernogolovka.
157. Bai, R.R., Zong, H.-J., He J.-G. (1984) *Macromol. Chem. Rapid Commun.*, **5**, 501.
158. Rakovskii, S., Neniev, L., Shopov, *et al.* (1987) *International Symposium Oxidation of Organic Compounds*, p. 56, Tallinn.
159. Belyakov, L.A., Kolotusha, T.P., Serova, T.E., *et al.* (1986) *Dokl. Akad. Nauk*, **288**, 1358.
160. Tavedyan, L.A., Tonikyan, A.K. (1987) *Arm. Khim. Zh.*, 40, 610; (1989) *Kinet. Katal.*, **30**, 1006.
161. Koroleva, N.V., Kamneva, A.I., Yukhnovets, L.B. (1985) *Tr. Mosk. Khim.-Tekhnol. Ins-t.*, **139**, 97.

162. Batyr, D.G., Reibel', I.M., Zubareva, V.E., Sandu, A.F. (1984) *Izv. Akad. Nauk Mold. SSR, Ser. Biol. Khim. Nauk*, **2**, 42.

163. Lunak, S., Vaskova, M., Lederer, P., Vepreksiska, J. (1986) *J. Mol. Catal.*, **34**, 321.

164. Berentsveig, V.V., Yuffa, A.Ya., Kudryavtsev, G.V., Baitova, L.V. (1980) *Vestn. Mosk. Univ., Ser.2: Khim.*, **21**, 495.

165. Yuffa, A.Ya., Berentsveig, V.V., Kudryavtsev, G.V., Lisichkin, G.V. (1981) *Kinet. Katal.*, **22**, 1469.

166. Batyr, D.G., Reibel', I.M., Sandu, A.F. (1979) *Izv. AN MSSR, Ser. Biol. Khim. Nauk*, **5**, 53.

167. Shklyayev, A.A., Terpugova, M.P., Amosov, Yu.I., Kotlyarovskii, I.L. (1980) *Izv. Akad. Nauk SSSR, Ser. Khim.*, **7**, 1526.

168. Frumkina, E.L. (1985) Effect of the content of coordination centers of ionite complexes on their catalytic activity in redox processes, *Cand. Sci. (Chem.) Dissertation*, Moscow.

169. Chandran, R.S., Ford W.T. (1988) *J. Chem. Soc. Chem. Commun.*, **2**, 104.

170. Ford, W.T., Chandran, R., Turk H. (1988) *Pure Appl. Chem.*, **60**, 395.

171. Artemov, A.V., Seleznev, V.A., Vainshtein E.F. (1983) *Izv. Vus. Khim. Khim. Tekhnol.*, **26**, 1298.

172. Inventor's Certificate 825138 USSR.

173. Artemov, A.V., Vainshtein, E.F., Seleznev, V.A., Tyulenin, Yu.P. (1982) *Voprosy Kinetiki i Kataliza* (Problems of Kinetics and Catalysis), p. 115, Ivanovo.

174. Seleznev, V.A., Artemov, A.V., Vainshtein, E.F., Tyulenin, Yu.P. (1981) *Neftekhimiya*, **21**, 114.

175. Seleznev, V.A., Tyulenin, Yu.P., Vainshtein, E.F., Artemov, A.V. (1984) *Kinet. Katal.*, **24**, 1085; (1984) *Izv. Vuz. Khim. Khim.Tekhnol.*, **27**, 64.

176. Tyulenin, Yu.P., Seleznev, V.A., Vainshtein, Yu.F., Artemov, A.V. (1983) *Proceedings of the V International Symposium on Heterogeneous Catalysis*, Vol. 1, p. 105, Sofia.

177. Yamagushi, S., Inoue, M., Enomoto, S. (1985) *Chem. Lett.*, **6**, 827; (1986) *Bull. Chem. Soc. Jpn.*, **59**, 2881.

178. Yamagushi, S., Shinoda, H., Inoue, M., Enomoto, S. (1986) *Chem. Pharm. Bull.*, **34**, 4467.

179. Gekhman, A.E., Makarov, A.P., Polstnyuk, O.Ya, Moiseev, I.I. (1986) *Abstracts of the VI All-Union Conference on Oxidation of Organic Compounds in the Liquid Phase* (in Russian), Vol. 2, p. 54, L'vov.

180. Men'shikov, S.Yu. (1988) Synthesis of heterogenised chelates of transition metals with azomethines and their catalytic activity in oxidation reactions, *Cand. Sci. (Chem.) Dissertation*, Ufa.

181. Lukashova, E.I. (1989) Insoluble polymeric catalysts containing manganese porphyrins in oxidation of unsaturated hydrocarbons, *Cand. Sci. (Chem.) Dissertation*, Moscow.

182. (1973) *Inorganic Biochemistry*, Amsterdam: Elsevier.

183. Pomogailo, A.D. (1988) *Polimernyeye Immobilizovannyye Metallokompleksnye Katalizatory* (Polymeric Immobilized Metal Complex Catalysts), Moscow: Nauka.

184. Adams, P.A., Berman, M.C., Baldwin, D.A. (1979) *J. Chem. Soc., Chem. Commun.*, **19**, 856.

185. Mochida, J., Yasutake, A., Fujitsu, H., Takeshita, K. (1982) *J. Phys. Chem.*, **86**, 3468.

186. Vorobyev, A.V., Lukashova, E.A., Solov'yeva, A.B., et al. (1988) Vysokomol. Soedin., Ser. B., **30**, 903.
187. Enikolopyan, N.S., Bogdanova, K.A., Karmilova, L.V., Askerov, K.A. (1985) Usp. Khim., **54**, 369; (1983) **52**, 20.
188. Enikolopyan, N.S., Solov'yeva, A.B. (1988) Zh. Fiz. Khim., **62**, 2289.
189. Roginskii, S.Z., Berlin, A.A., Kutseva, L.N. (1963) Dokl. Akad. Nauk, **148**, 118.
190. Levina, S.D., Andrianova, T.I., Sakharov, M.M., et al. (1966) Zh. Fiz. Khim., **40**, 118.
191. Solov'yeva, A.B., Karakozova, E.I., Karmilova, L.V., Timashev, S.F. (1984) Izv. Akad. Nauk SSSR. Ser. Khim., **11**, 2445.
192. Solov'yeva, A.B., Samokhvalova, A.I., Lebedeva, T.S., et al. (1986) Dokl. Akad. Nauk, **290**, 1383.
193. Ivanova, A.I. (1988) Catalytic activity of metal porphyrins and soluble copolymers containing metal porphyrins in reactions of olefin oxidation by molecular oxygen, Cand. Sci. (Chem.) Dissertation, Moscow.
194. Solov'yeva, A.B., Lukashova, E.A., Karakozova, E.I., et al. (1984) Dokl. Akad. Nauk, **274**, 858.
195. Weerarathna, S., Hronec, M., Malik, L., Vesely, V. (1983) React. Kinet. Catal. Lett., **22**, 7.
196. Inventor's Certificate 832316 SSSR.
197. Belousov, V.M., Kovtyukhina, N.I., Mikhailovskii, S.V., Novikov, Yu.N. (1980) Dokl. Akad. Nauk, **253**, 346.
198. Skorobogaty, A., Smith, T.D. (1984) Coord. Chem. Rev., **53**, 55.
199. Shpota, G.P., Tarkovskaya, I.A., Strazhenko, D.N. (1976) Ukr. Khim. Zh., **42**, 584.
200. Kopylova, V.D., Frumkina, E.L., Mochalova, L.A., Saladze, K.M. (1978) Kinet. Katal., **19**, 1356.
201. Michajliva, D., Andreev, A. (1977) React. Kinet. Catal. Lett., **6**, 443.
202. Kokorin, A.I., Frumkina, E.L., Kopylova, V.D., et al. (1989) Izv. Akad. Nauk SSSR. Ser. Khim., **9**, 1970.
203. Keptman, S.V., Keptman, G.M., Tomilova, M.V. (1993) Kinet. Katal., **34**, 1035.
204. Lobantseva, V.F., Kiseleva, N.N., Kurbanaliev, M.K. (1992) Vysokomol. Soedin. Ser. B., **34**, 20.
205. El.-Sheikh, M.Y., Ashmawy, F.M., Salem, I.A., Zaki, A.P. (1988) Z. Phys. Chem., 269, 126; (1989) Transition Met. Chem., **14**, 95.
206. Kreja, L., Plewka, A. (1982) Angew. Makromol. Chem., **102**, 45.
207. Kreja, L. (1987) Monatsh. Chem., **118**, 717.
208. Shirai, H., Maruyama, A., Kobayashi, K., et al. (1979) J. Polym. Sci., Polym. Lett. Ed., **17**, 661.
209. Shirai, H., Higaki, S., Hanabusa, K., et al. (1983) J. Polym. Sci., Polym. Lett. Ed., **21**, 157.
210. Skobeleva, V.D., Men'shikov, S.Yu., Molochnikov, L.M., et al. (1987) Izv. Akad. Nauk SSSR, Ser. Khim., **8**, 1728.
211. Men'shikov, S.Yu., Kolenko, I.P., Kharchuk, V.G., et al. (1989) Kinet. Katal., **30**, 742.
212. Dzhardimalieva, G.I., Pomogailo, A.D., Shul'ga, Yu. M. (1991) Izv. Akad. Nauk SSSR, Ser. Khim., **12**, 100.
213. Gokak, D.T., Ram, R.N. (1989) J. Mol. Catal., **49**, 285.
214. Gokak, D.T., Kamath, W.V., Ram, R.N. (1988) J. Appl. Polym. Sci., **35**, 1523.

215. Sigel, H., Wyss, K., Fisher, B.E., Prijs, B. (1979) *Inorg. Chem.*, **18**, 1354.
216. Arai, K., Ogiwara, Y. (1987) *Makromol. Chem.*, **188**, 1067.
217. Berkutov, E.A., Kudaibergenov, S.E., Sigitov, V.A. (1986) *Polymer. J.*, **27**, 1269.
218. Hatano, M., Norawa, T., Yamamoto, T., Kambara, S. (1968) *Makromol. Chem.*, **115**, 10; (1968) **111**, 73; (1966) **98**, 136.
219. Melnik, A.C., Kildahl, N.K., Rendina, A.R., Busch D.H. (1979) *J. Am. Chem. Soc.*, **101**, 3232.
220. Abuchowskii, A., McCoy, J.P., Palczuk, N.C., *et al.* (1977) J. *Biol. Chem.*, **252**, 3578.
221. Turkevich, J., Xu Z. (1984) *Proceedings of the VIII International Congress on Catalysis*, Vol. 4, p. 791, Weinheim.
222. Kapanchan, A.T., Pshezhetskii,V.S., Kabanov, V.A. (1969) *Vysomol. Soedin., Ser. B*, **11**, 5; (1970) *Dokl. Akad. Nauk*, **190**, 853.
223. Pshezhetskii, V.S., Lebedeva, T.A., Kabanov, V.A. (1972) *Dokl. Akad. Nauk*, **204**, 1492.
224. Pshezhetskii, V.S., Kuznetsova, T.A., Kabanov, V.A. (1975) *Bioorgan. Khim.*, **1**, 1224.
225. Pshezhetskii, V.S., Ikryannikov, S.G., Kuznetsova, T.A., Kabanov, V.A. (1976) *J. Polym. Sci., Polym. Chem. Ed.*, **14**, 2595.
226. Dryuk, V.G. (1985) *Usp. Khim.*, **54**, 1674.
227. Tolstikov, G.A. (1976) *Reaktsii Gidroperekisnogo Okisleniya* (Reactions of Hydroperoxide Oxidation), Moscow: Nauka.
228. Blumberg, E.A., Filippova, T.V., Dryuk V.G. (1988) *Itogi Nauki Tekh. Kinet. Katal.*, **18**, 3.
229. Kucher, R.V., Timokhin, V.I, Shevchuk, I.P., Vasyutin, Ya.M. (1986) *Zhidkofaznoye Okisleniye Nepredel'nykh Soedinenii v Okisi Olefinov* (Liquid Phase Oxidation of Unsaturated Compounds in Olefin Oxides), Kiev: Nauk. Dumka.
230. Sheldon, R.A. (1980) *J. Mol. Catal.*, **7**, 107.
231. Sheldon, R.A., Van Doorn, J.A. (1973 *J. Catal.*,**31**, 427; Stamenova, R.T., Tsvetanov, C.B., Vassilev, K.G., *et al.* (1991) *J. Appl. Polym. Sci.*, **42**, 807.
232. Minoum, H., Seree de Roch, Sajus, L. (1969) *Bull. Soc. Chim. Fr.*, **5**, 1481.
233. Yokoyama, T., Nishizawa, M., Kimura, T., Suzuki, T.M. (1983) *Chem. Lett.*, **11**, 1703.
234. Sobczak, J., Ziolkowski, J.J. (1979) *React. Kinet. Catal. Lett.*, **11**, 359
235. Inventor's certificates No. 34781 and 34827, Bulgaria.
236. Boeva, R.S., Kotov, S., Jardanov, N. (1984) *React. Kinet. Catal. Lett.*, **24**, 239.
237. Sherrington, D.C., Simpson, S. (1991) *J. Catal.*, **131**, 115; (1993) *React. Polym.*, **19**, 13.
238. Miller, M.M., Sherrington, D.C. (1994) *J. Chem. Soc. Chem. Commun.*, **32**, 55.
239. Grinenko, S.B., Konishevskaya, G.A., Alekseyenko, L.M. (1985) *Ukr. Khim. Zh.*, **51**, 822.
240. Inventor's Certificate 995854, USSR.
241. Sapunov, V.N., Riko, E.A., Litvintsev, I.Yu. (1980) *Kinet. Katal.*, **21**, pp. 529, 791, 794.
242. Sapunov, V.N. (1986) *Tr. Mosk. Khim.-Tekhnol. Ins-t.*, **141**, 42.
243. Ferruti, P., Ranucci, E., Tempesti, E. (1990) *J. Appl. Polym. Sci.*, **41**, 1923; *Makromol. Chem., Macromol. Symposium*, **59**, 381.

244. Bhaduri, S., Khwaja, H. (1983) *J. Chem. Soc., Dalton Trans.*, **3**, 415.
245. Filippov, A.P., Polishchuk, O.A. (1984) *Kinet. Katal.*, **25**, 1348.
246. Filippov, A.P., Salivan-Peskova, V.Ya. (1989) *Kinet. Katal.*, **30**, 668.
247. Kurusu, Y., Masyama, Y., Saito, M., Saito, S. (1986) *J. Mol. Catal.*, **37**, 235; Kurusu Y., Masyama Y. (1987) *J. Macromol. Sci.* A, **24**, 389.
248. Tanielyan, S.K., Kropf, H., Ivanov, S.K. (1987) *Proceedings of the VI International Symposium on Heterogeneous Catalysis*, p. 43, Sofia.
249. Gelbard G., Breton F., Benelmoundeni M., Charreyre M.-T., Dong D. (1994) *6th International Conference on Polymer Supported Reactions in Organic Chemistry*, p. 254, Venice.
250. Zhu, W., Ford, W.T. (1992) *J. Polym. Sci., Polym. Chem.*, **30**, 1305; Srinivasan, S., Ford, W.T. (1991) *Nouv. J. Chim.*, **15**, 693..
251. Yokoyama, T., Nishizawa, M., Kimura, T., Suzuki, T.M. (1985) *Bull. Chem. Soc. Jpn*, **58**, 3271.
252. Chapin, E.C., Twohig, E.F., Keys, L.D., Corski, K.M. (1982) *J. Appl. Polym. Sci.*, **27**, 811.
253. Lau, C.-P., Chang, B.-H., Grubbs, R.H., Brubaker, C.H. (Jr.) (1981) *J. Organomet.*, **214**, 325.
254. Farrall, M.J., Alexis, M., Trecarten, M. (1983) *Nouv. J. Chim.*, **7**, 449.
255. Chang, B.-H., Grubbs, R.H., Brubaker, C.H. (Jr.) (1985) *J. Organomet. Chem.*, **280**, 365.
256. Tabushi, T., Kurihara, K., Morimitsu, K. (1985) *J. Synth. Org. Chem. Jpn.*, **43**, 323.
257. Brill, W.F. (1986) *J. Org. Chem.*, **51**, 1149.
258. U.S. Patent 4480113, 1985.
259. Zasedatelev, S.Yu., Filippova, T.V., Blumberg, E.A. (1980) *Dokl. Akad. Nauk*, **250**, 132.
260. Moriarty, R.M., Penmasta, R., Prakash, I. (1985) *Tetrahedron Lett.*, **26**, 4699.
261. Berentsveig, V.V., Barinova, T.V. (1987) *Kinet. Katal.*, **28**, 850.
262. Denisov, E.T., Mitsevich, N.I., Agabekov, V.E. (1975) *Mekhanizmy Zhidkofaznogo Okisleniya Kislorod-Soderzhashchikh Soedinenii* (Mechanisms of Liquid Phase Oxidation of Oxygen-Containing Compounds), Minsk: Nauka I Tekhnika.
263. Virtahen, P.O.I., Samalkivi, R. (1985) *Finn. Chem. Lett.*, **12**, 47; (1985) **3**, 336.
264. Vijayasro, K., Rajaram, J., Kwriacose, J.C. (1987) *J. Mol. Catal.*, **39**, 203.
265. Cagne, R.R., Marks, D.N. (1984) *Inorg. Chem.*, **23**, 65.
266. Chou, T.-C., Hwang, B.J. (1986) *World Congress III Chem. Engineering*, Vol. 4, p. 1, Tokyo.
267. Galperina, A.I., Srednev, S.S., Gvozdovskii, G.N. (1986) *Abstracts of the All-Union Conference on Oxidation of Organic Compounds* (in Russian), Vol. 2, p. 105, L'vov.
268. Maternova, J., Satinek, K. (1980) *J. Polym. Sci., Polym. Symposium*, **68**, 239.
269. Cainelli, G., Gardillo, G., Orena, M., Sandri, S. (1976) *J. Am. Chem. Soc.*, **98**, 6737.
270. Frechet, J.M., Warnock, J., Farrall, M.J. (1978) *J. Org. Chem.*, **43**, 2618; (1981) **46**, 1728.
271. Frechet, J.M., Darling, P., Farrall, M.J. (1980) *J. Am. Chem. Soc. Polym. Prepr.*, **21**, 272.
272. Ishikawa, Y., Yoschida, M., Ando, T. (1982) *Bull. Covt. Ind. Res. Inst. Osaka*, **33**, 1.

273. Li, N.-H., Frechet, J.M.H. (1987) *React. Polym.*, **6**, 311.
274. Terlyanskaya, A.T., Toropkina, G.N., Sorokina, N.L., *et al.* (1980) *Plast. Massy*, **5**, 51.
275. Noureldin, N.A., Lee, D.G. (1981) *Tetrahedron Lett.*, **22**, 4889.
276. Huang, M.-Y., Ren, C.-Y., Jiang, Y.Y. (1987) *J. Macromol. Sci., Chem.*, **24**, 269.
277. Huang, M.-Y., Ren, C.-Y., Zhou, Y.-S., *et al.* (1985) *Organosilicon and Bioorganosilicon Chemistry. Proceedings of the VII International Symposium on Organosilicon Chemistry* (Kyoto, 1984), p. 275, New York.
278. Katayama-Aramata, A., Ohnischi, K. (1983) *J. Am. Chem. Soc.*, **105**, 658.
279. Mahajan, V.V. (1979) *Ind. Chem. Eng.*, **21**, 45.
280. Kozhevnikov, I.V., Taraban'ko, V.E., Matveev, K.I. (1980) *Kinet. Katal.*, **21**, 947.
281. Nomiya, K., Murasaki, N., Miwa, M. (1986) *Polyhedron*, **5**, 1031.
282. Magashima, H., Sakai, K., Tsuji, I. (1982) *Chem. Lett.*, **6**, 859.
283. Butina, N.P., Pshezhetskii, V.S., Stom, D.I. (1984) *Vysokomol. Soedin., Ser. B*, **26**, 813.
284. Panova, G.V., Solozhenko, E.G., Garbar, A.V., *et al.* (1984) *Zh. Obshch. Khim.*, **54**, 1391.
285. Casellato, U., Tamburini, S., Vagato, P.A., *et al.* (1983) *Inorg. Chim. Acta*, **69**, 45.
286. Challa, G., Mienders, H.C. (1977) *Relations Between Homogeneous and Heterogeneous Catalysis*, p. 309, Lyon.
287. Nishide, H., Tsuchida, E. (1982) *Makromol. Chem.*, **183**, 1883.
288. Tsuchida, E., Nishikawa, H., Terada, E. (1976) *J. Polym. Sci., Polym. Chem. Ed.*, **14**, 825.
289. Lin, Y., Zou, B. (1985) *Tsukhua Syuebao (J. Catal)*, **6**, 237.
290. Sharma, P., Vigee, G.S. (1984) *Inorg. Chim. Acta*, **88**, 29; (1984) **90**, 73.
291. Moore, K., Vigel, G.S. (1982) *Inorg. Chim. Acta*, **66**, 124.
292. Pshezhetskii, V.S. (1980) *Proceedings of the Symposium on Catalysts Containing Supported Complexes* (in Russian), p. 81, Novosibirsk.
293. Lin, Y.-B., Han, A.-B., Dai, Q.-Y. (1988) *Chin. J. Org. Chem.*, **1**, 45.
294. Matsumoto, M., Kuroda, K. (1981) *Tetrahedron Lett.*, **22**, 4437.
295. Zhu, W., Ford, W.T. (1993) *J. Mol. Catal.*, **78**, 367.
296. Dadze, T.P., Khidekel', M.L. (1970) *Izv. Akad. Nauk SSSR, Ser. Khim.*, **12**, 2722.
297. Hulsbergen, F.W., Manassen, J., Reedijk, J., Welleman, J.A. (1977) *J. Mol. Catal.*, **3**, 47.
298. Challa, G. (1981) *Makromol. Chem. Suppl.*, **182**, 70.
299. Verlaan, J.P.J., Bootsma, J.P.J., Challa, G. (1982) *J. Mol. Catal.*, **14**, 211.
300. Sato, M., Kondo, K., Takemoto, K. (1979) *J. Polym. Sci., Polym. Chem. Ed.*, **17**, 2729.
301. Sato, M., Shindo, M., Kondo, K., Takemoto, K. (1980) *J. Polym. Sci., Polym. Chem. Ed.*, **18**, 101.
302. Meinders, H.C., van Bolhuis, F., Challa, G. (1979) *J. Mol. Catal.*, **5**, 225.
303. Koning, C.E., Vierson, F.J., Challa, G., Reedijk, J. (1988) *J. Mol. Catal.*, **44**, 245.
304. Breemhaar, W., Meinders, H.C., Challa, G. (1981) *J. Mol. Catal.*, **10**, 33.
305. Nishinaga, A., Watanabe, K., Matsuura, T. (1974) *Tetrahedron Lett.*, **15**, 1291.
306. Drago, R.S., Gaul, J., Zomwack, A., Straub, D.K. (1980) *J. Am. Chem. Soc.*, **102**, 1033.
307. Bied-Charreton, C., Idoux, J.P., Gaudemer A. (1978) *Nouv. J. Chim.*, **2**, 303.
308. Turk, H., Ford, W.T. (1988) *J. Org. Chem.*, **53**, 460.

309. Yamagishi, A. (1982) *J. Phys. Chem.*, **86**, 229.
310. Wei Chen, Guangfu Ma, Zhaoying Lo, Mangin Gao (1987) *Macromol. Sci., Chem.*, **24**, 243.
311. Skobeleva, V.D., Men'shikov, S.Yu., Molochnikov, L.S., *et al.* (1987) *Zh. Prikl. Khim.*, **60**, 2160.
312. Lee, M.H., Dawson, C.R. (1973) *J. Biol. Chem.*, **248**, 6596.
313. Koning, C.E., Hiemstra, B.L., Challa, G., *et al.* (1985) *J. Mol. Catal.*, **32**, 309.
314. Koning, C.E., Brinkhuis, R., Wevers, R., Challa, G. (1987) *Polymer*, **28**, 2310.
315. Tsukube, H., Maruyama, K., Araki, T. (1983) *J. Chem. Soc., Perkin Trans. Pt II*, **10**, 1485.
316. Maruyama, K., Tsukube, H., Araki, T. (1980) *J. Polym. Sci., Polym. Chem. Ed.*, **18**, 753.
317. U.S. Patent 4176161, 1980.
318. Yukimasa, H., Sawai, H., Takizawa, T. (1979) *Makromol. Chem.*, **180**, 1681.
319. *Polymeric Reagents and Catalysts* (1986), Washington DC: American Chemical Society.
320. Materra, V.D., Barnes, D.M., Chaudhuri, S., *et al.* (1986) *J. Phys. Chem.*, **90**, 4819.
321. Risen, W.M. (1987) *Proceedings of the NATO Advances in Research, Workshop: Structure and Properties of Ionomers* (Villard, Dodrecht, 1983),p. 331.
322. Pol. Patent 119180, 1984.
323. Takanashi, N., Arakawa, H., Kano A., *et al.* (1990) *Chem. Lett.*, **2**, 205.
324. Narrison, C.R., Hodge, P. (1982) *J. Chem. Soc., Perkin Trans. Pt. I*, **2**, 509.
325. Gibietis, Ya.Ya., Putse, A.A., Zitsmanis, Ya.L. (1986) *Izv. Akad. Nauk Latv. SSR, Ser. Khim.*, **5**, 600.
326. Yaneda, F., Yamato, H., Nagamatsu, F., *et al.* (1982) *J. Polym. Sci., Polym. Lett. Ed.*, **20**, *667.*
327. Prokop, Z., Setinek, K. (1981) *Collect. Czech. Chem. Commun.*, **46**, 1237; **11**, 2657.
328. Miyazawa, T., Endo, T. (1986) *J. Polym. Sci., Polym. Chem. Ed.*, **23**, 2487.
329. U.S. Patent 4286068, 1982.
330. Yoshida, J., Sofuku, H., Kawabata, N. (1983) *Bull. Chem. Soc. Jpn.*, **56**, 1243.
331. Reisz, K., Inczedy, J. (1979) *J. Therm. Anal.*, **16**, 421.
332. Schutten, J.M., Beelen, T.P.M. (1980) *J. Mol. Catal.*, **10**, 85.
333. Brouwer, W.M., Piet, P., German, A.L. (1984) *Makromol. Chem.*, **185**, 363; (1984) *J. Mol. Catal.*, **22**, 297; (1985) **29**, 335, 347; (1985) **31**, 169.
334. Schutten, J.H., Piet, P., German, A.L. (1979) *Makromol. Chem.*, **180**, 2341; (1980) *Angew. Makromol. Chem.*, **89**, 201.
335. Brouwer, W., Traa, P.A.M., de Werd, T.J.W., *et al.* (1984) *Angew. Makromol. Chem.*, **128**, 133.
336. Brouwer, W., Piet, P., German, A.L. (1983) *Polym. Bull.*, **8**, 245.
337. Streun, K.H., Piet, P., German, A.L. (1987) *Eur. Polym. J.*, **23**, 941.
338. Welzen, J., Herk, A.M., German, A.L. (1987) *Makromol. Chem.*, **188**, 1923; (1989) **190**, 2477.
339. Babu, S.H., Ford, W.T. (1992) *J. Polym. Sci., Polym.Chem.*, **30**, *1917.*
340. Pratt, J.M. (1972) *Inorganic Chemistry of Vitamin B_{12}*, New York: Acad. Press.
341. Shinkai, S., Ando, R., Kunitake, T. (1978) *J. Chem. Soc. Perkin Trans.*, Pt II, **12**, 1271.

342. Hanabusa, K., Ye, X., Koyama, T., *et al.* (1990) *J. Mol. Catal.*, **60**, 127.
343. Hassanein, M., Ford, W.T. (1988) *Macromolecules*, **21**, 525.
344. Scorobogaty, A., Smith, T.D. (1982) *J. Mol. Catal.*, **16**, 131.
345. Streum, K.H., Meuldijk, A.L.J., German, A.L. (1989) *Makromol. Chem.*, **173**, 119; (1990) **191**, 2181.
346. U.S. Patent 4206043, 1981.
347. Kokorin, A.I., Astanina, A.N., Kopylova, V.D., Frumkina, E.L. (1989) *Izv. Akad. Nauk SSSR, Ser. Khim.*, **7**, 1491.
348. Kirik, N.P., Mashkina, A.V., Trofimchuk, A.K., Glushenko, L.V. (1981) *Ukr. Khim. Zh.*, **47**, 395.
349. Nikitayev, A.T., Pomogailo, A.D., Golubeva, N.D., Ivleva, I.N. (1986) *Kinet. Katal.*, **27**, 709.
350. Prokhov, L.T., Andreev, A.A., Shopov, D.M. (1986) *Khim. Tekhol. Topl. Masel*, **11**, 33.
351. Chanda, M., O'Driscoll, K.F., Rempel, G.L. (1980) *J. Mol. Catal.*, **7**, 389; (1980) **11**, 9; (1981) **12**, 197.
352. Gierzatowicz, R., Pawtawski, L. (1985) *Gaz, Woda Tech. Sanit.*, **59**, 26.
353. Chanda, M., O'Driscoll, K., Rempel, G.L. (1984) *Appl. Catal.*, **9**, 291.
354. Chanda, M., Grinshpun, A., O'Driscoll, K.F., Rempel, C.L. (1984) *J. Mol. Catal.*, **26**, 267.
355. Szszepaniak, W. (1963) *Chem. Anal. (Warsaw)*, **8**, 843.
356. Smirnova, G.L. (1989) Nature of Active Centers of Perfluorinated Polymer Films Modified by Transition Metal Ions, *Cand. Sci. (Chem.) Dissertation*, Moscow.
357. Svinova, L.I. (1989) Catalytic Activity of Hydrogels Modified by Transition Metal Ions in Oxidative Processes, *Cand. Sci. (Chem.) Dissertation*, Moscow.
358. Fam, M. T. (1989) Influence of the Content and Structure of Metal Complex Polymer Catalysts on Their Catalytic Activity in Oxidation of Sulfoanion by Molecular Oxygen, *Cand. Sci. (Chem.) Dissertation*, Moscow.
359. Smirnova G.L., Astanina A.N., Rudenko A.P., *et al.* (1989) *Zh. Fiz. Khim.*, **63**, 703.
360. Astanina, A.N., Smirnova, G.L., Skvortsova, L.N., *et al.* (1988) *Zh. Fiz. Khim*, **62**, 1786; (1987) **61**, 2802.
361. Astanina, A.N., Fom, M. T., Volkov, V.I., *et al.* (1989) *Zh. Fiz. Khim*, **63**, 615.
362. Golodov, V.A., Shindler, Yu. M. (1984) *Tr. Ins-t. Org. Katal. Elektrokhim. Akad. Nauk Kaz. SSR*, **23**, 190.
363. Matatov, Yu.I. (1987) *Abstracts of the VI All-Union Conference on Chemistry of Non Aqueous Solutions of Inorganic and Complex Compounds*, (Rostov-on-Don, 1987).
364. Kopylova, V.D., Astanina, A.N. (1987) *Ionitnyye Kompleksy v Katalize* (Ionite Complexes in Catalysis), Moscow: Khimiya.
365. Inventor's Certificate 978813, 1041142, 1181706, 1264974, USSR.
366. Matsuda Sinpei, Mari Tosikatsu, Takeuti Seidzi, *et al.* (1981) *Sekubai* (Catalyst), **23**, 293.
367. Sharipova, F.V., Leplyanin, G.V., Dzhemilev, U.M., *et al.* (1987) *Abstracts of the IV All-Union Symposium on Heterogeneous Catalysis in Chemistry of Heterocyclic Compounds* (in Russian), p. 42, Riga.
368. Carlini, C., Sbrana, G. (1981) *Macromol. Sci., Chem.*, **16**, 323.
369. Levitzki, A., Pechi, J., Berger, A. (1972) *J. Am. Chem. Soc.*, **94**, 6844.
370. Ueda, J., Hanaki, A. (1984) *Bull. Chem. Soc. Jpn.*, **57**, 3511.

371. Hatano, M., Nozawa, T., Ikeda, S., Yamamoto, T. (1971) *Makromol. Chem.*, **141**, pp. 11, 31.
372. Nozawa, T., Akimoto, Y., Hatano, M. (1972) *Makromol. Chem.*, **158**, 21; **161**, 289.
373. Strothkamp, K.G., Dawson, C.R. (1974) *Biochemistry*, **13**, 434.
374. Peterson, L.A., Trevor, A., Gastanoli, N. (1987) *J. Med. Chem.*, **30**, 249.
375. Barteri, M., Pispisa, B., Primiceri, M. (1979) *Biopolymers*, **18**, 3115; (1980) *J. Inorg. Biochem.*, **12**, 167.
376. Barteri, M., Pispisa, B. (1980) *Preprints from International Symposium on Macromolecules, Macro Florence* (Pisa, 1979), Vol. 4, p. 155.
377. Barteri, M., Pispisa, B. (1982) *J. Chem. Soc., Faraday Trans.*, **78**, pp. 2073, 2085.
378. Pispisa, B., Palleschi, A., Paradossi, G. (1987) *J. Mol. Catal.*, **42**, 269.
379. Aptekar', M.D., Garnovskii, A.D., Ismailov, Kh.M. (1983) *Koord. Khim.*, **9**, 487.
380. Pispisa, B., Palleschi, A. (1986) *Macromolecules*, **19**, 904.
381. Vengerova, N.A., Lukashina, N.N., Kirsh, Yu.E., Kabanov, V.A. (1973) *Vysokomol. Soedin.*, **15**, 773.
382. Scurlatov, Yu. I., Kovner, Y.Ya., Travin, S.O., *et al.* (1979) *Eur. Polym. J.*, **15**, 811.
383. Koyama, N., Ueno, Y., Sekiyama, Y., *et al.* (1986) *Polymer*, **27**, 293.
384. Jpn Patent 59–205343, 1985.
385. Sun, Z., Yan, C., Kitano, H. (1986) *Macromolecules*, **19**, 984.
386. Krupnenskii, V.I., Potapov G.P. (1987) *Abstracts of the XVI All-Union Chuguyev Conference on Chemistry of Complex Compounds* (Krasnoyarsk, 1986), Vol. 2, 594.
387. Astanina, A.N., Karpov, V.V., Trusov, P.Yu., Rudenko, A.P. (1985) *Zh. Fiz. Khim.*, **59**, 1641.
388. Pispisa, B., Paradossi, G., Palleschi, A., Desideri, A. (1988) *J. Phys. Chem.*, **92**, 3422.
389. Yoshioka, T., Shimamura, M. (1984) *Bull. Chem. Soc. Jpn.*, **57**, 334.
390. Belokon, Y.N., Tatarov, V.I., Savel'eva, T.F., *et al.* (1986) *Makromol. Chem.*, **187**, 1065.
391. Allen, N.S. (1986) *Chem. Soc. Rev.*, **15**, 373.
392. Kucher, P.V., Opeida, L.I., Nivikov, T.V., and Turovskii, A.A.(1987) *Dokl. Akad. Nauk UzSSR, Geol., Chem. Biology*, **5**, 37; (1989) *Zh. Prikl. Khim.*, **62**, 2781.
393. Yasman, Yu. B., Nel'kenbaum, E.M. (1985) *Tes. dokl. konf. mol. uchenykh* (Abstracts of a Conference of Young Scientists) (Ufa, 1995), p. 396.
394. Allara, D.L., White, C.W. (1978) *Stabilization and Degradation of Polymymers Symposium, 173 Meeting of the American Chemical Society* (New Orleans, 1978), p. 273, Washington DC,.
395. Ballard, D.G.H., Dawkins, J.M. (1974) *Eur. Polym. J.*, **10**, 829.
396. Inventor's certificate 23644, Bulgaria.
397. Abdel-Bary, E.M., Sarhan, A.A., Abdel-Razik, E.A. (1987) *Polym. Degrad. Stabil.*, **18**, 145.
398. Nikitin, A.V., Kholuiskaya, S.N., Rubailo, V.L. (1992) *Kinet. Katal.*, **33**, 58.
399. Huang, S.J., Wang, J.-F., Quinga, E. (1983) *Modified Polymers, Proceedings of the Symposium* (Las Vegas, 1982), p. 75, New York.
400. Mikheyev, Yu.A., Guseva, L.P. (1987) *Khim. Fizika*, **6**, 1259.
401. Inventor's certificate 234832, Czechoslovakia; Ger. Patent 3209434, 1984.
402. Paley, M.S., Harris, J.M. (1987) *J. Polym. Sci., Polym. Chem. Ed.*, **25**, 2447.
403. Jpn Patent 57–53331, 1984.

404. Mazaletskii, A.V., Vinigradova, V.G., Panova, G.V. (1989) *Neftekhimiya*, **29**, 414.
405. Zhdanov, A.A., Papkov, V.S. (1986) *Makromol. Chem. Makromol. Symp.*, **4**, 203.
406. Kolninov, O.V., Kolesnikova, V.V., Milinchuk, V.K. (1984) *Vysokomol. Soedin, Ser. B*, **26**, 9.
407. Staszewska, D.U. (1983) *Angew. Makromol. Chem.*, **118**, 1.
408. Quirk, R.P., Chen, W.-C. (1984) *J. Polym. Sci., Polym. Chem. Ed.*, **22**, 2993.
409. Khalil, M.J., Aglano, J., Habeish, A. (1982) *J. Appl.Polym.Sci.*, **27**, 2377.
410. Egorenkov, N.I., Kuzavkov, A.I., Doktorova, V.A. (1986) *Vysokomol. Soedin.*, **28**, 1525.
411. Pozdnyakov, O.F., Redkov, B.P., Tabarov, S.Kh. (1987) *Zh. Prikl. Khim.*, **60**, 1610.
412. Jellinek, H.H.G., Wei Y.C. (1986) *J. Polym. Sci., Polym. Phys. Ed.*, **24**, 503.
413. Eliseyeva, I.M., Sviridenko, V.G., Lin D.G. (1988) *Kauch. Rezina*, **2**, 7.
414. Irkhin, B.L., Spitsina, S.D., Ponomarenko, V.I., *et al.* (1989) *Vysokomol. Soedin., Ser. B*, **31**, 611.
415. Abou el Naga, H.H., Salem, A.E.M. (1984) *Wear*, **96**, 267.
416. Osawa, Z. (1984) *Develop. Polym. Stabilizat.*, **7**, 193.
417. Kishore, K., Verneker, V.R.P., Dharumaraj, G.V. (1983) *J. Polym. Sci., Polym. Lett. Ed.*, **21**, 627.
418. Maletin, Yu.A. (1985) *Teoreticheskaya I Prikladnaya Khimiya Diketonov* (Theoretical and Applied Chemistry of Diketones), p. 256, Moscow.
419. Gladyshev, G.P., Mashukov, N.I., Mikitayev, A.K., El'tsyn N.A. (1986) *Vysokomol. Soedin., Ser. B*, **28**, 62.
420. Yang, X.-Z., Chien, J.C.W. (1985) *J. Polym. Sci., Polym. Chem. Ed.*, **23**, 859.
421. Chiessi, E., Pispisa, B. (1994) *J. Mol. Catal.*, **87**, 177.
422. Bedioni, F., Devynck, J., Bied-Charreton, C. (1995) *Acc. Chem. Res.*, **28**, 30.
423. Maslinska-Solich, J., Szaton, A. (1993) *React. Polym.*, **19**, 191.
424. Parumova, A.M., Suleimanova, P.G., Sultanov, Yu.M., Molochnikov, L.S., Efendiev, A.A., Shakhtankhtinski, T.I. (1995) *Kinet. Katal.*, **36**, 523.
425. Tau Van Min, Rudenko, A.P. (1994) *Vest. Mosk. Universiteta, Seria 2, Khimia*, **35**, 367.
426. De, B.B., Lohray, B.B., Sivaram, S., Dhal, P.K. (1994) *Macromolecules*, **27**, 1291.
427. Dell' Anna, M.M., Mastrorilli, P., Nobile, C.F., Suranna, G.P. (1995) *J. Mol. Catal.: A:Chemical*, **103**, 17.
428. Naughton, M.J., Drago, R.S. (1995) *J. Catalysis*, **155**, 383.
429. Dupont, P., Vedrine, J.C., Paumark, E., Hecquet, G., Lefebvre, F. (1995) *Appl. Catal.: A: General*, **129**, 217.

CHAPTER 3

POLYMERIZATION PROCESSES CATALYZED BY IMMOBILIZED METAL COMPLEXES

In contrast with hydrogenation and oxidation reactions, n-merization (di-, *iso*-, oligo-, poly- and copolymerization) processes proceed via a single type of substrate activation, namely using reagents containing multiple bonds. Polymerization addition of one or several monomers, leads to the formation of one, sometimes huge, molecule. Polymerization reactions are not usually accompanied by the formation of any evolved species, whereas polycondensation releases simple product molecules, such as HCl, H_2O and NH_3. There are at least two entities in the transition metal ion coordination sphere. They are the growing macrochain and the coordinated monomer (for ionic coordinative polymerization).

If the immobilization of transition metal complexes under optimal conditions for hydrogenation and oxidation leads to an increase in catalyst activity and selectivity (as well giving the opportunity to regenerate the catalyst) this procedure causes even larger scale changes in the case of polymerization. This may affect not only catalyst activity but also the length of the macrochains formed, molecular-mass distribution, stereoregularity and the composition of polymer products in the case of copolymerization.

Metal complexes bound to a polymer support most frequently induce ionic polymerization of olefins, dienes and acetylenes, and less commonly radical polymerization of vinyl-type monomers, acting at all reaction stages: initiation, chain propagation and termination. Active sites for the addition of monomer molecules to the growing polymer chain can in many cases be regenerated yielding new polymer chains.

At present the application of metal complexes immobilized on polymer supports to catalysis of polymerization has been insufficiently generalized. At the same time the use of metal complexes bound to inorganic supports in polymerization catalysis, mainly of olefins, has been extensively developed. Such catalysts ranged from the first Solvay catalysts supported on silica, alumina and magnesia to the so-called titanium-magnesium catalysts (TiCl₄ on $MgCl_2$) [1–3].

Such a classification of polymer-immobilized catalysts is insufficient. Just metal complexes immobilized on polymer supports were used at first for polymerization (1965/1966). For instance, organoaluminium compounds as one of the catalyst components have been bound to macromolecular ligands with C=O, C≡N and C=N groups [4]. Stereoregulated catalytically active systems have been obtained by MX_n bonding with nitrogen-containing polymers [5], etc. Further investigations promoting development in the field will be summarized.

In this Chapter some peculiarities of the poly- and copolymerizations of olefins (mainly of ethylene and propylene), dienes (butadiene and isoprene), acetylenes (acetylene, phenylacetylene), etc., will be discussed. We will try also to summarize all the main results for copolymerization of α-olefins with polar monomers, and for radical polymerization of vinyl-type monomers. Catalysis of polycondensation processes in the presence of immobilized metal complexes is also included in the scope of this book.

3.1 Homo- and Copolymerization of Olefins Catalyzed by Metal-Containing Polymers

Olefin polymerizations in the presence of bound complexes is one of the most extensively investigated reactions and is also of commercial importance, especially in the case of ethylene and propylene together with their copolymers. Moreover, it is interesting to compare some kinetic peculiarities and mechanisms of catalysis by homogeneous, pseudo-homogeneous and heterogenized catalytic systems and so to clarify the role of the macroligand.

3.1.1 About the mechanism of catalytic polymerization of olefins

Homogeneous and pseudo-homogeneous catalysts usually show high activity for ethylene polymerization at modest temperatures (283–303 K). At the above processing temperatures (333–363 K) the activity of catalytic systems such as $TiCl_4$–AlR_3, Cp_2TiCl_2–$AlEt_2Cl$, $VOCl_3$–AlR_2Cl, $Ti(OR)_4$–AlR_2Cl, sharply reduces with time. This is the result of properties of metal–carbon bond formation and cleavage, because addition of the activated monomer (chain propagation stage) occurs at this bond. The reduction of alkylated transition metal ions ($Ti^4 \rightarrow Ti^3$, $V^5 \rightarrow V^4 \rightarrow V^3$, etc.) is one of the reasons for catalyst deactivation.

To a considerable degree this effect is essential for vanadium ions in high oxidative states (Figure 3.1), appearing in the fast attainment of maximum polymerization rates with subsequent sharp diminution with time. The reduction usually proceeds by bimolecular secondary alkylation with either further disproportionation of alkyl groups in the coordination sphere of the transition metal or the disproportionation of alkyl radicals at different reactive centers [6,7]:

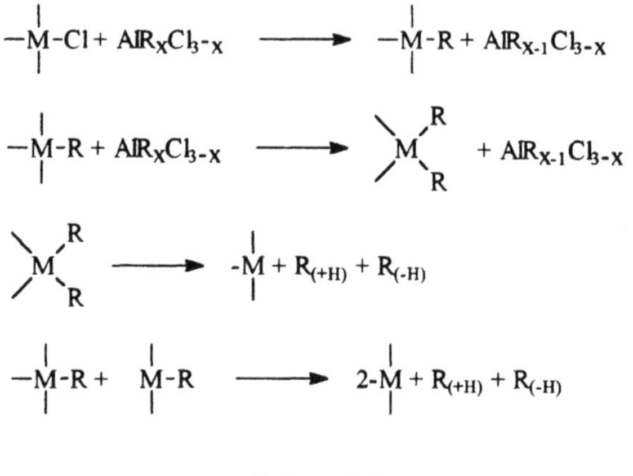

$$-\overset{|}{\underset{|}{M}}-Cl + AlR_xCl_{3-x} \longrightarrow -\overset{|}{\underset{|}{M}}-R + AlR_{x-1}Cl_{3-x}$$

$$-\overset{|}{\underset{|}{M}}-R + AlR_xCl_{3-x} \longrightarrow \overset{\diagdown \quad \diagup R}{\underset{\diagup \quad \diagdown R}{M}} + AlR_{x-1}Cl_{3-x}$$

$$\overset{\diagdown \quad \diagup R}{\underset{\diagup \quad \diagdown R}{M}} \longrightarrow -\overset{|}{\underset{|}{M}} + R_{(+H)} + R_{(-H)}$$

$$-\overset{|}{\underset{|}{M}}-R + \overset{|}{\underset{|}{M}}-R \longrightarrow 2-\overset{|}{\underset{|}{M}} + R_{(+H)} + R_{(-H)}$$

Scheme 3.1

However, both the reasons and the mechanism for homogeneous catalyst deactivation still remain unclear, so that the unstationary character of the polymerization may be the result of changes in either the concentration of propagation centers, n_{pr}, or their nature. It is impossible to maintain the constancy of active center concentration in solution due to their high reactivity[1].

The activity diminution may proceed either via active center deactivation by homo- and heterolytic metal–carbon bond cleavage (first-order reaction) or by disproportionation (second-order chain termination) In the first case the rate changes in n_{pr} follow the equations $-dn_{pr}/d\tau = k_d n_{pr}$ and $\ln n_{pr} = -k_d\tau$, while in the second case $-dn_{pr}/d\tau = k'_d n_{pr}^2$ and $1/n_{pr} = -k'_d\tau$. (k_d and k'_d in these equations contain the constant rates of deactivation for mono- and bimolecular active sites respectively).

Figure 3.2 shows that n_{pr} changes for VCl_4/Al^iBu_3 catalytic system cannot be described by one of the above deactivation equations. So, one can believe that after active site formation (w_p^{max}) and up to 10–15 min after initiation, the process is accompanied mostly with second-order deactivation of active sites, while at the above reaction time the first-order deactivation prevails.

By summation of all corresponding equations one can derive: $-dn_{pr}/n_{pr} = k_d n_{pr} + k'_d n_{pr}^2$ and $-dn_{pr}/n_{pr} = k_d - k'_d n_{pr}$. $k_d = 1.65 \times 10^{-3}$ s^{-1} and $k'_d = 0.48 \times 10^3$ l/(mol s) have been found from $(-dn_{pr}/n_{pr})$ dependence obtained by a radiochemical technique [9]. For homogeneous systems (at normal temperatures) n_{pr} values usually do not exceed 0.1–1.0 mol% of the total transition metal amount showing the great loss of transition metal in side processes. The value of the chain propagation rate constant, $k_{pr} = 7.5 \times 10^3$ l/(mol s), has been found from the equation $W_p = K_{pr} n_{pr} C_m$. By examination of the inverse dependence of P_n on the main agents of chain breaking concentration (monomer, C_m, organoaluminium compound, C_{Al}, etc.) the values of chain termination rates constant can be calculated:

$$1/P_n = (k_0^m C_m + k_0^{Al} C_{Al} + \sum_{1}^{j-2} k_{0_{j-2}} C_{0_{j-2}}^{\alpha_j})/k_{pr} C_m$$

To compare the orders of constant values let us notice that for the catalytic system VCl_4–$(isoBu)_2AlCl$ at 313 K in benzene at Al/V=10, $k_0^m = 37 \times 10^{-2}$, $k_0^{Al} = 2.6 \times 10^{-2}$ l/(mol), $n_{pr} = 0.8 \times 10^{-2}$ mol/mol of V, $E_a = 14.5 \pm 2$ kJ/mol [9].

Almost all catalytic systems derived from vanadium and titanium compounds with the metal ion in a high oxidation state can be attributed to pseudo-homogeneous (colloid-dispersed) catalysts [6]. They are well heterogenized simply by MX_n treatment with modified agents such as amines, esters, nitriles, etc., followed by separation of the metal complex catalysts [10–12]. Such an approach was the first route to metal complex catalyst heterogenization. On the one hand it essentially simplifies the kinetic regularities, and on the other hand it increases the efficiency of modified catalysts for ethylene polymerization (i.e. for the VCl_4·methyl methacrylate (MMA) complex (Figure 3.3)). The steadiness of active site concentrations and hence the stability of the polymerization process, is reached by active site regeneration. However the more

---

[1] Recently, highly active soluble catalysts of Ziegler–Natta type for ethylene polymerization have been developed. These are zirconium compounds bound to methylalumoxanes $(-Al-O-)_n$ (n=10–20; see for example [8]).
$\underset{CH_3}{|}$

promising method of heterogenization of homogeneous and pseudo-homogeneous olefin-polymerization catalysts is through their immobilization by oligomeric or polymeric supports [13].

3.1.2 Ethylene polymerization by oligomer-immobilized catalytic systems

Oligomeric supports occupy a position between low molecular mass materials (modified agents) and macromolecular ligands.

The products of MX_n interaction with oligomeric oxyethylated alcohols, carboxylic acids and anhydrous sorbates are of special interest (synthesis and characteristics are presented in [14]) [15,16]. The rate of ethylene polymerization catalyzed by immobilized on oxyethylated n-nonyl alcohol $MX_{n-1}-O(CH_2CH_2O)_n-C_9H_{19}-AlR_2Cl$ is affected both by the nature of the transition metal ion and the length of the oxyethylene oligomeric chains as well the degree of support functionalization (Figure 3.4). However, the stabilizing action of oligomeric ligands is not strong (the time of activity falls by a factor of two and is only two times higher (10 min) than that for homogeneous system (5 min).

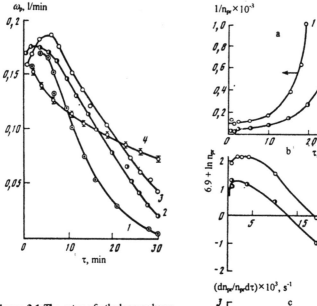

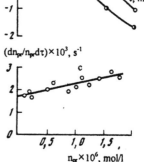

Figure 3.1 The rates of ethylene polymerization by catalytic systems with vanadium in high oxidation states: (1) VCl_4; (2) $VOCl_3$; (3) $VO(OEt)_3$; (4) $V(O\text{-}tert\text{-}Bu)_4$. Cocatalyst $isoBu_2AlCl$, [Al]/[V]=10, 313 K, $C_V=5\times10^{-5}$ mol, ethylene concentration 0.04 mol/l.

Figure 3.2 Active center deactivation with time in accordance with kinetics measurements (1) and by the radiochemical method (2).

A rise of catalyst activity for ethylene polymerization has been achieved by $TiCl_4$ immobilization on oligomeric supports, and also with alcohols with long hydrocarbon chains.

The relation between the nature of the transition ion bond with the support and catalyst activity [17] can be illustrated by the example of oligomeric ligands, particularly for oxyethylated diesters of pentaerythritol. These contain both end group carboxyl and carbonyl groups in ester units as well as oxygen atoms in oxyethylene residues. Interaction with $TiCl_4$, VCl_4, $VOCl_3$, etc., is a multistage process. First is the formation of oligo-metal esters. Second is bonding with carbonyl groups, and last is connection with the oxyethylene residue, each MX_n molecule being linked with three oxyethylene groups.

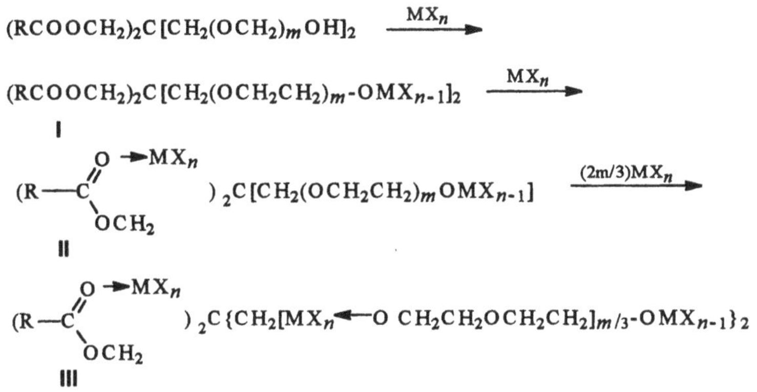

$$(RCOOCH_2)_2C[CH_2(OCH_2)_mOH]_2 \xrightarrow{MX_n}$$

$$(RCOOCH_2)_2C[CH_2(OCH_2CH_2)_m\text{-}OMX_{n\text{-}1}]_2 \xrightarrow{MX_n}$$

I

$$\left(R-C\overset{O \rightarrow MX_n}{\underset{OCH_2}{\diagdown}}\right)_2 C[CH_2(OCH_2CH_2)_mOMX_{n\text{-}1}] \xrightarrow{(2m/3)MX_n}$$

II

$$\left(R-C\overset{O \rightarrow MX_n}{\underset{OCH_2}{\diagdown}}\right)_2 C\{CH_2[MX_n \leftarrow O\,CH_2CH_2OCH_2CH_2]_{m/3}\text{-}OMX_{n\text{-}1}\}_2$$

III

Scheme 3.2

Time-dependent changes in specific catalytic activity (SCA) for complexes I–III are plotted in Figure 3.5. The type of VCl_4 connection with the oligomeric ligand (σ or nv) affects the ethylene polymerization rate ((isoBu)$_2$AlCl as cocatalyst). Vanadium-containing esters (I) showed the largest SCA value, while all the products of VCl_4 interaction with OH, C=O, CH$_2$-O-CH$_2$– groups, etc., revealed noticeably low activity. Therefore strong restrictions should be imposed on the functional uniformity of the polymer support.

It has been shown [18,19] that ethylene polymerization in the presence of both oligo-immobilized and modified catalysts proceeds at uniform Ziegler-type active centers similar to those of homogeneous systems (MX_n–$AlR_xCl_{3\text{-}x}$). The differences are of degree rather than of kind[2]: both n_{pr} and k_{pr} factors have similar values to those of the systems with which they are compared (Tables 3.1 and 3.2).

The increase of catalyst activity in olefin polymerizations has also been observed in the presence of systems modified with a number of other oligomeric compounds [21–23].

---

[2] The so-called oligomeric effect has been observed in several catalytic processes, for instance in oxidative dimerization of acetylenes [20]. The effect is displayed by the fact that metal complexes with oligomeric ligands are more effective than those with monomeric and polymeric analogues. In the case mentioned the effect has not been observed.

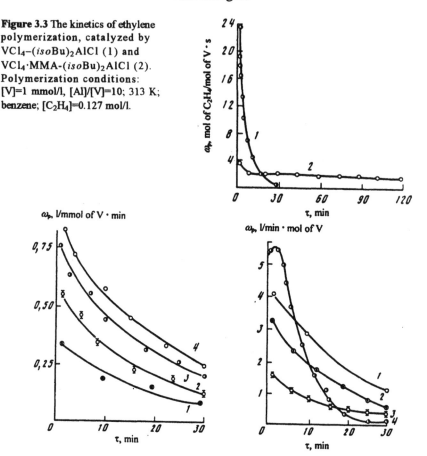

Figure 3.3 The kinetics of ethylene polymerization, catalyzed by $VCl_4-(isoBu)_2AlCl$ (1) and $VCl_4 \cdot MMA-(isoBu)_2AlCl$ (2). Polymerization conditions: $[V]=1$ mmol/l, $[Al]/[V]=10$; 313 K; benzene; $[C_2H_4]=0.127$ mol/l.

Figure 3.4 The influence of degree of n-nonyl alcohol oxyethylation on the catalytic properties of the products of its interaction with $VOCl_3$ (1) or VCl_4 (2–4) for ethylene polymerization. Reaction conditions: $(isoBu)_2AlCl$ as cocatalyst; benzene; 313 K; $[Al]/[V]=14$; $[C_2H_4]=0.04$ mol/l. The value of oxyethylation is: (2) 4.5, (3) 5.2, (4) 9.

Figure 3.5 The influence of the bond nature between VCl_4 and the oxyethylated ($n=20$) divalerianate of pentaerythritol on the rate of ethylene polymerization. Products of interaction: (1) end group hydroxyls; (2) end groups of both hydroxyls and carbonyls; (3) at all accessible groups (III), and (4) in the presence of conventional catalysts $VCl_4-(isoBu)_2Cl$.

3.1.3 Ethylene polymerization in the presence of metal complexes bound to homo- and copolymers

Both carbon and heterochain polymers with carboxyl, amino, amido and hydrosulfide groups, as well phosphorus-containing substances are widely used as supports. There are several methods of macroligand functionalization [17]. The simplified scheme of MX_n immobilization can be performed as follows [13]:

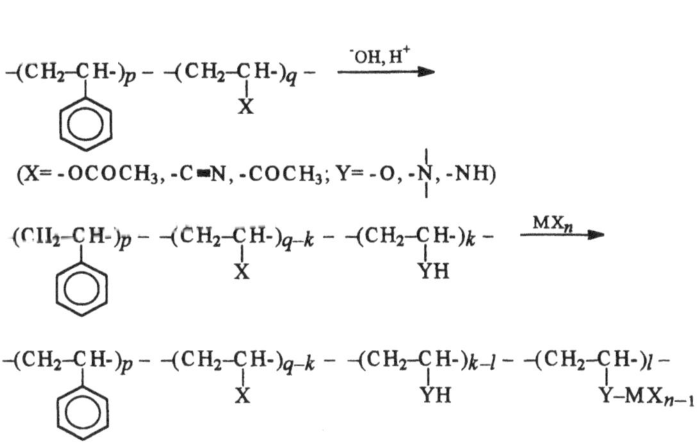

$$-(CH_2-CH-)_p - -(CH_2-CH-)_q - \xrightarrow{^-OH, H^+}$$

(ring) | X

$$(X= -OCOCH_3, -C\!\equiv\!N, -COCH_3; \; Y= -O, -\!\overset{|}{N}, -NH)$$

$$(CH_2-CH-)_p - -(CH_2-CH-)_{q-k} - -(CH_2-CH-)_k - \xrightarrow{MX_n}$$

(ring) | X | YH

$$-(CH_2-CH-)_p - -(CH_2-CH-)_{q-k} - -(CH_2-CH-)_{k-l} - -(CH_2-CH-)_l -$$

(ring) | X | YH | Y-MX_{n-1}

Scheme 3.3

The value of the $l/(k-l)$ ratio can change across a broad range (up to 10–20 in the case of a saponifiable copolymer of styrene with vinylacetate [13]) depending on both the nature of the functional groups and the reaction conditions; the degree of loading of polymer functional groups (with $TiCl_4$) for such supports can reach 80% [24]. Some other variants of MX_n immobilization are also known [25–40].

The data on polymerization reactions catalyzed by macrocomplexes are not numerous; they are mainly concentrated in patents. The conditions explored for such catalysts are not comparable. Moreover, it is rather difficult to find a universal parameter for SCA evaluation. The most commonly used one is determined as "polymer yield per hour related to 1 g of fixed transition metal on a polymerized monomer at a pressure of 1 atm.". However, this parameter does not correctly display in many cases the true value of the catalyst activity because the process is nonstationarity.

Analysis of results presented in Figure 3.6 shows that elongation of the ligand chain also increases catalyst activity, as it does for oligomer-immobilized systems. The overall values of polyethylene (PE) yield are in the ratio 1:1.5:5.0 for complexes of VCl_4 with MMA, oligomeric MMA and PMMA, respectively (Table 3.2).

It is significant that covalently bound MX_n are more active than coordinated macrocomplexes [13]. Similar regularities have been observed [24] in comparisons of catalysts obtained by the interaction of $TiCl_4$ with a hydrolyzed copolymer of ethylene, or with vinylacetate or its analogue, n-butoxy(titanium) trichloride (Figure 3.7b).

Among all the kinetic regularities for ethylene polymerization in the presence of all the systems compared, following are worthy of note. The same npr values are reached at greater Al/M ratios (10–15) for immobilized catalysts than those for homogeneous catalysts (Al/M=2) or modified systems (Al/M=6–8). In all cases the increase of Al/M ratio is accompanied with a reduction of PE molecular mass. At optimal Al/M ratios, polymerization is a first-order reaction with respect to both monomer and transition metal concentrations. The chain transfer reaction involving monomer competes with this with participation of an organoaluminum compound [41]. The magnitudes of activation energies for the chain propagation step are similar. This indicates that the same active centers are responsible for catalysis in both systems. However, the rise in the effective value of the activation energy, E_a, and temperature variations of the number of active sites, E_{anpr}, both suggest a complicated mechanism for active site formation in immobilized systems. In addition, both the rate constant values for monomolecular, k_d, and bimolecular deactivation of active sites are reduced by a factor of 5 for immobilized catalysts. This is connected with differences in both mechanism and kinetics of active site formation in comparable systems. While for homogeneous VCl_4–AlR_2Cl, the value of n_{pr} reduced by a factor of 45 within 30 min of polymerization, it reduced only by 5 for immobilized system, remaining essentially constant for 3 hours (Figure 3.8). Some increase of the n_{pr} value has even been observed for $VCl_4 \cdot MMA$.

a_p, mol of C_2H_4/mol of V · s

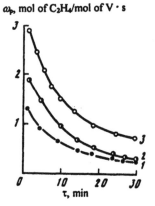

Figure 3.6 The effect of the ligand chain length on ethylene polymerization rate by $VCl_4 \cdot L$ complexes. Ligands: (1) MMA, (2) $(PMMA)_0$, (3) PMMA.

Figure 3.7 The influence of the type of immobilization on catalytic behavior of VCl_4–$(isoBu)_2AlCl$ (a) and $TiCl_4$–$AlEt_3$ (b) systems. a: (1) vinylacetate (VA), (2) polyvinyl-acetate (PVA), (3) hydrolized copolymer of styrene with VA. Polymerization conditions: benzene; 313 K; [Al]/[V]=50; [C_2H_4]=0.04 mol/l. b: (1) hydrolized copolymer of ethylene with VA, (2) BuOH.

a_p, l/min · g of V

a_p, ml of C_2H_4/min

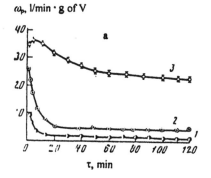

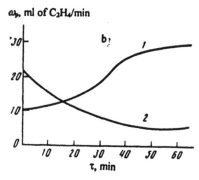

Another type of macromolecular catalyst for ethylene polymerization is also of great interest. They are macromolecular carbonium salts of transition metals [42,43]. Tritylium salts of general formula $[Ph_3C]^+[MX_nHal]^-$ (Hal – halogen) are the active components of ethylene polymerization catalysts. Their polymer analogues based on copolymers of styrene (CS) with monomers containing an active halogen atom (α-bromostyrene (CSBr-S), α,β,β'-trifluorostyrene (CS3FS), chloromethylated polystyrene (CMPS), etc.) are several times as active as their homogeneous analogues (Figure 3.9). Moreover, these catalysts are soluble in aromatic solvents, at least during the initial stage of the polymerization process. However, these systems are non-stationary even at 281 K. This is the outcome of significant lability of active sites bound to mobile chain segments, and results in bimolecular deactivation as in the case of homogeneous catalysts.

However, a polymer chain can have a protective effect on the active sites. While, k_d and k_d' values for the homogeneous system VCl_4–AlR_2Cl are equal to 2.2×10^{-3} s^{-1} and 1.93×10^3 l/(mol s), these values for polymer-immobilized systems $VCl_4 \cdot CS3FS$ and $VCl_4 \cdot CSBrS$ are equal to 0.9×10^{-3}, 1.2×10^{-3} s^{-1} and 0.95×10^3, 1.3×10^3 l/(mol s), respectively. The progressive fall of ethylene polymerization rate can also be the result of the phase discontinuity which appears just after the initiation of polymerization. This effect is accompanied with occlusion, or microcapsulation of a significant part of the active sites in the volume of the polymer, thus providing an additional diffusion barrier for polymerized ethylene.

Immobilized catalysts derived from block copolymers, (polystyrene)$_{100}$–(butadiene)$_3$–Li–$TiCl_4$, are soluble and effectively catalyze ethylene polymerization [44]. The nature of the active sites in such systems is, probably, of Ziegler type.

Immobilization of organoaluminium compounds, the second catalyst component, is also promising for an increase of catalyst activity. It can be realized via interaction with the polymers, containing C=O, C=N and C≡N groups [4,35] or by adding AlR_2H or AlR_3 to the multiple bonds of the polymers [45,46]:

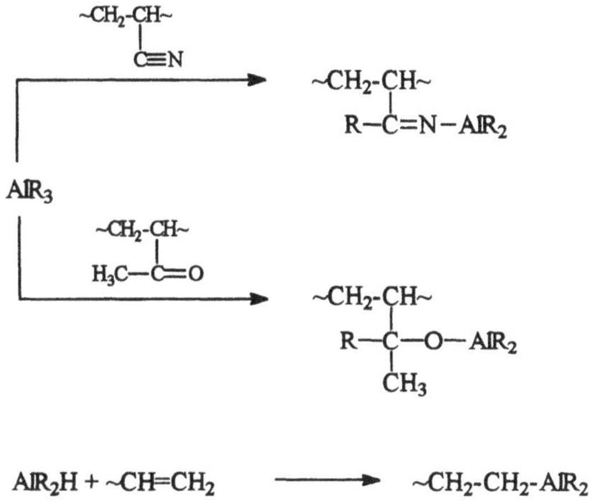

Scheme 3.4

However, in these systems the unbound transition metal compound is used for ethylene polymerization. Mono-component organo-transition metal systems [47] bound to polymers are also used for ethylene polymerization. Specifically, tetrabenzylzirconium, $Ti(CH_2SiMe_3)_4$ and π-allyl compounds of Ni^{2+}, Cr^{3+} and Mo^{6+}, as well as a $Cr(norbornyl)_4$ step in 2-addition to poly(ethylene terephthalate) or nylon-66 [48] providing an effective catalyst for ethylene polymerization.

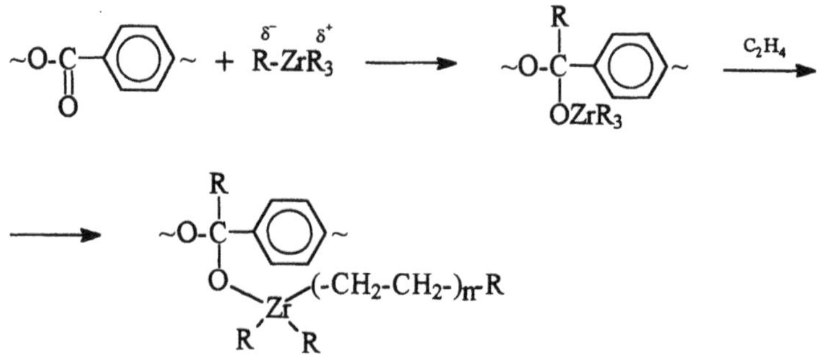

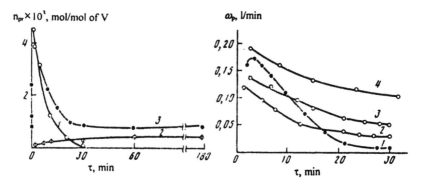

Scheme 3.5

If the catalyst is bound to a cord fiber, a protective cover of polyethylene is formed around polyamide or poly(ethylene terephthalate) thread proofing it against the action of moisture. The significant technological demand from the process is a high polymerization rate at the commencement of the process (in the first several milliseconds) at the instant of the thread passing through a special spinneret (the rate of thread winding is about 600 m/min). The activity of the catalyst at that moment reaches the value of 10^6 g/(mmol of Zr atm h) and sharply decreases during several seconds due to the sharp fall of monomer access. Uniform polyethylene covers on the fibers of poly(vinyl acetate) (PVAc), polyamides, cotton, flax, viscose, etc., containing OH, NH and NHR groups for the bonding of Ti, Zr, V and Cr can also be obtained. However, for this additional activation with organoaluminium compounds is required [36].

Figure 3.8 The dependence of active site number on the time of ethylene polymerization by catalytic systems: (1) VCl_4-$(isoBu)_2AlCl$; (2) $VCl_4 \cdot MMAc/(isoBu)_2AlCl$; (3) $VCl_4 \cdot PMMAc/(isoBu)_2AlCl$. Reaction conditions as for Figure 3.70.

Figure 3.9 The kinetics of ethylene polymerization by macromolecular carbonium salts of VCl_4 in benzene. Polymer supports: 1 – VCl_4-$AlEt_2Cl$, 2 – CSBrS, 3 – CMPS, 4 – CS3FS. Reaction conditions: $[C_2H_4]=0.04$ mol/l, $[V]=0.5$ mmol/l, $[Al]/[V]-10$, 313 K.

Table 3.1 Values of n_{pr} and k_{pr} in ethylene polymerization[a] catalyzed by oligomeric immobilized systems obtained by interaction of VCl4 with the oxyethylated (n=20) ester of the anhydrosorbite of lauric acid (complex nomenclature as given for Scheme 3.2).

Polymerization time (s)	I			II			III		
	w_p	$n_{pr}\times10^3$	$k_{pr}\times10^{-3}$	w_p	$n_{pr}\times10^3$	$k_{pr}\times10^{-3}$	w_p	$n_{pr}\times10^3$	$k_{pr}\times10^{-3}$
60	2.85	1.27	8.1	2.26	0.66	12.3	1.25	0.46	9.7
120	2.74	1.25	7.9	2.15	1.0	7.7	1.15	0.30	14.1
180	2.60	1.38	6.8	2.04	0.87	8.4	1.10	0.32	12.5
300	2.35	1.13	7.5	1.85	0.94	7.1	1.05	0.24	15.9
600	1.98	0.80	9.0	1.37	0.32	11.4	0.80	0.25	11.4
1200	1.27	0.86	5.3	0.85	0.28	10.9	0.45	0.12	13.4
1800	0.88	0.50	6.4	0.51	0.26	7.1	0.39	0.15	9.9
Average			7±2			9±3			13±3

[a] w_p – mol of C_2H_4/mol of V; n_{pr} – mol/mol of V; k_{pr} – l/(mol s); reaction conditions see at the Figure 3.4; $[C_2H_4]$=0.28 mol/l.

Table 3.2 Kinetic parameters for ethylene polymerization catalyzed by the systems $MX_n\cdot L$–$Al(isoBu)_2Cl$ [41].

$MX_n\cdot L$	$k_{pr}\times10^{-3}$ (l/mol s)	$n_{pr}\times10^3$ (mol/mol)	k_0^M (l/mol s)	k_0^{Al} (l/mol s)	$k_d\times10^3$ (s⁻¹)	$k_d'\times10^{-3}$ (l/mol s)	E_a^{eff} (kJ/mol)	E_a^k (kJ/mol)	$E_a^{n_{pr}}$ (kJ/mol)
VCl4	(7.5±0.5)[a]	(12.5)	(0.345)	(1.68)	(2.9)	(1.76)	(16.4±2)	5.8±1	11.3±2
VCl4PMMAc	18.5±1	4.58	0.85	4.14	2.2	1.93	(24±4)	7±2	17.2±2
VCl4MMAc	16.2±2	3.8	0.61	2.90	0.54	0.56	(2.1±1)	9.7±2	28.6±2
VCl4PBCarb[c]	10.5±2[b]	0.2	0.72	2.02				6.3±2	21.9±2
VCl4VCarb	14±2	1.2						8±2	31.5±2
	12±2	0.19							

[a] Data obtained by kinetic method, the rest of the data obtained by radiochemical analysis (^{14}CO).

[b] The value at corresponding n_{pr}^{max}.

[c] PVCarb and VCarb are polyvinylcarbynol and vinylcarbynol, respectively.

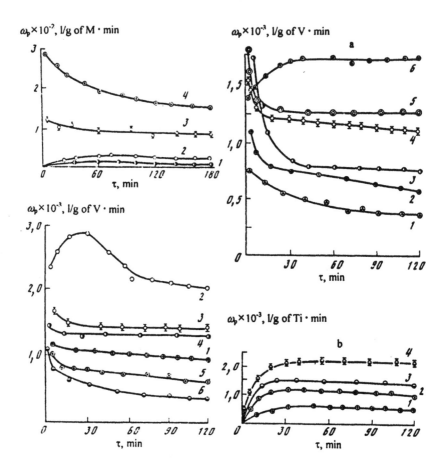

Figure 3.10 Dependences of ethylene polymerization rate on the type of metal halide immobilized on PE-gr-PAAl. 1, MoCl₅; 2, ZrCl₄, 3, TiCl₄, 4, VCl₄. Polymerization conditions: heptene, 343 K, [Al]/[M]=50, ethylene pressure 0.4 MPa.

Figure 3.11 Ethylene polymerization kinetics by catalytic systems derived from VCl₄ (a) and TiCl₄ (b) bound to polymer supports by donor-acceptor interaction. Polymer supports: (a) 1, PE-gr-poly(diisobutyl ester of vinyl phosphonic acid), 2, PE-gr-poly(diallyl sulfide), 3, PE-gr-PAN, 4, PE-gr-PMMA, 5, PE-gr-poly(methyl vinyl ketone), 6, PE-gr-PVP; (b) 1, PE-gr-poly(diallyl sulfide), 2, PE-gr-PVP, 3, PE-gr-poly(vinyl acetate), 4, PE-gr-PMMA. Polymerization conditions: heptene, 343 K, AlEt₂Cl as cocatalyst, [Al]/[M]=50, ethylene pressure 0.4 MPa.

Figure 3.12 Ethylene polymerization kinetics by catalytic systems derived from VCl₄ immobilized on polymer supports by covalent bonding. Polymer supports: 1, PE-gr-poly(allyl sulfide), 2, PE-gr-PAAl, 3, PE-gr-poly(allyl amine), 4, PE-gr-poly(diallyl amine), 5, PE-gr-poly(acrylic acid), 6, PE-gr-poly(α,β,β′-trifluorostyrene).

We have considered systems that are merely variants when MX_n or AlR_xCl_{3-x} are chemically bound to the polymer support. However, there are numerous ways for the physical bonding of metal complex components of ethylene polymerization catalysts. Below we have not attempted to be comprehensive but rather to identify the most commonly used. These are impregnation of PE granules [37], ion-exchange resins [38], macroporous copolymer of styrene with divinylbenzene (CSDVB) species [39] with MX_n or Al–organic compounds (either from solutions or the gas phase). The properties of catalysts obtained by interaction of titanium chlorides and macroreticular polymer sorbents can be regulated on a large scale by changing reaction conditions [49].

The ease of VCl_3 precipitation on the surfaces of different materials (sometimes, directly by the reduction of vapors, $2VCl_4 \rightarrow 2VCl_3 + Cl_2$) is well known. This technique is often used to obtain ethylene-polymerization immobilized catalysts. Polyethylene membrane species can be obtained by using these catalytic systems [50]. Some other mechanochemical methods of MX_n chemical immobilization in the solid phase accompanied by metal complex microcapsulation within polymer matrixes [40] (see Section 3.1.7) can also be attributed to this group of techniques.

3.1.4 Immobilized catalysts on the base of grafted complexes

Numerous studies have shown that to increase the catalytic activity of immobilized systems by 1–2 orders of magnitude, it is necessary to replace the active sites on the surface of the support or within a thin presurface layer (due to the heterophase character of the reaction).

As the topochemistry of bound MX_n is specified by the polymer functional group distribution [17, pp.92–99], special polymer carriers with functional groups, donor centers, or chelate nodes grafted to the surface of inert polymer materials, such as PE, polypropylene (PP), copolymer of ethylene with polypropylene (CEP), polystyrene (PS), poly(vinyl chloride) (PVC), have been developed [51,52]. The general scheme of formation of such supports is powder polymer is subjected to some mechanical, chemical or irradiation action in the presence of the desirable monomer. Grafting polymerization proceeds on active sites (free radicals, ions) yielding the grafted polymer layer.

$$] \xrightarrow{\text{initiation}}]^* \xrightarrow{\underset{Y}{\overset{CH_2=CH}{|}}}]\text{-(-CH}_2\text{-CH-})_n\text{-}$$

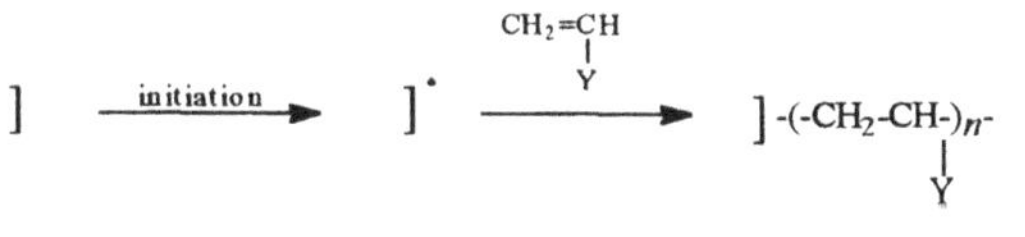

] - polymer surface;

$Y = $ -OH, -NHR, -CH$_2$NH$_2$, -SR, -CH$_2$OH, -OCOCH$_3$, -COOH, -C≡N, -C$_5$H$_4$N, ≡P=O, ‾PR$_2$, etc.

Scheme 3.6

It has been shown that the grafted polymer layer (at the optimal content of functional groups, which is equal to 0.1–1.3 mmol/g), of thickness   approx. 10–30 nm, is completely accessible to the reagents. PE-grafted-polyallyl alcohol (PE-gr-PAAl), PE-gr-poly(4-vinylpyridine) (PE-gr-P4VP), PE-gr-poly(acrylic acid) (PE-gr-PAAc), PE-gr-poly(allyl amine) (PE-gr-PAA), PE-gr-poly(diallyl amine) (PE-gr-PDAA), etc., have proved to be active. The fixating of catalytic components to such grafted layers follows the same procedure as for homo- and copolymer supports [53,54]. The transition metal compound can penetrate and bind with functional groups at a depth of 10 monolayers of grafted monomer. For comparison let us notice that the depth of a mono-component (Ti^{4+} or Zr^{4+} compounds) catalyst penetration into the fibers is estimated to be equal to 20 monolayers [48].

By analogy with heterogeneous catalysts that contain the active component in thin presurface layer, these ones can be called "cork" or "crust" catalysts. Let us analyze and compare the catalytic properties of such systems.

The ethylene polymerization rate significantly depends on the nature of the transition metal. Compounds of vanadium in high oxidation states also show the greatest activity (Figure 3.10). The high stability of vanadium-containing catalysts (even at elevated temperatures, 343 K) is worth consideration. The $\tau_{1/2}$ value is 180 min, whereas for a homogeneous system it is equal to 5 min even at 313 K. Any organoaluminium compound can serve as cocatalyst for these metal polymers; however, AlR_2Cl is the most active in the sequence: $AlEt_2Cl>AlPr_2Cl>Al(isoBu)_2Cl$. Furthermore, the highest activity is reached at larger Al/M ratios (50–100) than those for homogeneous systems (2–10). However, the overall concentration of organoaluminium compound in the reactive volume is, as a rule lower than that for the soluble systems (0.1–0.5 mmol/l).

The nature of the polymer support, especially of the polymer grafted layer, has an inhibiting action on the activity of such catalysts. The dependence of ethylene polymerization rate on the nature of the donor centers in immobilization of VCl_4 and $TiCl_4$, as well on the type of M–L (L–polymer ligand) σ-bond is illustrated by the Figures 3.11 and 3.12. In all cases immobilization of MX_n yields active and stable catalytic systems. Usually, σ-bonding of transition metal with functional cover increases both catalyst activity and stability [55].

It is difficult to compare kinetic parameters for ethylene polymerization by homogeneous and grafted metal complexes. This results from differences in the operating temperatures. Thus, a working temperature range for VCl_4–$(isoBu)_2AlCl$ is 283–313 K. The stage of chain propagation is defined by the equation [9] $k_{pr}=2.2\times10^8\exp(-4600/RT)$. The extrapolation to polymerization temperature in the presence of immobilized vanadium compounds (343 K) gives the value $k_{pr}^{343}=4.1\times10^4$ l(mol s). The values of n_{pr}, k_{pr}^{343}, and effective values of activation energy for ethylene polymerization by immobilized VCl_4 systems are given in Table 3.3.

One can see that k_{pr} values for both systems are of the same order of magnitude within experimental error. However, the concentration of specific sites for immobilized systems (PE-gr-PAAl) is by 2–3 times greater than the maximum value for soluble systems, and the n_{pr} value is 80–250 times (after 30 min polymerization) greater than for the homogeneous system. This is the main reason for the high specificity of immobilized catalysts in ethylene polymerization[3].

This effect becomes clearly evident by developing heterogeneous analogues of the homogeneous catalytic systems (Cp_2TiCl_2, Cp_2VCl_2, $Ti(OBu)_4$, $VO(OEt)_3$, etc.). Thus, immobilized Cp_2TiCl_2–$AlEt_2Cl$ shows significantly increased activity and stability (Figure 3.13). Under optimal conditions $Ti(OBu)_4$ activity can rise by a factor of 100–150 (Figure 3.14) . Some kinetic peculiarities of such systems which have no explanation at the present time are as follows. The decrease of reaction order with respect to titanium from 1.4

---

[3] Similar data [1] both on the nature of active sites and the values of rate constants have been obtained for systems based on $TiCl_4$ anchored to inorganic supports; data sets remained similar after the fixation.

(homogeneous system) to 0.66 (immobilized system) (similar to the monomer (0.5)) and the sharp reduction (to 11 kJ/mol) of the activation energy in comparison with that for the homogeneous system (58 kJ/mol). It is most probable that immobilization of the bulky Ti(OBu)$_4$ molecule is favorable for she substrate entering structures having a reducing insertion barrier. The constancy of k_{pr} values for immobilized systems (Table 3.3) indicates, in turn, the sharp drop of the rate constant pre-exponential factor. The observed compensating effect indicates that adjacent to the active site polymer fragments are involved in the reaction, resulting in a decrease of entropy for active site reactions [56].

The stabilization of activities of immobilized catalysts can be demonstrated by their behavior under aging (exposure in the absence of monomer). The aging of soluble (homogeneous and colloid) systems leads to a significant specificity decrease. This is the result of side catalyst interactions with components of the reaction medium. These are accompanied by transition metal ion reductions (see Scheme 3.1), aggregation of active forms, etc. In immobilized catalysts deactivation of active sites is controlled by their formation; the surface of the polymer support or grafted functional layer, in their turn, prevent deactivation by separation of active sites (Figure 3.15). It has been shown that monomer does not influence the deactivation processes.

Let us consider two interesting phenomena characteristic of ethylene polymerization catalyzed by immobilized complexes which are also an additional means of controlling it [57]. The first one can be observed through the extreme dependence of activity on the surface density (concentration) of fixed metal complexes. Thus, the highest activity in ethylene polymerization by VCl$_4$·PE-gr-P4VP–AlEt$_2$Cl has been observed [58] at the fixed vanadium content of 0.6 mass% (18 atoms/nm^2). Similar dependences ("bell-shaped") have also been observed for a number of other catalytic systems (Figure 3.16) as well as for some other processes catalyzed by immobilized metal complexes.

There is no doubt that such dependence is the result of peculiarities of active component distribution, particularly V^{4+} in the above example. Rather reliable data have been obtained on the dual character of metal ion distribution on the surface of supports [17, pp.253–260]. Some of these are in the form of isolated centers (C_{iso}), while others are cluster-type aggregations with high local concentrations of fixed particles, C_{cl}. The increase of overall concentration of fixed MX$_n$ results in changes of the ratio between *iso*lated ions and those in cluster-type formations. Neither C_{iso} nor C_{cl} are directly responsible for the activity changes. However, it is evident that just a number of small associates (dimers, trimers, etc.) should exceed the maximum value.

On the other hand, it is reasonable to assume that active sites in immobilized systems are isolated transition metal ions stabilized throughout cluster formation. Their fraction also should pass the maximum value. However, it is rather difficult to evaluate the number of these stabilized species as it may change both under organoaluminium compound action and in the course of polymerization.

In addition to these examples, and also mentioned above (in Chapters 1 and 2), there are some other examples of "bell-shaped" SCA dependency on the surface density of transition metal ions. Thus, the rising branch of the plot of PE yield against amount of vanadium bound to alumoxanes has been connected with cooperative actions of vanadium ions [59]. On the other hand, a progressive decrease of specificity of TiCl$_4$ bound to inorganic supports has been observed until each active center becomes well separated from others [60]. The area per active center was estimated to be equal to 0.4 nm^2 [6]. The proposal that active sites can be stabilized through cluster formation has also been made in the studies of *n*-butane oxidation to acrolein catalyzed by vanadium-containing aerosols [61] as well for some other catalytic processes [62].

Table 3.3 Characteristics of MX_n immobilized on polymers with grafted functional layers and kinetic parameters of ethylene polymerization (343 K).

Polymer carrier	Content of grafted fragments $\times 10^4$ (mol/g)	MX_n	Content of fixed metal $\times 10^4$ (g-at/g)	n_p (mmol mol)	$k_{pr} \times 10^{-4}$ (l/(mol s))	E_a^{eff} (kJ/mol)
PE-gr-poly(4-vinylpyridine) (PE-gr-P4VP)	6.0	TiCl$_4$	2.04			
	6.0	VCl$_4$	0.70			
PE-gr-poly(acrylonitrile) (PE-gr-PAN)	6.4	TiCl$_4$	0.40			
	11.9	VCl$_4$	0.70	4.0	2.8±0.6	38±4
PE-gr-poly(methylmethacrylate) (PE-gr-PMMAc)	12.0	VCl$_4$	1.14	4.5	3.4±0.7	38±4
	12.0	TiCl$_4$	0.77			
PE-gr-poly(methylvinylketone) (PE-gr-PMVK)	6.0	VCl$_4$	1.50	4.5	4.0±0.8	31±4
	6.0	TiCl$_4$	2.70			
PE-gr-poly(vinylacetate) (PE-gr-PVAc)	12.8	TiCl$_4$	1.70			
PE-gr-poly(diallylsulfide) (PE-gr-PDAS)	1.1	TiCl$_4$	0.18			
PE-gr-poly(dibutyl)ester of vinylphosphonic acid (PE-gr-PDBEVPAc)	1.1	VCl$_4$	0.50	3.8	2.9±0.6	32±4
PE-gr-poly(trifluorostyrene) (PE-gr-P3FS)	2.3	VCl$_4$	0.45	5.0	4.8±1.0	31±4
PE-gr-poly(vinylpyrrolidone) (PE-gr-PVPr)	8.6	VCl$_4$	2.6	4.2	3.1±0.6	29±4
PE-gr-polydiallylamine (PE-gr-PDAA)	2.6	VCl$_4$	2.50	5.0	4.2±0.8	27±4
PE-gr-polyacrylic acid (PE-gr-PAAc)	11.0	VCl$_4$	1.90	2.1	5.1±1.0	38±4
PE-gr-poly(allyl alcohol) (PE-gr-PAAl)	6.2	TiCl$_4$	1.20			
	6.2	VCl$_4$	3.00	12.5	3.4±0.7	34±4
	6.2	Cp$_2$TiCl$_2$	0.10			
	3.4	Ti(OBu)$_4$	0.11			
	3.4	VO(OEt)$_3$	0.84			

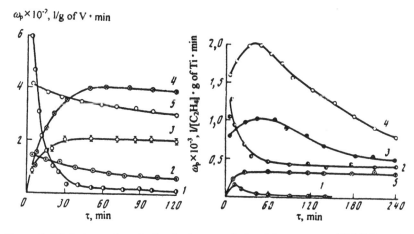

Figure 3.13 Ethylene polymerization by heterogenized and homogeneous VO(OEt)$_3$ systems. Polymer supports: (1) homogeneous system, (2) hydrolysated copolymer of styrene with vinylacetate, (3) PE-gr-PAAl, (4) reduced PE-gr-PMVK, (5) PE-gr-PAA. Polymerization conditions: toluene, C$_2$H$_4$ pressure 1.0 MPa, 343 K, AlEt$_2$Cl as cocatalyst, [Al]/[V]=100.

Figure 3.14 Ethylene polymerization by Ti(OBu)$_4$–AlEt$_2$Cl heterogenized systems. Polymer supports: (1) homogeneous system, (2) PE-gr-PDAA, (3) PE-gr-PAA, (4) PE-gr-PAAl, (5) PE-gr-PAAc. Polymerization conditions: toluene, C$_2$H$_4$ pressure 1.0 MPa, 343 K, [Al]/[Ti]=100.

Another unusual phenomenon, perhaps also general for catalysis by immobilized complexes, is the temperature behavior of polymerization rate. With a rise in temperature, the ethylene polymerization rate follows the Arrhenius law only in a certain temperature range: up to the glass (or softening) temperature of the polymer support. It has been shown for gas-phase ethylene polymerization catalyzed by TiCl$_4$·P4VP–AlEt$_2$Cl, that Arrhenius dependence of the polymerization rate is valid up to 383 K. The effective activation energy in this range is typical for anionic-coordinate polymerization, 16±2 kJ/mol (Figure 3.17a and b). At these temperatures the polymer segments show higher mobility and relaxation relative to reorientation of support macromolecules which becomes the controlling factor. The temperature coefficient becomes negative (–130±10 kJ/mol). If we suppose that the decrease of polymerization rate in this temperature range is the result of irreversible active sites deactivation caused by enhanced mobility of polymer segments, then $w_p=Kn_{pr}(1-\Delta n/n_{pr})=Kn_0(1-w_d\Delta t)$, where w_d is the specific value of the deactivation rate by any of the possible mechanisms. It therefore follows that $w_d\Delta t=1-w_p/w_0$, where w_0 is the value of polymerization rate without deactivation which can be calculated from the temperature dependence (see Figure 3.17b, 1). The shape of the curve corresponds to those observed for time relaxation dependences of segment mobility; the apparent value of activation energy sharply increases in the approach to the temperature of mobility freezing. The temperature dependence of the relaxation rate is linearized on the coordinates lgw_d against $(\Delta T)^{-1}$, i.e. assumes the form $w_d=A\exp(-B/(T-T_0))$.

It is known [63] that the B value is connected with both glass transition temperature, T_g, and the fraction of free volume in glassing, α_g, by the relation $B=(T_g-T_0)/2.3\alpha_g$. If we assume that $\alpha_g=0.025$ [63], then $T_g=360$ K is the temperature which releases the brake on mobility leading to a decrease of polymerization rate (see Figure 3.17a). The lower T_g

values for surface layer indicated a more unconsolidated packing of polymer molecules in the grafted covering which appears to be acceptable. Specifically, such an approach can be used as a kinetic method for evaluation of mobility in grafted layers.

Thus, the restriction of transport diffusion of active sites is a significant factor responsible for the activity and stability of immobilized catalysts.

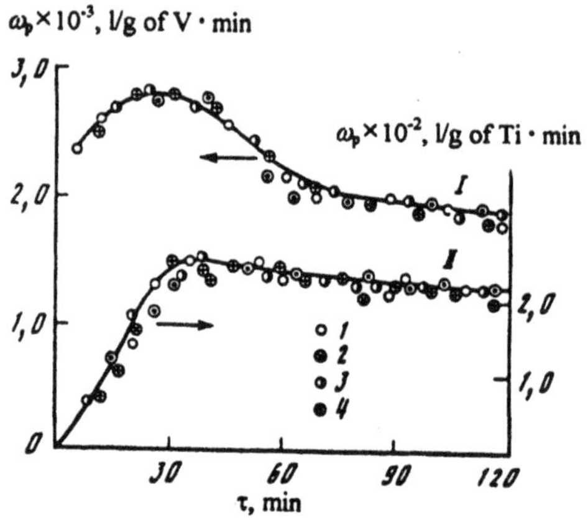

Figure 3.15 The influence of VCl$_4$ (I) and TiCl$_4$ (II)-immobilized catalysts aging on the ethylene polymerization rate. Time of exposure in the absence of monomer, min: (1) 0, (2) 10, (3) 15, (4) 20. Polymerization conditions: heptene, 343 K, AlEt$_2$Cl, [Al]/[M]=50, C$_2$H$_4$ pressure 0.4 MPa.

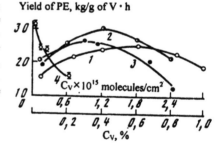

Figure 3.16 The dependence of immobilized catalysts activity in ethylene polymerization on the surface density of bound transition metal ions. (1) VCl$_4$·PE-gr-P4VP, (2) VO(OEt)$_3$–PE-gr-PAAl, (3) VO(OEt)$_3$–PE-gr-PAAc, (4)   –   Ti(OBu)$_4$-PE-gr-PAAl. Polymerization conditions as at the Figure 3.15.

3.1.5 Ethylene polymerization by titanium–magnesium catalysts

At present, immobilized titanium–magnesium catalysts (TMC), which are titanium compounds (TiCl$_4$, TiCl$_3$(OR), etc.) bound to inorganic magnesium-containing supports, are widely used for ethylene polymerization. These are obtained either by reduction of titanium compounds with Grignard's reagents or by supporting on inorganic carriers of Ti^{4+} complexes from solutions as well by joint grinding of MgCl$_2$ with TiCl$_4$, TiCl$_3$, etc. [1]. Reduction yields bulky TMC with 10–20% of titanium, while attached catalysts (MgO, MgCl$_2$, Mg(OH)$_2$, Mg(OH)Cl, MgCO$_3$, Mg(OR)$_2$, etc., as supports) contain only 1–2 mass% of titanium. Organomagnesium compounds are known to exhibit polymer-like structures themselves [64].

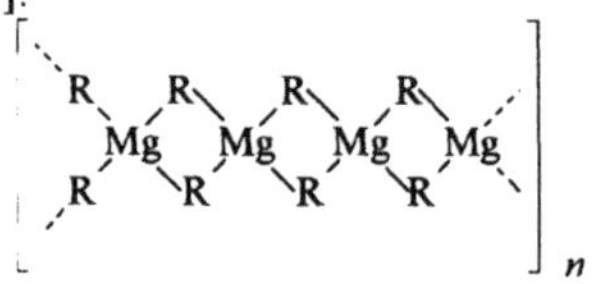

$$\left[\begin{array}{cccc} R & R & R & R \\ \diagdown & \diagdown & \diagdown & \diagdown \\ Mg & Mg & Mg & Mg \\ \diagup & \diagup & \diagup & \diagup \\ R & R & R & R \end{array} \right]_n$$

Scheme 3.7

Data on the synthesis and activity of polymer-immobilized TMC are rather scarce [65–67]. For instance, TMCs grafted to polymers with functional groups covering have been obtained according to the general scheme (with the example of PE-gr-PAAl):

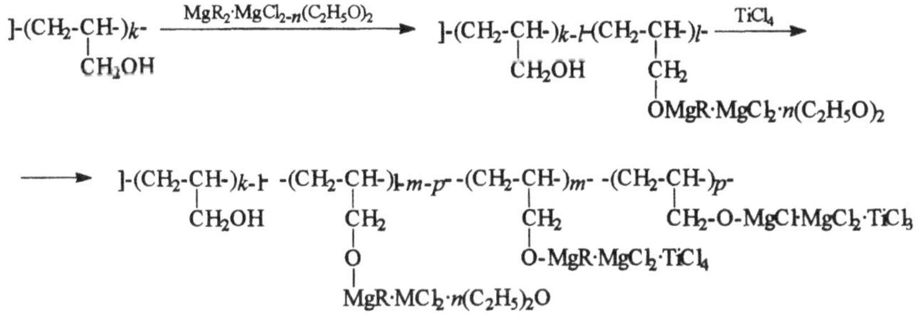

$$]\text{-(CH}_2\text{-CH-)}_k\text{-} \quad \xrightarrow{\text{MgR}_2 \cdot \text{MgCl}_{2-n}(\text{C}_2\text{H}_5\text{O})_2} \quad]\text{-(CH}_2\text{-CH-)}_k\text{-}\text{-(CH}_2\text{-CH-)}_l\text{-} \quad \xrightarrow{\text{TiCl}_4}$$
$$\qquad\; \text{CH}_2\text{OH} \qquad\qquad\qquad\qquad\qquad\qquad \text{CH}_2\text{OH} \quad\;\; \text{CH}_2$$
$$\qquad\qquad\qquad\qquad\qquad\qquad\qquad\qquad\qquad\qquad\qquad\qquad \text{OMgR} \cdot \text{MgCl}_2 \cdot n(\text{C}_2\text{H}_5\text{O})_2$$

$$\xrightarrow{\qquad} \;\;]\text{-(CH}_2\text{-CH-)}_k\text{-}\text{-(CH}_2\text{-CH-)}_{k\text{-}m\text{-}p}\text{-}\text{-(CH}_2\text{-CH-)}_m\text{-} \text{-(CH}_2\text{-CH-)}_p\text{-}$$
$$\qquad\qquad \text{CH}_2\text{OH} \qquad\; \text{CH}_2 \qquad\qquad\; \text{CH}_2 \qquad\qquad \text{CH}_2\text{-O-MgCl} \cdot \text{MgCl}_2 \cdot \text{TiCl}_3$$
$$\qquad\qquad\qquad\qquad\qquad\; \text{O} \qquad\qquad\; \text{O-MgR} \cdot \text{MgCl}_2 \cdot \text{TiCl}_4$$
$$\qquad\qquad\qquad\qquad \text{MgR} \cdot \text{MCl}_2 \cdot n(\text{C}_2\text{H}_5)_2\text{O}$$

Scheme 3.8

There are a number of different functional sites on such catalysts, such as unreacted functional groups of the polymer support, resting Mg centers as well bound Ti^{4+}- and Ti^{3+}- species. Both the amount of bound titanium and the value of the Ti^{4+} to Ti^{3+} ratio depend on the nature of the functional groups and on the amount of attached magnesium. Mixed immobilized complexes of $MgCl_2$ and $TiCl_3$ have not been produced with low titanium and magnesium content (<0.3 mmol/g), probably, due to strong interaction with the functional group cover. Highly dispersed active crystallites of size 4.5–7.5 nm, growing mainly along the polymer chain, have been produced at higher metal concentrations. Kinetic studies showed a first-order reaction with respect to monomer in the range of concentrations 0.12–0.48 mol/l (ethylene pressure 0.2-0.8 MPa), the effective values of activation energy being equal to 53.6 kJ/mol (323–343 K) and 19.3 kJ/mol in the temperature range 343–363 K. Species containing about 50% titanium in the form of isolated Ti^{3+} ions showed the highest activity for ethylene polymerization (550 kg of PE/(g of Ti MPa h)). The strong dependence of catalyst activity on isolated Ti^{3+} ion content (Figure 3.19) leads to the conclusion that these (probably stabilized with $TiCl_3$ or $TiCl_2$ crystallites) appear in active centers[4].

If we take the value of $k_{pr}=1.2\times10^4$ l/(mol s), defined by radiochemical analysis for bulky TMCs [69], the fraction of active sites in immobilized catalysts reaches 6% of overall titanium content (4–7 mol% for bulky TMCs based on $TiCl_4$ and $MgCl_2$). If we take into account that the polymer support surface area is about 4 m^2/g, it follows that each active site may occupy 2.7 nm^2, and this leas to the conclusion that all active sites may be located in a thin surface layer of the support. It is also important that in general the same kinetic parameters and activities as those for bulky TMCs are reached at small magnesium contents, comparable to the amount of titanium. This feature allows the properties of the polyethylene formed to be easily regulated (molecular mass characteristics, dispersion composition, etc.). However, additionalstudies are necessary before reaching definite conclusions.

---

[4] A proposal that isolated Ti^{3+} ions do not take part in polymerization has been made [68]. By the hypothesis, the active sites were alkylated titanium ions incorporated in surface associates (with the example of $MgCl_2$), thus unobserved by ESR spectroscopy.

Recently, the same approach has been used [70] to obtain TMC from both CSDVB (DVB 5.5–55%, S_{sp}=5–410 m^2/g) species and chloromethylated derivatives (Cl content 9.8–20.3%). The activity of such catalysts (Mg/Ti=2–8) reaches the values of 13–76 kg of PE/(g of Ti h atm) or in terms of catalyst weight 0.7–5.4 kg/g.

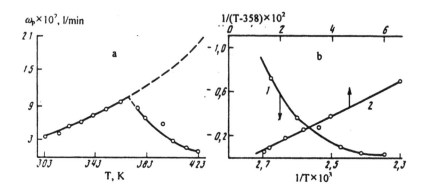

Figure 3.17 Temperature dependence (a) of gas-phase ethylene polymerization rate for catalytic system TiCl$_4$·P4VP–AlEt$_2$Cland this (b) in coordinates $1-w_p/w_0$ – $1/T$ (1) and – $1/(T-358)$. The calculated polymerization rate (a) is shown by dashed line, the observed polymerization rate is represented by solid line.

However at present alkoxydes [71] and carboxylates [72] of magnesium as well NdCl$_3$-supported Ziegler–Natta catalysts [73] are more often used for olefins polymerization than organomagnesium compounds.

It was mentioned above that zirconium-containing catalysts can also be immobilized, for example [74] by polybutadiene hydrozirconation with Schvartz' reagent (Cp$_2$Zr(H)Cl). The treatment yielded soluble zirconium-containing polymers in which zirconium complexes were bound to the polymer with σ-bonds. Recently the reactions of organozirconium compounds have been reviewed [75].

3.1.6 Gel-immobilized catalysts for ethylene polymerization

If the main goal of the formation of grafted polymer catalysts was to dispose active sites on the surface of the support so as to increase its specificity, it is opposite for gel immobilized ones. In catalysis by gel-immobilized systems one should aim to use all active centers dispersed throughout the total polymer volume. This can be achieved by the use of swelling, rather than insoluble polymers in the reaction mixture, that are permeable to catalyst components, substrate and solvent molecules, so-called "mosaic" gel polymers [76–79]. Basically these are rubbers (ethylene–propylene, tercopolymers of ethylene, propylene and unconjugated dienes, siloxanes, etc.) which are functionalized with radical grafting of different monomers (vinylpyridines, acrylic acid, butadiene, MMA, allyl alcohol, etc.). The following rubber linkage gives three-dimensional networks preventing the removal of catalyst particles. Thus, the structure of gel-immobilized catalysts prepared (in the form of granules, fibers, films, etc.) is favorable for diffusion of reagents into (and out of) the polymer volume and so stabilizing the bound active centers (Figure 3.20).

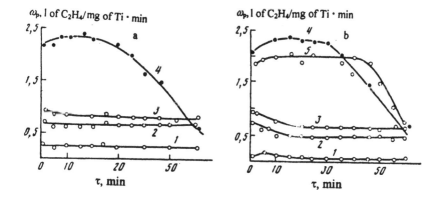

Figure 3.18 Ethylene polymerization kinetics in the presence of polymer TMCs. The dependence of ethylene polymerization rate on the type of support (a) and Mg content in PE-gr-PAAc (b) (343 K, ethylene pressure 0.2 MPa). (a) 1, PE-gr-PMVK; 2, PE-gr-PAAl; 3, PE-gr-PDAA; 4, PE-gr-PAAl. (b) 1, [Mg]=0.13 mmol/g; 2, 0.17; 3, 0.32; 4, 0.57; 5, bulky TMC on $MgCl_2$. [$AlEt_3$]=4.4 mmol/l.

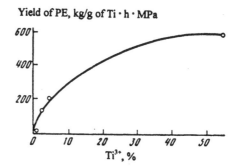

Figure 3.19 The dependence of the specificity of TMC supported on PE-gr-PAAc on Ti^{3+} content.

Gels can also be used for TMC bonding as well as soluble catalytic systems for ethylene polymerization at elevated temperatures. In the last case, PE growing inside the swelling polymer granule steadily enters the outer solvent [80].

Temperature dependences of polymerization rate in the presence of these catalytic systems also exhibit extreme points. However, these are completely reversible, i.e. the decrease of polymerization rate in the range of elevated temperatures is not the result of active center deactivations. Kinetic curves of ethylene polymerization catalyzed by gel-immobilized systems are substantially different above and below the polyethylene melting temperature (Figure 3.21). For the temperature range (303–363 K) polymerization proceeds mainly inside the catalyst granules and the titanium content remains constant with time (Figure 3.21 curve 5), the polymerization rate being steady even at 363 K. At temperatures above T_m (408 K), the PE being produced is extracted by the solvent from the polymer granules. Active centers associated with PE macromolecules can also be removed from the support [81] and further suffer from bimolecular deactivation. In this case the polymerization rate fall is accompanied by a decrease of the content of titanium inside the support granules (see Figure 3.21 curves 4,6).

Microsyneresis of the polyhydrocarbon gel in the medium of hydrocarbon solvent leading to a decrease in monomer concentration in the vicinity of active centers is the important feature of gel-immobilized systems providing them with structural–conformational adaptive ability. As a result, a temporary retardation of the heat-liberating reaction takes place in the vicinity of the catalyst. Such a feedback gives catalysts self-protected from irreversible thermal deactivation [82].

In general terms, gel catalysts are similar to biomembranes with embedded enzymes. Under optimization they can be converted from single use catalysts into systems of prolonged action. In the Section 3.3 we shall review the processes of monomer dimerization, including ethylene, catalyzed by similar systems.

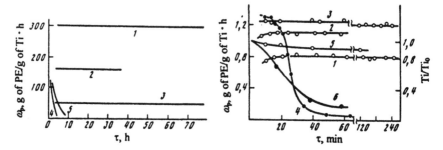

Figure 3.20 The kinetics of ethylene polymerization by both gel catalytic systems and their analogues. (1) triple copolymer of ethylene, propylene and diene (TCEPD)–PAAl–Mg^{2+}–TiCl$_4$–AlR$_2$Cl (ethylene pressure 1.5 MPa, 473 K); (2) TCEPD–PAAl–AlR$_3$–TiCl$_4$–AlR$_2$Cl (433 K); (3) TCEPD–P4VP–TiCl$_4$–AlR$_3$ (388 K); (4) TiCl$_4$–AlR$_3$ (418 K); (5) Py–TiCl$_4$–AlR$_3$ (418 K); (2)–(5) ethylene pressure 11 MPa.

Figure 3.21 The dependences of ethylene polymerization rate (gel catalytic system Cp$_2$TiCl$_2$–TCEPD–Al(*iso*Bu)$_2$Cl in *n*-heptene) and the portion of unextracted titanium (Ti/Ti$_0$) (5, 6) on time. *T*, K: (1) 303, (2) 323, (3) 363 (ethylene pressure 0.4 MPa), (4) 408 (ethylene pressure 0.6 MPa). Order of reagents addition: gel catalytic system (22 g/l), heptane, Al(*iso*Bu)$_2$Cl ([Al]/[Ti]=7.6), ethylene.

3.1.7 Stereospecific propylene polymerization

The creation of stereospecific propylene polymerization catalysts is a complex problem. The formation of submicrocrystals of MCl$_3$ (M=Ti, V) on polymer supports is one of the ways of solving it. Coprecipitation of MCl$_3$ in situ, MCl$_4$ reduction in the presence of macroligands, including chemisorbed species, as well as joint grinding of MCl$_3$ with polymers are some of the suitable approaches. For instance, the treatment of PS or CSDVB with VCl$_4$ with subsequent reduction under vacuum at 353 K yields a highly stable catalyst with a SCA of 0.5 kg of PP/(g of V h atm) [83]. In a similar way the effective titanium catalysts with a SCA of 0.3–0.6 kg of PP/(g of TiCl$_3$ h) have been obtained by TiCl$_3$ supported on a porous polymer carrier, as well on PE and PP [39,84]. The ease of VCl$_3$ precipitation on polymer surfaces especially on those with excess electrons is well known [50]. Electron donors favor powdered VCl$_3$ crystals and significantly affect the morphology of the metal–containing polymer produced. Stereoregular PP (93 % insoluble in boiling heptane) has been obtained [38] in the presence of δ–TiCl₃ supported on ion-exchange resins, both on anionite in the form of NH_4^+ salt and on cationites containing carboxylic and sulfoacid groups.

Titanium trichloride incorporated within polymer pores (CSDVB, 40% DVB, total pore volume 1.56 cm³/g) was 2.5 times more active in propylene polymerization than commercial TiCl₃ species (TAS–101) (Figure 3.22). Polypropylene of relative molecular mass 10^6 and with a yield of 24 kg/(g of Ti h) has been produced by this catalyst in liquid propylene [49]. Dispersed TiCl₃ chemically bound to the polymer molecule, however, only yields oligomer products, and has been produced by complex formation of TiCl₄ with a copolymer of styrene, DVB and MMA (60:30:10 mass%) and subsequent reduction with AlEt₂Cl.

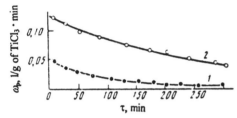

Figure 3.22 Kinetics of propylene polymerization in the presence of TiCl₃ catalytic systems. (1) TAS–101, (2) TiCl₃ supported on CSDVB (40% of DVB). Polymerization conditions: liquid propylene, 343 K, AlEt₂Cl as cocatalyst.

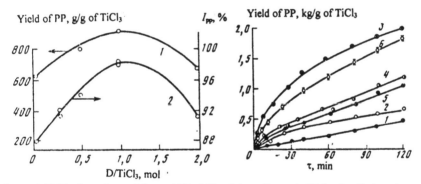

Figure 3.23 The dependence of the total PP yield and that of isotactic PP fraction (I_{PP}) on the molar ratio [β-TiCl₃]/[MMA units] in the catalyst produced by grinding together of TiCl₃ and PP-gr-PMMA. Polymerization conditions: condensed propane–propylene fraction containing 90% of propylene, [propylene]=9.3 mol/l, [TiCl₃]=0.74 mmol/l, 343 K, polymerization time 120 min.

Figure 3.24 Kinetics of propylene polymerization in the presence of catalyst produced by grinding together of β-TiCl₃ and polymers. Polymer supports: (1) TiCl₃ without support; (2) TiCl₃/PP (10%) (in parentheses the % of TiCl₃ mass is given); (3) TiCl₃·PP-gr-P4VP (7.5%); (4) TiCl₃·PP-gr-PAAc (5.0%); (5) TiCl₃·PP-gr-PAN (7.5%); (6) TiCl₃·PP-gr-PMMA (6.0%). Conditions: pentane, 343 K, propylene pressure 0.4 MPa, AlEt₂Cl.

Methods used for mechanochemical MCl₃ dispersion on polymers, especially with grafted functional layers, are widely employed. Thus grinding together of TiCl₃ and PP-gr-PMAc (polypropylene grafted poly(methyl acrylate)) yields catalysts [85] for which plots of activity and stereospecificity give a curve having a maximum at [MAc]/TiCl$_3$=1:1 (Figure 3.23). The greatest activity value of ~900 g of PP/g of TiCl$_3$ is 2.8 times greater than that of the standard system and 1.4 times that of TiCl$_3$ supported on PP without grafted functional groups. It is also important that such systems combine enhanced activity with high stereospecificity. It is rather unusual that polymerization of the condensed propane–propylene fraction in the presence of H$_2$ (as a regulator of polymer molecular mass) is accompanied by a 2-fold increase of specific activity and not the expected decrease. The nature of the grafted functional groups significantly influences the catalyst activity (Figure 3.24).

The size of the MCl$_3$ crystals depends on the grinding time as well as the L/MCl$_3$ ratio. The excess of functional groups can induce the total decomposition of the MCl$_3$ crystal phase. Two processes accompany MCl$_3$ immobilization at L/MCl$_3$<1. These are also (i) breakdown of the crystal phase due to complex formation and (ii) proper grinding and aggregation of MCl$_3$ crystallites [86].

As a result, a phase of MCl$_3$ with particle size independent of grinding time has been observed by X-ray analysis of this catalyst. The combination of these two factors allows verification of crystallite size down to total homogenization in the polymer support. The diminution of crystallite size causes a significant rise of activity but it is not the route to enhanced stereospecificity (Table 3.4).

Table 3.4 Some characteristics of PP obtained[a] in the presence of bound TiCl$_3$ [86].

Catalyst	PP yield (kg/g)	Stereoisomeric composition of the fraction soluble in heptane (%)		Crystallinity (%)		Index of stereo-specificity (IR data)
		Cold	Boiling	X-ray analysis	IR spectroscopy	
TiCl$_3$[b]	0.45	4.2	3.8	54	64	87
TiCl$_3$	0.78	4.0	5.2	54	65	–
TiCl$_3$·PE	1.0	4.0	5.2	–	–	–
TiCl$_3$·PE-gr-P4VP[c]	1.5	8.2	9.4	–	–	–
TiCl$_3$·PP-gr-PMMA	1.8	1.0	1.0	61	65	–
TiCl$_3$·PP-gr-PAN	1.1	0.8	3.2	56	64	89
TiCl$_3$·PP-gr-P4VP	2.0	0.9	1.1	58	68	–
TiCl$_3$·PP-gr-PAAc	1.2	1.3	3.7	54	65	–

[a]AlEt$_2$Cl as cocatalyst, propylene pressure 0.4 MPa, τ=2 h.
[b]Non-grinding, for other entries the time of grinding was 1 h.
[c]Al/Ti=20, for all others 50.

In a similar way typical TMCs (for example, obtained by interaction of anhydrous MgCl$_2$ with TiCl$_4$) can in addition be mixed in a vibrating mill together with polymer powder, such as poly(tetrafluoroethylene) (PTFE) [87]. During this procedure even the smallest particles of Ti-containing product are stabilized with a PTFE envelope, and the catalyst produced is characterized by a narrower distribution of particle size.

Specified granule size and polymer particle morphology can also be obtained by support treatment with phosphorus-containing compounds also leading to particles stabilization [88].

The use of macroligands as electron-donor modifying agents for additional growth of metal-complex catalyst stereospecificity such as polyesters and poly(ester)acrylates seems to be in prospect [89].

There are no reports in the literature concerning the stereospecific polymerization of higher α-olefins (butene, 4-methylpentene, etc.) catalyzed by immobilized systems.

3.1.8 Copolymerization of α-olefins catalyzed by immobilized metal complex catalysts

The comparisons of copolymerizations of α-olefins, in particular of ethylene with propylene, catalyzed by either homogeneous or immobilized complexes is expected to provide information concerning both general and specific features of the processes and reveal additional features of the polymerization mechanism. This expectation is based on the high sensitivity of different monomers to the nature of the active sites.

A rise of propylene content in ethylene–propylene mixtures results in a progressive drop of reaction rate in the presence of both homogeneous and immobilized VCl$_4$ catalysts [55], the active site concentrations being independent of the composition of monomers mixtures. The fall of polymerization rate is the result of only steric factors increasing the time of propylene addition at the stage either of adsorption by the active site or during insertion into growing chain.

Both mobile and immobilized systems have close values of copolymerization parameters corresponding to reactivities of used monomers (Table 3.5). The values of the tabulated $r_1 r_2$ products indicate that both monomer units are dispersed along the copolymer chain; the probability of block-copolymer formation is low.

Factors concerning not only reactivity but also adsorption are of great importance in catalytic copolymerization [90]. In view of this, the differential equation expressing copolymer composition may be represented as follows:

$$\frac{dM_1}{dM_2} = \frac{[M_1]\dfrac{k_{11}k_1[M_1]}{(k_{12}k_2[M_2])}+1}{[M_2]\dfrac{k_{22}k_2[M_1]}{k_{21}k_1[M_2]}}$$

which can be transformed into the copolymerization equation with the following copolymerization constant $r_1^* = k_{11}k_1 / k_{12}k_2$ and $r_2^* = k_{22}k_2 /k_{21}k_1$ (where k_1 and k_2 are adsorption coefficients, $k_1 = k_1'/k_1'' + k_{11}$ and $k_2 =k_2'/k_2'' + k_{22}$; and k_1', k_1'', k_2'' are adsorption and desorption rates constant for the monomers M_1 and M_2, respectively). Both are derived from constant relations so that r_1 and r_2 will be the resultant copolymerization rate constants. For the above systems adsorption is not the controlling step, so the reactivity of active sites will be dependent on the type of end unit bound with centers of chain propagation: $k_{11}/k_{21} = 2.7–6.3$; $k_{22}/k_{12} = 0.058–0.135$. In addition, if monomer adsorption would significantly affect the total rate of both ethylene and propylene addition, then $k_{11} = k_{21}$, $k_{22}=k_{12}$ and $k_{11}/k_{21}=k_{21}/k_{22}$ so that $r_1=1/r_2$ and $r_1 r_2=1$.

This inversion of activity may be connected with selective sorption of more reactive monomer by active centers and should be considered as one of the criteria for copolymerization with intermediate monomer coordination by active sites. The fact that in comparative systems the product $r_1 r_2 < 1$ (see Table 3.5) means that copolymerization of ethylene with propylene both by mobile and immobilized systems is characterized by the comparable values of the adsorption coefficients.

The equation for copolymer composition requires the condition of equilibrium (proportionality) between active sites including different end units $n_{M1} = x n_{M2}$ so that $n_{M1} + n_{M2} = n_{pr}$. The value of n_{pr} sharply drops with time for the non-stationary catalytic system VCl_4–AlR_2Cl. Nevertheless, the proportionality between n_{M1} and n_{M2} holds true in this case, the copolymer composition being independent of the number of chain propagation active sites. This is a consequence of the fact that the decrease of

copolymerization rate with time is the result of a decrease of n_{pr} but not of a change of their nature. From here, the selectivity of active sites with respect to both ethylene and propylene remains constant.

The overall copolymerization rate for different compositions of monomer mixtures ($F = M_1/M_2$) follows the equation

$$w_{copol} = \frac{k_{11}k_{22}(r_1 F + r_2 F + 2)}{r_1 k_{22} F + r_1 k_{11} + r_1 k_{22} + r_2 k_{11}/F} n_{pr}(M_1 + M_2)$$

From the analogy with the equation for homopolymerization rate ($w = k_{pr} n_{pr} C_M$) the expression in the form of a fraction can be considered as the formal copolymerization rate constant k_{pr}^{copol}. This depends on the four constants which characterize homo- and hetero- (crossing) addition of monomers as well on their relation to F in the reaction volume. As can be seen from Figure 3.25, k_{pr}^{copol} increases proportionally with ethylene content in the range of F values 0.05–1.0; the calculated values of k_{pr}^{copol} for several systems agree with those found by experiment. Similar results have been obtained for ethylene copolymerization with α–butene.

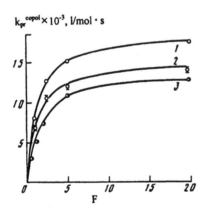

Figure 3.25 The calculated (solid line) and experimental (symbols) values of copolymerization rate constants in the presence of V-containing systems as a function of composition of monomers mixture. Catalysts: (1) VCl_4, (2) VCl_4 immobilized on PE-gr-PAAl, (3) VCl_4 immobilized on PE-gr-poly(diisobutyl ester of vinylphosphonic acid).

Other characteristic examples are as follows. The complex catalyst, macroligand–$TiCl_4$, yields amorphous ethylene–propylene elastomers [91]. Catalysts on $Ni(AcAc)_2$ and phosphorus-containing polymer supports [92], as well gel-immobilized metal-complex systems [93] show high activity in copolymerization of ethylene with propylene. Vanadium complexes bound to polymer supports with ester units, such as poly(acrylates), poly(vinyl ester), polyester derived from dicarbonic acid and diols [94] are effective in the production of tercopolymers ethylene–propylene–ethylidenenorbornadiene (TCEPD). Olefin copolymerization with supported metallocene–methyl-aluminoxane has been studied [95].

Copolymerization of ethylene with α-butene has been studied in detail. The interest in this problem is connected with the possibility of modification of linear polyethylene properties (crystallinity, strength, improvement of rheological characteristics and increase of the resistance to cracking, etc.). Incorporation of higher α-olefins units into the PE chain usually proceeds with maximum effect for α-butene.

The introduction of α-butene into the reaction mixture usually leads to a decrease of copolymerization rate which is similar to that seen with propylene. The output of copolymers with high C_4H_8 content (above 10 mol%) is accompanied by high catalyst expandable coefficient; copolymerization parameters are equal to $r_1=29.5\pm5$, $r_2=0.014\pm0.003$, $r_1\times r_2=0.413$ [93].

It is worth noting the impossibility of both ethylene copolymerization with trans–butene–β and cis–butene–β and their isomerization in the presence of immobilized V–containing catalysts. Some problems of ethylene–butene copolymerization and also butene isomerization will be considered in Section 3.3.

3.1.9 Diffusion restrictions in catalytical polymerization in the presence of metal-containing polymers

The characteristic dependence of polymerization rate on concentration ratios and functional group covering parameters, as well as on the type of MX_n immobilization, show that diffusion restrictions in catalysis by metal complexes bound to the polymer carrier surface (even in the case of highly effective TMCs) do not usually dominate over other factors.

One of the criteria for diffusive limitation in coordinative polymerization (for the processes proceeding within the kinetic range) is the correspondence of calculated and experimental values for M_n and the coefficient of molecular mass distribution (γ). However, it is often difficult to apply this necessary and sufficient condition to catalysis by immobilized metal complexes. The second necessary criterion for reaction rate limitation can be presented by the relation [96]

$$\alpha = \left(\frac{k_{pr}n_{pr}}{D_m}\right)^{\frac{1}{2}} r_0 \ll 1$$

where the Thiele modulus α is the characteristic diffusion time to characteristic reaction time ratio, D_m determines monomer ability to diffuse and r_0 is the initial radius of catalyst particles. For diffusion-controlled systems α reach values of the order of 10. Otherwise, they are in the range 1×10^{-4} to $4.8\div10^{-2}$. Immobilized systems, especially those of grafted type, usually fulfill the latter requirement ($\alpha\ll1$). The model of multigranular porous catalyst microparticles [97] applied to olefin polymerization by heterogenized systems describes rather well kinetic regularities, molecular mass and polydispersity values, as well as limitations of heat and mass transfer through the external layer of the microparticles.

For ethylene polymerization by gel-immobilized systems, the contribution of inner diffusion limitations to the chain propagation stage in gels should be taken into account [81]. The values of the Thiele modulus are $\alpha\gg1$ for such systems, so that in the case of strong intramolecular retardation, the reaction rate per catalyst volume is expressed by the equation

$$w = \frac{1}{r_0}\sqrt{w_{\text{eff}}D_{\text{m}}[M]^{n+1}}$$

where w is intrinsic polymerization rate per catalyst volume, $[M]$ is monomer concentration on the external surface of the catalyst granules and n is the reaction order. The inverse dependence of initial polymerization rates on the size of catalyst granules is a specific feature of ethylene polymerization by gel-immobilized systems.

Under equilibrium swelling, the apparent sizes of catalyst particles may further rise wit solvent uptake after contact with the monomer; and in addition, the polyethylene produced is accumulated inside each particle. Thus, super-equilibrium swelling of such cross-linked catalyst particles is caused by polymer accumulations within the preformed polymer carrier network. If V_{st} is the particle volume in the stationary working state and V_{eq} is that under the conditions of equilibrium swelling by pure solvent, then $\Delta V = (V_{\text{st}} - V_{\text{eq}})$ is the volume increment due to "super-equilibrium" swelling of the three-dimensional structure of initial gel. Values of V_{st} can be large, and generally are greater by some orders of magnitude than V_{eq} values.

The expansion work of catalyst particles by the ΔV increment is defined as

$$A = \int_{V_{\text{eq}}}^{V_\bullet} P(V)\mathrm{d}V$$

and is the result of ethylene polymerization free energy release followed by generation of additional tension within the elastic support network. Moreover, the specific mechanism of physical chain terminations is often displayed for such systems; the swelling pressure acting on growing macromolecules squeezes it out of the catalyst granule. It is also possible that some additional tension appears at the point of chain growth, and this may be the reason for the elementary activation step leading to macromolecular chain cleavage from the active site. The rate constant for such an event, k, is defined by Eyring's equation: $k \sim \exp[-(E - \sigma v / RT)]$, where E is activation energy value without mechanical tension; σ is the tension at the point of chain growth; v is the volume element (on the scale of one monomer unit). Obviously, molecular mass parameters for the polymers produced can be affected by the degree of cross-linking.

3.1.10 Mechanism of active site formation and stabilization

The processes which underlie the formation of metal complex catalyst active sites are often complicated and unclear even for comparatively simple homogeneous systems. The possibility of directly identifying active site composition and their action is small. However, comparative kinetic analysis of mobile and immobilized catalysts gives the opportunity of characterizing the types of active centers in analogues systems with certain reliability. Moreover, it is possible in some favorable cases to discern and characterize by physico-chemical methods the intermediates that are the precursors of the active centers.

For instance, the ligand-exchange (recoordination) reaction has been characterized [19] by studying active site formation (n_{pr}) in modified $VCl_4 \cdot MMA-AlR_xCl_{3-x}$ systems

Table 3.5 Rate parameters for ethylene–propylene copolymerization catalyzed by homogeneous and immobilized VCl₄ systems.

n_{pr} (mmol/mol)	$r_1\pm3.0$	$r_2\pm0.004$	$r_1\times r_2$	$k_{11}\times10^{-3}$ (l/mol s)	$k_{22}\times10^{-3}$ (l/mol s)	$k_{12}\times10^{-3}$ (l/mol s)	$k_{21}\times10^{-3}$ (l/mol s)	k_{11}/k_{21}	k_{22}/k_{21}
VCl₄									
3.0	15.1	0.021	0.317	18±5	77±10	1.2	3.7	5.0	0.064
VCl₄–AOC[b]									
0.4	13.4	0.022	0.295	7.2	60±8	0.5	2.7	2.7	0.12
PE–gr–PAAl–VCl₄									
10.8	17.2	0.020	0.344	15.1	75±10	0.9	3.8	4.0	0.083
PE–gr–PDBEVPAc–VCl₄									
3.2	16.7	0.026	0.434	12.8	108±15	0.8	4.1	3.1	0.135

[a] Reaction conditions: benzene, 313 K, AlEt₂Cl as cocatalyst, Al/V=50, flow–through system.
[b] Oxyethylated alcohol.

$$VCl_4 \times MMAc + AlR_xCl_{3-x} \leftrightarrow VCl_4 + MMAc \times AlR_xCl_{3-x}$$

The released VCl_4 with the excess of organoaluminium compound yields active sites of the Ziegler type, similar in composition to those in mobile systems. All the facts [89] concerning olefin poly- and copolymerization by immobilized systems give evidence of the same mechanism of active site formation.

Generally, the following reaction routes are feasible by interactions of the catalyst components:

1. the formation of transition metal alkyl derivatives bound to the polymer support;

2. transition metal ion reduction with subsequent attachment to the polymer support;

3. transition metal ion substitution by an organoaluminium compound with conservation of the bond with support through the organoaluminium fragment;

4. transition metal ion substitution by an organoaluminium compound yielding mobile forms of transition metal ions in different oxidation states.

These processes can be generalized by the scheme:

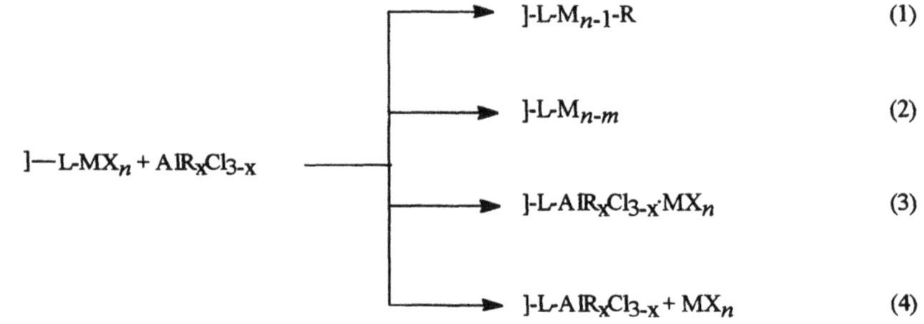

$$]\text{-L-M}_{n-1}\text{-R} \qquad (1)$$

$$]\text{-L-M}_{n-m} \qquad (2)$$

$$]\text{—L-MX}_n + AlR_xCl_{3-x}$$

$$]\text{-L-A}lR_xCl_{3-x}\cdot MX_n \qquad (3)$$

$$]\text{-L-A}lR_xCl_{3-x} + MX_n \qquad (4)$$

Scheme 3.9

The specific weight of the above reactions depends on the nature of the reagents, the type of MX_n bond with the support and reaction conditions. For instance, in the case of an MX_n covalent bond with the support, the contribution of the processes leading to transition metal complex increased lability is not usually above 10–20%,. whereas, donor–acceptor bonding favors the processes and the amount of labile MX_n can reach 40%. These effects are usually observed during the initial polymerization stage (15–20 min) corresponding to the non-stationary part of the kinetic curves. The lack of correlation between Al^{3+} content on the support and amount of labile transition metal points to participation of AlR_2Cl not only in MX_n substitution but also in the interaction with available functional groups.

These reactions can be decreased by MX_n chelation. Macrochelates are less active in ethylene polymerization than monodentate MX_n bound to the support and show higher stability (Figure 3.27). No lability of transition metal compounds has been observed for such systems. For different reasons, the mobile forms of eluted transition metals are not capable of catalyst activity.

The products of reactions (1) and (3) above might form potential active centers of polymerization. However, if active centers were alkylated transition metal derivatives immobilized by the support (reaction (1)), this would certainly lead to a dependence of k_{pr} on the nature of polymer support (or on the nature of its functional groups). This is inconsistent with experimental data (see Tables 3.3 and 3.5).

Thus, it is most probable that active site formation in immobilized systems follows an S_N1 addition mechanism via recoordination in the "cage" of polymer support

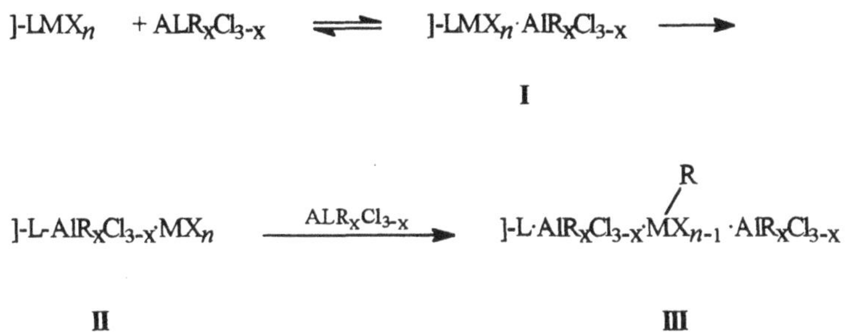

$$]\text{-LMX}_n \quad + \text{ALR}_x\text{Cl}_{3-x} \quad \rightleftharpoons \quad]\text{-LMX}_n \cdot \text{AIR}_x\text{Cl}_{3-x} \quad \longrightarrow$$

I

$$]\text{-L-AIR}_x\text{Cl}_{3-x'}\text{MX}_n \quad \xrightarrow{\text{ALR}_x\text{Cl}_{3-x}} \quad]\text{-L-AIR}_x\text{Cl}_{3-x'}\overset{\overset{\displaystyle R}{|}}{\text{M}}\text{X}_{n-1} \cdot \text{AIR}_x\text{Cl}_{3-x}$$

II   **III**

Scheme 3.10

Compound (**III**) is a trinuclear, bridged metal complex[5] composed of two aluminum ions and one transition metal ion. Both the structure and the reactivity of the proposed active centers (or their precursors) will not be directly influenced by the polymer support (its functional groups); however their concentration dependents o its nature. The last effect is often observed.

According the proposed scheme, active sites (**III**) formation proceeds through the following stages: the formation of intermediate complex (**I**), its rearrangement into complex (**II**) with further alkylation. Thus, the multistage pathway is accompanied by several side reactions which are the reason for the low degree of transition metal compound activation (see Table 3.3). Consequently, purposeful presynthesis of product **II** (pretreatment of polymer support with organometallic compound with following

---

[5] The idea of trinuclear, bridged active site structure is fairly realistic; a complex of the type

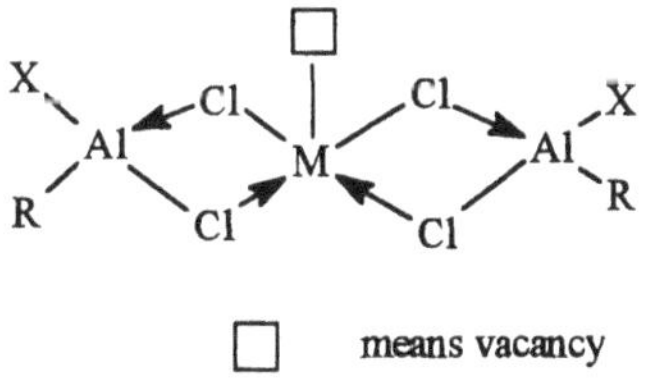

☐   means vacancy

is identified as the active center in analogous homogeneous catalysts [98]. A trinuclear bicyclic Ti^{3+} complex

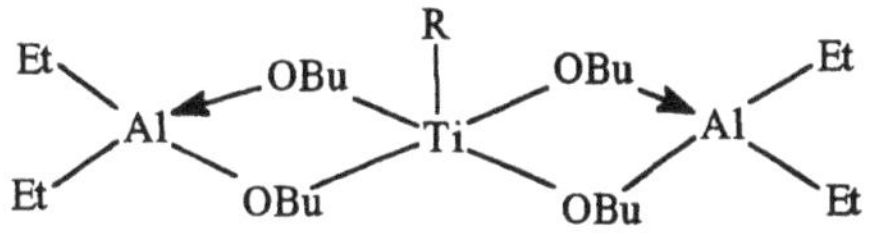

has been identified by ESR for the catalytic system $Ti(OBu)_4$–$AlEt_3$ [99]. There are also a number of other similar examples of such active site structures.

deposition of MX_n) would contribute to the increase of active site content. This approach has been successfully used in the synthesis of bimetallic immobilized metal complexes, including titanium–magnesium catalysts.

Under optimal conditions, multiple regeneration of active sites (n_{pr}) can be observed. For instance, in ethylene polymerization by the catalytical system $Ti(OBu)_4$–PE-gr-PAAl–$AlEt_2Cl$ the reproducibility of active sites (even assuming the participation of each titanium atom, determined from the number of PE moles per mole of titanium) was shown to be equal to 21 at 343 K and 1200 at 363 K [89]. For the homogeneous system this value is equal to 0.25.

Transition metal ion reduction to lower oxidation states is known to be the main reason for metal-complex catalyst deactivation [6]. Rather low equilibrium concentrations of Ti^{3+} ions due to their subsequent aggregation is usually observed for the catalytic systems $TiCl_4$–AlR_xCl_{3-x} [100]. The migration of Ti^{3+} metal ions along the surface or within thin a presurface layer[6] is hindered in immobilized systems. Thus, the thermodynamically favorable process for an increase in particle numbers ($nTiCl^{3+} \rightarrow [Ti^{3+}]_n$) is hindered for these systems and so the equilibrium Ti^{3+} concentration is not reached even after a prolonged reaction time (Figure 3.28). Moreover, even the simple adsorption of $TiCl_4$–$AlEt_2Cl$ interaction products on unfunctionalized polyethylene causes a rise in the equilibrium concentration of Ti^{3+} ions. Chemical bonding of $TiCl_4$ on oligo-like support $C_{12}H_{25}OH$ or the hydrolyzed copolymer of styrene with vinyl acetate, and to a greater extent on PE-gr-PAAl, leads to a sharp decrease of both Ti^{4+} and Ti^{3+} ions reduction with the progressive accumulation of the latter.

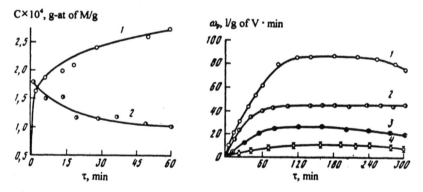

Figure 3.26 The changes in Al(1) and V(2) concentrations on the support with time of catalyst components interaction. Catalytic system $VO(OEt)_3$ on PE-gr-PAAl–$AlEt_2Cl$; solvent n-heptane; 343 K; [Al]/[V]=50.

Figure 3.27 Ethylene polymerization kinetics catalyzed by chelated VCl_4 systems. Polymer supports: (1) PE-gr-PMVK–*ortho*–hydroxybenzoic acid hydrazide, (2) PE-gr-PAAl–alizarin, (3) PE-gr-PAA–β–diketonate, (4) PE-gr-poly(monomethyl methacrylic ester of ethylene glycol).

These experimental facts have also been confirmed by electron spectroscopy for chemical application (ESCA) data [101]. In these studies changes in the electron bond

---

[6] As will be shown in Sections 3.3.2. and 3.3.3. retardation of such processes is also characteristic of the bound complexes of a variety of metals, say for Co^{2+} and Ni^{2+}.

energy of $2p_{3/2}$ shell depending considerably on the ligand surroundings have been studied. In the interaction of $TiCl_4 \cdot PE$-gr-PAN with $AlMe_2Cl$ (Al/Ti=50) in heptane at 293 K a small shift of $2p$-signal from 459.6 eV to 459.4 eV has been observed. This is probably attributed to the stage of immobilized Ti^{4+} alkylation (reaction a). However, interaction of both $TiCl_4 \cdot PE$-gr-PAN and $TiCl_4 \cdot PE$-gr-PAAl with $AlEt_2Cl$ under the same conditions results in appearance of Ti^{3+} ions with electron bond energy E_b=458.8 eV (this value for bulky $TiCl_3$ is equal to 459±0.5 eV) [102] (Figure 3.29). The ESR spectra of these species show broad singlet signal with g-values equal to 1.93. It should be noted that for the similar homogeneous system all titanium ions are in the form Ti^{2+} under the same conditions [6]. Only at 343 K (the temperature of ethylene polymerization by immobilized metal complexes) does the more complete reduction of Ti^{4+} occur. The ESCA spectrum of the system examined had a broad signal composed of at least of two peaks, the first attributed to Ti^{3+} (Eb=458.3 – 458.9 eV) and this probably of Ti^{2+} (E_b=456.2–456.8 eV).

Cooperative stabilization is another probable reason for the rise in activity and stability of immobilized catalyst. A notable number of polymer chain segments can take part in MX_n bonding. The rupture of one of the bonds between a metal ion and polymer chain unit requires the same energy as the cleavage of the similar low molecular mass complex. However, in the first case the energy expended is not immediately compensated by the appearance of additional translatory movements of the complex (i.e. the increase of configuration entropy does not take place). For this reason the probability of deactivation of active sites in immobilized systems is usually less than that for homogeneous systems under the same conditions. The stability of intermediates on (or within) polymer supports can also be considerably increased by spontaneous conformational transitions of the neighboring polymer units as well by fluctuations of structure, etc.

To understand the reasons for the rise in stability of immobilized catalysts let us compare the decomposition of organometallic transition metal compounds (the principal element of the active centers) in both homogeneous and heterogeneous systems.

Metal-carbon σ-bond energies are not of great value [103] however sufficient to provide these compounds stability in the case of simple M–C bond decomposition. This bond energy value is comparable with those for carbon–main groups metals and is not the reason of high reactivity and low stability of alkylated transition metal compounds well known in solutions. The main processes controlling the above properties are those proceeding in the inner coordination sphere of the transition metal, such as β-elimination and disproportionation. The activation energy for these processes is considerably less than that for M–C bond rupture. Thus to increase the stability of these compounds one must exclude this possibility for the alkyl group. The general scheme of these reactions with participation of electron donor D can be presented as follows [56]:

$$RMX_n + D \rightleftharpoons RMX_n \cdot D$$

$$RMX_n \cdot D \longrightarrow \begin{cases} MX_n \cdot D + R^{\bullet} \\ RH + MX_n \cdot D^{\bullet} \end{cases}$$

Scheme 3.11

In spite of differences in the character of the reagents nature, there are many common features between the above reactions and those of active centers in olefin polymerization because olefins can be considered to be weak organic bases while active sites of metal complex catalysts certainly have an M–R bond.

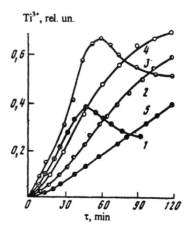

Figure 3.28 The kinetics of Ti^{3+} accumulation during interaction of Ti^{4+} with $AlEtCl_2$. Ti^{4+} compounds: (1) $TiCl_4$, (2) $TiCl_4$ bound to hydrolyzed copolymer of styrene with vinyl acetate, (3) $C_{12}H_{25}OTiCl_3$, (4) $TiCl_4/PE$, (5) $TiCl_4$ bound to PE-gr-PAAl. Reaction conditions: benzene, 293 K, [Al]/[Ti]=50.

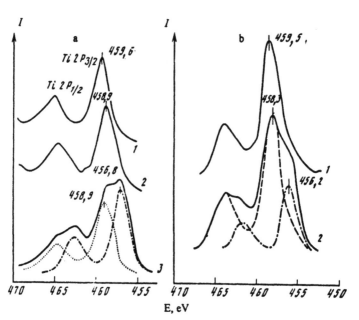

Figure 3.29 ESCA spectra of immobilized Ti^{4+} compounds and the products of their interaction with $AlEtCl_2$. (a) 1, PE-gr-PAN·$TiCl_4$; 2, as 1 exposed at 293 K during 30 min; 3, as 1, but exposed at 343 K. Al/Ti=50. (b) 1, PE-gr-PAA·$TiCl_4$, 2, as 1, exposed at 293 K during 30 min.

As a model active site fixed at a polymer surface, CH_3TiCl_3 bound to PE-gr-P4VP or to PE-gr-PAN has been used [104–106]. CH_3TiCl_3 complexes with 4-substituted pyridines break down in homogeneous systems below ambient temperatures yielding methane. The process follows the concerted mechanism in cyclic complexes and proceeds in the titanium coordinative sphere; Ti^{4+} is not reduced in the course of the reactions. An increase of complex stability has been observed for immobilized complexes. Only at temperatures above 283 K has the progressive breakage of the Ti–C bond been observed followed by the evolution of a gas (methane. ethane and C_4 hydrocarbons). A rise of temperature above 353 K led to an increase of homolytic Ti–CH$_3$ bond cleavage with the formation of CH$_3$ radicals Accompanying reduction of Ti^{4+} to Ti^{+3} has been observed:

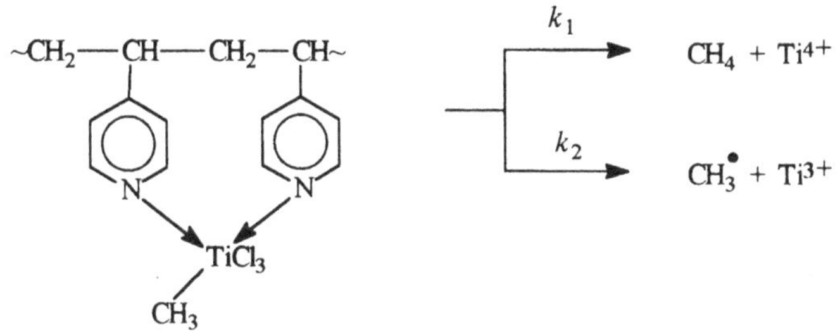

Scheme 3.12

The value of the rate constant of concurrent cleavage of the Ti–CH$_3$ bond, k_1, is 2 orders of magnitude greater than that for decomposition of CH$_3$TiCl$_3$ complex with monomer ligand. This is the result of an extremely low value of the preexponential factor, 10^{-2}–10^{-3} s^{-1} (for homogeneous systems this value is within the range of 10^3–10^5 s^{-1}). In other words, the "alignment" (deformation) of the polymer ligand which provides the opportunity for the concurrent reaction needs a significantly greater change in entropy for the formation of the intermediates in comparison with that for the homogeneous complex. It is also important that the rates of Ti–CH$_3$ homolysis yielding free radicals (rate constant k_2) are constant both for homogeneous and heterogeneous systems.

Thus, because the Ti–C bond energy values exhibit similar behavior for homogeneous and immobilized systems, then the increased stability and activity of the immobilized systems are to be identified with a sharp retardation of the concurrent processes in the transition metal coordinative sphere. Another important circumstance also should be noted. Obviously, deactivation processes, such as β-elimination, disproportionation and transition metal ion reduction, have to proceed with a rearrangement of significantly greater number of ligands than for monomer insertion at M–C bonds.

So, the rise of complex stability can be observed on these examples: soluble complexes<modified complexes<immobilized catalysts.

3.2 Diolefin Polymerizations in the Presence of Immobilized Catalysts

Interest in this problem is connected with the possibility of observing (and under optimal conditions influencing) the effect of immobilization not only on activity but also on stereospecificity of metal-complex catalysts. Dienes give additional opportunities for such changes in comparison with α-olefins. Immobilization of transition metal compounds yields unpredictable effects in butadiene and isoprene polymerization. For instance inversion of the stereospecific action in isoprene polymerization of π-allyl chromium compounds immobilized on aluminosilicates has been shown [107]. One can assume that polymer supports will also considerably affect these processes.

3.2.1 cis-Stereoregular polymerization by cobalt compounds

Homogeneous systems based on cobalt compounds ($CoCl_2 \cdot Py_2$, $CoCl_2 \cdot R'OH$, $Co(AcAc)_2$, $Co(R''COO)_2$, etc.) in combination with AlR_2Cl are effective and stereospecific catalysts for butadiene polymerization to cis-1,4-polybutadiene [108, 109].

To make a correct comparison of the actions of these catalysts, the main requirement is similarity of the ligand environment in both homogeneous and heterogeneous systems. Otherwise, the nature of the ligand environment can significantly affect the structure of the polydienes formed.

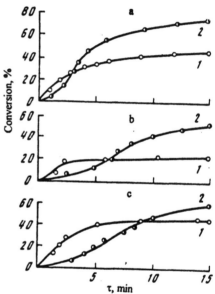

Figure 3.30 Kinetics of polybutadiene polymerization by homogeneous (1) and heterogeneous (2) systems formed from $CoCl_2 \cdot Py_2$ (a), $CoCl_2 \cdot (VIm)_4$ (b) and $Co(AcAc)_2$ (c). Polymer supports: (a) PE-gr-P4VP, (b) PE-gr-PVIm, (c) – PE-gr-PAAc. Polymerization conditions: 323 K, [butadiene]=2.0 mol/l, $[Co^{2+}]=2\times10^{-4}$ mol/l, [Al]/[Co]=200, toluene.

Table 3.6 Characteristics of polybutadienes obtained in the presence of homogeneous and heterogeneous catalytic systems (polymerization conditions as in the Figure 3.30; butadiene conversion not above 10%).

Co^{2+} compound	$\overline{M}_w \times 10^3$	$\overline{M}_n \times 10^3$	γ	Units content (%)		
				1,4-cis	1,4-trans	1,2-
$CoCl_2 \cdot Py_2$	16.2	8.1	2.00	70.2	6.8	23.0
PE-gr-P4VP·$CoCl_2$	28.4	9.7	2.92	75.5	7.8	19.7
$CoCl_2 \cdot (VIm)_4$	15.4	6.1	2.52	72.3	7.7	20.0
PE-gr-PVIm·$CoCl_2$	52.8	19.2	2.75	79.1	7.1	13.8
$Co(AcAc)_2$	15.3	5.7	2.66	66.6	7.4	26.0
PE-gr-PAAc·$Co(AcAc)_2$	40.6	16.2	2.50	77.6	6.2	16.2

A comparison of homogeneous catalytic systems ($CoCl_2 \cdot Py_2$, $CoCl_2 \cdot (VIm)_4$, $Co(AcAc)_4$) with heterogeneous showed (Figure 3.30) [110–112] that a considerable inductive period for active sites formation has been observed for the last catalysts. Besides that, retardation of deactivation reactions has also been observed for immobilized systems in some cases giving rise to an increase of both total yield and molecular mass of the polybutadiene (Table 3.6). This last observation may be connected with a number of objectives. One of them is possible adsorption of the polybutadiene molecules formed on the surface of the immobilized catalyst leading to the slowing down of chain termination reactions ("gel"-effect).

The microstructures of polymers obtained by comparative catalytic systems are similar although the production of 1,4-*cis*– units increases by 2–11% in heterogeneous systems at the expense of a diminution in 1,2-structure content. Their presence is connected [108] with the appearance in the systems of cobalt in low oxidation states.

A reduction in the polymerization activation energy for the heterogeneous systems (30 kJ/mol) should also be mentioned (for homogeneous systems this value is 55 kJ/mol).

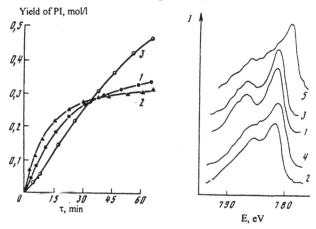

Figure 3.31 Isoprene polymerization kinetics in the presence of homogeneous ($CoCl_2 \cdot Py_2$– Et_2AlCl) (1) and heterogeneous ($CoCl_2 \cdot PE$-gr-P4VP–Et_2AlCl) (2), $Co(CH_3COO)_2 \cdot PE$-gr-poly(*N*-acryloylacetylamide)–Et_2AlCl) (3) systems. Polymerization conditions: 323 K, [isoprene]=0.7 mol/l, [Co^{2+}]=2.0×10^{-4} mol/l, [Al]/[Co]=30, toluene.

Figure 3.32 ESCA $Co(2p_{3/2})$ signals for immobilized and individual cobalt complexes. (1), $CoCl_2$; (2) $Co(AcAc)_2$; (3) $CoCl_2 \cdot P4VP$; (4) $Co(AcAc)_2 \cdot PAAc$ (3.2×10^{-4} Co^{2+} ions/g); (5) $CoCl_2 \cdot P4VP$–$AlEt_2Cl$.

As homogeneous "cobalt" catalytic systems are insufficiently stereospecific in isoprene polymerization, immobilization has been supposed to increase the 1,4-*cis* unit content in the polymer. However, the compared systems show high activity (Figure 3.31), the desirable increase of stereospecificity due to the heterogeneous catalysts not being achieved (Table 3.7).

It has been shown by the combined kinetic-radiochemical method[7] that rates constant of chain propagation reactions are of similar values for the systems compared. However, the heterogeneous processes show higher efficiency; each Co atom generates an active center, while only 38% of cobalt atoms yield active sites in the homogeneous systems.

Polymers obtained in the presence of the catalytic systems compared have molecular masses close to ($\overline{M}_n$=137 000−370 000), and show a narrow unimodal polydispersity factor although $\overline{M}_n$ that of polyisoprene catalyzed by the heterogeneous system under similar conditions is somewhat higher.

The main factor concerning the properties of the active site responsible for diene polymerization consists of formation of π-allyl complexes with the monomer involved [113]. Their formation can be presented by following scheme [114]:

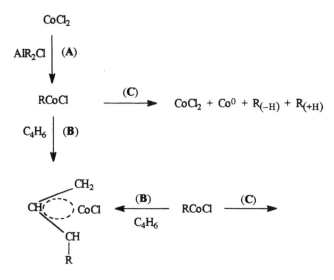

Scheme 3.13

Immobilization of $CoCl_2$ on polymer supports allows the formation of the complex to be monitored during the course of these reactions (by means of ESCA, measuring the species magnetic susceptibility, etc.) [115]. For instance, it has been shown that magnetic moments of octahedral Co^{2+} complexes (d^7 configuration) are within the range 5.0–5.2 μ_B at room temperature. These values for high-spin tetrahedral complexes are within 4.4-4.8 μ_B. Low spin octahedral Co^{2+} complexes show magnetic moments of 1.8–1.85 μ_B, close to the magnetic moment of a single unshared electron. These data allow on the one hand the coordination state of bound Co^{2+} to be characterized and on the other hand the study of the complex transformation in the course of the catalyzed reaction.

The dependence of the magnetic susceptibility on magnetic field strength shows the presence of ferromagnetic cobalt species in the $PE\text{-}gr\text{-}P4VP\cdot CoCl_2\text{-}AlEt_2Cl$ reaction

---

[7] In concept the method consists of chain termination at low monomer conversion degree by use of $^{14}CH_3OH$; k_{pr} is determined on $\overline{M}_n$ changes with time and polymer radioactivity gives the values of n_{pr}.

products. Tabulated in Table 3.8 are values of ferro-   [Co(0)], dia-   [Co^{1+}] and paramagnetic [Co^{2+}] contents that have been calculated on the basis of this and the temperature dependences of magnetic susceptibility. Ferromagnetic cobalt represents about 50% in these products, the remainder being diamagnetic. In ESCA spectra Co (2p$_{3/2}$) a response (Figure 3.32) at 779.9 eV, attributed to metallic cobalt [115] has been observed. Furthermore, an increase of N/Co ratio in comparison with the initial PF-gr-P4VP·CoCl$_2$ product demonstrates the gradual increase in the number of uncoordinated pyridine rings during the course of the reaction. The peak N$_{1s}$, being strongly broaden shows the presence of several bonding states of nitrogen atoms in pyridine fragments. One of these is characterized by E$_b$=399.3 eV and may be attributed to the uncoordinated state; a second (at 401.9 eV) may be attributed to nitrogen coordinated with different alkylated aluminum chlorides and stronger Lewis acids than CoCl$_2$.

Table 3.7 The comparative kinetics of isoprene polymerization by homogeneous and heterogeneous "cobalt" systems.

Factor	System	
	Homogeneous	Heterogeneous
E$_a$, kcal/mol	16.5±1.0	13.0±1.0
Reaction order		
with respect to monomer	0.4	0.8
with respect to cobalt	1.4	1.0
k$_{pr}$ (l/mol s) (kinetic method)	6.5±0.5	3.5±0.5
k$_{pr}$ (l/mol s) (radiochemical method)	10.7	2.0
n$_{pr}$ (mol/mol of Co)	0.38	1.0
Polymer structure (%)		
1,4-cis-		40–46
1,4-trans		30–32

Table 3.8 Magnetic state of cobalt ions in the products of interaction of catalyst components.

System	Color	Co ion magnetic state (%)		
		ferromagnetic	paramagnetic	diamagnetic
PE-gr-P4VP·CoCl$_2$+Al(C$_2$H$_5$)$_2$Cl	gray	50		50
in air after 0.5 h	blue	7	35	58
after 2 weeks	blue	3.5	60	37.5
after 3 week	blue	–	100^a	–
PE-gr-P4VP·CoCl$_2$+isoprene+ Al(C$_2$H$_5$)$_2$Cl	gray	8	–	92
in air after 0.5 h	gray	3	20	77
after 3 weeks	blue	–	100^b	–

^{a}The share of paramagnetic Co has been calculated in accordance with μ_{eff}=4.8; that of the diamagnetic form by subtraction of the paramagnetic and ferromagnetic contributions from the total amount of cobalt present.

$^b\mu_{eff}$=5.18 μ_B for these species.

Thus, the interaction between components of catalytic cobalt systems in the absence of monomer is accompanied by the essential reduction of Co^{2+} to free metal products, the magnetic moment of ferromagnetic part being fixed (to an accuracy of 10%) by decreasing the temperature to 4.2 K and increasing the magnetic field strength to 10^4 G. This means that at least 90% of ferromagnetic cobalt is in the form of large (>10 nm) particles [116]. Aggregation of free metal particles is probably the main reason for loss catalyst activity on aging. However, the interactions of catalyst components in the presence of isoprene (in the course of polymerization) prevents this process. Only 6–8% of ferromagnetic cobalt has been detected in the reaction mixture after polymer separation. The main part of cobalt was shown to be in diamagnetic form (Co^{+1}). Most probably, the monomer[8] stabilizes Co^{+1} ions in *statu nascendi* through π-complex formation (active in polymerization) as shown for the complex of $CoCl_2$ with $Al_2Et_3Cl_3$ [117]:

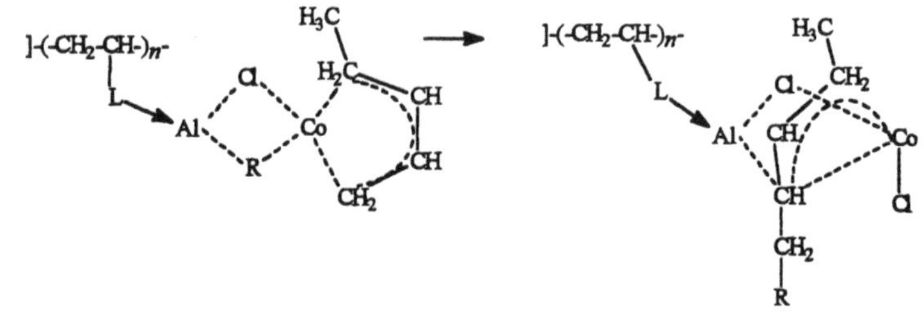

Scheme 3.14

Without discussing the detailed structures of active centers[9], let us only point out that the main role of the polymer support for cobalt-immobilized catalytic systems probably consists of the regulation of redox processes leading to active site regeneration (for instance by oxidation of Co^0 by path **C** in scheme 3.13). Similar possibilities will be discussed below for ethylene dimerization by nickel catalysts[10] (Section 3.3).

It is well known that the essential component in Co^{2+} homogeneous systems for diene polymerization is the electron donor. Electron donors are added to provide a high polymerization rate, a high molecular mass of polymers produced and *cis*-stereoregulation: water additives are of special interest for these catalysts and particularly influence homogeneous systems (Figure 3.33). In addition, a sharp increase

---

[8] Simple Co–C and/or Ni–C σ-bonds are significantly less stable than Ti–C or V–C. However their stability increases under the action of π-allyl structures.

[9] Up to now there have been no data on both cobalt coordination number and active center stereochemistry. As was mentioned above, cobalt complexes may exist in the form of a 6-coordinated octahedron, 4-coordinated tetrahedron or square planar structures, each of these being a potential active site or a precursor. Inactive in butadiene polymerization, tetrahedral $[CoCl_4]^{2-}$ complex was shown to form at ratios Al/Co<50 [118]. However, both coordination number and stereochemistry may change in the course of polymerization.

[10] The value of "nickel" catalyst activity correlates with neither nickel cluster concentration nor with Ni^+ [119]. Active centers are supposed to be formed either of free metal particles of Ni(0) (may be of clusters of the size less than 10 nm) or directly of Ni^{2+} species.

of catalyst activity is usually observed within a narrow range of H_2O concentrations (10 mol% against $AlEt_2Cl$ for homogeneous systems and 15 mol% for heterogeneous systems). Water additives also essentially affect the stereoregularity of the polymer products by increasing 1,4-structures at the expense of 1,2-units. Under optimal H_2O concentrations polybutadiene yields reach 30 kg/(g of Co h); the molecular mass of the polymer increases under the action of water additives for both systems (Table 3.9).

Table 3.9. The effect of water additives on the characteristics of polybutadiene obtained in the presence of homogeneous and heterogeneous systems[a]

$H_2O \cdot 10^4$, mol/l	$CoCl_2 \cdot Py_2$							
	$\overline{M}_w$	$\overline{M}_n$	$\overline{M}_w / \overline{M}_n$	$\overline{M}_z$	$\overline{M}_z / \overline{M}_w$	1,4-cis	1,4-trans	1,2-
1.4	247 000	94 000	2.62	445 000	1.80	76	6	18
2.2	268 000	100 000	2.68	480 000	1.78	89	5	6
7.0	460 000	225 000	2.04	640 000	1.39	92	5	3
13.0	464 000	268 000	1.73	610 000	1.31	94	5	1

$H_2 \cdot 10^4$, mol/l	$CoCl_2 \cdot PE\text{-}gr\text{-}P4VP$							
	$\overline{M}_w$	$\overline{M}_n$	$\overline{M}_w / \overline{M}_n$	$\overline{M}_z$	$\overline{M}_z / \overline{M}_w$	1,4-cis	1,4-trans	1,2-
1.4	136 000	58 000	2.35	230 000	1.67	81	6	13
2.2	224 000	68 000	3.30	415 000	1.85	90	5	5
7.0						95	4	1
13.0	440 000	250 000	1.76	600 000	1.36	95	4	1

[a]Reaction conditions: benzene, [butadiene]=2.0 mol/l, $[Co^{2+}]=7 \times 10^{-5}$ mol/l, [Al]/[Co]=200, 323 K.

Table 3.10. The magnetic state of cobalt ions in the products of interaction of catalyst components [120].

System	Reaction time	Magnetic state of Co,%		
		ferro-(Co^0)	dia-(Co^+)	para-(Co^{2+})
PE-gr-P4VP·$CoCl_2$+$AlEt_2Cl$	2 h	50	50	–
with H_2O	2 h	25	75	–
PE-gr-P4VP·$CoCl_2$+$AlEt_2Cl$+butadiene	2 h	–	100	–
with H_2O	2 h	–	100	–
PE-gr-P4VP·$CoCl_2$+alumoxane+butadiene	2 h	–	-	100
PE-gr-P4VP·$CoCl_2$+alumoxane+butadiene	7 days	–	50	50
with H_2O	7 days	–	65	35
PE-gr-P4VP·$CoCl_2$+alumoxane+$AlEt_2Cl$	1 h	25	75	–

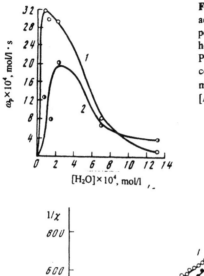

Figure 3.33 The influence of water additives on the rate of butadiene polymerization in homogeneous (1) and heterogeneous (2) (support=PE-gr-P4VP) systems. Polymerization conditions: 323 K, [butadiene]=0.7 mol/l, $[Co^{2+}]=2.0\times10^{-4}$ mol/l, [Al]/[Co]=30, toluene.

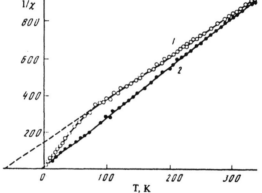

Figure 3.34 The dependence of $1/\chi$ on T for immobilized complexes PE-gr-P4VP·CoCl$_2$ in (1) inert atmosphere; (2) in air.

Generally, the promoting action of water additives appears concurrently with its activity in respect of organoaluminium molecules yielding alumoxanes. This reaction shows diminishing reducibility in alkylation of Co^{2+} by stabilizing the Co^{1+} ions needed for *cis*-1,4-polybutadiene formation.

Immobilization gives opportunity to determine the mechanism of action of the water in both diene polymerizations and the stereochemistry of the cobalt complexes. For this purpose, the products of interaction in the system PE-gr-P4VP·CoCl$_2$–AlEt$_2$Cl both in the presence of monomer and without it have been separated and studied at different reaction stages in an inert atmosphere [120].

These studies show the significant influence of H_2O on Co^{2+} reduction only in the absence of monomer. It has been revealed that approximately half of the ferromagnetic species are formed without water additives, the reminder cobalt being in the form of Co^{+} (Table 3.10). At the same time during the first stage of butadiene polymerization, H_2O additives do not change the electronic state of the bound cobalt ions; it completely reduces to Co^{+} and is stabilized by the monomer in the form of π-allyl complexes which block further reduction. Since mild hydrolysis of organoaluminium compounds yields alumoxanes and their formation is believed to be the result of the water additives action, studies of the activities of alumoxanes in butadiene polymerization have been carried out.

It has been shown under model conditions [120] that butadiene polymerization by the system PE-gr-P4VP·CoCl$_2$–(isoBu)$_2$Al–O–Al(isoBu)$_2$ proceeds at a low rate. Only modest amounts of polybutadiene are formed after several days reactions, while under otherwise similar conditions 100% conversion of butadiene achieved in 1 h using the system PE-gr-P4VP·CoCl$_2$–AlEt$_2$Cl–H$_2$O. Addition of AlEt$_2$Cl to the mixture increases activity, the polymerization rate nevertheless remaining low (10% conversion was reached after 7 days). Studies of the magnetic properties of the reaction products showed that low polymerization rates in the absence of added water were the result of very low reduction of Co^{2+} species providing catalytically active Co$^+$ complexes.

However, as illustrated by Table 3.10, water additives up to concentrations of 20 mol% do not essentially influence the cobalt oxidation state in the course of polymerization. Therefore, accelerated water action is not the result of the diminution of organoaluminum compound reducibility due to the formation of alumoxanes.

In this connection, one should assume the direct action of water molecules (as a proton source) on the active site leading to M–C bond activation as was supposed for titanium-containing compounds in ethylene polymerization [121]. On the other hand, the effect of water may be the result of active site rearrangement accompanied by a rise of cobalt oxidation state number favorable to monomer coordination[11].

Finally, measurements of magnetic susceptibility (within the broad temperature range 4.2–300 K) showed that Co^{2+} complexes aging in air significantly change their magnetic properties. The temperature dependence $1/\chi = f(T)$ for macrocomplexes PE-gr-P4VP·CoCl$_2$ in inert atmosphere show abnormal character, testifying to the presence on the polymer support of polynuclear structures with strong antiferromagnetic exchange activity (Figure 3.34). This, for complexes exposed in air, exactly follows the Curie–Weiss law indicating decomposition of these polynuclear species (μ_{eff}=5.18 μ_B corresponds to the magnetic moment of Co^{2+} octahedral complexes). Since, the contact with air does not change catalyst activity, one can assume the mononuclear species to be the active sites.

Kinetic studies of the action of water on diene polymerization in combination with physico-chemical data on the structure of formed polydienes lead to the conclusion that the active sites are similar in both homogeneous and heterogeneous systems. However, isolation of intermediates is possible only in the latter.

Finally, it should be noted that gel-immobilized Ti^{4+}, Fe^{3+}, Ni^{3+}, Co^{2+} complexes are also active in butadiene polymerization [122]. The structure of the polymer formed is predetermined by both the structure of the transition metal ion and that of the cocatalyst. The systems based on Fe^{3+} yield polymers containing 30–40% of 1,4-$trans$ units, with 1,2-structures constituting the reminder. Cobalt systems preferentially give 1,4-cis units (87–97%). Another feature is the formation of strong cross-linked polybutadiene no matter what cobalt salt has been used. Ti^{4+} catalytic systems yield 1,4-$trans$ and 1,2-units.

---

[11] The presence of two types of active sites both being products of AlR$_2$Cl interaction with water and capable of interchanging (the amounts of each depending on water concentration) has also been assumed [114].

3.2.2 Macromolecular organolithium compounds in the polymerization of dienes

Organolithium compounds (mainly LiBu, LiMe, LiEt and their complexes with electron donor additives) initiate anionic polymerization of dienes and styrene. Insoluble in hydrocarbon solvents, organolithium compounds sometimes transform into soluble form by addition of monomer units. Through a series of organometallic reactions these mixtures yield polydienes composed of 90–97% 1,4-*cis*-units. Macromolecular organolithium compounds produced either by direct *trans*-lithiation or by indirect lithiation (by halogen exchange with lithium) have recently found use for this purpose ([17], pp.61–65). Polystyryllithium, polystyrylbenzyllithium, etc., are more widely used. For illustration, macromolecular initiators soluble in aromatic solvents have been produced by metalation of polybenzyl (relative molecular mass of 2300), polydiphenylethyl (relative molecular mass of 1170) with LiBu. Proceeding from the observation that the microstructure of polydienes produced by these initiators is similar to that formed in the presence of LiBu it is concluded that the mechanism is similar to that of polydiene formation. The key stage of the mechanism is identified as electron transfer to monomer following recombination of anion radicals; chain propagation proceeds further through the dianions.

3.2.3 Palladium-immobilized complexes in diene polymerizations

The data on these catalysts are rather scarce. The comparison of homogeneous and heterogeneous Pd^{2+} π-complexes in polymerization of bicyclohepta-2,5-diene has been carried out [123]. Immobilization of $Pd(PhCN)_2Cl_2$ on phosphorylated CSDVB in the presence of β-pinene yielded catalysts of low activity (Figure 3.35) as a result of coordinative saturated complex (I) formation. Straight bonding of Pd^{2+} to polymer supports carrying partly reduced benzene rings (1,4-cyclohexadienes) yields active coordinative unsaturated polymer π-complexes (complex II):

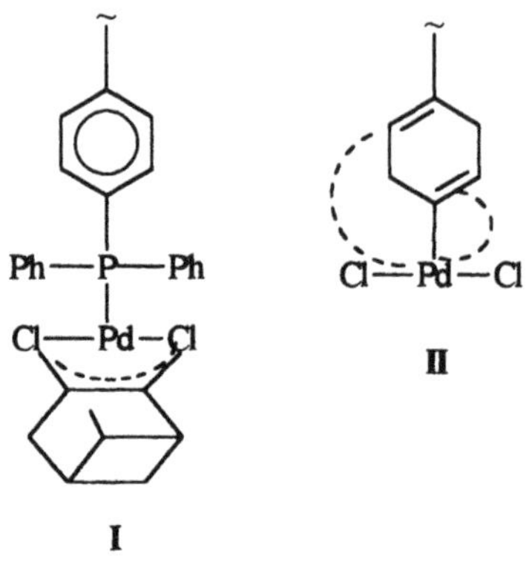

Scheme 3.15

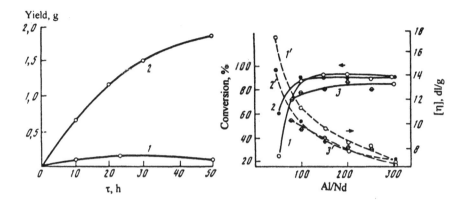

Figure 3.35 Polymerization of bicyclohepta-2,5-diene in the presence of Pd-bound catalysts. Catalysts: (1) complex **I**, (2) complex **II**, T=353 K.

Figure 3.36 The influence of [Al]/[Nd] ratio on the yield and viscosity of polybutadiene at 323 K. (1,1′) copolymer of styrene with acrylic acid (CSAAc)·Nd–Ph₃CCl–Al(isoBu)₃, Cl/Nd=3.5, reaction time 6 h; (2,2′) CSAAc·Nd–Ph₃CCl–Al(isoBu)₃, Cl/Nd=3.5, 4 h; (3,3′) CSAAc·Nd–Al(isoBu)₃, Cl/Nd=3, 8 h.

3.2.4 Diene polymerizations in the presence of immobilized lanthanides and actinides

Rare earth metals have found use as catalyst components for butadiene and isoprene polymerization [124, 125]. Polymerization of diolefins under the action of $PrCl_2$ and $NdCl_2$ even in the absence of cocatalyst has been reported [126]. Usually, these catalysts yield products with a low content of gel-fractions reached with 1,4-*cis* units. Apart from chlorides, carboxylic acid salts (more often Nd(naphthenate)₃) are sometimes used. A number of techniques for immobilization of the rare earths have been developed [127,128]. For instance, interaction of rare earth elements (Nd, Pr, La, Eu, etc.), or chlorides with functionalized polymers, such as the copolymer of styrene with methaacrylic acid, PP-gr-PAAc, or the copolymer of styrene with 2-(methylsulfynyl) ethyl methacrylate, produce highly effective catalysts for stereospecific polymerization of butadiene and isoprene into 1,4-*cis*-polymers (98% in the case of butadiene and 95–96% for isoprene) [129,130]. The activity of immobilized systems has proved to be 2–3 times more active than the homogeneous NdCl₃–4DMSO–Al(isoBu)₃ system; the special additives of CCl₄, alkyl- and benzyl chlorides, and most notably Ph₃Cl (polymerization in toluene at 323 K, the content of bound Nd = 0.66 mmol/g) enhance catalyst activity (up to 170 kg of polybutadiene/(g of Nd h)) (Table 3.11). Catalytic properties of these macrocomplexes depend on a number of factors such as the polymer properties (of prime importance), molar ratio [Al]/[Nd], solvent nature, organoaluminum compound (decreasing in the sequence Al(isoBu)₃>(isoBu)₂AlH>AlEt₃), the nature of the rare earth compound (carboxylic, naphthenate, octanoate can be used instead of chloride).

However, the microstructure of the polymers formed does not depend on the above factors; polymer molecular mass clearly drops with an increase of [Al]/[Nd] ratio (Figure 3.36). The activity of polymer-immobilized Nd complexes (bidentate group bound through the carboxylic groups of the copolymer of styrene with acrylic acid (12%) (CSAAc)) is dependent on the [Nd]/[COOH] ratio [132]. The optimal activity has been observed when this ratio is 0.2. New catalysts for 1,4-*cis*-butadiene polymerization based on bimetallic polymer-immobilized complexes [133] of the type CSAAc/Nd/Na or CSAАc/Nd/Fe should also be noted. Their activity is higher at lower [Nd]/[COOH] ratios than those of monometallic catalysts and increases with the rise of Fe or metallic sodium content. The relation between the structure of polymer-immobilized lanthanide complexes and their activity in butadiene polymerization has been analyzed [134].

A promising method for the production of rare earths-containing polymers is poly- and copolymerization of lanthanide-containing monomers [135–140]. These metal copolymers in combination with AlR_3 or AlR_2Cl are effective in 1,4-*cis*-polymerization of dienes. Catalysts can be regenerated under optimal conditions.

Data on diene polymerization by immobilized actinides complexes are rather scarce. 1,4-*cis*-Polymerization of isoprene by uranium halides or oxyhalogenides bound to higher ($C_8 \div C_{20}$) alcohols, trialkylphosphates and triphosphynoxides [141] should be mentioned as such catalysts (AlR_3 as cocatalyst), yielding polymers containing 99% *cis*-units. The uranium oxidation state is supposed to be +3 in these systems. The first act of interaction is the reversible recoordination of the "microsupport" from uranium to aluminum (as it usually observed in modified systems for olefin polymerization, see Section 3.1.10) followed by uranium alkylation and reduction. Unfortunately these studies have had no further development although the review "Developments in the surface and catalytic chemistry of supported organoactinides" [142] has recently been published.

Table 3.11 Comparison of activities of neodymium catalysts in butadiene polymerization [131].

Catalytic system	Nd (mmol/g)	Al/Nd	τ (min)	Polybutaiene yield (kg/(g of Nd h)
Nd–CSAAc[a]–Ph₃CCl–Al(*iso*Bu)₃	0.1	2000	20	137
	0.2	1500	10	169
	0.3	500	15	93
NdCl₃·CSMS[b]–Al(*iso*Bu)₃	0.3	1000	15	56
NdCl₃·4DMSO–Al(*iso*Bu)₃	0.3	1000	30	16
	0.3	300	30	8

Catalytic system	[η] (dl/g)	Microstructure (%)		
		1,4-cis	1,4-trans	1,2-
Nd·CSAAc–Ph₃CCl–Al(*iso*Bu)₃	3.5	95.8	0.7	0.5
	2.2	98.4	1.2	0.4
	3.4	98.8	0.7	0.5
NdCl₃–CSMS–Al(*iso*Bu)₃	2.7	98.7	0.8	0.5
NdCl₃·4DMSO–Al(isoBu)₃	5.9	99.0	0.6	0.4

[a] CSAAc = copolymer of styrene with acrylic acid.

[b] CSMS = copolymer of styrene and 2-(methylsulphynyl)ethylmethacrylate.

3.2.5 Synthesis of 1,4-trans-isoprene

1,4-*trans*-Polyisoprene being the analogue of gutta-percha and balata recently has attracted the attention of investigators. Complex vanadium-containing systems doped with special additives (including $TiCl_4$) are effective for its production (*trans*-stereospecific polymerization of butadiene, isoprene and *trans*-piperylene) [143]. The *trans*-1,4 unit content of these polymers is above 90%; their properties are similar to those of α- and β-gutta-percha.

Since polymerization in the presence of homogeneous systems proceeds at moderate temperatures, gives low polymer yields, needs specific additives (amines, sulfides, phosphynes), one can suppose that immobilized catalysts would be effective in 1,4-*trans* isoprene polymerization. Immobilized Ti^{4+} and V^{4+} compounds showed high activity in the process up to 320 K (polymer yield increased by 1.5–2 times) [89]. The use of bimetallic Ti–V immobilized systems (molar ratio Ti/V=0.1–0.5) has also proved successful. The mechanism of isoprene polymerization by bimetallic systems still remains unclear. However active centers probably include both transition metal ions. This assumption is based on the comparative analogy between α- and β-$TiCl_3$–(π-$C_3H_5)_nM$ (M=Ni, Zr, Cr, n=2–4) systems, the former being non stereospecific in 1,4-*cis*-polyisoprene formation and the latter yields high stereospecific polymers [144]. This indicates that the formation of stereoregular polymer proceeds at M–C bridged bonds in bimetallic complexes of the type

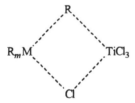

Scheme 3.16

According to other opinions, competitive monomer addition to Ti–C and M–C bonds may be the reason for the dependence of polyisoprene microstructure on the properties of the metal nature in π-allyl complexes. The use of mixed ligands in the metal ion environment (of the type $TiCl_2I_2$, etc.) may also have an enhancing effect on catalyst activity.

3.2.6 The joint polymerization of di- and mono-olefins by immobilized catalysts

Copolymerization of monomers of different types yields polymers with unusual properties, so called chemical hybrids. Copolymerization of dienes with olefins has been catalyzed by binuclear complexes of d- and f-metals (e.g. $NdCl_3·3D$–$TiCl_4$–$Al(isoBu)_3$). Analysis of the reaction products (microstructure of the diene part of the copolymer, the presence of an oligo-butadiene fraction) indicates that the Ti-containing centers are responsible for copolymerization. In the presence of modified additives and under certain conditions alternating copolymers of butadiene with ethylene or propylene can be synthesized.

Copolymerization of butadiene with ethylene catalyzed by bimetallic Co^{2+}–V^{4+} bound to PE-gr-PVPr possibly follows another mechanism [145]. Active sites of two types, the first for ethylene polymerization (V-containing) and the second for butadiene polymerization (Co-containing) can be formed under the activation of the bound macrocomplex with AlR$_2$Cl. Butadiene is known to be a strong inhibitor of ethylene polymerization and there is only moderate copolymerization with ethylene in the presence of vanadium-containing catalysts ([90] pp.133–160). Although, the rate of ethylene being used up decreases in the presence of butadiene in the above system, inhibiting action of butadiene is considerably less than in vanadium-containing systems (Figure 3.37). The rate of butadiene polymerization on cobalt-containing active centers also decreases to some extent in the presence of ethylene [146]. This fact probably indicates that α-olefins take part in chain termination[12] (by addition of 1 or 2 end ethylene molecules) but do not copolymerize with butadiene. However, due to the close proximity of both active sites, copolymeric materials can be formed by a migration processes, such as interchain exchange [147], polybutadiene chain transfer to polyethylene[13] (the top path in Scheme 3.17), or by chain recombination (lower path 2).

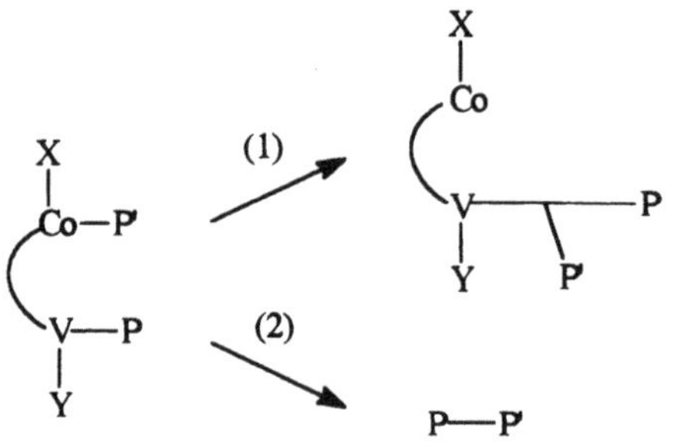

Scheme 3.17

A two-block copolymer P-P′ is formed in the case of chain disproportionation, whereas tri- and poly-block copolymers can be obtained when chain transfer occurs with

---

[12] The mechanism of the process is similar to chain transfer to the polymer molecule and proceeds via decomposition of the less stable σ-alkyl cobalt derivative formed after ethylene addition to the π-allyl complex [146]:

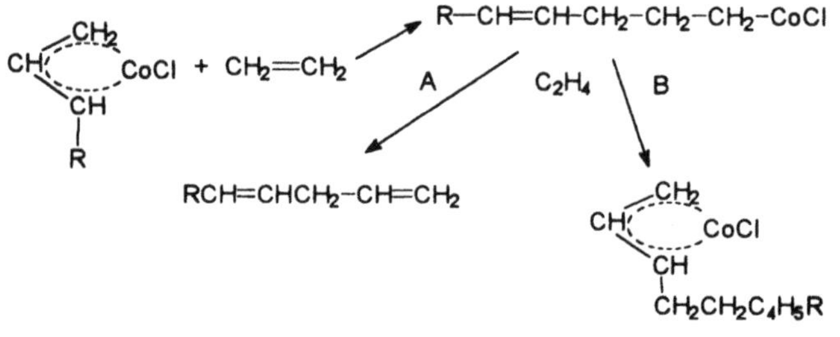

The path A is accompanied by a chain transfer reaction; path B leads to copolymerization of ethylene with butadiene.

[13] The branching of polybutadiene obtained in the presence of cobalt systems increases with the rise of the degree of polymerization [148]. This is the result of chain transfer to polymer molecules.

ethylene polymerization. The mechanism described is unusual[14] and has never been mentioned for the formation of polyolefins and polydienes.

Structural analysis of these chemical hybrids is based on characterizing in the copolymers the content of butadiene units (^{14}C technique). The radioactivity of the copolymer of ethylene with butadiene obtained in the presence of immobilized monometallic (V^{4+}) catalyst is assumed to be a measure of the statistical copolymer. The characteristics of the polymers obtained are presented in Table 3.12. The increase of the amount of ethylene being used up in polymerization is accompanied by a decrease of butadiene unit content. Polybutadiene blocks are mainly composed of 1,4-*cis* units pointing out the lack of influence of the monomers.

Table 3.12 Characteristics of block PE–polybutadiene (PBD) copolymer obtained in the presence of bifunctional Co^{2+}–V^{4+}-immobilized catalyst[a]

Polymerized ethylene amount (g)	Units content (%)	
	Butadiene in product	1,4-cis in PBD fraction
0.5[b]	2.7	-
0.5	38	95
1.0	24	90
2.0	14	90

[a]Polymerization conditions: AlEt$_2$Cl as cocatalyst, 323 K, gas pressure 0.4 MPa, [butadiene]=0.1 mol/l.

[b]Copolymerization of ethylene with butadiene catalyzed by monofunctional catalyst VCl$_4$·PE-gr-PVP-AllEt$_2$Cl.

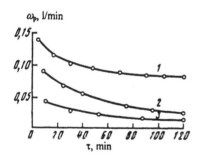

ω_p, l/min

τ, min

Figure 3.37 Kinetics of ethylene polymerization: (1) VCl$_4$·PE-gr-PVPr; (2) with addition of 1×10^{-2} mol of butadiene; (3) with addition of butadiene to the catalytic system CoCl$_2$–VCl$_4$ bound to PE-gr-PVPr. Polymerization conditions: ethylene pressure 0.4 MPa, 343 K, AlEt$_2$Cl, benzene, [Al]/[Co+V]=100.

[14] In a study of copolymerization of butadiene with styrene catalyzed by organolithium compounds in combination with potassium alkoxides the equilibrium between two active sites has been characterized [148]. Lithium centers were shown to be responsible for butadiene polymerization, whilst K reacts with styrene. The product is not random but is a microblock copolymer, the size of the microblocks correlating with the frequency of ⇨C-Li ↔ ⇨C-K transitions. A study of anionic isoprene polymerization catalyzed by the Li–Na system revealed the presence of active sites of two types differing in the nature of the M–C bond. However, it has been shown that under certain conditions chain propagation proceeds preferentially on one of these [149].

To explain the exchange mechanism the concept of active site for alternating copolymerization of dienes and olefins [150] can be used. In this active site vanadium is bivalent; there is a coordination vacancy for the olefin in π-allyl–vanadium complex. After olefin insertion vanadium ion appears in the position favorable for the bidentate diene molecule coordination

3.3 Di-, Iso- and Oligomerization of Olefins

The interest in these reactions catalyzed by immobilized complexes arises not only from the commercial value of the resulting products (butenes, pentenes, hexenes, and other starting materials for production of higher alcohols, butadiene, isoprene, high-octane benzyne, etc.) but also from their opportunities to obtain information about the mechanism of catalyst action, the nature of the active sites, the principles of their deactivation, as well understanding both general and particular features of catalysis of n-merization.

3.3.1 Dimerization of ethylene to isomeric butenes catalyzed by Ni-containing immobilized systems

Dimerization of α-olefins can be catalyzed by different systems [151–154]. These are catalysts for cationic oligomerization (inorganic acids, including those immobilized on the support), those for anionic dimerization (alkali metals or their bound organometallic compounds), complexes of transition metal ions and Ziegler–Natta systems (especially Ni-containing).

Three possible mechanisms of ethylene dimerization in the presence of Ni catalysts (with more or less reliable experimental results) are now under polemic. These are restricted or degenerated polymerization when chain termination reactions predominate over chain propagation, synchronous coupling (stepwise addition) of two ethylene molecules to a metal ion with a C–C bond being made between them and reductive dimerization via nickel hydride formation.

The last mechanism of ethylene dimerization seems to be the most probable. The presence of hydride–nickel complexes *in situ* and their interaction with ethylene have been indicated by several physico-chemical methods. σ-Alkyl–nickel complexes have been characterized before [153].

Hydride–nickel complexes active in dimerization are generated from nickel compounds in the usual oxidation state (+2) through the following series of transformations

$$Ni^{2+} + AIR_xCl_{3-x} \longrightarrow Ni^+ + Ni^0 \longrightarrow Ni^0 \text{ (colloid particles)}$$

$$\downarrow AIR_xCl_{3-x}$$

$$Ni^0 + Ni^{2+}$$

$$\downarrow AIR_xCl_{3-x}$$

$$Ni^+ + AIR_xCl_{3-x} \longrightarrow Ni^{2+} + Ni^0$$

Scheme 3.18

The average value of active complex output (chain length) is 2.3×10^3 mol of monomer per mol of σ-alkyl (or hydride)–Ni^{2+} complex. Regeneration of active centers proceeds via a redox cycle with participation of Ni^0. However, in homogeneous systems (for example $Ni(napht)_2$–$AlEtCl_2$) complete catalyst deactivation occurs at 333–343 K after 2 h of ethylene dimerization. Under otherwise similar conditions the heterogeneous analogue, $Ni(OCOCH_3)$, bound to PE-gr-PAAc–$AlEt_2Cl$, shows higher activity and long-term stability (during 40 h) (Figure 3.38) [155,156]. Dimerization kinetics depend to a large extent on the nature of the immobilized nickel compound (Figure 3.39). Immobilized $NiCl_2$ provides the largest rate at the beginning of ethylene dimerization with a subsequent decrease of reactivity albeit to a smaller extent than that for the homogeneous system. The maximum value of turnover number (TN) in the starting period for the homogeneous system is 300 min^{-1} while in heterogeneous systems this is 600–1500 min^{-1}, those values being equal to 70 and 400–600 min^{-1}, respectively, after 1 h of the reaction.

Dimerization products have a similar composition for compared systems, mol%: α-butene 3–6; trans-β-butene 54; cis-β-butene 29; hexenes ~8 (α-hexene 0.5–3.5; methyl-pentenes 0.1–5.5; dimethylbutenes 1–3). The reactivities of Ni^{2+} compounds can be enhanced by their immobilization on chelating polymers, including those with grafted azomethine β-dicarbonylic compounds [157,158]. Strong nickel fixation in chelate units prevents metal leaching in the course of the catalyzed reaction and provides complex stability even at 363 K (Table 3.13). Certainly the specificity of nickel chelates (N, O–) considerably affects their reactivity [159].

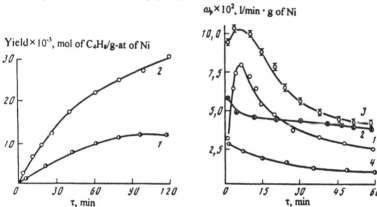

Figure 3.38 Kinetics of ethylene dimerization by nickel-containing systems: (1) homogeneous $Ni(napht)_2$–$AlEtCl_2$, (2) $Ni(CH_3COO)_2$–PE-gr-PAAc-$AlEtCl_2$. Dimerization conditions: n-heptane, 333 K, [Al]/[Ni]=30, ethylene pressure 1.0 MPa.

Figure 3.39 The influence of the type of Ni^{2+} immobilization on ethylene dimerization rate. Catalytic systems: (1) $Ni(napht)_2$–PE-gr-PAAc, (2) $Ni(CH_3COO)_2$–PE-gr-PAAc, (3) $NiCl_2$–PE-gr-PAAc (hydrolyzed with NaOH), (4) homogeneous $Ni(napht)_2$. Dimerization conditions as in Figure 3.38, T=343 K.

The ethylene dimerization rate increases in aromatic solvents for both of the systems compared; however reactions of solvent alkylation are of great importance together with ethylene di- and oligomerization.

All experimental data (including analysis of dimerization products) indicate a similar or very close resemblance of active sites in both homogeneous and immobilized nickel systems. Active site formation probably also develops in similar ways. The role of the polymer support consists of stabilization of the active sites by hindering their association in the course of the catalyzed reaction.

Magnetochemical measurements showed [160] (Table 3.14) the dependence of magnetic susceptibility on the intensity of the magnetic field, probably due to presence of ferromagnetic nickel moieties along with the predominant paramagnetic forms. Moreover, the share of the ferromagnetic species increases on going to low temperatures. This is caused by the presence of some finely dispersed superparamagnetic particles of size 1–10 nm together with large ferromagnetic particles (>10 nm).

Ethylene dimerization by homogeneous systems is accompanied by considerable nickel reduction. About 30% of Ni^{2+} ions are reduced to finely dispersed, free metal particles after 30 min of reaction, which can be the main cause of catalyst deactivation. Under otherwise similar conditions, in immobilized systems less than 1% of metal nickel has been identified in solid products. Metal complexes bound to the polymer support are magnetically diluted, so migration of reduced metal particles along the support has to precede their association. Functional cover of the polymer support prevents these processes so that the isolated Ni^0 complexes formed may be reoxidized with AlR_nCl_{3-n} to Ni^+ (even without ethylene) and are involved in a redox cycle according to Scheme 3.18.

Table 3.13 Dimerization and oligomerization of ethylene[a] by immobilized Ni^{2+} chelates with azomethynes of β-dicarbonylic compounds.

Polymer support and the way of its modification	Ni^{2+} compound	Ni^{2+} content on the support $(g\text{-}at\times10^4/g)$	The stationary rate of ethylene transformations (kg/g of Ni h)
PE-gr-PMVK+o-amino-phenol	$Ni(CH)_3(COO)_2$	1.01	12.0
PE-gr-PMVK+o-amino-phenol	$Ni(AcAc)_2$	1.20	30.0
PE-gr-PMVK+o-amino-phenol	$NiCl_2\cdot6H_2O$	1.40	19.0
Pe-gr-PMVK+ aniline hydrochloride	$NiCl_2\cdot6H_2O$	0.70	10.0
PE-gr-PMVK+o-toluidine	$Ni(CH_3COO)_2$	0.80	14.0
Homogeneous system (313 K)	$Ni(napht)_2$	–	8.0

	Reaction product composition (%)			
	Butenes	α-Hexene	Methyl-pentenes	Dimethyl-butenes
PE-gr-PMVK+o-aminophenol	93.3	3.5	1.5	1.7
PE-gr-PMVK+o-amino-phenol	91.7	1.6	3.9	2.8
PE-gr-PMVK+o-amino-phenol	92.3	2.8	1.2	3.7
Pe-gr-PMVK+ aniline hydrochloride	90.8	3.1	4.2	1.9
PE-gr-PMVK+o-toluidine	93.4	1.5	2.1	3.0
Homogeneous system (313 K)	89.5	2.2	4.6	3.7

[a] Reaction conditions: heptane, 343 K, ethylene pressure 1.0 MPa, cocatalyst $AlEtCl_2$, [Al]/[Ni]=50.

In contrast to the strong influence of the diene mentioned above on the formation of ferromagnetic Co (see Table 3.8) in PE-gr-P4VP·CoCl$_2$–AlEt$_2$Cl systems (see Section 3.2.1) ethylene does not affect Ni^0 ferromagnetic particle formation in nickel systems.

Table 3.14 Content of ferromagnetic nickel (%) in the products of interaction of Ni(CH$_3$COO)$_2$ on both PE-gr-PAAc (A) and Ni(napht)$_2$ (B) with AlEt$_n$Cl$_{3-n}$ [160].

System	Measurement temperature (K)		
	293	80	4.2
A + Al(C$_2$H$_5$)$_3$	0.34	0.56	2.4
A + Al(C$_2$H$_5$)$_3$ + C$_2$H$_4$	0.48	0.50	1.4
B + Al(C$_2$H$_5$)$_3$	1.2	2.1	
A + Al(C$_2$H$_5$)$_2$Cl	0.55	1.2	6.3
A + Al(C$_2$H$_5$)$_2$Cl + C$_2$H$_4$	0.18	0.37	
B + Al(C$_2$H$_5$)$_2$Cl	27	34	37.5
B + Al(C$_2$H$_5$)$_2$Cl + C$_2$H$_4$	34	53	
A + Al(C$_2$H$_5$)Cl$_2$	0.34	0.31	
A + Al(C$_2$H$_5$)Cl$_2$ + C$_2$H$_4$	0.35	0.41	3.8
B + Al(C$_2$H$_5$)Cl$_2$	31.2	31.6	51.4
B + Al(C$_2$H$_5$)Cl$_2$ + C$_2$H$_4$	14.5	17.6	43.8

Gel-immobilized nickel catalysts [161] as well those supported on porous polymers [162] have also been used for liquid phase ethylene dimerization. NiCl$_2$ [163] or Ni(AcAc)$_2$ bound to swollen but insoluble vinylpyridine-containing resins are effective in both ethylene and propylene dimerization [164].

Nickel complexes bound to polyvinyl- , polyvinylidene- and polyallylpyridines are effective in olefin dimerization under mild conditions [165]. The activity of NiCl$_2$ immobilized on a copolymer of 2-methyl-5-vinylpyridine with styrene in combination with AlEtCl$_2$ or Al(isoBu)$_2$Cl is the same as that of systems derived from bound Ni^{2+} complexes on P2VP or P4VP; the steady-state rate of propylene dimerization reaches 1.8–3.1 kg of propylene/(g of Ni h).

Phosphorous-containing polymer additives are usually used as stabilizers in homogeneous systems [166] and heterogenized by inorganic oxides (γ-Al$_2$O$_3$ or SiO$_2$) [167,168]. Ni^{2+} catalysts immobilized by macroporous phosphynated CSDVB initiate ethylene dimerization to butene with a selectivity of 90% [169]. Phosphorous-containing polymers of the gel-type have been used for propylene dimerization [170].

The "bell-shaped" dependence of the SCA on the surface density of bound nickel complexes has been observed in ethylene dimerization by nickel-containing systems [156] and has been reproduced in this form for different thickness of grafted layer[15] (Figure 3.40).

---

[15] A similar bell-shaped dependence of catalyst activity on the polymer ligand content (units of 4-vinylpyridine) has been observed also for gel-immobilized systems [171]; the optimal range for grafted 4-vinylpyridine fragments is about 8–15 mass%.

These peculiarities are probably caused by structure specificity of bound complexes.

For instance, it has been shown that mainly disubstituted complexes $\overset{\displaystyle L\quad L}{\underset{\displaystyle \overset{\backslash/}{N}\,2+}{|\quad|}}$ are formed

at the polymer surface for low polymer loading with a Ni^{2+} compound, while

monosubstituted complexes $\underset{N^{2+}\ X}{L}$ are essentially formed at higher nickel contents

[172]: both are able to form associates. Probably these complexes show different activities in ethylene dimerization showing specificity in redox cycles.

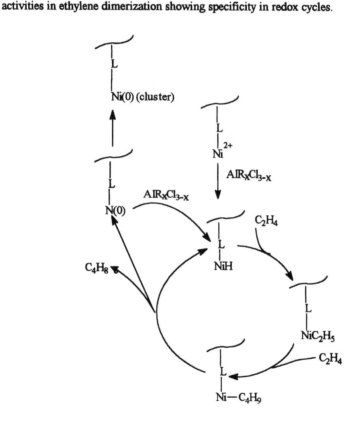

Scheme 3.19

The measurements of magnetic susceptibility at high magnetic field intensity (up to 70 kG) and low temperature (4.2 K) showed the specificity of nickel ion distribution to depend on their surface concentration [174]. Under these conditions the susceptibility of different magnetic species changes in a non-linear manner with the rise of magnetic field intensity. This gives the opportunity to characterize the distribution of nickel states from values of the magnetic moments. Thus, a comparison of two catalysts, one containing 0.8 (specimen I) and the second cotaining 1.6 (specimen II) mass% of bound nickel treated with AlEtCl$_2$ (Figure 3.40) showed that both contained nickel ions in several magnetic states (diamagnetic complexes of Ni^0, complexes of Ni$^+$ of spin value $s=0.5$, Ni^{2+} complexes of $s=1.0$, as well finely dispersed Ni^0 particles of different spin values). The calculations indicate the presence of Ni^0 aggregates with various average sizes in each specimen. Microparticles of the size < 1 nm (composed of 10–30 metal atoms) formed by nickel reduction in specimen **I** are approximately half the size of those in specimen **II**.

Therefore, the rise of Ni^{2+} surface density results in the formation of enlarged clusters. Consequently specimen **B** (with higher Ni^{2+} surface density) will have relatively smaller (in number) boundary nickel atoms which are identified as responsible for ethylene dimerization. This is an illustration of the bell-shaped dependence of specific activity on the surface density of bound metal ions. A favorable factor is the presence in the polymer support of moieties with high local concentration of transition metal ions originated in the course of transition metal compound immobilization ([17], pp. 255–260).

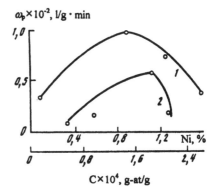

Figure 3.40 Dependence of initial specific ethylene dimerization rate on the surface concentration of bound nickel. PAAc grafting degree (%): (1) 7.2; (2) 4.0. Reaction conditions: PE-gr-PAAc, 343 K, [Al]/[Ni]=50, n-heptene, AlEt$_2$Cl, ethylene pressure 0.4 MPa.

Experimental data show that in the course of ethylene dimerization in the presence of immobilized nickel complexes, a reduction of Ni^{2+} ions to Ni^0 takes place. The latter aggregate as small size clusters[16]. Active sites, probably located on these cluster boundaries are stabilized by the entire electron system of the cluster[17]. The bell-shaped dependence of SCA on the surface density of Ni^{2+} ions correlates with the cluster size distribution function [174].

It is worth noticing that even simple impregnation of Al$_2$O$_3$ or SiO$_2$ with nickel salts prevents catalysts from rapid deactivation during ethylene oligomerization or propylene dimerization [175], increasing both catalyst lifetime and product yields by 5–7 times. The same effect has been observed in selective dimerization of ethylene to α-butene by Ti(OBu)$_4$ supported on SiO$_2$ or Al$_2$O$_3$ in combination with an organoaluminum compound [176]. An increase of ethylene dimerization rate by several orders of magnitude by covalent bonding of RhCl$_3$ on SiO$_2$ has also been reported [177]. However these effects on ethylene dimerization have not been observed when RhCl$_3$ has been immobilized on polymer supports, such as PE-gr-PAAl or PE-gr-P4VP [89] under the

---

[16] Superparamagnetic nickel particles have also been detected in [173] for oxide-immobilized catalysts.

[17] See footnote 10.

above conditions. Immobilized complexes of Nb and Ta (support = spherical particles of CSDVB modified with cyclopentadienyl ligands) are effective in ethylene dimerization [178]. Thus, the activity of immobilized CpTaCl$_4$ treated with neopentylithium in ethylene dimerization was shown to be 18 mmol of butene mmol^{-1} of Ta h^{-1}.

In principle, immobilized metal complexes may be useful for catalysis of olefin co-dimerization, i.e. ethylene with propylene in the presence of gel-immobilized systems [179], alkenes with vinyl- and allyl ethers [180], ethylene oligomerization [181], etc.,. It is important that cationic centers formed in these systems due to evolutionary transformations catalyze co-dimerization of lower olefins with reaction products.

Table 3.15 Ethylene oligomerization[a] catalyzed by Ni(COD)$_2$ bound to phosphorous-containing supports [183].

Catalyst	TN^b (s^{-1})	Oligomer yield (%)		β-factor^c (C_6-C_{16})
		Linear	α-Olefins	
]—PPh$_2$ Ni (Ph$_2$P—CH=CH—Ph···O)	~1	99	96	0.2
]—CH$_2$·PPh$_2$ Ni (Ph$_2$P—CH=CH—Ph···O)	~3	99	96	0.1
]—P(Ph)(Ph) Ni (CH=C(Ph)—O) PPh$_3$	~25	95–97	93–97	0.5
]—CH$_2$—P(Ph)(Ph) Ni (CH=C(Ph)—O) ·PPh$_3$	~60	99	99	0.4
Ni(COD)$_2$/SiO$_2$–Al$_2$O$_3$	~0.1	92	56	1.4

[a] Reaction conditions: 348 K, toluene, ethylene pressure 4.0 MPa.
[b] Moles of ethylene per mole of polymer-bound nickel per second.
[c] β-factor (mol of C_n fraction/mol of C_{n+2}) – 1.

Homogeneous nickel catalysts with bidentate chelating ligands (usually phosphorus-containing) are widely used for alkenes oligomerization [182]. Their commercial value has been demonstrated in the SHOP-process (Shell development of higher α-olefins manufacturing process) for manufacturing of oligomers containing up to 95% linear α-olefins. The same approach can be realized through the use of immobilized catalysts.

As demonstrated in Table 3.15, Ni(COD)$_2$ bound to phosphynated CSDVB or to CSDVB modified with benzoylmethylenephosphorane (with additives of PPh$_3$) show high activity and selectivity (without a cocatalyst) for ethylene oligomerization[18] [183]. The ratio between the proportions of olefin oligomerization and polymerization catalyzed by organometallic nickel complexes has been shown to be dependent on the reaction conditions [187]. One can see that polymer-immobilized catalysts are more active and selective than those bound to oxides.

3.3.2 Structural olefin isomerizations catalyzed by homogeneous and immobilized metal complexes

As it has been shown above, ethylene dimerization in the presence of immobilized catalysts is often accompanied by the formation of mixtures of isomeric butenes (with small amounts of hexenes). Structural or skeleton isomerization changes of substrate under action of catalysts can be used as models for the study of catalyst mechanisms. Moreover one should take into account these reactions as competetive ones to numerous catalytic processes (e.g. hydrogenation, see Chapter 1).

All prototropic transformations in butenes catalyzed by Ni-containing systems can be generalized as follows:

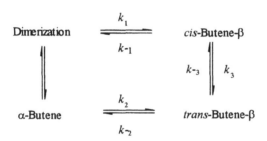

Scheme 3.20

Both butene dimerizations and their skeleton isomerization (yielding isobutene) are not usually induced by these catalysts.

A comparative analysis of α-butene isomerization by homogeneous (Ni(napht)₂–AlEtCl₂) and heterogeneous systems has been carried out [188]. The first system provides the higher reaction rate which reaches a fixed value after 1 h (Figure 3.41). The equilibrium composition of isomeric butenes was shown to be, mol%: α-butene 52; cis-butene-β 18; trans-butene-β 30. However, these equilibrium values differ from the thermodynamic values (at 300 K, α-butene 3, cis-butene-β 75; at 373 K, 7.26 and 67%, respectively [188]). Probably only step-by-step equilibria between cis- and trans-butene-β is significant. In isomerization of cis-butene-β under identical conditions a steady state is reached more rapidly (α-butene 6, cis-butene-β 30 and trans-butene-β 64) and the system almost reaches thermodynamic equilibrium.

At the same time isomerization in the presence of Ni²⁺-immobilized catalysts proceeds at a notably low rate. To illustrate, the value of α-butene conversion to isomers is only about of 4 mol% after 2 h, the ratio of cis-butene-β to trans-butene-β being close to

---

[18] Ethylene dimerization proceeds in the presence of Ti(OBu)₄ bound to PE-gr-PAAl [184] as well as on gel-immobilized systems [185,186].

thermodynamic equilibrium. It is worth noticing that *cis*-butene-β is not found in any prototropic transformations.

Butene isomerization is probably an intramolecular process [189] and under the assumption of hydride formation can be represented by a sequence of additions to nickel hydride and further eliminations. The double bond position is predetermined either by butene insertion (addition) or by proton transfer. Isomerization inertness of immobilized catalysts is caused by steric hindrances for butene addition to immobilized nickel hydride. In addition, it is known that the volume of phosphyne ligand even for homogeneous Ni-containing systems significantly affects the rates of both olefin dimerization and olefin isomerization.

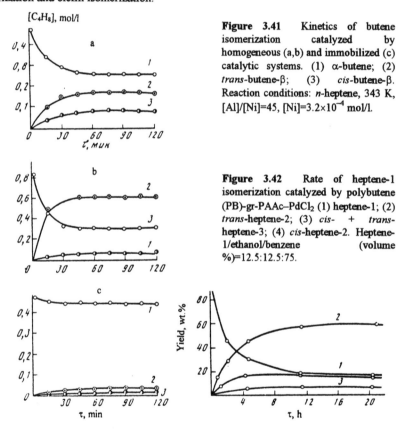

Figure 3.41   Kinetics of butene isomerization   catalyzed   by homogeneous (a,b) and immobilized (c) catalytic systems. (1) α-butene; (2) *trans*-butene-β;   (3)   *cis*-butene-β. Reaction conditions: *n*-heptene, 343 K, [Al]/[Ni]=45, [Ni]=3.2×10^{-4} mol/l.

Figure 3.42   Rate of heptene-1 isomerization catalyzed by polybutene (PB)-gr-PAAc–PdCl$_2$ (1) heptene-1; (2) *trans*-heptene-2; (3) *cis*- + *trans*-heptene-3; (4) *cis*-heptene-2. Heptene-1/ethanol/benzene   (volume %)=12.5:12.5:75.

Thus, immobilized catalysts are sensitive to steric hindrance in the steps of butene addition as already shown for hydrogenation reactions (geometrical regioselectivity).

Somewhat similar regularities have been observed [89] from comparisons of butene isomerization in the presence of monocomponent catalysts, homogeneous PdCl$_2$(PhCN)$_2$ and the same species bound to PE-gr-P4VP. α-Butene isomerization is effectively catalyzed by PdCl$_2$(PhCN)$_2$, however at a lower rate than that in the presence of Ni-containing systems, yielding all types of butenes[19]. Heterogenized Pd^{2+} is inactive in these reactions.

At the same time gel-like catalysts, $PdCl_2$ bound to PBD, are strongly reactive in double bond isomerization [171]. More than 70% of heptene isomerizes after 4 h, the product consisting of 46% of trans-heptene-2, 5.7% of cis-heptene-2, 19% of cis- and trans-heptene-3 is the reminder (Figure 3.42). The SCA of this catalyst depends on its swelling ability (in other words, on the accessibility of active sites inside the gel) and the value is 22.6 g/(g of Pd h), being approximately twice that of $PdCl_2$ alone[20] (13 g/(g of Pd h)) [188]. The alcohol moieties in the solvent mixture are supposed to be responsible for Pd^{2+} to Pd^+ reduction, Pd^+ being active in isomerization [192].

It is also interesting that these systems also catalyze cis–trans-isomerization of heptene-2: 2.5% of cis-heptene-2 isomerized to trans-heptene-2 in 24 h of reaction, neither heptene-2 nor heptene-3 being formed.

We have already mentioned above the activity, for allylbenzene isomerization to cis- and trans-propenyl benzene [193,] of both $C_2H_5TiCl_3$ and $(C_2H_5)_2TiCl_2$ bound to CSDVB modified with cyclopentadienyls. The rise of catalyst SCA caused by a decrease of the titanium content was attributed to the activity of isolated titanium centers. 1,5-Cyclooctadiene isomerization yields the more stable 1,3-cyclooctadiene through intermediate 1,4-cyclooctadienes.

Olefin hydrogenations, including butene-1, is often accompanied by substrate isomerization. For instance, $Os_3(CO)_{12}$ clusters bound to CSDVB modified with $HO(CH_2)_2$ groups are effective in α-butene isomerization [194]. The TN for each cluster molecule reaches 30. The reaction is considerably promoted by hydrogen, the isomerization rate being remarkably low in the absence of H_2. Thus the interaction of the coordinatively unsaturated groups $HOs_3(CO)_{10}-O(CH_2)_2$ with H_2 was assumed to yield the active sites for isomerization. The activity of osmium clusters bound to polymers is only modest in comparison with those bound to SiO_2 or Al_2O_3 in spite of the fact that α-butene isomerization is catalyzed by several polymers, such as ion-exchange resins containing sulfonic acid residues [195].

Isomerization is often an accompanying reaction in olefin metathesis. Thus the relation between α-octene isomerization and disproportionation products in the presence of tungsten carbonyls bound to strong basic anionic exchange resin[21] (in combination with $AlEtCl_2$) is proportional to the ratio Al/W within the range 7–10; introduction of O_2 suppresses the isomerization [197]. The catalyst for pentene-2 disproportionation is obtained either by polymerization of different monomers (ethylene, butadiene) in the presence of WCl_6 or by bonding of $Na[W(CO)_3(\eta\text{-}C_5H_5)]$ to CMPS, as well as $W(CO)_6$ to phosphynated PS [198]. The processes of quadricyclane–norbornadiene isomerization will be discussed in Chapter 4.

---

[19] Both ethylene dimerization and butene isomerization catalyzed by Pd^{2+} as well as by nickel catalysts follow the mechanism of hydride formation [190].

[20] The activity of $PdCl_2(PhCN)_2$ in heptene-1 isomerization can be increased by complex bonding to inorganic supports with weak acid centers (SiO_2, MgO, etc.) [191].

[21] The yield of α-hexene metathesis catalyzed by Mo and W bound to oxides is increased by 6 times in comparison with homogeneous analogues and was accompanied by sharp decrease of isomerization products yield (hexene-2 and hexene-3) from 15 to 1.5% [196]; this has been ascribed to be the result of strong dependence on surface properties (the formation of Si-O-M units).

3.3.3 Ethylene dimerization and the processes of "relay-race" copolymerization for low density linear polyethylene (LDLPE) production

The development of bifunctional catalysts for specified catalytic sequences of reactions when the product of the first reaction can serve as substrate for the second is of great importance. These reactions can be widely exemplified. They are, for instance monomer-isomerizing polymerization of heptene-2, heptene-3 [199] and 4-methyl-2-pentene [200] catalyzed by Ziegler–Natta systems $(AlEt_3–TiCl_3–Ni(AcAc)_2$ in xylene, and the combination of propene disproportionation with oligomerization in the presence of tungsten catalysts [201], etc.

Usually the overall action of a bifunctional catalyst is the consequence of non-additive actions by both of its components. The ability of polymer supports to combine both active centers in the same matrix to minimize diffusive restrictions in comparison with those for two separate monofunctional catalysts (see Chapter 5) is inexhaustible.

Bifunctional catalysts are most widely used for ethylene copolymerization with α-butene *in situ* in the production of so called linear polyethylene of low density (LDLPE) [202–206]. All general methods for LDLPE production are based on incorporation into a PE backbone of short chain branches, which can be made by catalytic copolymerization of ethylene with α-olefins $C_3–C_{10}$ in the gas phase at low pressure (0.2–0.4 MPa) or in suspension, or in solution in the presence of highly active catalysts. Comonomer concentration is usually not above 1–8%; the density of the formed LDLPE is ordinarily within the range 0.94–0.92 g/cm^3. With respect to the last parameter these polymers are somewhat in between high density polyethylene and low density polyethylene, being closer to the second.

The essence of "relay-race" (or concerted) copolymerization thus lies in the coexistence of two different bound active sites in the same polymer matrix, the first being responsible for dimerization and the second for copolymerization. The general scheme of the process can be represented as follows:

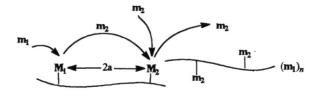

Scheme 3.21

The first monomer m_1 (say ethylene) is converted into m_2 by some simple reaction , such as di- , oligo- , isomerization, etc. (in this case ethylene is converted into α-butene) in the active center M_1 (for instance, Ni^{2+}). Copolymerization of m_1 and m_2 proceeds in an active center M_2 $(Ti^{4+}, V^{4+}, V^{5+}, Zr^{4+},$ etc.). The microcell containing both active sites is the size of about half the average distance between two neighboring centers of the bifunctional catalyst. In turn, the distance between M_1 and M_2 can be specified at the stage of catalyst preparation. The simplest way to prepare heterometallic complexes of transition metals randomly distributed along a polymer chain is the gradual deposition of metal pairs, such as $Ni^{2+}-V^{4+}$, $Ni^{2+}-V^{5+}$, $Ni^{2+}-Ti^{4+}$, etc. Such an approach has been realized for PE-gr-PAAc [207].

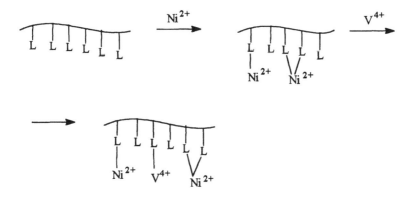

Scheme 3.22

However these heterometallic complexes bound to the polymer support should not be assumed to be electronically disconnected systems. Cooperative interaction between paramagnetic centers has been identified for this system [207,208].

The stationary mass-transfer in an individual microcell of this type is determined by the set of equations [209]:

$$D_i \frac{d^2 C_i}{dx^2} = 0 \quad (i = 1,2)$$

where C_1, D_1 is the concentration and effective diffusion coefficient of monomer m_1, whilst C_2, D_2 are those for dimer m_2. The branching degree (BD) of LDLPE is specified as the number of side branches per 1000 carbon atoms of polymer backbone (giving the relative share of butene in the copolymer) and is characterized by the ratio of the effective rates of co- and homopolymerization:

$$BD - \left.\frac{W_c}{W_p}\right|_{x=a} = \frac{k_c C_2}{k_p C_1}$$

where k_c and k_p are effective rate constants for the respective processes and a is the distance between catalytic centers.

Apart from the usual factors (such as concentrations of m_1 and m_2, those of M_1 and M_2, the average distances between microcells, k_c and k_p values), the BD value can also be regulated by variation of the distance between centers of different nature and diffusion factors. Under given diffusion conditions the BD value is completely specified by catalyst topology. The degree of branching decreases with the rise of diffusion coefficient. This is caused by an increase of the dissociation of the formed dimer m_2.

Generally, the dependence of BD on intercenters distances at fixed macrounit size follows the bell-shaped form (Figure 3.43). This dependence for $Ni^{2+}-V^{4+}$-immobilized system is illustrated by Table 3.16. Without discussing details of the BD value dependence on both mass transfer between two active sites and the kinetic parameters, we shall only point to the experimental evidence of the upper limit for the degree of branching obtained by means of macrokinetic analysis [209].

Figure 3.43 The calculated dependence of branching degree on the distances between catalytic centers (a) at a fixed size of macrounit. This is discussed further in the text.

Table 3.16 The dependence of the degree of branching for LDLPE obtained in the presence of immobilized catalystsa on the ratio of Ni^{2+}/V^{4+}.

V^{4+} on the support (mmol/g)	Ni^{2+}/V^{4+}	$\bar{r}_{V-V}$ (nm)	$\bar{r}_{V-M}$ (nm)	BD, $C_2H_5/1000\ C$
0.03	5.7	1.44	0.90	7.1
0.06	2.8	1.26	0.72	65
0.10	1.7	1.06	0.77	53
0.22	0.8	0.75	0.67	47

aPolymerization conditions: support PE-gr-PAAc (7.2 mass% of PAAc), S_{sp}=4 m^2/g, [Ni^{2+}]=0.17 mmol/g, 343 K, ethylene pressure 0.4 MPa, AlEtCl$_2$, n-heptane.

This is also confirmed by analysis of BD for LDLPE obtained in the presence of immobilized bifunctional catalysts with fixed (both geometrical and mass) distribution of transition metal ions. These catalysts have been obtained [210] by bonding of three ring azomethine Ni^{2+} chelates on polymer supports after preliminary treatments with Ti^{4+}, V^{4+} or V^{5+} compounds according to the scheme

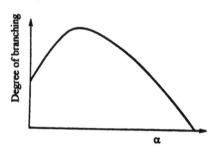

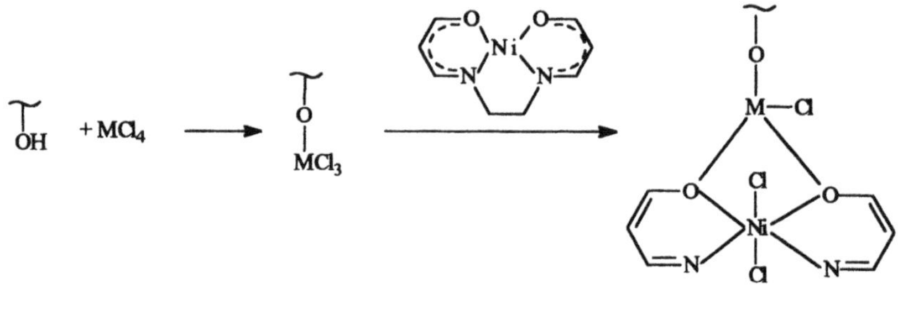

Scheme 3.23

As estimated in [211], the average distance M_1-M_2 in monomeric azomethine heterometallic complex is 0.22 nm and the BD value is not less then 30. Thus distance M_1-M_2, and consequently BD can be regulated by changing both the chelate unit and the nature of the polymer support.

The cage effect in an immobilized bifunctional center can appear either just as an increase in the frequency of first encounters or as an increase in the contact time of reacting molecules (leading to an increase in the number of repeating encounters with the active centers). Surface diffusion of butene molecules formed to the copolymerizing center is also possible in this case.

An increase in butene content in copolymers has never been observed for homogeneous systems. Thus the efficiency of ethylene with α-butene copolymerization can be enhanced by means of immobilized heterocomplexes. According to authors descriptive expression [212], the work of the bimetallic center an be regarded as "assembly belt where each worker completes his task with the highest professional skill and rapidity". In our case nickel dimerizes ethylene whilst vanadium copolymerizes the butene formed with ethylene. The closeness of both active sites permits rapid component transfer from each to the other as on a well-coordinated conveyer belt. The opportunity for highly precise synthesis in the production of immobilized heterocomplexes also favors this effect.

Meanwhile, some failures in the work of the conveyer are also possible. Irregular composition of gaseous products that does not correlate with reaction conditions makes "relay-race" copolymerization difficult to control and evaluate.

With regard to α-butene accumulation, isomerization and consumption due to copolymerization, all the processes discussed can be generalized[22] by the scheme:

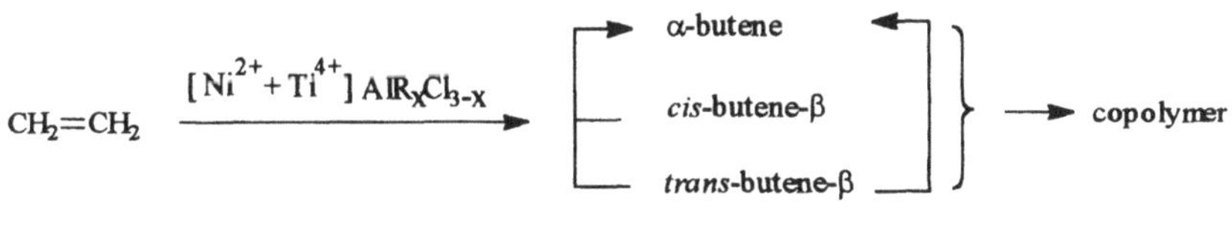

$$CH_2{=}CH_2 \xrightarrow{[\,Ni^{2+}+Ti^{4+}\,]\,AlR_xCl_{3-x}} \left[\begin{array}{c} \text{α-butene} \\ \textit{cis}\text{-butene-β} \\ \textit{trans}\text{-butene-β} \end{array}\right] \longrightarrow \text{copolymer}$$

Scheme 3.24

All these reactions in their turn also introduce complications into the kinetics of the sequential bifunctional process.

It is worth mentioning that the degree of branching of LDLPE obtained by immobilized bifunctional catalysts (especially for those with controlled bonding of M_1 and M_2) is greater than that obtained from introduction of "exterior" butene in the presence of monometallic immobilized complexes (Ti^{4+} or V^{4+}). This effect is probably caused by higher local concentrations of butene molecules in the vicinity of the active site (cage concentration) in the case when they are generated *in situ*.

---

[22] Monomer-isomer polymerization of 2-butene catalyzed by Ziegler–Natta systems (including those with NiX_2 as the third component):

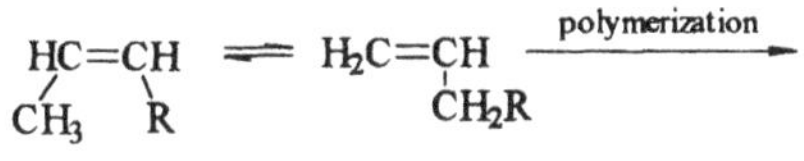

$$\underset{\substack{| \\ CH_3}}{HC}{=}\underset{\substack{| \\ R}}{CH} \rightleftharpoons \underset{\substack{| \\ CH_2R}}{H_2C}{=}CH \xrightarrow{\text{polymerization}}$$

has been reported in [213].

It is probably that the effect can be reduced through a combination of "relay-race" copolymerization with subsequent homogeneous isomerization of the β-butenes formed or by their abstraction.

One of the methods for LDLPE production is the combination of ethylene dimerization by a selective catalytic system making only α-butene (Ti(OBu)$_4$–AlR$_3$) with its copolymerization by high effective catalysts, such as TMC. A lot of catalytic systems have been studied in this process. For example, they are Ti(OBu)$_4$–PE-gr-PAAc–Mg^{2+}·Ti^{3+}–AlEt$_3$ [205,206], AlEt$_3$–Ti(O-isoC$_3$H$_7$)$_4$–TiCl$_4$/MgCl$_2$/PE [214,215]. Synthesis of LDLPE can also be achieved by gas-phase polymerization [216,217]. Properties of LDLPE depend not only on branching degree but also on the character of the distribution of branches (random or block-like) [197, 218–222]. The distribution of comonomer sequences in LDLPE, including those obtained by immobilized metal complexes, has been studied by different methods, such as IR [223–226], ^{13}C NMR [227], etc. [228–230]. It worth mentioning that supported bimetallic catalysts are sometimes used to regulate the molecular-mass characteristics of polyethylene formed [208].

3.3.4 The processes of di-, iso-, tri- and oligomerization of dienes

Butadiene oligomerization in the presence of metal complex catalysts yields a set of different products, such as 3-methylheptatriene (3-MHT), 4-vinylcyclohexene (VCH), octatriene-1,3-E, 6E (OTE), cyclooctadiene-1,5 (COD), cyclododecatriene-1E, 5E, 9Z (CDT):

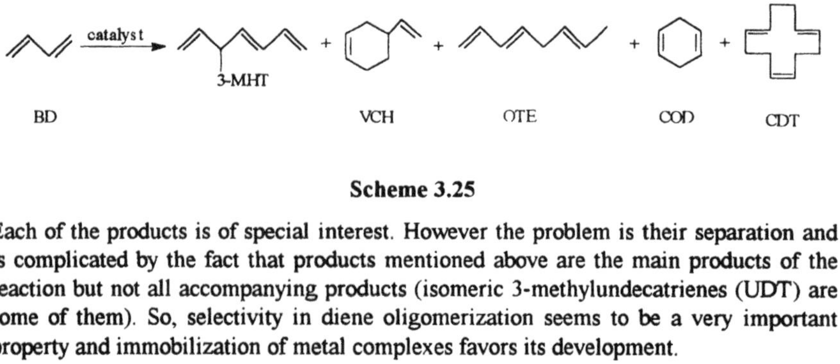

BD   VCH   OTE   COD   CDT

Scheme 3.25

Each of the products is of special interest. However the problem is their separation and is complicated by the fact that products mentioned above are the main products of the reaction but not all accompanying products (isomeric 3-methylundecatrienes (UDT) are some of them). So, selectivity in diene oligomerization seems to be a very important property and immobilization of metal complexes favors its development.

For example, Fe(AcAc)$_3$ [231] bound to cross-linked elastomer containing butylphosphytes or this chromium-containing polymer [232] in combination with AlEt$_3$ are effective in linear dimerization of butadiene yielding mainly 3-methylheptatriene-1,4,6. The same product is dominant (95–98%) under certain conditions (AlEt$_3$, 323 K, n-heptane, Al/Co=6) in the presence of TCEPD–PVP·CoCl$_2$ [122]. However the activity of this catalyst[23] is lower than that in butadiene polymerization.

CDT is the main product of butadiene oligomerization catalyzed by Ti^{4+}-immobilized complexes (under optimal conditions up to 97–100%), the selectivity of the process

---

[23] The specific features of butadiene-1,3 dimerization catalyzed by cobalt catalytic systems are represented in [233] while all previous studies are reviewed in the monograph [2].

increasing with the rise of donor properties of the binding ligands. This effect is supposed [234] to be caused by the growth of isolated Ti^{3+} ion content.

As with olefin dimerization, immobilized nickel complexes are effective in butadiene oligomerization. Specifically, Ni^0 (the product of $Ni(AcAc)_2$ reduction with $EtOAlEt_2$ in the presence of polymers) stabilized by P2VP and P4VP is active in butadiene cyclotrimerization [235] to mixtures of *trans-, trans-, cis-* and *trans-, cis-, cis-*CDT-1,5,9. However, complexes bound to cross-linked PVPs show no activity in this reaction.

Macrocomplexes of $NiCl_2$ or $Ni(OAc)_2$ immobilized on CSDVB modified with dipyridyl (Dipy) groups are effective (in combination with AlR_3) in butadiene trimerization, the yield of CDTs increasing from 9 to 70% with a temperature increase from 313 to 343 K [236].

The selectivity of the nickel-containing catalysts discussed above of gel-like type in butadiene cyclotrimerization is essentially influenced by a set of the factors, such as the ratio Al/Ni, phosphyte ligand arrangement in the polymer matrix as well reaction temperature:

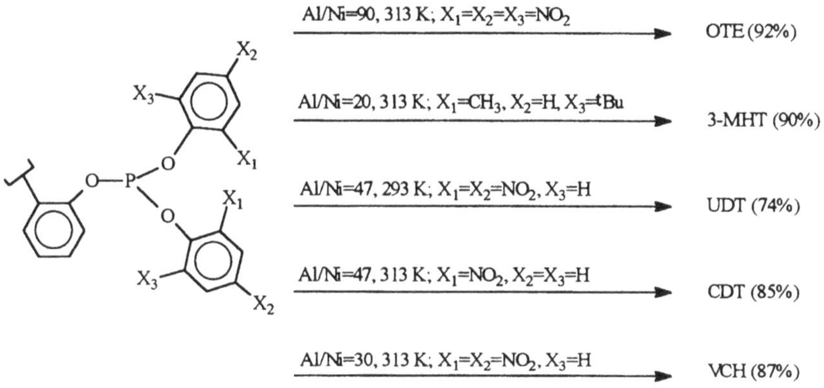

Scheme 3.26

The view of these mechanisms gives additional opportunities for control of the selected process control and improvement of catalyst selectivity.

There are only limited data on isoprene and chloroprene [237] oligomerization in the presence of immobilized complexes. However, the product compositions are too complex for analysis. There is a detailed study of the complex reaction of dimerizing acetoxylation in the system butadiene–acetic acid under the action of $PdCl_2$, $Pd(PPh_3)_4$ and $Pd(OAc)_2$ bound to phosphynated CSDVB [238]. The product compositions obtained under otherwise identical conditions by both homogeneous and heterogenized catalysts were shown to be rather similar.

3.4 Polymerization of Vinyl Monomers in the Presence of Immobilized Metal Complexes

Some aspects of vinyl polymerization by the use of metal complexes have been survived in the review [239]. Most commonly this dwells on radical polymerization and rarely on ionic mechanism. Here are some typical examples.

Metal chelates in combination with promoters (usually halogenated hydrocarbons) are known as initiators of homo- and copolymerization of vinylacetates. The polymerization mechanism is not well understood, but believed to be not exclusively radical or cationic (as polymerization proceeds in water).

Similar polymer-bound systems are also known. The macrochelate of Cu^{2+} with a polymeric ether of acetoacetic acid (synthesized on PVAl and phenol followed by treatment with Cu^{2+}) effectively catalyze acrylonitrile polymerization [240]. Meanwhile, this monomer is used as an indicator for the radical mechanism of polymerization. Another polymer chelate bound to a copolymer of vinyl alcohol with ethylene is active in polymerization of MMA, styrene, etc. [241,242].

Trisacethylacetonate–Mn^{3+} complexes bound to carboxylated (co)polymers have been used [243] for emulsion polymerization of a series of vinyl monomers. Mn^{3+} chelates in these systems act not only as polymerization initiators but also as emulsion stabilizers.

The rate of MMA polymerization in both water and CCl_4 in the presence of complexes formed by binding of $Cu(NO_3)_2$ and $Fe(NO_3)_3$ with diaminocellulose increases proportionally with monomer concentration and with the square root of metal complex content [244]. This is typical for radical polymerization. This mechanism is also confirmed by the value of the activation energy, E_a=49 kJ/mol (333–363 K). This initiator also shows activity in polymerization of styrene and acrylamide. In a similar way polymerization of MMA in water is initiated by a catalytic system composed of a copolymer of MVK with acrylamide, $CuSO_4$ and CCl_4 [245]. Oligomeric polyethylene glycol (PEG) can also be used as a macroligand (in system PEG–$CuCl_2$–H_2O) [246].

Complexes of Cu^{2+}, Fe^{3+}, V^{5+}, etc., immobilized on polymers, are often used as components of redox systems which are effective as initiators of vinyl-type monomer polymerization. For instance, the system vanadylpolycarboxylate–thiocarbamide–water solution of HNO_3 initiates acrylonitrile polymerization [247]. The process is accompanied by oxidation of bound V^{4+} to V^{5+}. Chain termination follows a bimolecular mechanism but not chain transfer to V^{5+}. At the same time cobalt complexes of etioporphyrin immobilized on polymer gel via azomethine bond keep their catalytic activity as chain transferers to monomer [248]. The chemical inertness of the polymer support with respect to radical attack is an important requirement in these processes. It has been shown [249] that Cu^{+}–porphyrin complexes immobilized in a volume of polyacrylamide gel cause joint polymerization of acrylamide and bisdiazonium salts in water. Ferrocenic polymers are effective initiators for polymerization of p-substituted chloroformylated vinylic monomers: α-chloro-β-formyl-p-methoxystyrene and α-chloro-β-formyl-p-iodostyrene [250]. The formed products are oligomers with average polymerization degree of 5–12.

We will now mention another practically useful approach. Macrocomplexes of Cu^{2+} with PVAl are irreversibly and rapidly adsorbed by inorganic supports containing OH groups giving a monolayer with an average density of 7.2 structural units per nm^2. Adsorbed PVAl–Cu^{2+} acts as an initiator of (co)polymerization of allyl-type and methacrylic monomers proceeding on the surface of the support by a radical mechanism (due to $Cu^{2+}\rightarrow Cu^{+}$ reduction). The process is considerably accelerated in the presence of carbon tetrahalides [251] making polymer cover on mineral fillers (both micro- and macroscopic up to 10–20 mass%) sometimes forming grafted polymer layers. Such an approach can be widely used for production of composites with enhanced properties (modulus of elasticity, adhesion, etc.).

Ionic polymerization catalyzed by immobilized complexes, in particular Lewis acids on polymer supports, so-called solid superacids (see Section 4.6), was now been extensively developed. For example, aluminum chloride catalysts bound to a sulphocation exchanger are active in olefin (isobutene) [252] polymerization. $AlCl_3$ immobilized on PS (up to 1.7–2.6% of $AlCl_3$ in gel-like polymers and about 3.3–6.1% on macroporous supports) effectively initiates carbonium/ionic polymerization of styrene, α-methylstyrene and β-pinene [253]. Cationic polymerization of styrene proceeds in the presence of $SnCl_4$–PS [254], as well as Lewis acids immobilized on flaky graphite compounds, polymerization efficiency decreasing in the sequence of Lewis acids: $SbF_5 > AlCl_3 > ZnCl_2$ [255]. Cationic polymerization of α-methylstyrene is initiated by $CaCl_2$ bound to PS [256]. The mechanism of the reaction was shown to be typical for cationic polymerization:

$$]-GaCl_3 + H_2O \longrightarrow]-(GaCl_3OH)-H^+$$

Scheme 3.27

Polymerization of *N*-vinylcarbazole catalyzed by Mn^{2+}, Co^{2+}, Ni^{2+}, Cu^{2+} or Zn^{2+} immobilized on PVC in the form of dimethylglyoxime complexes [257]

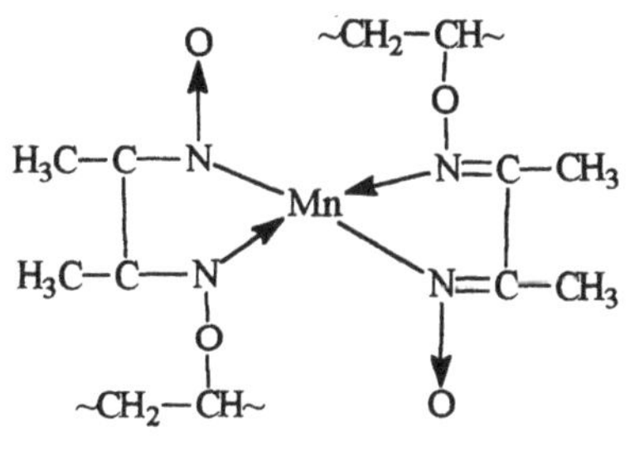

Scheme 3.28

follows the same mechanism.

Macrochelates of this kind are more active than Mn^{2+} supported on sieves. The relative molecular mass of the polymers formed reaches values of 20 000–30 000 with an isotacticity of 37% instead of 30% obtained by sieves.

Lewis acids (BF_3, $SnCl_4$, $SnCl_2$, $TiCl_4$, $FeCl_3$) immobilized in a volume of swollen polymer gel (slightly cross-linked copolymer of butyl acrylate, acrylonitrile and acrylic acid) catalyze cationic polymerization and oligomerization of vinyl ethers [258]; bound $Al(OR)_3$, BF_3 and $Ti(OBu)_4$ initiate oligomerization of furfuryl alcohol, 2-furfurylidene acetone and tetrachlorofurane by the mechanism of cycloaddition [259]. The rare earth catalyst ($NdCl_3$) bound to a copolymer of ethylene with vinyl alcohol (in combination with $Al(isoBu)_3$–CCl_4) catalyzes styrene polymerization most probably by both free radical and coordinative mechanisms [260]; tetrabenzyl titanium with methylalumoxane causes syndiospecific polymerization of styrene [261].

Ziegler–Natta catalysts (including immobilized ones) can be used as initiators of vinyl monomers (MMAc, vinyl chloride, etc. [262,263]) polymerization, radical polymerization with cyclic opening [264], etc. Thus, the catalytic system PMMA·VCl₄–AlR₂Cl (as well as its pseudohomogeneous analogue [265]) initiates homolytic disruption of the M–C bond in the active site with following homo- and grafting polymerization of MMA:

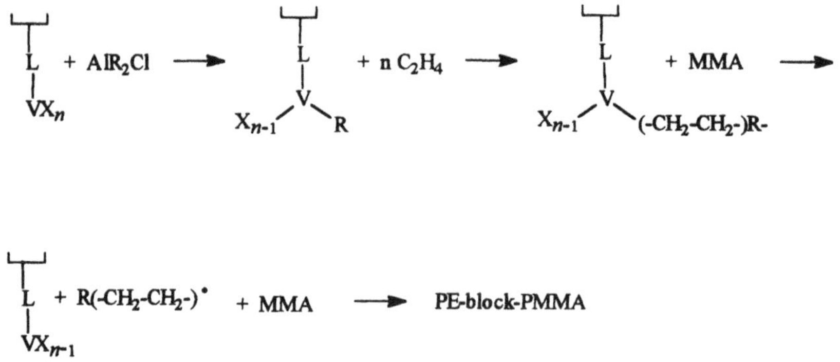

$$L \quad + AlR_2Cl \longrightarrow \quad L \quad + n\ C_2H_4 \longrightarrow \quad L \quad + MMA \longrightarrow$$

$$VX_n \qquad\qquad X_{n-1}\diagup\diagdown R \qquad\qquad X_{n-1}\diagup\diagdown (\text{-}CH_2\text{-}CH_2\text{-})R\text{-}$$

$$L \quad + R(\text{-}CH_2\text{-}CH_2\text{-})^\bullet \quad + MMA \quad \longrightarrow \quad \text{PE-block-PMMA}$$

$$VX_{n-1}$$

Scheme 3.29

Such an approach gives an opportunity of forming polymer–polymer compositions just at the stage of catalytic polymerization (Section 3.7).

3.5 A Comparative Analysis of Ethynyl Monomer Polymerization Catalyzed by Mobile and Immobilized Metal Complexes

Polymerization of acetylene monomers catalyzed by immobilized complexes of W, Mo and Pd is interesting in many respects. Firstly, monocomponent catalysts provide very suitable models for studying the mechanism of catalyst action. Besides, ethynyl monomers provide extensive opportunities in comparison with those of diolefin-type for studying the influence of immobilization on the stereoregularity of the polymers produced as well as their molecular mass characteristics.

So far as the structure of substituted polyacetylenes is specified with chain position with respect to both double and single bonds, four isomeric helicoid chain conformations with three double bonds in each segment are possible [266,267]:

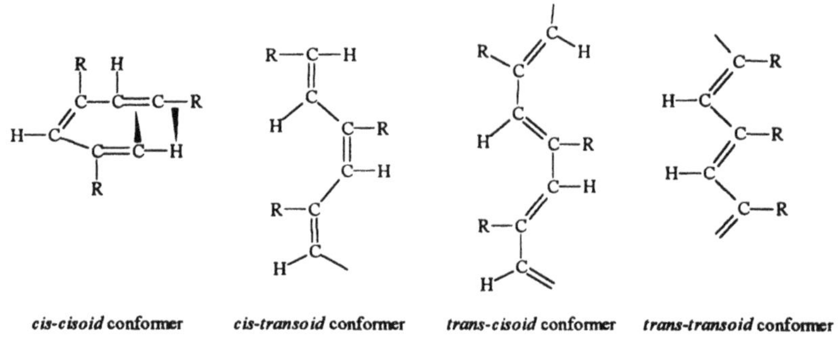

cis-cisoid conformer   *cis-transoid* conformer   *trans-cisoid* conformer   *trans-transoid* conformer

Scheme 3.30

Of all the structures of compounds soluble in organic solvents, polyphenylacetylene can be most easily recognized by IR and NMR spectroscopic methods.

3.5.1 Heterophase polymerization of acetylenes by molybdenum complexes

Macrocomplexes obtained by interaction of $MoCl_5$ and WCl_6 with PE-gr-PAAl, PE-gr-poly(propargyl alcohol) (PE-gr-PPAl), PE-gr-PAAc, PE-gr-P4VP, PE-gr-PAN, etc., are effective in phenylacetylene (PhA) polymerization in hydrocarbon or halocarbon solvents [268–270]. The process of complex fixation is accompanied with Mo^{5+} to Mo^{4+} reduction (on polymeric amines and nitriles [271]). The same processes have also been observed in interactions of molybdenum pentachloride with low molecular mass compounds (Py, Dipy, CH_3CN, etc.) yielding complexes of the type $MoCl_4 \cdot 2D$ with monodentate ligands and $MoCl_4 \cdot D$ with bidentate ones [272]. Immobilization of $MoCl_5$ on PE-gr-PAN at low temperatures does not change the electronic state of molybdenum (the value $\mu_{eff}=1.78$ μ_B points to the oxidation number of molybdenum as being five) [273]. The molybdenum ion distribution on such supports is rather complex. There are simultaneously two types of molybdenum centers, first isolated metal ions and second in the form of "clusters" of different sizes. The general tendency is toward enlargement of "clusters" with an increase in the polymer loading with complexes.

The rate of heterophase PhA polymerization strongly depends on the nature of the polymer support (Figure 3.44). $MoCl_5$ immobilized on PE-gr-PAAl, PE-gr-PAN and PE-gr-PPAl is 1.5–2 times as effective as the homogeneous system under otherwise identical conditions. Although the PhA polymerization rate in the presence of molybdenum halides bound to PE-gr-P4VP or PE-gr-PAAc is less than that catalyzed by $MoCl_5$, these catalysts show long-term stability and can more effectively catalyze the process after long reaction times than the homogeneous system. The effective values of PhA polymerization rate constants at 343 K catalyzed by complexes bound to PE-gr-PAAl, PE-gr-PAN, PE-gr-PPAl, PE-gr-P4VP and PE-gr-PAAc are respectively 10, 4, 89, 54, 28 and 18 mol of PhA/(mol of Mo min), while for the homogeneous system is 52 mol of PhA/(mol of Mo min). The reaction follows the Arrhenius law, $E_{eff}^a=38\pm2$ kJ/mol (for homogeneous polymerization $E_{eff}^a=33\pm2$ kJ/mol).

Immobilized molybdenum halides show relatively high activity in polymerization of some other monosubstituted acetylenes, such as propargyl alcohol and propargyl bromide (Figure 3.45). However, their rates of polymerization are usually less than that of PhA. The ratio of initial rates is equal to 1:0.6:0.4. Moreover, the polymers formed are insoluble in organic solvents making their study difficult, in particular the analysis of molecular mass characteristics.

Meanwhile just this information can be very useful for investigation of characteristics of PhA polymerization by Mo-containing catalysts. Heterophase polymerization of PhA usually yields polymers with higher molecular mass than that of polymers obtained in the presence of homogeneous system. Molecular mass of PPhA usually decreases with temperature rise both for homogeneous and heterogeneous processes (Table 3.17).

A typical chromatogram of PPhA obtained in the presence of immobilized catalysts (after separation of the oligomeric fraction soluble in methanol) shows two peaks, the left one (Figure 3.46, 1) attributed to the polymer fraction with $\overline{M}_n=6000$–14 000 and the right peak characteristic for fraction soluble in the mixture of THF/methanol=50:50

vol% with $\overline{M}_n$=1000–1400. The share of this fraction is only 5–8% of the high molecular mass component and it is mainly composed of linear, low molecular mass products and the cyclic trimer, triphenylbenzene (Figure 3.46, curve 3).

Table 3.17 The influence of reaction temperature on molecular mass of PPhA, obtained in the presence of homogeneous and heterogeneous catalytic systems (support PE-gr-PAN).

Reaction temperature, K	Polymerization					
	Homogeneous			Heterogeneous		
	$\overline{M}_w \times 10^{-3}$	$\overline{M}_n \times 10^{-3}$	γ	$\overline{M}_w \times 10^{-3}$	$\overline{M}_n \times 10^{-3}$	γ
283	44.4	11.6	3.83	–	–	–
303	30.1	7.7	3.90	29.1	13.4	2.17
323	22.4	7.2	3.11	22.9	9.7	2.35
333	–	–	–	23.0	9.3	2.47
343	17.6	6.7	2.63	15.0	7.5	2.00
353	–	–	–	18.2	6.3	2.89
363	–	–	–	14.8	6.2	2.38

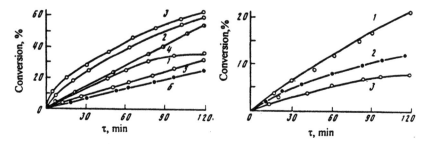

Figure 3.44 A comparison of the degree of conversion in PhA polymerization catalyzed by homogeneous (1) and heterogeneous (2–6) catalytic systems. Polymer supports: (2) PE-gr-PAN, (3) PE-gr-PAAl, (4) PE-gr-PPAl, (5) PE-gr-P4VP, (6) PE-gr-PAAc. Polymerization conditions: [PhA]=1.6 mol/l, [PhA]/[Mo]=400, 343 K.

Figure 3.45 Polymerization of various acetylene monomers: (1) PhA; (2) propargyl alcohol; (3) propargyl bromide. Polymerization conditions: polymer support PE-gr-PAAl, [Mo]=2.5 mmol/l, [monomer]=1.6 mol/l, 343 K.

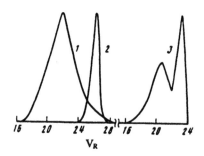

Figure 3.46 Typical gel-chromatograms of poly(phenylacetylene) (PPhA) obtained at 343 K in the presence of a heterogeneous system (1) and its oligomeric fractions (2,3). (1,2) the curves obtained on columns of set B; (3) of set A.

Thus, heterophase polymerization of PhA also gives four polymer products as well as homogeneous polymerization [274], the molecular mass characteristics of the high molecular mass fraction being in the ranges $\overline{M}_n=(18\div30)\times10^3$ and $\overline{M}_n=(7\div14)\times10^3$. These polymers contain polyconjugated double bonds, possess a linear polyene structure and are characterized by a high content of *cisoid*-transoid fraction (75–80%), as is true for the polymers obtained by the homogeneous system under comparable conditions.

The content of unpaired electrons in these polymers depends on both polymerization conditions and the type of catalyst, being within the range $(1\div8)\times10^{17}$ spin/g. They can be attributed to insulators as their conductivity (σ) is less than 10^{-13} $\Omega^{-1}cm^{-1}$, whereas poly(propargyl alcohol) ($\sigma=1\times10^{-7}$ $\Omega^{-1}cm^{-1}$) and poly(propargyl bromide) ($\sigma=5\times10^{-6}$ $\Omega^{-1}cm^{-1}$) are typical semiconductors. The double bond contents in PPhAs are less than 1 mol per mole of polymer units. This may be the result of partial cyclization or some other accompanying processes (see Table 3.18).

The concept of Mo^{5+} reduction under the action of PhA at the stage preceding that of active site formation [274] has been experimentally established in catalysis by immobilized molybdenum complexes. Several intermediates have been separated and characterized by different physico-chemical methods.

Table 3.18 Some characteristics of PPhA obtained in the presence of immobilized molybdenum complexes[a].

Polymer support	T (K)	PPhA yield (g/g of Mo)	$D_{870}{}^b/D_{910}$	Double bond content (mmol/g of PPhA)	Unpaired electron content $\times10^{17}$ (spin/g)	$\overline{M}_w\times 10^{-3}$	$\overline{M}_n\times 10^{-3}$	γ
PE-gr-PAAl	323	48	1.26	7.74	3.5	27.7	9.7	2.86
PE-gr-PAAl	343	95	1.33	6.32	3.5	23.4	8.5	2.74
PE-gr-PAAl	353	120	1.39	7.90	3.7	15.9	6.7	2.39
PE-gr-PAN	353	105	–	6.84	3.9	18.2	6.3	2.89
PE-gr-P4VP	353	32	–	–	–	19.7	7.8	2.52
PE-gr-PAAc	353	25	1.41	–	–	20.8	7.9	2.63
Without support	323	37	1.12	7.24	3.0	22.4	7.2	3.11
Without support	343	44	–	–	–	17.6	6.7	2.63

[a] Polymerization conditions: benzene, [Mo]=6.7 mmol/l; [PhA]=1.67 mol/l.
[b] Factor of relative content of *cis*-structures.

Measurements of magnetic susceptibility showed [273] that practically all bound Mo^{5+} ions became diamagnetic just at the initial stage of polymerization though the process continues for some hours. This evidence indicates rapid molybdenum reduction to Mo^{4+} or Mo^{2+} in the course of polymerization, as only these ions can be diamagnetic. However, the formation of Mo^{2+} in the course of polymerization is less probable. In the case of PhA interaction with bound Mo^{4+} ions strong changes in complex coordination appearing via Mo^{4+} transformation from a high-spin ($\mu_{eff}=2.4$ μ_B) to a low-spin diamagnetic state have been revealed. The lack of an induction period in PhA polymerization by immobilized Mo^{5+} and Mo^{4+} catalysts testifies that these transformations proceed very rapidly. These conclusions have also been confirmed by

ESCA data [275]. The reduction $Mo^{5+} \rightarrow Mo^{4+}$ (the support PE-gr-PAN) has appeared as the shift of the $Mo_{3d5/2}$ peak from 232.5 eV to 232 eV (which corresponds to the Mo^{4+} state [276]). The progressive reduction $Mo^{5+} \rightarrow Mo^{4+} \rightarrow Mo^{3+}$ has also been observed in the course of polymerization since the growth of the $Mo_{3d5/2}$ at 231 eV peak intensity and its further broadening (Figure 3.47) have been detected. The electron binding energy in Mo^{3+} has been estimated to be equal 229.7 eV [277] (similar to that for Mo_6Cl_{12} clusters). Due to partial overlapping of all these peaks in ESCA spectra it is difficult to evaluate the share of each molybdenum valent state. The fact that the molybdenum concentration did not change in the course of polymerization points to all the above reduction processes not being accompanied by molybdenum leaching.

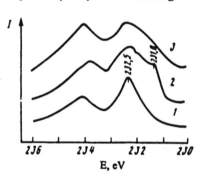

Figure 3.47 ESCA spectra of $MoCl_5$ (77 K) (1), PE-gr-PAN+$MoCl_5$ (2) and that of the catalyst after 30 min of PhA polymerization (3).

Thus, immobilization of molybdenum halides on polymer supports[24] provides opportunities for experimental confirmation of a multicenter mechanism of PhA polymerization which can be generalized as follows:

1. High molecular mass PPhA ($\overline{M}_n = 7\,000$–$14\,000$) is formed by a coordinative mechanism; the active center contains a Mo^{4+} ion. This valency state is probably necessary for the appearance of additional coordinative vacancies. The following stages of metathesis and isomerization of the formed ladder-type polymers [278] yield linear polyconjugated PPhA:

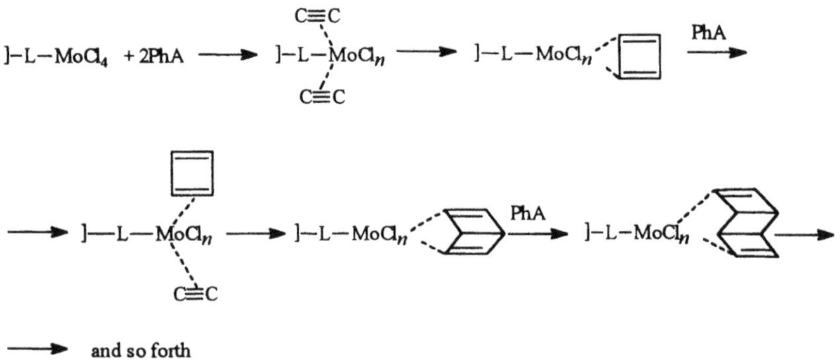

Scheme 3.31

---

[24] This is also substantial for both $MoCl_5$ binding to inorganic supports and for PhA polymerization in the presence of these catalysts.

The reduction $Mo^{5+} \to Mo^{4+}$ proceeds at this stage, the latter being stabilized in the π-complex, and is responsible for high molecular mass PPhA formation.

2. Bound molybdenum halides can probably produce molybdenum-carbon bonds with subsequent PhA incorporation at this bond (metal–carbon mechanism). Active sites of this type make PPhA of $\overline{M}_n = 1400$–2000 (due to chain degradation):

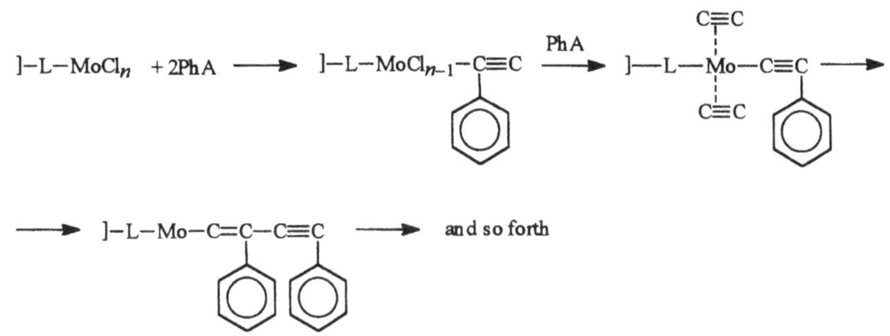

Scheme 3.32

The data on PhA polymerization quenching by radioactive inhibitors[25], such as $^{14}CH_3OH$, CH_3O^3H and ^{14}CO (which are usually used as tests of polymerization type) also confirm this conclusion. PPhA showed radioactivity only in the case when $^{14}CH_3OH$ was used as an inhibitor. This indicates sufficient Mo–C bond polarization so that the positive charge is mainly concentrated at the end of the growing chain (which can be classified as coordinative-cationic polymerization)

3. Oligomer formation proceeds by both an organometallic mechanism and at reduced molybdenum ions. Actually, PhA polymerization catalyzed by $MoCl_4$ yielded mainly oligomers with $\overline{M}_n \approx 400$; only a trace amount of PPhA of $\overline{M}_n \approx 7000$ has been detected.

4. The product of PhA tricyclomerization is probably formed at strongly reduced molybdenum compounds[26], say $MoCl_3$. This compound is unique of all molybdenum chlorides[27] and gives octahedral complexes with three molecules of monodentate ligand,

---

[25] The method of radioactive inhibitors (CH_3O^3H) has also been used for quenching of acetylene polymerization catalyzed by the system $Ti(OBu)_4$–$AlEt_3$ with the aim to determine the molecular mass of formed PA $(\overline{M}_n = 500$–36 000).

[26] M^{5+} reduction under the action of PhA is follows by cyclotrimerization of PhA with MX_5 (M=Nb, Ta; X=halogen) to 1,2,4- and 1,3,5-triphenylbenzene [279]. The oligomeric fraction (soluble in methanol), the product of PhA copolymerization with acetylene or 3,3-dimethyl-1-butyne catalyzed by a $MoCl_5$ solution in toluene at 298 K, was also shown to be composed of mainly cyclic trimers [280].

[27] Cyclic oligomers are usually formed in the coordination sphere of the transition metal ion from coordinated monomer molecules. The ring size is predetermined with the number of coordinative vacancies of the metal complex. Thus, cyclooctatriene is formed from acetylene coordinated with Ni^{2+} which has four coordinative vacancies [281]. The polymerization induction period which is observed in alkyne tricyclomerization in the presence of $NbCl_5$ [282] can also be

MoD$_3$Cl$_3$. The possibility of cyclotrimer formation caused by radical initiation should not be excluded; these particles can be formed in the system in parallel with reduction processes[28].

The reactions of chain termination and restriction also follow several pathways. For example, they can be caused by chain transfer to monomer at the stage of both ladder polymer and product formation (by an organometallic mechanism), by molybdenum reduction and catalyst leaching out and by chains degradation (chain termination as a result of reactive center deactivation due to unpaired electron delocalization along the chain of conjugation). It has been shown that chain transfer to solvent molecules does not occur in these processes.

Thus, heterophase PhA polymerization under the action of immobilized molybdenum complexes is accompanied by a strong interaction of monomer with molybdenum ions which in their turn are reduced to lower oxidation states. These processes cause changes in active site nature and mechanism of chain growth and therefore in the structure of the products formed. Immobilization affects all stages of the polymerization process, coordinative polymerization the most, which appears as increases of both polymer yield and their molecular mass. These effects are similar to those observed in polymerization of mono- and diolefins.

Immobilization of molybdenum complexes gives additional improvements. As already mentioned above the process is not accompanied by catalyst leaching. Thus no trace of catalyst has been detected in PPhA solution after the first filtration, whereas a two or even threefold precipitation of PPhA obtained in the presence of the homogeneous system is required to obtain the same effect. Besides, the catalysts can be reused by chlorination of the reduced molybdenum. The appearance of the ESR signal characteristic of the presence of Mo^{5+} has been observed after treatment of the catalyst that has leached out with Cl$_2$.

Finally, let us note some reactions of PhA, for example with CO, in the presence of immobilized molybdenum complexes. Thus, a catalyst prepared by interaction of Mo(CO)$_6$ at 383 K with PS (with all CO ligands being released) yields derivatives of (2+2)-cycloaddition, containing strained four-membered rings (bicyclo- and tetracycloadditions) [286].

3.5.2 Stereospecific PhA polymerization by immobilized Pd^{2+} complexes

Homogeneous cationic polymerization of PhA under the action of PdCl$_2$ complexes (for example, PdCl$_2$(PPh$_3$)$_2$) [287] yields PPhA[29] with $\overline{M}_n \approx 2000$, totally composed of a trans-cisoid structure.

---

caused by Nb^{5+} reduction processes. PhA converts to triphenylbenzene, while octadiene undergoes intramolecular ring closure with the formation of 1,4-bis(tetraaline)butane.

[28] Gel-immobilized catalytic systems derived from nickel halides or acetylacetonate vanadium cause PhA cyclotrimerization; depending on the reaction conditions the mixture of 1,2,4- and 1,3,5-triphenylbenzenes is usually formed [283]. Ni(CO)$_2$PPh$_3$ bound to

phosphynated PS catalyzes trimerization of HC≡CR [284] to , R=C$_2$H$_5$OOC ,

while triphenylphosphyne complexes of PdCl$_2$ induce its polymerization [285].

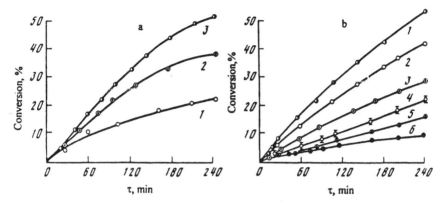

Figure 3.48 PhA polymerization by Pd^{2+} complexes. (a) 1, $PdCl_2$(4-vinylpyridine)$_2$; 2, Pd^{2+} on PE-gr-PAAc; 3, $PdCl_2 \cdot$PE-gr-P4VP. (b) Pd^{2+} obtained by exchange interaction on polymer support: 1, PE-gr-polyvinylcarbazole; 2, PE-gr-PVPr; 3, PE-gr-PAAI; 4, PE-gr-PAN; 5, PE-gr-poly(vinylimidazole) (PVIm); 6, homogeneous system $PdCl_2(C_6H_5CN)_2$. Polymerization conditions: 343 K, benzene, [PhA]=3.0 mol/l, $[Pd^{2+}]$=5 mmol/l.

Immobilization of Pd^{2+} on the polymer support results in a rise of PhA polymerization rate in comparison with the homogeneous systems (Figure 3.48). Attempts to establish a correlation between Pd^{2+} macrocomplex stability and their activity in PhA

---

[29] The mechanism of acetylene-type monomer polymerization has been studied in detail (see [288,289]) and consists of a series of intermediate stages:

(i) monomer incorporation at the Pd–Cl bond:

$$Pd(PhCN)_2Cl_2 \xrightarrow{\ CH\equiv CY\ } Pd(CH\equiv CY)(PhCN)Cl_2 \ \rightleftharpoons \ \left[Pd(CH\equiv CY)(PhCN)Cl_2 \right] \rightleftharpoons$$

(ii) the formation of cyclobutenyl complexes of different composition:

Probably, these process are also accompanied by Pd^{2+} to Pd^+ and Pd^0 reductions.

polymerization (as has been achieved for nitrocompound hydrogenation, see Section 1.3) have failed. All polymers were completely composed of *trans*-cisoid structure; molecular mass characteristics were also insensitive to immobilization (Table 3.19).

Stereoregular PPhA with relative molecular mass 5000–9000 can be obtained [290] in the presence of chelate Rh^{1+} complexes; *cis*-structure contents reached 100% in these polymers. Data on acetylene polymerization by these complexes has also been reported [291].

Among other processes with participation of ethynyl compounds and palladium complexes we shall note co-dimerization of acetylenes with alkyl halides catalyzed by $PdCl_2$ bound to phosphynated PS [292]:

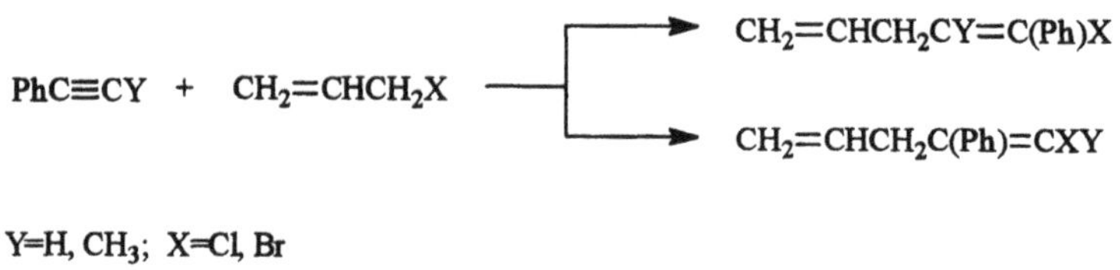

$$PhC{\equiv}CY \;+\; CH_2{=}CHCH_2X \longrightarrow \begin{array}{l} CH_2{=}CHCH_2CY{=}C(Ph)X \\[2mm] CH_2{=}CHCH_2C(Ph){=}CXY \end{array}$$

Y=H, CH_3; X=Cl, Br

Scheme 3.33

Table 3.19 Molecular mass characteristics of PPhA obtained in the presence of Pd^{2+} catalysts.

Pd^{2+} compound	$\overline{M}_w \times 10^{-3}$	$\overline{M}_n \times 10^{-3}$	γ	$\overline{M}_z \times 10^{-3}$	$\dfrac{\overline{M}_z}{\overline{M}_w}$
$PdCl_2(PhCN)_2$	2.4	1.68	1.43	4.2	1.75
$PdCl_2 \cdot$(PE-gr-PVP)	1.8	1.20	1.50	5.6	3.11
$PdCl_2 \cdot Py_2$	2.1	1.75	1.20	3.9	1.86
$PdCl_2 \cdot$(PE-gr-PAN)	2.9	2.1	1.38	5.1	1.76
$PdCl_2 \cdot$(PE-gr-PVIm)	2.5	1.95	1.30	4.8	1.92
$PdCl_2 \cdot$(PE-gr-PAA)	2.85	1.80	1.58	3.7	1.30
$PdCl_2 \cdot$(PE-gr-polyvinylcarbazole)	2.50	1.75	1.43	5.4	1.76

3.5.3 Catalytic cis-trans-isomerization of PPhA

Polyphenylacetylene polymerization is often accompanied by competitive processes. Thus, as already mentioned above the double bond deficiency in the PPhA obtained can be the result of partial chain cyclization. Another reaction which also accompanies polymerization and is probably catalyzed by the same metal complexes is chains isomerization. Thermal *cis–trans*-isomerization of PhA [293] as well *cis–trans*-isomerization catalyzed by electron π-acceptors, for example, by chloranil in dichloroethane [294] are only some examples. The main stages of the mechanism are the fast formation of a charge transfer complex as an intermediate and further low-speed one-electron transfer yielding a cation-radical polymer. Electrochemical *cis–trans*-isomerization of PhA yielding stable polymers is also known [295].

The opening of the triple bond in the course of PhA polymerization leads to *cis*-structure formation, while *trans*-structures are the result of *cis–trans*-isomerization. In homogeneous systems (based on Pd^{2+} derivatives), *cis–trans*-isomerization proceeds via cyclization and yields 1,3-cyclohexadiene structures.

This stage is followed by double bond migrations and changes in the structural configuration of the units from *cis* to *trans* without π-bond cleavage[30]. Further stages of this complex process are cyclohexadiene fragment aromatization, chain decomposition with triphenylbenzene abstraction and recombination of the radicals formed.

Isomerization of PPhA containing 87% *cis*-transoidal structures (obtained in the presence of $MoCl_5$) catalyzed by homogeneous palladium complexes yields *trans*-cisoidal PPhA and is accompanied by chain degradation. It is important that immobilized Pd and Mo complexes cause neither *cis–trans*- nor *trans–cis*-isomerization of PPhA. This PPhA behavior is quite similar to comparative butene isomerization catalyzed by either heterogeneous or homogeneous complexes (see Section 3.3.2.).

3.5.4 Polymer compatibility as a way of producing electroconductive polymers with a system of conjugated double bonds

Immobilized $Ti(OBu)_4$ (for example on PE-gr-PAAl) in combination with AlR_3 can be used effectively in acetylene polymerization for the production of electroconductive polymer composites on doped polyacetylene (PA). Pure PA films have two main drawbacks. They are inconvenient for application due to their friability and instability. The simple scattering of PA powder (obtained by acetylene polymerization catalyzed by $Ti(OBu)_4$–$AlEt_3$) in the matrix of PE did not produce a material with good properties [296], because PA shows a tendency to form massive agglomerates (10–100 μm). To achieve good mixing of the polymers, another approach has been developed. This is *in situ* polymerization of acetylene on PE films (of thickness 0.3 mm) containing $Ti(OBu)_4$ preactivated with AlR_3 [296, 297]. The material products showed good mechanical and electrical properties up to a PA content of 6–10 mass%. The polymer support provides flexible and stable PA covering. Polyacetylene obtained by polymerization on a PE matrix at 383 K has a *trans*-structure $-(CH-)_x$. It can also be cross-linked with the formation of three-dimensional structures incorporated into the PE matrix. The electroconductivity of these composite materials doped with iodine vapor was shown to be within the range 1–5 Ω/cm.

To improve the metal complex catalyst component binding to the support (of desirable shape) these films can be preliminary modified by grafting polymerization of different monomers, such as acrylic acid or allyl alcohol (0.5–11.0 mass%). Dark, strong, bound PA film of thickness 0.01–0.05 mm has been obtained by acetylene polymerization on these films with immobilized $Ti(OBu)_4$ [298]. Doping with iodine vapor provides these films with electroconductivity (Table 3.20). It is interesting that only the surface film is electroconductive, the support (PE, PP, PS, PVAl, PTFE) remaining dielectric. Multilayer polymer–polymer composites (PPC) with only one conductive layer can be obtained in this way. These PPC are easily made and preserve their properties on storage. The increased stability of the materials may well cause difficulties in O_2 penetration of PA.

---

[30] Both catalytic intermolecular cyclization (with the preservation of the *cis*-configuration) and double bond migration accompanied by *cis–trans*-isomerization are characteristic of secondary reactions proceeding in polydiene chains (including catalysis by cobalt-containing systems) [114].

Polymer–polymer composites of polyolefins and polymers with a system of conjugated double bonds can also be obtained by polymerization of PhA (electroconductivity of doped PPhA reaches 1.8×10^{-5} Ω/cm, the mechanism of conductivity is ionic [299]), propargyl alcohol, propargyl bromide, propargyl amine, all catalyzed by Mo complexes immobilized on PE, PP, PS, etc. [270].

Materials with very interesting properties (both mechanical and conductive) have been obtained by mixing PA and PB [300, 301] as well triple copolymers, for instance of diene, ethylene and propene [302]. Acetylene polymerization in the matrix of the last mentioned copolymer yields a product characterized by both high electroconductivity and resistance to the action of an aggressive medium, as well as good suitable for applications; cross-linking leads to PA immobilization in the copolymer matrix. Grafted polymers polydiene–PA can be obtained by polymerization *in situ*. The first stage is catalyst formation through the incorporation of a titanium–alkyl bond at 1,2-units of PBD; the second stage is acetylene polymerization by this catalyst [303]:

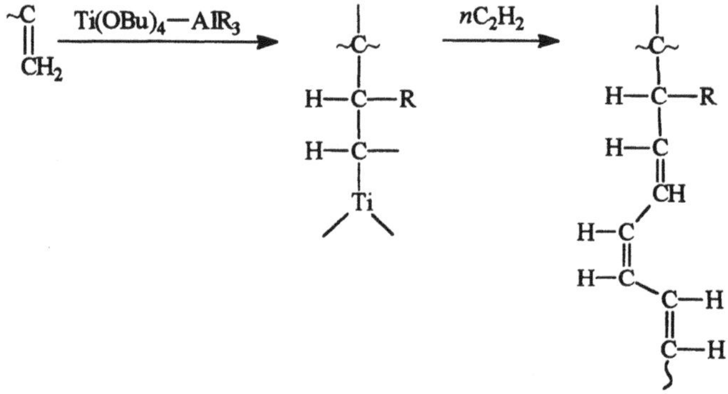

Scheme 3.34

Table 3.20 Some characteristics of electroconductive polymer materials doped with iodine and obtained by acetylene polymerization catalyzed by immobilized metal complexes[a] [298].

Polymer support	Ti^{4+} (mmol/g)	Polymerization time (h)	PA film thickness (μm)	σ (Ω/cm)
PE-gr-PAAl (0.5 %)	0.04	2	0.05	6.25
PE-gr-PAAl (0.5 %)	0.06	2	0.016	4.16
PP-gr-PAAl (0.8 %)	0.10	1	0.02	14.38
Teflon-gr-PAAc (5 %)	0.40	3	0.01	2.40
PVAl	0.20	2	0.02	3.90
PE	traces	3	-	1×10^{-7}
PS	0.15	3	0.025	6.67

[a]Polymerization conditions: 343 K, cocatalyst $AlEt^3$, acetylene pressure 0.1 MPa.

Some other monomers, such as pyrrole can also be used to transfer charges. Thus, highly transparent and electroconductive films have been obtained by polymerization of pyrrole on PVAc [304]. The same properties were found in the films of polymer melts derived from polypyrrole, PVC, etc., which have been produced by electrochemical polymerization [305] (see Section 4.9).

Electroconductive PPC obtained at the stage of polymerization can also be synthesized by retracing the above steps. In this way catalyst or initiator are bound to the surface of the electroconductive polymer and then the monomer yielding non-conducting layer is polymerized. Thus, interaction of PA with sodium hydronaphthalyde gave charge transfer complexes $[Na_{0.57}-(CH)-]_x$ responsible for MMA polymerization[31] [307].

Filler electroconductive composites have been obtained by treatment of carbon black with LiBu with subsequent anionic MMA polymerization [308], or by polymerization of α-olefins by metal complexes bound to graphite [309].

The use of rare earth naphthenates (La, Pr, Nd, etc.) as coordinative catalysts for stereospecific acetylene polymerization is a new approach to the polymerization of acetylenes. Probably, their heterogenized analogues will soon be used for the same purpose.

There are no data on allene polymerizations catalyzed by immobilized metal complexes. These monomers are known to be strongly adsorbed by heterogeneous metal-complex catalysts and to completely inhibit α-olefin polymerization. This property is used for quantitative examination of active site content in these systems [310]. At the same time complexes of Ni, Rh, Pd, etc., for example bis-π-allylhalogeno nickel complexes adsorbed on the support (formed from nickel carbonyl and allyl halide), are active in allene cyclodimerization and polymerization [311]. The nature of the support considerably affects both the structure and the activity of the Ni complex.

3.6 Catalysis of Polycondensation

Polycondensation is formally an exchange or substitution reaction in contrast to polymerization which is an addition reaction. This property predetermines the main features of the process.

Polycondensation reactions can be generally attributed to one of three groups: (i) non-catalytic; (ii) accelerated by catalyst; (iii) occurring particularly in the presence of catalyst.

The role of catalyst in these reaction is to form active intermediates with one of the monomers [312]. Thus, catalyst activity is supposed to be proportional to its ability to form active intermediates; however it is the result not only of reagent activation but also of the rise of interaction probability due to monomers drawing together and favorable orientation. Acid-base catalysts initiating ionic polycondensation are most widely used. For example, tertiary amines are able to initiate cationic, anionic, ionic coordinative and even radical condensations.

Although catalysis has important practical uses in polycondensation, its theoretical basis is rather scanty. Effective catalysts are usually selected empirically. To a considerable extent this can be attributed to metal-complex catalysis and especially to immobilized catalysts which are only beginning to find a use in this field.

---

[31] Similar composites can be obtained in several ways. Thus in the study [306] PA doped with AsF5 has been further covered with the film of poly-n-xylylene.

3.6.1 Oxidative coupling of 2,6-disubstituted phenols catalyzed by metal-containing polymers

Oxidative phenol polycondensation is a very valuable route to the production of aromatic polyethers and is the most extensively studied field of polycondensation in the presence of immobilized metal complexes.

One of the first studies to have been completed was the synthesis of poly(2,6-dimethylphenyl oxide) by the oxidation of 2,6-dimethylphenol with oxygen from air in the presence of a copper complex of pyridine [313]. The reaction is expressed by the scheme:

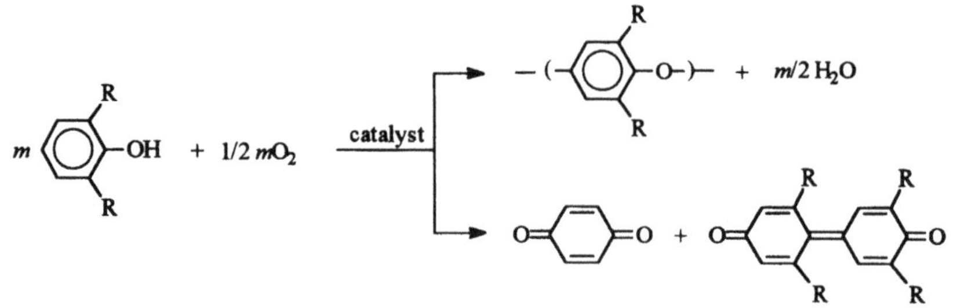

R=H, CH₃, *tert*-Bu, Ph, etc.

Scheme 3.35

The oxidation of these phenols catalyzed by immobilized metal complexes has already been discussed above in the Section 2.7.3. Diphenoquinone as a product of coupling at C–C bonds (see Scheme 2.8) is formed in the presence of low molecular mass catalysts in the absence of amines [314]. In the presence of amines the main product is poly(2,6-dialkyl-1,4-phenylene oxide) which is formed by coupling at C–O bonds. By different methods it has been shown that the active sites responsible for polymer product formation are hydroxo-bridged binuclear copper complexes (which are formed in the presence of bases), while chlorine-bridged dimers account for the C–C coupling.

The kinetic scheme for the reaction can be represented as follows [315]:

$$\text{Cu}^{2+}\text{—complex} + \text{DAPh} \underset{k_{-1}}{\overset{k_1}{\rightleftharpoons}} \text{Cu}^{2+} \bullet \text{DAPh}$$

$$\text{Cu}^{2+} \bullet \text{DAPh} \xrightarrow{k_2} \text{Cu}^{2+}\text{—complex} + [\text{C—O}]/[\text{C—C}]$$

$$\text{Cu}^{2+}\text{—complex} + \text{H}_2 \xrightarrow{\text{O}_2} \text{Cu}^{2+}\text{—complex} + \text{H}_2\text{O}$$

Scheme 3.36

where DAPh is dialkylphenol and [C–O] and [C–C] are the coupling products at the C–O and C–C bonds respectively.

The process is expressed by an equation of the form

$$\frac{1}{w} = \frac{1}{w_0} + \frac{K_M}{w_0[DAPh]}$$

where K_M is the Michaelis constant; $K_M=(k_1+k_2)/k_1$.

It has been assumed that this process proceeds by a radical mechanism [316].

Immobilized complexes, mainly of copper in the medium of acetonitrile, alcohols, 1,2-dichlorobenzene/methanol (13:2) and other solvents are those most widely used in the oxidative coupling of disubstituted phenols. These subjects are also interesting for they allow one to observe chain effects in catalysis. Copolymers of styrene with 4-vinylpyridine (its quaternized derivatives), N-vinylimidazole and dimethylaminostyrene with different parts of nitrogen-containing monomers, as well as macromolecular analogues of tetramethyl ethylene diamine (TMED) are generally used as supports.

From the example of Cu^{2+} complexes with P4VP [317] it has been shown that the rate of oxidative coupling of xylenol-2,6 and o-cresol and their mixture with oxygen is lower than that catalyzed by the Cu^{2+} complex with pyridine. This is probably caused by hindrances for both monomer coordination and electron transfer from coordinated monomer to metal ion. The activity can be changed by a few tens or even hundreds of times.

The rate of 2,6-dimethylphenole polymerization increases with a rise of Cu^{2+} concentration (up to 0.3 mmol/l) (Figure 3.49) and then reaches a plateau [318]. The molecular mass of poly(2,6-dimethylphenyl oxide) formed in the presence of complexes immobilized on CSDVB–4VP (14 200) is as low as that obtained by homogeneous complexes (20 500-24 500). The reaction follows the same mechanism as with the homogeneous system:

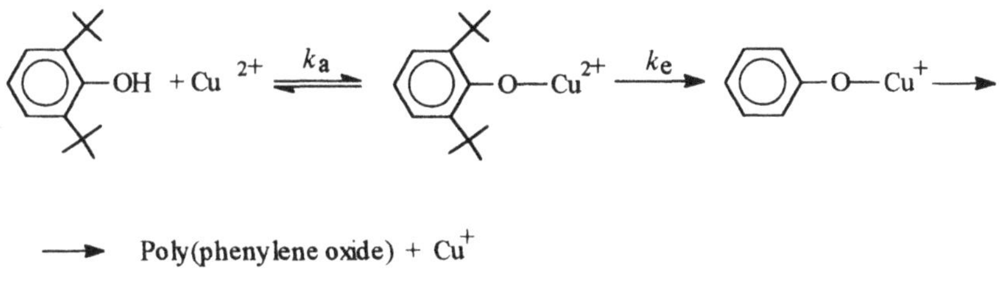

$$\longrightarrow \quad Poly(phenylene\ oxide) + Cu^+$$

Scheme 3.37

Monomer coordination by metal-containing polymers insoluble in the reactive mixture has been shown experimentally (by the measurement of its diminution in liquid phase under argon atmosphere). The observed substrate coordination rate constant was within the range $K=1.9-2.8$ l/(mol min) being considerably less than that for homogeneous systems (10–10^2 l/(mol min)) [319] as a consequence of steric factors characteristic of insoluble polymer matrixes.

The portion of Cu^{2+} ions of the total copper content ($[Cu^{2+}]/\Sigma Cu$) has been determined from magnetic susceptibility measurements which decrease in the course of polymerization and is the result of copper ion reduction under the action of the monomer (Figure 3.50). The electron transfer rate constant, K_e, is sensitive to the nature of the solvent being the largest for methanol which is capable of penetration inside the polymer matrix. K_e and $[Cu^{2+}]/\Sigma Cu$ are equal to 10 l/min and 0.35 for methanol, while those for water are 5.3 l/min and 0.87, and for benzene are 15 l/min and 0.62. K_e also has lower values than in homogeneous system $(Cu^{2+} \cdot Py)$ [317]. The same features have been observed for reaction activation energy (30.7 and 46.2 kJ/mol, respectively). The variances in E_a are caused by the energetic factor of the electron-transfer controlled stage. The Cu^{2+} complex rearrangement from square planar to tetragonal with significant changes of the bond angles (see Scheme 2.9) requires more energy for insoluble macrocomplexes.

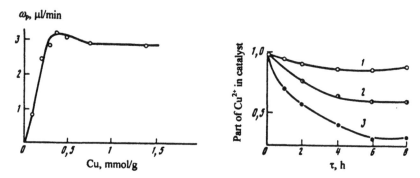

Figure 3.49 The dependence of 2,6-dimethylphenol polymerization rate on copper content in the complex Cu^{2+}–CSDVB–4VP. Reaction conditions: $[Cu^{2+}$–CSDVB–4VP]=150 mg, [monomer]=0.05 mol/l, benzene as solvent, 303 K, in air.

Figure 3.50 Changes of the Cu^{2+} portion with time in 2,6-dimethylphenol polymerization. Solvents: (1) H_2O, (2) benzene, (3) CH_3OH. Reaction conditions: catalyst PVP–Cu; [Cu]=1 mmol/l; [monomer]/[Cu]=5; 303 K, argon.

All these processes proceed faster in the presence of soluble metal-containing polymers. In the case of copolymers, catalytic activity depends not only on the nature of the comonomers but also on functional group distribution along the copolymer chain. It was conclusively demonstrated for the examples of Cu^{2+} complexes with copolymers of styrene with 4-(N-methyl-N-n-vinylbenzyl)aminopyridine, etc. [320, 321]. The spatial interaction between coordinated substrate and polymer segment between two amino groups bound with different copper ions increases with a decrease of segment length. This interaction promotes reaction due to the facility of tetragonal intermediate complex formation from the initial square-planar. The tetragonal state induced by complex deformation is more favorable for electron transfer from substrate to the copper ion. It has been shown on models that for binuclear cooper complex formation at least six styrene units are required. The increase of catalyst stereospecificity with excess of NaOH causes partial substrate deprotonation under an increase of medium basicity.

The reaction rate increases with a rise of the degree of support functionalization, α, up to its value of 0.4. This is the result of a change of activation parameters as well as low spatial hindrances for binuclear Cu^{2+} complex formation and Cu^{2+}–macroligand intermediate complex formation.

The activity of the Cu^{2+} complex with the copolymer of styrene with N-vinylimidazole depends on the proportion of amines to Cu^{2+} and reaches its largest value when this ratio equals 2, while the same effect for a low molecular mass ligand analogue is reached at [amine]/[Cu^{2+}]=3. The reaction rate being a function of α has two maxima [322]: at $\alpha=0.14$ and $\alpha=0.25$ in the case when amino groups are separated by three or six inert styrene units (Figure 3.51). The increased stability of metal-polymer catalysts can also be the result of enhanced local concentration of metal complexes in the polymer globule.

The intrinsic viscosity of partially quaternized P4VP decreases from 6.23 to 0.84 dl/g by introduction of Cu^{2+} ions indicating intramolecular chelation. Copper reduction in turn causes a viscosity rise from 0.84 to 1.27 dl/g due to considerable expansion of the polymer globules. In the course of the catalytic action the immobilized catalyst fluctuates thus accelerating electron transfer [323].

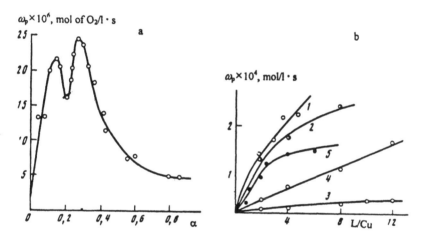

Figure 3.51. The dependencies of initial rate of condensation on the macroligand functionalization degree, α (copolymer of styrene with N-vinylimidazole) (a) and the rate of O_2 absorption on the ratio L/Cu (b). (b) 1, 4-pyrrolidinepyridine; 2, 4-dimethylaminopyridine; 3, pyridine; 4, 1,2-bis(dimethylamino)ethane; 5, dimethyl-aminopyridine; $\alpha=0.056$.

It should be noted that Cu^{2+} complexes with polyaminodiamines (polymer analogues of TMED) are 4 times as active as the corresponding low molecular mass catalysts (CH_3OH, 298 K). Binuclear Cu^{2+} complexes include one TMED unit per copper ion and only one bridging group binding two copper ions, [TMED·Cu(OH)Cl–Cu·TMED]·2Cl, instead of two bridging OH groups in the usual complexes.

Oxidative coupling (for example of 2,6-di-*tert*-butylphenol) is also catalyzed by Cu^{2+} macrocomplexes bound to nonporous particles of silica gel [321]. Although this approach does not change significantly the catalyst SCA (and is only 70% of that for soluble polymer complexes), it considerably facilitates both catalyst separation and regeneration for multiple reuse (more than 10 recycles).

Metal polymers derived from bis(ethylene-1,2-dimethylato)Cu^{2+} are also active in oxidative polymerization of 2,6-dimethylphenole [324]. Heterometallic complexes (including immobilized ones), such as copper chelates with Schiff bases in combination with nickel halides, are also used for synthesis of polyphenylene oxides of the high molecular mass [325]. For instance, the heterometallic complex of P4VP with Cu^{2+} and Mn^{2+} is more active than the monometallic one and yields poly(phenylene oxide) of higher molecular mass [326]. Monomer in these catalysts is also activated by coordination by Cu^{2+}, while redox cycle $Cu^+ \rightarrow Cu^{2+}$ is provided by Mn^{3+} ions:

$$Cu^+ + Mn^{3+} \xrightarrow{\text{P4VP}} Cu^{2+} + Mn^{2+}$$

The reduced Mn^{2+} is oxidized by O_2 under very mild conditions. Thus, the route of the processes is composed of two cycles:

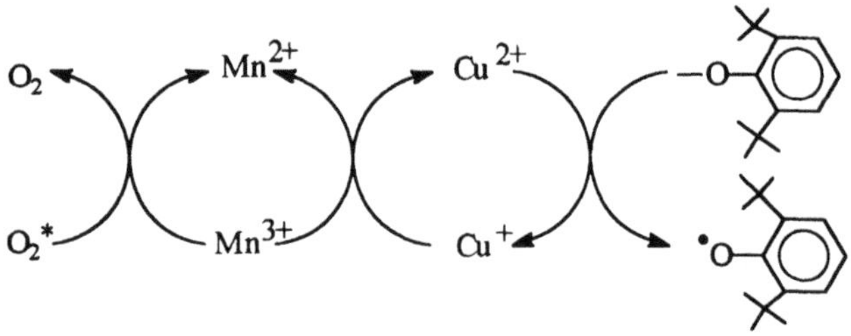

Scheme 3.38

Oxidative polymerization of phenols is also catalyzed by Mn^{2+} chelates, while poly-Schiff bases of Co^{2+}, Ni^{2+}, Cu^{2+} initiate only oxidative dimerization of 2,6-di-tert-butylphenole [327].

Apart from 2,6-disubstituted phenols, compounds containing mobile a hydrogen atom in the 1,4-position, of the type [328]

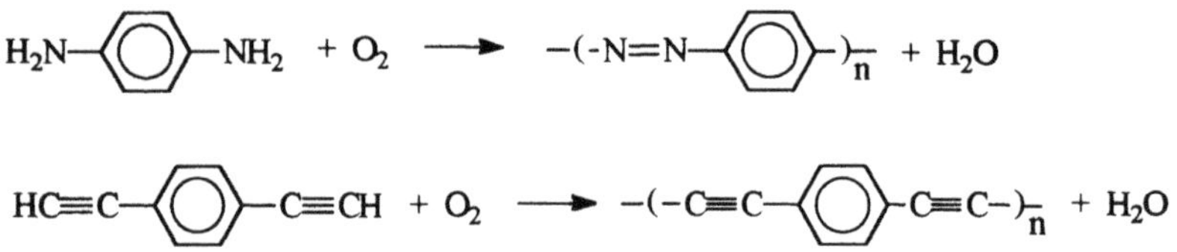

Scheme 3.39

can also follow oxidative coupling reactions.

Oxidative polycondensation of acetylene [329] yields a linear carbon compound, "carbyne":

Scheme 3.40

3.6.2 Catalytic curing of reactive oligomers catalyzed by metal polymers

Complexes of some metal ions (Ni, Co, Zn, etc.) and especially their acetyl acetonates effectively accelerate solidification of several reactive oligomers. The possibility of epoxy–anhydrous composites curing in the presence of metal-containing oligoethers has been demonstrated [330]. The activity of metal ions in these processes follows the sequence $Ca^{2+} < Mn^{2+} < Mg^{2+}$ and correlates with the value of metal ion radii. Ni and Co complexes show higher activity than Zn-containing in solidification of epoxynovolac block copolymers [331]. The promoting action of the product of salicyl aldehyde condensation with $TiCl_3$ and aqueous ammonia on the properties of epoxy-polymers has been observed [332]. Acetyl acetonates of polyvalent metals, due to coordinative unsaturation at the central metal ion (provided by substitution of labile acetyl acetonate rings), are able to activate both the curing agents and the epoxy-compound promoting solidification [333]. These chelate curing agents (latent accelerators of solidification of diane epoxy resins with dicyandiamide) can be modified by several polymers, for example rubbers. Epoxy binders containing chemically bound copper show long term stability and low heat of solidification [334]. It is important that curing agents containing a metal ion as a structural element of the matrix do not deteriorate the dielectrical properties of the materials.

Immobilized catalysts are also used for catalytic solidification of phenol–formaldehyde oligomers (PhFO) in the production of resols and novolacs:

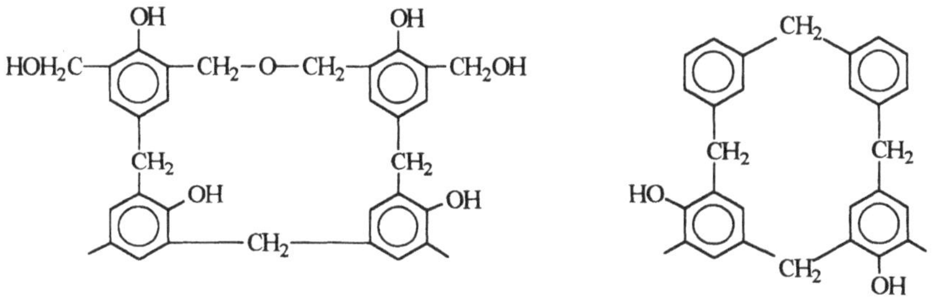

Scheme 3.41

This curing process (433-453 K) is accompanied by an increase of bridging methylene group content; 2,4-bonds dominate, the content of 2,2'-fragments is small [335]. However, the conditions of PhFO solidification are extreme for convenient metal complex catalysts. Thus, their stability can be enhanced by immobilization. This can be achieved by binding Ni^{2+}, V^{4+}, V^{5+}, Co^{2+}, Ti^{4+} and Pd^{2+} complexes to several carbo- and heterochain polymers [336]. They show comparatively high activity in PhFO solidification (Table 3.21). It is important that the catalytic properties of metal-containing polymers appear only at temperatures above the polymer support melting temperature. Under these conditions catalysis in the melt takes place. Under otherwise identical conditions, $Co(OCOCH_3)_2$ accelerates the reaction by only 25%, while the same species bound to a support increases activity by approximately twice. It is most probable that unbound Co^{2+} complexes in the course of epoxidation aggregate into cluster particles, while immobilized ones results in groups remaining separated on the polymer support.

Table3.21 The influence of immobilized catalyst metal ion nature on the time of both PhFO and phenoplasts solidification.

Catalyst	Time (s)		
	Gel formation (423 K)	Hardening in the form (443 K)	Viscousoplastic state (393 K)
Without catalyst	120	250	380
Co(OCOCH$_3$)$_2$	90	200	360
PE-gr-PAAc–Co^{2+}	50	120	570
PE-gr-PAAc–Ni^{2+}	70	150	575
PE-gr-PAAc–Cu^{2+}	75	170	570
PE-gr-PAAc–Zn^{2+}	82	180	578
PE-gr-PAAc–V^{4+}	95	192	574
PE-gr-PAAc–Cr^{3+}	100	204	570
PE-gr-PAAc–Mn^{2+}	105	210	570
CSMMAc–Ti^{4+}	85	185	420

On the model system CoCl$_2$–dimethylole-n-cresol it has been shown by several physico-chemical methods [337] that the process of polycondensation involves a stage of Co^{2+} complex formation with substrate (K_{eq}=13.1 l/mol at 353 K). Complex formation causes proton activation in the methylole groups coordinated by Co^{2+} and consequently the acceleration of PhFO solidification.

It is also interesting that even under extreme conditions (443 K) the curing rate as a function of surface density of bound metal ions follows the bell-like shape (Figure 3.52) with the maximum effect at a Co^{2+} content of ~1.0 mass%.

Solidification of resins derived from furfuryl alcohol proceeds basically in a similar way:

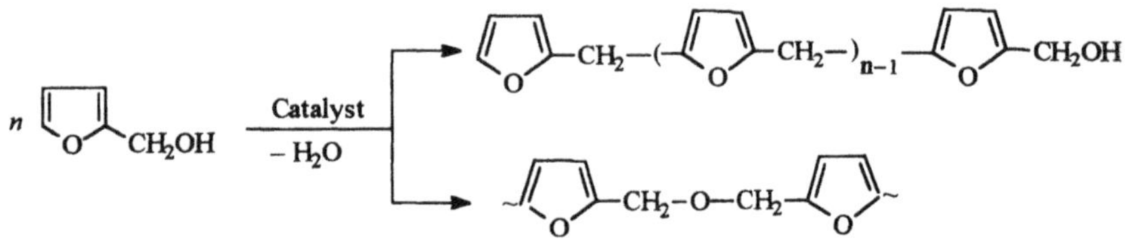

Scheme 3.42

Ti^{4+} chelates, including oligoester-type [(bis(chelate)-bis(oligoester)titanates] are effective in reesterification and polycondensation [338]. Introduction of glycols residues or oligoethers into the molecule of the titanium compound causes a rise of catalytic activity.

The synthesis of aliphatic–aromatic polyamides by polycondensation of aliphatic diamines, aromatic dibromides and carbon oxide catalyzed by palladium complexes is also of great interest [339]:

$$H_2N-R-NH_2 \ + \ Br-Ar-Br \ + \ 2CO \ \xrightarrow{\text{Catalyst}} \ \left[NH-R-NH-\overset{\overset{O}{\|}}{C}-Ar-\overset{\overset{O}{\|}}{C} \right]_n$$

Scheme 3.43

Polycondensation can also be catalyzed by the polymer support. Thus, polymeric triphenylphosphyne initiates (in the presence of Py, Et_3N) condensation of terephthalic acid with 4,4-diaminodiphenyl ether, 4-aminobenzoic acid, iso- and terephthalic acids with diphenylpropane, etc.

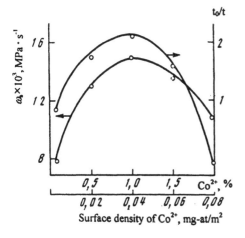

Figure 3.52 The dependencies of phenoplast solidification rate and the value of the t_0/t ratio on the surface density of Co^{2+} in PE-gr-PAAc–Co^{2+}. Cobalt content 0.01 mass%, 343 K.

3.6.3 Catalysis of urethane formation[32]

Metal complexes also catalyze formation of ethyl carbamates [342]. Metals of the platinum group, including Pt blacks [343], are most widely used as catalysts. The same metals can be immobilized on different supports, e.g. polymeric ones, in particular ion-exchange resins. In the reaction of phenylisocyanate with methyl alcohol, the catalytic activity of mono- and binuclear Cu^{2+} complexes depends on the spatial composition of the chelate node [344], the Cu^{2+} complexes being more active. The length of the PEG oligomer chain (of relative molecular mass 400–4000) considerably effects the reaction of urethane formation catalyzed by Cu^{2+} chelates [345, 346]. The stage of PEG coordination by Cu^{2+} ions precedes urethane formation reducing the activation energy of the process in comparison with the non-catalyzed reaction by ~12 kJ/mol. The increase of PEG molecular mass decreases the extent of OH⋯HO bonding and favors the more

---

[32] Formally, these reactions can be attributed to condensation processes [340, 341]. Their consideration in this section results from (i) a similar composition of the products formed, although the reaction follows another mechanism, and (ii) a desire to attract investigators to this growing and progressive field.

effective interactions between the oxygen atoms of ester glycol groups with the catalyst, i.e. causing catalyst deactivation. All these processes have been studied in detail [347].

Polyurethane formation in reactions of 2,4-toluylene- or hexamethylenediisocyanate with polyols, polyethyleneadipinate and poly(propylene glycol), both of molecular mass 2000, is catalyzed by Sn and Pd complexes, the catalyst activities being dependent on the spatial composition of reagents or in other words, on the structure of intermediate complexes [348].

Organotin compounds are the most active catalysts for urethane formation. As with so many other reactions, immobilization is effective in promoting their activity as it prevents the formation of non-reactive complexes (polymolecular heteroassociates incorporating several catalyst molecules and binding either by hydrogen or donor–acceptor bonds) [349–351]. Modified PEGs, copolymers of maleic anhydride with styrene, etc., are usually used as supports.

Secondary reactions that appear at certain reaction conversions (~60%) are of great importance for catalysis by immobilized oligomeric systems. These reactions can lead to polymer branching and even to gel or elastic networks formation. This is the result of high local concentrations of reactive groups in these systems. In the absence of species capable of complex formation only linear polymers are produced. These reactions in turn promote the enhancement of local effects [342] due to additional immobilization of monomer molecules or segments of oligomer support within the cross-linked polymer ranges. The structural organization of polyurethane carbamides cross-linked with chelates of three-valent metals [352] is worth mentioning. The bound metal ion action on the structural characteristics of the metal-containing polymer is caused not by the nature of the metal–polymer bonds but by its ability to coordinate functional groups of different flexible blocks of the macromolecule (or macromolecules). This leads to distention of the polymer structure and some increase in uniformity. Materials derived from transition metal compounds bound to isocyane polymer supports (the ratio [isocyane units]/[M] reaches 1:1) are stable up to 300°C [353]. The synthesis of condensation-type metal-containing polymers not only by polyurethane modification with metal complexes but also by direct condensation of metal-coordinated monomers [354] is also a promising route.

Polymers or oligomers promoting catalyst activity by several times can also be used as catalysts for the synthesis of polyurethane [355].

3.6.4 Some other condensation reactions

Aldol condensation (see Section 5.6) of benzaldehydes and acetophenone is catalyzed by macrocomplexes of Co^{2+} with P4VP or copolymers of 4VP with styrene and 1,4-DVB [356]. Metal-polymer catalysts increase the product yield and can be reused without loss of activity (353 K, dimethyl formamide). Surely, there is some analogy with aldol condensation in the presence of cobalt-containing catalysts and metal enzymes of enzymohexase class II incorporating Zn^{2+} ions in the active center.

Immobilized complexes are good for benzoin condensation [357] as well as for cross-aldol reactions [358]; supported Pd^{2+} complexes are used [359] in synthesis of condensation monomers and also in several other processes [360] including condensation of carboxylic compounds with CH acids of general formula XCH_2Y and XYCHZ, where X, Y and Z are atoms or groups promoting carbanion stabilization [361].

As mentioned above, polymers without metal complexes can also be used as condensation catalysts (see [362]). Ion-exchange resins (sulfonated CSDVB) are active in the condensation of acetone with phenol [363], and polymeric thiazonium salts are effective in reactions similar to benzoin condensation, etc.

3.7 The Formation of Polymer–Polymer Composites in the Course of Polymerization and Polycondensation Catalyzed by Metal Polymers

To continue examination of the catalyzed reaction's influence on the polymer support started above for hydrogenation (Section 1.14) and oxidation (Section 2.10) let us discuss this effect for polymerization and polycondensation.

In the course of the consideration of olefin, diene and acetylene polymerization catalyzed by immobilized metal complexes we have already mentioned the possibility of forming macrochains by grafting to the polymer support. This process can proceed by several routes [364]. Primarily, multiple bonds of polymer support functional groups can be opened at the stage of polymerization initiation and become the end groups of the growing macrochain. At the stage of chain propagation these bonds are capable of copolymerization addition. Finally, grafting can also proceed by chain transfer to the support surface; the end chain point or functional group reacts with an appropriate functional group of the support with the formation of covalent bond. Both the mechanism and participation by the above reactions are dependent on the nature of the support and the monomer.

The most important result is that new types of PPCs with enhanced mechanical and operating properties in comparison with those of mechanical blends can be synthesized by such polymerization *in situ* [365]. It is also interesting for the following reason. Most of the systems considered in this section are heterophase polymerization processes without catalyst leaching. So, it is important to pass the macroligand effect (assumed as PPC ingredient of low content) onto the properties of the obtained polymers. Such analysis has been made for systems (PMMA–PE [366], PVC–PE [367], PVAc–PE [368], PS–block-(ethylene-co-propene) [369], etc.) obtained by ethylene poly- and copolymerization catalyzed by macrocomplexes of VCl_4, $TiCl_4$, $VOCl_3$ [370] in combination with AlR_2Cl. In all cases the availability of the interphase (boundary) layer between components of synthesized PPC has been demonstrated. These layers have been formed not only by several physical bonds or by polymer chains interweaving but also by polyolefin grafting to a polar polymer component. Both grafting degree and the structure of the interphase layer are dependent on PPC composition and reaction conditions. This interphase layer is responsible for the strength of component binding and in the long run for the main properties of PPC [371], such as strength, impact strength, heat resistance (PVC–PE), etc.

The ethylene polymerization considered above (Section 3.1.3) catalyzed by organometallic complexes bound to fibers [50] leads to the formation of original "envelopes" around fibers: PE intergrowing with the oriented fiber of PTFE or nylon protecting them from wearing down due to attrition and decomposition due to moisture absorption, as well as enhancing their resistance to wear, etc..

Synthetic PPC can also be obtained by polycondensation, for example by introduction of novolacs (0.1–15 volume%) into PVAl, PE, etc. [372].

3.8 Conclusions

The comparative analysis of poly-, co-, and dimerization of olefins, dienes and acetylenes by homogeneous and immobilized metal complexes shows the similarity of active center nature in these systems. The effect of immobilization is quantitative rather than qualitative though it affects all stages of the polymerization process. Sometimes it is possible by several methods to disclose all complex transformations in the course of active site formation: isolated ions→the ranges with high local concentration of transition metal ions→associations composed of exchanged, bound particles. Under optimal conditions one can clarify the mechanism of each polymerization stage and study intermediates.

The active role of polymer supports consists of regulation of reactions of active sites formation and deactivation and sometimes promoting their regeneration. The processes of cooperative stabilization of active sites promoting thermal stability of these catalysts are also of great importance.

Immobilized bifunctional catalysts are very promising systems for realization of cooperative reactions, such as the "relay race" mechanism of ethylene dimerization with parallel ethylene copolymerization with the α-butene formed catalyzed by active sites of two types immobilized in a single polymer cage (LDLPE production). It is also possible to combine in the same polymer matrix active sites of different types, such as σ-alkyl and π-allyl. These considerations can be applied to a broad number of catalytic systems, not only for polymerization, and may be useful for a comprehensive analysis of catalytic action.

Macrocomplexes for use as redox systems in initiating polymerization and copolymerization of vinyl-type monomers is also of great interest although these studies are rather rare.

Although immobilized metal complexes are mainly used for oxidative coupling of phenols with steric hindrance in the production of polyethers, the dynamics in polycondensation studies promise fast development in the field.

Catalysis by immobilized metal complexes gives opportunities for the production of new composite materials. By selection of polymer supports and optimization of reaction conditions one can obtain materials with unique properties. Polymer–polymer composites of incompatible polymers, chemical "hybrids" of polyolefins and polydienes and electroconductive materials can be obtained in such a way.

Each of the problems mentioned requires further study, however the number of practical advantages which are provided by immobilization of metal complexes will lead to further development by investigation of polymerization processes catalyzed by metal-containing polymers.

This statement is corroborated by the set of new data in this field. The results on catalytic properties of polymer-immobilized Ti–Mg catalysts for ethylene polymerization [373], Cp_2ZrCl_2–polymethylalumoxane systems immobilized on α-cyclodextrin [374], immobilized catalysts for stereospecific propylene polymerization [375] and polymer-immobilized lanthanide complexes for butadiene polymerization [376] have been recently published. The new developments in tailored cationic palladium compounds as catalysts for highly selective dimerization of vinylic monomers [377], immobilized catalysts for propargyl alcohol polymerization to high molecular mass products (2×10^3 mol/g) [378], heterogenization of soluble polymer-supported organotin catalysts of urethane formation [379] are also the evidences for the further progress in this field. Recent developments in catalytic chemistry of supported organoactinides, including complexes immobilized on polymer supports are summarized recently [380].

References

1. Ermakov, Yu.I., Zakharov, B.A. and Kuznetsov, B.N. (1980) *Zakreplennye Kompleksy na Okisnykh Nositelyakh v Katalize* (Complexes Immobilized on Oxides in Catalysis). Novosibirsk: Nauka.
2. Hartley, F.R. (1985) *Supported Metal Complexes*. Reidel: Dordrecht.
3. *Catalytic Polymerization of Olefins* (1986) edited by T.Keii and K.Soga. Amsterdam: Elsevier.
4. Belg. Patent 716,375, 1969.
5. Br. Patent 1,148,650, 1970.
6. Chirkov, N.M., Matkovskii, P.E., Dyachkovskii, F.S. (1976) *Polymerizatsiya na Kompleksnykh Metalloorganicheskikh Katalizatorakh* (Polymerization by Complex Organometallic Catalysts). Moskow: Khimiya.
7. Mazurek, V.V. (1974) *Polymerizatsiya pod Deistviem Soedinenii Perekhodnykh Metallov* (Polymerization Catalyzed by Transition Metal Complexes). Leningrad: Nauka.
8. Kaminsky, W., Kulper, K., Brinstinger, M.H., Wild F.P. (1985) *Angew. Chem.*, **97**, 507.
9. Pomogailo, A.D., Baishiganov, E., Mambetov, U.A., Khvostik, G.M. (1980) *Vysokomol. Soedin.*, **22**, 248.
10. Pomogailo, A.D., Matkovskii, P.E., Zavorokhin, N.D., *et al.* (1967) *Dokl. Akad. Nauk SSSR*, **176**, 1347.
11. Pomogailo, A.D., Matkovskii, P.E., Beikhol'd, G.A. (1971) *Vysokomol. Soedin.*, **13**, 1931.
12. Pomogailo, A.D. (1970) In: *Khimiya Nefti i Neftekhimicheskii Sintez* (Petrochemistry and Petrochemical Synthesis), vol. 2, p.76. Alma-Ata: Nauka.
13. Pomogailo, A.D., Lisitskaya, A.P., Gor'kova, N.S., Dyachkovskii, F.S. (1974) *Dokl. Akad. Nauk SSSR*, **219**, 1375.
14. Pomogailo, A.D., Mambetov ,U.A., Gluzman, E.M., Sokol'skii, D.V.(1978) *Zh. Obshch. Khim.*, **48**, 2101.
15. Mambetov, U.A., Pomogailo A.D., Sokol'skii D.V., *et al.* (1972) In: *Kompleksnaya Pererabotka Mangyshlakskoi Nefti* (Complex Utilization of Mangyshlak's Petroleum), vol. 4, 127. Alma-Ata: Nauka.
16. Mambetov, U.A., Pomogailo, A.D. (1980) *Catalysts Containing Immobilized Complexes*, (Proceedings of the Symposium), (Tashkent, 1980), Part.1, p.143. Novosibirsk: Institute of Catalysis, Siberian Division, USSR Academy of Sciences.
17. Pomogailo, A.D. (1988) *Polimernye Immobilizovannye Metallokomplrksnye Katalizatory* (Polymer-Immobilized Metal Complex Catalysts). Moscow: Nauka.
18. Mambetov, U.A. (1981) Synthesis, Investigations and Catalytic Properties of Immobilized Titanium and Vanadium Complexes in Poly- and Copolymerization of Olefins. *Cand. Sci. Dissertation*, Alma-Ata.
19. Pomogailo, A.D., Baishiganov, E., Khvostik, G.M. (1980) *Kinet. Katal.*, **21**, 1535.
20. Maruyama, K., Tsukube, H., Araki, T. (1980) *J. Polym. Sci., Polym. Chem. Ed.*, **18**, 753.
21. Grudova, V., Radenko, F., Petkov, L., Kircheva, R. (1983) *Neftekhimiya*, **17**, 7.
22. Karayannis, N.M., Lee, S.S., Mangan, D.J. (1987) *J. Appl. Polym. Sci.*, **34**, 1329.
23. Azizov, A.G. (1987) *Usp. Khim.*, **61**, 2098.

24. Suzuki, T., Izuka, S.-I., Kondo, S., Takegami, Y. (1977) *J. Macromol. Sci.* A, **1**, 633.
25. Forte, M.M.C., de Souza, G.A., Quiyada, R. (1984) *J. Polym. Sci., Polym. Lett. Ed.*, **22**, 25.
26. Br. Patent 1,135,976, 1970.
27. Br. Patent 1,055,404, 1969.
28. Fr. Patent 1,405,371, 1969; Fr. Patent 1,518,052 ,1969.
29. U.S. Patent 3,600,367, 1974; U.S. Patent 3,396,155 1973.
30. Belg. Patent 706,659, 1969.
31. Br. Patent 834,217 1968.
32. USSR Inventor's Certificate no. 442,187, 422,255, 420,330, 1976.
33. USSR Inventor's Certificate no. 395,407; 1973.
34. U.S. Patent 3,882,046, 1975; Ind. Patent 137,834, 1973; Br. Patent 1,420,760, 1973; Belg. Patent 794,486 1973; Fr. Patent 2,208,919 1973.
35. Br. Patent 1,161,286, 1970; Br. Patent 1245507, 1972.
36. Jpn. Patent 27242, 1969.
37. Jpn. Patent 38572, 1970; 1,274,697, 1973; Br. Patent 1,381,605, 1975; U.S. Patent 3,642,748, 1973.
38. U.S. Patent 3,595,849, 1973.
39. Fr. Patent 1,550,186, 1970; USSR Inventor's Certificate no. 682262, 1978.
40. USSR Inventor's Certificate no. 925965, 1982.
41. Baishiganov, E., Pomogailo, A.D. (1980) *Complex Organometallic Catalysts for Olefins Polymerization*, p.46. Chernogolovka: Institute of Chemical Physics.
42. Pomogailo, A.D., Lisitskaya, A.P., Dyachkovskii, F.S., (1980) *Complex Organometallic Catalysts for Olefins Polymerization*, p.66. Chernogolovka: Institute of Chemical Physics.
43. Pomogailo, A.D., Lisitskaya, A.P. (1980) *Catalysts Containing Immobilized Complexes*, (Proceedings of the Symposium), (Tashkent, 1980), Part 1, p.147. Novosibirsk: Institute of Catalysis, Siberian Division, USSR Academy of Sciences.
44. Siove, A., Fontanille, M. (1984) *J. Polym. Sci., Polym. Chem. Ed.*, **22**, 3877.
45. Greiber,G., Egle, G. (1962) *Makromol. Chem.*, **53**, 206; **64**, 68.
46. Hulot, H., Landler, Y. (1969) *Kinetics and Mechanism Polyreacts.*, Budopest, 1969, **22**, p.215 (Prepr.).
47. Yamamoto, A., Yamamoto, T. (1978) *J. Polym. Sci., Macromol. Rev.*, **13**, 161.
48. Dobbs, B., Jackson, R., Gaitskell, J.N., *et al.*, (1976) *J. Polym. Sci., Polym. Chem. Ed.*, **14**, 1429.
49. Uvarov, B.A., Tsvetkova, V.I., Litvinova, V.D., *et al.*, (1980) *Catalysts Containing Immobilized Complexes*, (Proceedings of the Symposium), (Tashkent, 1980), Part 1, p.225. Novosibirsk: Institute of Catalysis, Siberian Division, USSR Academy of Sciences.
50. Marchessault, R.H., Chanzy, H.D. (1970) *J. Polym. Sci. C.*, 311.
51. Kritskaya, D.A., Pomogailo, A.D., Ponomarev, A.N., Dyachkovskii, F.S. (1979) *Vysokomol. Soedin., Ser. A*, **21**, 1107.
52. Kritskaya, D.A., Pomogailo, A.D., Ponomarev, A.N., Dyachkovskii, F.S. (1980) *J. Appl. Polym Sci.*, **25**, 349; *J. Polym. Sci., Polym. Symp.*, 23.
53. Pomogailo, A.D., Kritskaya, D.A., Lisitskaya, A.P., *et al.* (1977) *Dokl. Akad. Nauk SSSR*, **232**, 391.

54. Dyachkovskii, F.S., Pomogailo, A.D. (1978) *Catalysts Containing Immobilized Complexes*, Proceedings of the All-Union Conference, p.25. Novosibirsk: Institute of Catalysis, Siberian Division, USSR Academy of Science.

55. Pomogailo, A.D., Dyachkovskii, F.S. (1984) *Acta Polym.*, **35**, 41.

56. Dyachkovskii, F.S., Pomogailo, A.D. (1986) *Proceedings of the V International Symposium on the Connection between Homogeneous and Heterogeneous Catalysis*, vol. 2, Part 1, p.134. Novosibirsk: Institute of Catalysis, Siberian Division, USSR Academy of Science.

57. Pomogailo, A.D., Dyachkovskii, F.S. (1994) *Vysokomol. Soedin.*, **36**, 651.

58. Pomogailo, A.D., Nikitaev, A.T., Dyachkovskii, F.S. (1984) *Kinet. Katal.*, **25**, 166.

59. Ushakova, T.M., Dubnikova, I.L., Meshkova, I.N. (1979) *Vysokomol. Soedin., Ser. A*, **21**, 2713.

60. Chien, J.C.W., Hsieh, J.T.T. (1975) *Coordination Polymerization*, edited by J.C.W. Chien, p.305. New York: Akademic Press.

61. Golub, A.A., Zaitsev, Yu.P., Zazhigalov, V.A. (1981) *Ukr. Khim. Zh.*, **47**, 821.

62. Perkins, P., Vollhardi, K.P.C. (1979) *J. Am. Chem. Soc.*, **101**, 3985.

63. Vinogradov, G.V., Malkin, A.Ya. (1977) *Reologiya Polymerov* (Polymers Reology). Moscow: Khimiya.

64. Markies, P.R., Schat, G., Akkerman, O.S., *et al.* (1989) *J. Organomet. Chem*, **375**, 11.

65. Bochkin, A.M., Pomogailo, A.D., Dyachkovskii, F.S. (1986) *Kinet. Katal.*, **27**, 914.

66. Bochkin, A.M., Pomogailo, A.D., Dyachkovskii, F.S. (1987) *Vysokomol. Soedin., Ser. A*, **29**, 1353.

67. Bochkin, A.M., Pomogailo, A.D., Dyachkovskii, F.S. (1988) *React. Polym.*, **9**, 99.

68. Makhtarulin, S.I., Zakharov, V.A., Moroz, E.M., Maksimov, N.G. (1980) *Catalysts Containing Immobilized Complexes*, (Proceedings of the Symposium), (Tashkent, 1980), Part.2, p.205. Novosibirsk: Institute of Catalysis, Siberian Division, USSR Academy of Sciences.

69. Zakharov, V.A., Bukatova, Z.K., Makhtarulin, S.I., *et al.* (1979) *Vysokomol. Soedin., Ser. A*, **21**, 496.

70. Fuhrmann, H., Wilcke, F.-W., Bredereck, I., *et al.* (1990) *Plaste Kautsch.*, **37**, 145.

71. Jeong, Y., Lee, D., Shiano, T., Soga, K. (1991) *Makromol. Chem.*, **192**, 1727.

72. Gupta, V.K. (1992) *Makromol. Chem.*, **193**, 1043.

73. Xiao, S., Wang, H., Cai, S. (1991) *Makromol. Chem.*, **192**, 1059.

74. Baukova, E.Yu., Bronshtein, L.M., Valetskii, P.E., *et al.* (1992) *Metalloorg. Khim*, **5**, 1386.

75. Negishi, E., Takahashi, T. (1994) *Acc. Chem. Res.*, **27**, 124.

76. Potapov, G.P., Fedorova, E.I., Malygina, O.A., *et al.* (1982) *Vysokomol. Soedin., Ser. B*, **24**, 181.

77. Kabanov, V.A., Smetanyuk, V.I., Popov, V.G., *et al.* (1977) *Complex Organometallic Catalysts for Olefins Polymerization*, Part 8, p.18. Chernogolovka: Institute of Chemical Physics.

78. Kabanov, V.A., Smetanyuk, V.I. (1981) *Makromol. Chem. Suppl.*, **182**, 121.

79. Kabanov, V.A., Smetanyuk, V.I. (1984) *Itogi Nauki Tekhn., Kinet. Katal.*, **13**, 213.

80. Kabanov, V.A., Ivanchev, S.S., Smetanyuk, V.I., *et al.* (1980) *Vysokomol. Soedin., Ser. A*, **22**, 345.

81. Popov, V.G., Fomina, N.M. (1991) *Complex Organometallic Catalysts for Olefins Polymerization*, Part 11, p.59. Chernogolovka: Institute of Chemical Physics.

82. Volodin, V.V., Shupik, A.N., Shapiro, A.M. (1987) *Vysokomol. Soedin., Ser. B*, **29**, 496.

83. USSR Inventor's Certificate no. 682262, 1979.

84. Belg. Patent 719,650, 1969.

85. Dubinina, L.M., Amerik, V.V., Kleiner, V.I., Pomogailo, A.D. (1980) *Catalysts Containing Immobilized Complexes*, (Proceedings of the Symposium), (Tashkent, 1980), Part 1, p.181. Novosibirsk: Institute of Catalysis, Siberian Division USSR Academy of Sciences.
86. Saratovskikh, S.L., Pomogailo, A.D., Babkina, O.N., Dyachkovskii, F.S. (1984) *Kinet. Katal.*, **25**, 464.
87. Br. Patent 1,547,409 1974.
88. Fr. Patent 2,591,602, 1975; Fr. Patent 2,596,397 1975.
89. Pomogailo, A.D. (1981) Immobilization of Metal Complexes on Macromolecular Supports. Catalytic Properties of Immobilized Systems in Polymerization Processes, *Doctorate Dissertation*, Moscow.
90. Chirkov, N.M.,, Matkovskii, P.E. (1974) *Sopolimerizatsiya na Kompleksnykh Katalizatorakh* (Copolymerization Catalyzed by Metal Complexes). Moscow: Nauka.
91. Hermans, J.P. (1970) *III Conference on European Plast. P.*, (Paris, 1970) p.35.
92. USSR Inventor's Certificate no. 687083, 1978.
93. Lunin, A.F., Vaizi, Z.S., Ignatov, V.M., *et al.*, (1981) *Neftekhimiya*, **21**, 199.
94. Appl. 3326839 BRD, 1976.
95. Chien, J.C.W., Dawei, H. (1991) *J. Polym. Sci., Polym. Chem. Ed.*, **29**, 1603.
96. Chien, J.C.W. (1979) *J. Polym. Sci., Polym. Chem. Ed.*, **17**, 2555.
97. Floyd, S., Heiskanen,T., Taylor, T.W., *et al.* (1986) *J. Appl. Polym Sci.*, **32**, 5451; **33**, 1021.
98. Boor, J. (Jr.), Youngman, E.A. (1970) *J. Polym. Sci. A-1*, **4**, 1861.
99. Dzhabjev, T.S., Dyachkovskii, F.S., Shishkina, I.I. (1973) *Izv. Akad. Nauk SSSR, Ser. Khim.*, 1238.
100. Ademy, E.H., Hersberg, J.C., Steinberg, H. (1961) *Recl. Trav. Chim. Pays-Bas. Belg.*, **80**, 173; **81**, 73.
101. Karklin', L.N., Pomogailo, A.D., Lisitskaya, A.P., Borod'ko, Yu.G. (1983) *Kinet. Katal.*, **24**, 657.
102. Furuta, M. (1981) *J. Polym. Sci., Polym. Phys. Ed.*, **19**, 135.
103. L'vovskii, V.E., Fushman, E.A., Dyachkovskii, F.S. (1982) *Zh. Phys. Chem.*, **56**, 1864.
104. Serebryanaya, I.V., Khrushch, N.E., Pomogailo, A.D., Dyachkovskii, F.S. (1986) *Kinet. Katal.*, **27**, 389.
105. Khrushch, N.E., Chukanova,O.M., Dyachkovskii, F.S., Serebryanaya, I.V. (1984) *Proceedings of IV International Symposium on Homogeneous Catalysis*, Part 1, p.195.Leningrad: Nauka.
106. Serebryanaya, I.V., Khrushch, N.E., Leonova, A.G., *et al.* (1990) *Kinet. Katal.*, **31**, 540.
107. Shmonina,V.D., Stefanovskaya, N.N., Tinyakova, E.I., Dolgoplosk, B.A. (1973) *Vysokomol. Soedin., Ser. A*, **15**, 647.
108. Dolgoplosk, B.A., Makovetsskii, K.L., Tinyakova, E.I., Sharaev, O.K. (1968) *Polimerizatsiya Dienov pod Vliyaniem π-Allil'nykh Kompleksov* (Polymerization of Dienes by π-Allyl Complexes). Moscow: Nauka.
109. Dolgoplosk, B.A., Tinyakova, E.I. (1977) *Vysokomol. Soedin., Ser. A*, **19**, 2444.
110. Golubeva, N.D., Pomogailo, A.D., Kuzaev, A.I., Dyachkovskii, F.S. (1977) *Catalysts Containing Immobilized Complexes*, p.87. Novosibirsk: Institute of Catalysis, Siberian Division, USSR Academy of Science.

111. Golubeva, N.D., Pomogailo, A.D., Kuzaev, A.I., *et al.* (1979) *Dokl. Akad. Nauk SSSR*, **244**, 89.
112. Golubeva, N.D., Pomogailo, A.D., Kuzaev, A.I., *et al.* (1980) *J. Polym. Sci., Polym. Symp.*, 33.
113. Dolgoplosk, B.A., Tinyakova, E.I. (1982) *Metallorganicheskii Kataliz v Protsessakh Polimerizatsii* (Organometallic Catalysis in Polymerization Processes). Moscow: Nauka.
114. Kropacheva, E.N. (1973) Transformations in Diene Poymer Chains under Action of Ionic-Type Catalysts, *Doctorate Dissertation*, Leningrad.
115. Ivleva, I.N., Pomogailo, A.D., Echmaev, S.B., *et al.* (1979) *Kinet. Katal.*, **20**, 1282.
116. Vonsovskii, S.V. (1971) *Magnetizm. Magnitnye Svoistva dia-, ferro- and ferrimagnetikov* (Magnetism. Magnetic Properties of Dia-, Ferro- and Ferrimagnetics). Moscow: Nauka.
117. Cooper, W. (1970) *Ind. Eng. Chem., Prod. Res. Dev.*, **91**, 457.
118. Ueno, H., Hakatomi, S., Inoue, T. (1968) *Kogyo Kagaku Zasshi (J. Chem. Soc. Jpn. Ind. Chem. Sect.)*, **71**, 293.
119. Murachev, V.B., Kuz'kina, I.M., Makarov, I.G., *et al.* (1985) *Kinet. Katal.*, **26**, 502.
120. Ivleva, I.N., Echmaev, S.B., Golubeva, N.D., Pomogailo, A.D. (1983) *Kinet. Katal.*, **24**, 663.
121. Fushman, E.A., Borisova, L.F., Shupik, A.N., *et al.* (1982) *Dokl. Akad. Nauk SSSR*, **264**, 651.
122. Kalinina, L.P., Popov, V.G., Smetanyuk, V.I., Kabanov, V.A. (1991) *Complex Organometallic Catalysts for Olefins Polymerization*, Part 11, p.51. Chernogolovka: Institute of Chemical Physics.
123. Hojabre, F. (1976) *Appl. Chem. Biotechnol.*, **26**, 382.
124. Hsieh, L., Yeh, H.C. (1985) *Rubber Chem. Technol.*, **58**, 117.
125. Duvakina, N.V., Marina, N.G., Monakov, Yu.B., Rafikov, S.R. (1989) *Vysokomol. Soedin., Ser. A*, **31**, 227.
126. Sanyagin, A.A., Kormer, V.A. (1985) *Dokl. Akad. Nauk SSSR*, **283**, 1209.
127. Wolff, N.E., Pressley, R.J. (1963) *Appl. Phys. Lett.*, **2**, 152.
128. Metlay, M. (1964) *J. Electrochem. Soc.*, **111**, 1253.
129. Bergbreiter, D.E., Chen, L.-B., Chandran, R. (1985) *Macromolecules*, **18**, 1055.
130. Li, Y., Quang, J. (1988) *Gaofenzi Xuebao (Acta Polym. Sin.)*, 39.
131. Li, Y., Quang ,J. (1987) *J. Macromol. Sci. A*, **24**, 227.
132. Liu, G., Li, Y. (1990) *Acta Polym. Sin.*, 136.
133. Yu, G., Li, Y., Qu, Y., Li, X., (1993) *Macromolecules*, **26**, 6702.
134. Li, Y., Yu, G., Yao, J. (1992) *Acta Polym. Sin.*, 572.
135. Banks, E., Okamoto, Y., Ueba, Y. (1980) *J. Appl. Polym Sci.*, **25**, 359.
136. Okamoto, Y., Ueba, Y., Dzhanibekov, N.F., Banks, E. (1981) *Macromolecules*, **14**, 17.
137. Okamoto, Y., Ueba, Y., Nagata, I., Banks, E. (1981) *Macromolecules*, **14**, 807.
138. Ueba, Y., Banks, E., Okamoto, Y. (1982) *J. Polym. Sci., Polym. Chem. Ed.*, **20**, 1271.
139. Selenova, B.S., Pomogailo, A.D., Raspopov, L.N. (1990) *Complex Organometallic Catalysts for Olefins Polymerization*, Part 11, p.115. Chernogolovka: Institute of Chemical Physics.
140. Xu ,W.-Y., Wang, Y.-S., Zheng, D.-G., Xia, S.-L. (1988) *J. Macromol. Sci. A*, **25**, 1397.

141. Sokolov, V.N., Grebenshchikov, G.K. (1980) *Catalysts Containing Immobilized Complexes,* (Proceedings of the Symposium), (Tashkent, 1980), Part.1, p.221. Novosibirsk: Institute of Catalysis, Siberian Division, USSR Academy of Sciences.
142. Eisen, M.S., Marks, T.J. (1994) *J. Mol. Catal.,* **86**, 23.
143. Shatalov, V.P., Yudin, V.P., Krivoshein, V.V. (1977) *Complex Organometallic Catalysts for Olefins Polymerization,* Part 6, p.98. Chernogolovka: Institute of Chemical Physics, Russian Academy of Sciences.
144. Garmonov, I.V., Kormer, V.A. (1974) *Zh. Vses. Khim. Obva.,* **19**, 628.
145. Pomogailo, A.D. (1991) *Complex Organometallic Catalysts for Olefins Polymerization,* Part 11, p.116 Chernogolovka: Institute of Chemical Physics, Russian Academy of Sciences.
146. Sterenzat, D.E., Davydova, L.M., Dolgoplosk, B.A., *et al.* (1970) *Vysokomol. Soedin., Ser. B,* **12**, 388.
147. Rozenberg, B.A., Irzhak, V.I., Enikolopyan, N.S. (1975) *Mezhtsepnoi Obmen v Polymerakh* (Interchain Interactions in Polymers). Moscow: Khimiya.
148. Poddubnyi, I.Ya., Grechanovskii, V.A. (1964) *Vysokomol. Soedin., Ser. A,* **6**, 64.
149. Novikova, E.V., Balashova, N.I., Polyakova, G.R., *et al.* (1986) *Vysokomol. Soedin., Ser. A,* **28**, 1629.
150. Furukawa, J., Hirai, R., Nakaniva, M. (1969) *J. Polym. Sci. B,* **7**, 671.
151. Lefebvre, G., Chauvin, Y. (1970) Dimerization and Codimerization of Olefins in the Presence of Transition Metal Complexes. In: *Aspects of Homogeneous Catalysis,* edited by R. Ugo, vol.1. Milano: Carlo Manfredi.
152. Fel'dblyum, V.Sh. (1978) *Dimerizatsiya i Disproportsionirovanie Olefiniv* (Dimerization and Disproportionation of Olefins). Moscow: Khimiya.
153. Shmidt, F.K. (1986) *Kataliz Kompleksami Metallov Pervogo Perekhodnogo Ryada Reaktsii Gidrirovaniya i Dimerizatsii* (Catalysis by First Transition Row Metal Complexes. Reactions of Hydrogenation and Dimerization), pp. 158-230. Irkutsk: University Press.
154. Muthukumaru, P.S., Ravindranathan, M., Sivaram, S. (1986) *Chem. Rev.,* **86**, 353.
155. Echmaev, S.B., Ivleva, I.N., Bravaya, N.M., Pomogailo, A.D. (1980) *Kinet. Katal.,* **21**, 1530.
156. Pomogailo, A.D., Khrisostomov, F.A., Dyachkovskii, F:S. (1985) *Kinet. Katal.,* **26**, 1104.
157. Uflyand, I.E., Pomogailo, A.D., Gorbunova, I.O., *et al.* (1987) *Kinet. Katal.,* **28**, 613.
158. Uflyand, I.E., Pomogailo, A.D. (1989) *J. Mol. Catal.,* **55**, 302.
159. Uflyand, I.E., Pomogailo, A.D., Kurbatov, V.P. (1988) *Koord. Khim.,* **14**, 378.
160. Borod'ko, Yu.G., Ivleva, I.N., Echmaev, S.B., *et al.* (1980) *Catalysts Containing Immobilized Complexes,* (Proceedings of the Symposium), Part.1, p.123. Novosibirsk: Institute of Catalysis, Siberian Division, USSR Academy of Sciences.
161. USSR Inventor's Certificate no. 681036; 692822; 827152; 827151, 1979.
162. Carlu, J.-C., Caze, C. (1990) *React. Polym.,* **13**, 153.
163. Kurbatov, V.A., Grishin, G.A., Martynova ,M.A., *et al.* (1985) *Kinet. Katal.,* **26**, 1427.
164. Volodin, V.V., Kalinina, L.P., Shepelin, V.A., *et al.* (1988) *Vysokomol. Soedin., Ser. B,* **30**, 888.
165. U.S. Patent 3737474, .
166. Shmidt, F.K., Mironova, L.V., Proidakov, A.G., *et al.* (1978) *Kinet. Katal.,* **19**, 150.

167. Shvets, V.A., Makhlis, L.A., Vasserberg, V.E., Kazanskii, V.B. (1984) *Kinet. Katal.*, **25**, 1496.
168. Angelescu, E., Angelescu, A., Undrea, M., *et al.* (1987) *Proceedings of the VI International Symposium on Heterogeneous Catalysis*, (Sofia, 1986), Part 2, p.55. Sofia: Academia.
169. Carlu, J.C., Caze, C. (1987) *Proceedings of the Meeting on Macromolecules, XXX Microsymposium· Polymer Supported Organic Reagents and Catalysis*, p.40. Praque: Institute of Macromolecular Chemistry, Czechoslovak Academy of Sciences.
170. Potapov, G.P., Shgelepin, V.A. (1985) *React. Kinet. Catal. Lett.*, **28**, 287.
171. Kabanov, V.A., Martynova, M.A., Pluzhnov, S.K., *et al.* (1979) *Kinet. Katal.*, **20**, 1012.
172. Bravaya, N.M., Pomogailo, A.D., Vainshtein, E.F. (1984) *Kinet. Katal.*, **25**, 1140.
173. Umeda, S. (1980) *Nippon Kagaku Kaishi (J. Chem. Soc. Jpn., Chem. Ind. Chem.)*, 1202.
174. Echmaev, S.B., Ivleva, I.N., Golubeva, N.D., Pomogailo, A.D. (1986) *Kinet. Katal.*, **27**, 394.
175. Tkach, V.S., Shmidt, F.K., Sergeeva, T.N. (1976) *Kinet. Katal.*, **17**, 242.
176. Smith, P.D., Kiendworth, D.D., McDaniel (1987) *J. Catal.*, **105**, 7489.
177. Takahashi H., Okura, J., Keii, T. (1975) *J. Am. Chem. Soc.*, **97**, 7489.
178. Chang, B.-H., Lau, C.-P., Grubbs, R.H., Brubaker, C.H. (1985) *J. Organomet. Chem*, **281**, 213.
179. Kabanov, V.A., Smetanyuk, V.I., Popov, V.G. (1975) *Dokl. Akad. Nauk SSSR*, **225**, 1377.
180. Bersellini, U., Gateilani, M., Chiusoli, G.P., *et al.* (1979) *Fundamental Research on Homogeneous Catalysis, Proceedings of the I International Conference*, (Corpus Chirsti, TX, 1978), vol. 3, p.893, New York: Plenum Publ. Corp.
181. Popov, V.G., Vasil'chenko, S.V. (1988) *Kinet. Katal.*, **29**, 981.
182. Azizov, A.G. (1976) Gomogennye Katalizatory na Nositelyakh: *Review of the Institute of Petroleum Chemical Processes Akad. Nauk Azerb. SSR*, Baku, p.65.
183. Peuckeri, M., Keim, W. (1984) *J. Mol. Catal.*, **22**, 289.
184. Pomogailo, A.D., Gor'kova, N.S., Lokhov, A.A., Dyachkovskii, F.S. (1980) *Complex Organometallic Catalysts for Olefins Polymerization*, Ser.1, Part 8, p.46. Chernogolovka: Institute of Chemical Physics, Russian Academy of Sciences.
185. Kabanov, V.A., Smetanyuk, V.I., Prudnikov, A.I., *et al.* (1987) *Oligomerization of Unsaturated Hydrocarbons*, p.86. Moscow: Topchiev Institute of Petrochemical Synthesis Russian Academy of Sciences.
186. Potapov, G.P., Punegov, V.V., Dzhemilev, U.M. (1985) *Izv. Akad. Nauk SSSR, Ser. Khim.*, 1468.
187. Braca, G., Di Girolamo, M., Raspolli, A.M., Sbrana, G. (1992) *J. Mol. Catal.*, **74**, 421.
188. Zhorov, Yu.M., Panchenkov, G.M., Volokhova, G.S. (1977) *Izomerizatsiya Olefinov* (Isomerization of Olefins), Moscow: Khimiya.
189. Chuvin, Y., Phung, N.-H., Guichard-Loudet, N., Lefebre, G. (1966) *Bull. Soc. Chim. Fr.*, 3223.
190. Barlow, M.G., Bryant, M.J., Naszeldine, R.N., Mackie, A.G. (1970) *J. Organom. Chem*, **21**, 215.
191. Zhorov ,Yu.M., Shelkov, A.V., Panchenkov, G.M., Belov, B.A. (1977) *Neftekhimiya*, **17**, 230.

192. Grigor'ev, A.A., Klimenko, M.Ya., Moiseev, I.I. (1971) *Neftekhimiya*, **11**, 182.
193. Lau, C.-P., Chang, E.-H., Grubbs, R.H., Brubaker, C.H., Jr. (1981) *J. Organomet. Chem*, **214**, 325.
194. Lieto, J., Prochazka, M., Gates, B.C. (1985) *J. Mol. Catal.*, **31**, 89.
195. Ahn, J.H., Ihm, S.K., Park, K.S. (1988) *J. Catal.*, **113**, 434.
196. Guo, L.-P., Liao, S., Yu, D.-R., Guo, H.-F. (1986) *Cuihua Xuebao (J. Catal.)*, **7**, 177.
197. Pienaar, J.J., Plesisis, J.A.K. (1985) *S. Afr. J. Chem.*, **38**, 95.
198. Warmel, S., Buschmeyer, P. (1978) *Angew. Chem., Int. Ed. Engl.*, **17**, 131.
199. Otsu, T., Nagahara, H., Endo, K. (1975) *J. Macromol. Sci. Chem.*, **9**, 1245.
200. Otsu, T., Endo, K., Nagahara, H., Simizu, A. (1975) *Bull. Chem. Soc. Jpn.*, **48**, 2470.
201. Shmidt, F.K., Grechkina, E.A., Levkovskii, Yu.S., *et al.* (1978) *Neftekhimiya*, **18**, 23.
202. USSR Inventor's Certificate no. 594127, 1974.
203. Pomogailo, A.D., Khrisostomov, F.A., Lisitskaya, A.P., *et al.* (1980) *Catalysts Containing Immobilized Complexes*. (Proceedings of the. Symposium), (Tashkent, 1980), Part.1, p.151. Novosibirsk: Institute of Catalysis, Siberian Division, Russsian Academy of Sciences.
204. Kabanov, V.A., Smetanyuk, V.I., Popov, V.G., *et al.* (1980) *Vysokomol. Soedin.*, *Ser. A*, **22**, 335.
205. Enikolopyan, N.S., Raspopov, L.N., Pomogailo, A.D., *et al.* (1984) *Dokl. Akad. Nauk SSSR*, **278**, 1392.
206. Enikolopyan, N.S., Raspopov, L.N., Pomogailo, A.D., *et al.* (1989) *Vysokomol. Soedin., Ser. A*, **31**, 2624.
207. Pomogailo, A.D., Golubeva, N.D., Ivleva, I.N., *et al.* (1984) *Kinet. Katal.*, **25**, 1145.
208. Barazzoni, L., Menconi, F., Ferrero, C., *et al.* (1993) *J. Mol. Catal.*, **82**, 17.
209. Surkov, N.F.,Davtyan, S.P., Pomogailo, A.D., Dyachkovskii, F.S. (1986) *Kinet. Katal.*, **27**, 714.
210. Uflyand, I.E., Pomogailo, A.D., Golubeva, N.D., Starikov, A.G. (1988) *Kinet. Katal.*, **29**, 885.
211. Sinn, E., Narris,C.M. (1969) *Coord. Chem. Rev.*, **47**, 391.
212. Likholobov, V.A., Lisitsyn, A.S. (1989) *Zh. Vses. Khim. O-va*, **34**, 340.
213. Endo, K., Ueda, R., Otsu, T. (1991) *Macromolecules*, **24**, 6849; *Polymer J.*, **23**, 1173.
214. Beach, D.L., Kissin, Y.V. (1984) *J. Polym. Sci., Polym. Chem. Ed.*, **22**, 3027.
215. Nowlin, T.E., Kissin, Y.V., Wagner, K.P. (1988) *J. Polym. Sci., Polym. Chem. Ed.*, **26**, 755.
216. Yang, X. (1984) *Cuihua Xuebao (J. Catal.)*, **5**, 185.
217. Baulin, A.A., Chernykh, A.I., Rezznikova, O.N., Stanotkin, A.M. (1985) *Plast. Massy*, 6.
218. Mattise, W.L., Stehling, F.C. (1981) *Macromolecules*, **14**, 1479.
219. Kunn, R., Krömer, H. (1982) *Colloid. Polym. Sci.*, **260**, 1083.
220. Valenza, A., La.Mantia, F.P., Acierno, D. (1988) *Eur. Polym. J.*, **24**, 81.
221. Krause, J. (1984) *Acta Polym.*, **35**, 515.
222. Briston, J.H. (1982) *Converter*, **19**, 11.
223. Hsich, E.T., Randall, J.C. (1982) *Macromolecules*, **15**, 353.

224. Usami, T., Takayama, S. (1984) *Polym. J.*, **16**, 731.
225. Chukanov, N.V., Kumpanenko, I.V., Grigoryan, E.A., *et al.* (1989) *Vysokomol. Soedin., Ser. B*, **31**, 423.
226. Housaki, T., Saton, K. (1988) *Makromol. Chem. Rapid. Commun.*, **9**, 525.
227. Kimura, K., Shigemura, T., Yuasa, S. (1984) *J. Appl. Polym. Sci.*, **29**, 3161.
228. Yamamoto, K. (1982) *J. Macromol. Sci. Chem.*, **17**, 415.
229. Maderek, E., Strobl, G.R. (1983) *Colloid. Polym. Sci.*, **261**, 471.
230. Bubeck, R.A., Baker, H.M. (1982) *Polymer*, **23**, 1680.
231. USSR Inventor's Certificate no. 1117294, 1984.
232. Potapov, G.P., Koshkarova, A.A., Punegov, V.V., *et al.* (1987) *Khimiya Elementoorganicheskikh Soedinenii* (Chemistry of Organoelemnt Compounds), p.43. Gorky: Gorky State University.
233. Kabanov, V.A., Smetanyuk, V.I., Kalinina, L.P., *et al.* (1988) *Kinet. Katal.*, **29**, 984.
234. Potapov, G.P., Poluboyarov, V.A., Punegov, V.V. (1988) *Kinet. Katal.*, **29**, 679.
235. Schuchardt, U., Santos, D.F. (1985) *J. Mol. Catal.*, **29**, 145.
236. Moberg, C., Rakos, L. (1987) *J. Organomet. Chem*, **335**, 125.
237. Bouchal, K., Pokorny, S. (1975) *Makromol. Chem.* , **176**, 1679.
238. Pittman, C.U. (Jr.), Wuu, S.K., Jacobson, S.E. (1976) *J. Catal.*, **44**, 87.
239. Inaki, Y. (1975) *Prog. Polym. Sci. Jpn.*, **8**, 244.
240. Mizaki, T., Otsu, T. (1971) *J. Chem. Soc. Jpn.*, **74**, 1005.
241. Nishikawa, Y., Otsu, T. (1969) *Makromol. Chem.*, **128**, 276.
242. Otsu, T., Minamii, N., Nishikawa, Y. (1968) *J. Macromol. Sci., Chem.*, **2**, 905.
243. Brotter, M.A., Beilin, V.V., Andreeva, E.D. (1981) *Tekhnologiya Vysokomolekulyarnykh Soedinenii* (Technological Aspects of Macromolecular Compounds). Leningrad: Teknological Institute (Available from VINITI, no. 1-XP481).
244. Nishiuchi, T., Segawa, S., Yasuoka, S. (1980) *Nippon Kagaku Kaishi (J. Chem. Soc. Jpn. Chem. Ind. Chem.)*, 1269.
245. Narita, H., Shintani, I., Araki, M., Machida, S. (1977) *Bull. Fac. Textile Sci. Kyoto Univ. Ind. Arts Textile Fibers*, **8**, 83.
246. Ouchi, T., Beika, N., Imoto, M. (1982) *Eur. Polym. J.*, **18**, 725.
247. Wu, J.-Y., Yang, C.-X., Liang, Z.-M. (1987) *Cuihua Xuebao (J. Catal.)*, **8**, 293.
248. Pashchenko, D.I., Vinogradova, E.K., Bel'govskii, I.M., *et al.* (1982) *Dokl. Akad. Nauk SSSR*, **265**, 889; (1983) *Proceedings of III All-Union Conference on Chemistry and Biochemistry of Porphyrines, (Samarkand*, 1982), p.77. Moscow: Nauka.
249. Potapov, G.P., Alieva, M.I., Fedorova, E.I., Mitusov, A.A. (1984) *Vysokomol. Soedin., Ser. B*, **26**, 819.
250. Simionesku, Cr., Mazulu, I., Cocirla, I., Tataru, L., Lixandru, T. (1988) *Eur. Polym. J.*, **24**, 221.
251. Godard, P., Wertz, J.L., Biebuyck, J.J., Mercier, J.P. (1989) *Polym. Eng. Sci.*, **29**, 127.
252. Sangalov, Yu.A., Gladkikh, I.F., Minsker, K.S. (1982) *Vysokomol. Soedin., Ser. B*, **26**, 356.
253. Naik, K.M., Sivaram, S. (1977) *Proceedings of International Symposium on Macromolecules,* (Dublin, 1977), vol. 2, p. 481. Amsterdam: Elsevier.
254. Ran, R., Jia, Y., Jiang, S. (1987) *Shiyou Huagong (Petrochem. Technol.)*, 15.

255. Kakuliya, Ts.V., Khananashvili, L.M., Tsomaya, N.I., *et al.* (1988) *Izv. Akad. Nauk Gruz.SSR*, **130**, 557.
256. Ran, R.-C, Jia, X.-R., Li, M.-Q., Jiang, S.-J. (1988) *J. Macromol. Sci., Chem.*, **25**, 907.
257. Biswas, M. (1989) *Eur. Polym. J.*, **25**, 981.
258. Potapov, G.P. (1983) *React. Kinet. Catal. Lett.*, **23**, 307.
259. Potapov, G.P., Shishova, T.L. (1982) *React. Kinet. Catal. Lett.*, **21**, 361.
260. Zhao, J., Yang, M., Zheng, Y., Shen, Z. (1991) *Macromol. Chem.*, **192**, 309.
261. Chien, J.C.W., Salajka, Z. (1991) *J. Polym. Sci., Polym. Chem. Ed.*, **29**, 1243; 1253.
262. Solf, I., Herrmann, L. (1971) *Plaste Kautsch.*, **18**, 243.
263. Mazurek, V.V. (1980) *Kataliz i Mekhanizm Reaktsii Obrazovaniya Polymerov* (Catalysis and Mechanism of the Reactions of Polymers Formation), p. 41. Kiev: Naukova Dumka.
264. Iwhau, C., Soo, S.S. (1989) *Makromol. Chem. Rapid. Commun.*, **10**, 85.
265. Pomogailo, A.D., Baishiganov, E., Dyachkovskii, F.S. (1981) *Vysokomol. Soedin., Ser. A*, **23**, 220.
266. Simonesku, Kh., Dumitriesku, S., Negulesku, I., *et al.* (1974) *Vysokomol. Soedin., Ser. A*, **16**, 790.
267. Chauser, M.G., Rodionov, Yu.M., Misin, V.M., Cherkashin, M.I. (1976) *Usp. Khim.*, **45**, 695.
268. Kiyashkina, Zh.S., Pomogailo, A.D., Kuzaev, A.I., *et al.* (1977) *Catalysts Containing Immobilized Complexes*, p. 111. Novosibirsk: Institute of Catalysis, Siberian Division, USSR Academy of Science,
269. Kiyashkina, Zh.S., Pomogailo, A.D., Kuzaev, A.I., *et al.* (1980) *J. Polym. Sci., Polym. Symp.*, 13.
270. Pomogailo, A.D., Kiyashkina, Zh.S., Kuzaev, A.I., *et al.* (1985) *Vysokomol. Soedin., Ser. A*, **27**, 707.
271. Ivleva, I.N., Echmaev, S.B., Pomogailo, A.D., *et al.* (1977) *Dokl. Akad. Nauk SSSR*, **233**, 903.
272. Opalovskii, A.A., Tychinskaya, I.I., Kuznetsova, Z.M. (1972) *Galogenidy Molibdena* (Molybdenum Halides). Novosibirsk: Nauka.
273. Echmaev, S.B., Ivleva, I.N., Pomogailo, A.D., *et al.* (1983) *Kinet. Katal.*, **24**, 1428.
274. Kiyashkina, Zh.S., Pomogailo, A.D., Kuzaev, A.I., *et al.* (1979) *Vysokomol. Soedin., Ser. A*, **21**, 1796.
275. Karklin`, L.N., Pomogailo, A.D. (1984) *Izv. Akad. Nauk Latv. SSR, Ser. Khim.*, 218.
276. Zingg, D.S., Makovsky, L.E., Tischer, R.E., *et al.* (1980) *J. Phys. Chem.*, **84**, 2898.
277. Hamer, A.D., Walton, R.A. (1974) *Inorg. Chem.*, **13**, 1446.
278. Woon, P.S., Farona, M.F. (1974) *J. Polym. Sci., Polym. Chem. Ed.*, **12**, 1749.
279. Masuda T., Mouri T., Higoshimura T. (1980) *Bull. Chem. Soc. Jpn.*, **53**, 1152.
280. Kaminska, G., Hocker, H. (1988) *Makromol. Chem.*, **189**, 1447.
281. Temkin, O.N., Flid, R.M. (1968) *Kataliticheskie Prevrashcheniya Atsetilenovykh Soedinenii v Rastvorakh Kompleksov Metallov* (Catalytical Transformations of Acetylene Compounds in Solutions of Metal Complexes). Moscow: Nauka.
282. Toit C., J., Plessis, J.A.K., Lachmann G. (1985) *S. Afr. J. Chem.*, **38**, 8.
283. Luksha, V.G., Potapov, G.P., Koshkarova, A.A., *et al.* (1981) *Khimiya Elementoorganicheskikh Soedinenii* (Chemistry of Organoelement Compounds), p.80. Gorky: Gorky State University.

284. Editors comments (1973), *Chemtech.*, **3**, 560.
285. Simionescu, C.I., Bulacovschi, V., Grovu-Ivanoui, M., Stanciu, A. (1987) *J. Macromol. Sci., Chem.*, **24**, 611.
286. Vatanathamm, S., Farona, M.F. (1980) *J. Mol. Catal.*, **7**, 403.
287. Simionescu, C.I., Percec, V., Dumitrescu, S. (1977) *J. Polym. Sci., Polym. Chem. Ed.*, **15**, 2497.
288. Maitlis, P.M. (1972) *Zh. Vses. Khim. Ova.*, **17**, 403.
289. Maitlis, P.M. (1980) *J. Organomet. Chem*, **200**, 161.
290. Furlani, A., Napoletano,C., Russso, M.V. *et al.* (1989) *J. Polym. Sci., Polym. Chem. Ed.*, **27**, 75.
291. Cataldo, E. (1992) *Polymer*, **33**, 3073.
292. Kaneda, K., Uchiyama, T., Terasawa, M., *et al.* (1976) *Chem. Lett.*, 449.
293. Simonescu, C.I., Percec, V. (1980) *J. Polym. Sci., Polym. Chem. Ed.*, **18**, 147.
294. Orlov, I.G., Kutafina, N.S., Golubenko, L.I., Cherkashin, M. (1981) *Izv. Akad. Nauk SSSR, Ser. Khim.*, 291.
295. Chung, T.-C., Mac Diarmid, A.G., Feldblum, A., Heeger, A.J. (1982) *J. Polym. Sci., Polym. Lett. Ed.*, **20**, 427.
296. Galvin, M.E., Wnek, G.E. (1983) *J. Polym. Sci., Polym. Chem. Ed.*, **21**, 2727.
297. Galvin, M.E., Wnek, G.E. (1982) *Polym. Prepr. Am. Chem. Soc., Div. Polym. Chem.*, **23**, 99; (1984) **25**, 229.
298. Kozub, G.I., Pomogailo, A.D., Tkachenko, L.I., *et al.* (1988) *Proceedings of the VII All-Union Conference Complexes with Charge Transfer and Ion-Radical Salts*, p.31. Chernogolovka: Institute of Chemical Physics, Russian Academy of Sciences.
299. Cukor, P., Krugler, J.J., Rubner, M.F. (1980) *Polym. Prepr. Am. Chem. Soc., Div. Polym. Chem.*, **21**, 161.
300. Rubner, M.F., Tripathy, S.K., Georger, J (Jr.), Choleva, P. (1983) *Macromolecules*, **16**, 870.
301. Sichhel, E.K., Rubner, M.F. (1985) *J. Polym. Sci., Polym. Phys. Ed.*, **23**, 1629.
302. Lee, K.I., Jopson, H. (1983) *Makromol. Chem. Rapid. Commun.*, **4**, 375.
303. Destri, S., Catellani, M., Bolognesi, A. (1984) *Makromol. Chem. Rapid. Commun.*, **5**, 353.
304. Ojio, T., Miyata, S. (1986) *Polym. J.*, **18**, 95.
305. Niiwa, O., Kakuchi, M., Tamamura, T. (1987) *Macromolecules*, **20**, 749.
306. Osterholm, J.E., Yasuda, H.K., Levenson, L.L. (1982) *J. Appl. Polym. Sci.*, **27**, 931.
307. Kminek, I., Trekoval, J. (1984) *Makromol. Chem. Rapid. Commun.*, **5**, 53.
308. Miyauchi, S., Sorimachi, Y., Mitsui, M., *et al.* (1984) *J. Appl. Polym. Sci.*, **29**, 251.
309. Enikolopyan, N.S., Galashina, N.M., Shklyarova, E.I., *et al.* (1988) *Vysokomol. Soedin., Ser. B*, **30**, 867; (1984) *Dokl. Akad. Nauk SSSR*, **278**, 620.
310. Petts, R.W., Wengh, K.C. (1982) *Polymer*, **23**, 897.
311. Dykh, Zh.L., Lafer, L.I., Yakerson, V.I., Rubinshtein, A.I. (1980) *Catalysts Containing Immobilized Complexes*, (Proceedings of the Symposium), (Tashkent, 1980), Part.1, p. 41. Novosibirsk: Institute of Catalysis, Siberian Division, Russian Academy of Sciences.
312. Korshak, V.V. (1982) *Usp. Khim.*, **51**, 2096.
313. Kopylov, V.V., Pravednikov, A.N. (1968) *Vysokomol. Soedin., Ser. A*, **10**, 1794.
314. Tsuruya, S., Kuse, T., Masai, M., Imamura, S.-I. (1981) *J. Mol. Catal.*, **10**, 285.

315. Challa ,G. (1981) *Makromol. Chem. Suppl.*, **182**, 70.
316. Tsuchida, E., Nishikawa, H., Terada, E. (1976) *J. Polym. Sci., Polym. Chem. Ed.*, **14**, 825.
317. Tsuchida, E., Kaneko, M., Nishide, H., Nishi, S. (1971) *Kogyo Kagaku Zasshi (J. Chem. Soc. Jpn. Ind. Chem. Soc.)*, **74**, 2370; (1973) *Makromol. Chem.*, **164**, 203.
318. Tsuchida, E., Nischida, H. (1980) *Preprints of International Symposium on Macromolecules* (Pisa, 1979) vol. 4, p. 147.
319. Tsuchida, E., Nishide, H., Nishikawa, H. (1974) *J. Polym. Sci., Polym. Symp.*, 47.
320. Verlaan, J.P.J., Alferink, P.J.T., Challa, G. (1984) *J. Mol. Catal.*, **24**, 235.
321. Verlaan, J.P.J., Bootsma, J.P.C., Challa, G. (1982) *J. Mol. Catal.*, **20**, 203.
322. Breemhaar, W., Meinders, H.C., Challa, G. (1981) *J. Mol. Catal.*, **10**, 33.
323. Tsuchida, E., Nishide, H. (1977) *Adv. Polym. Sci.*, **24**, 1.
324. Kaneko, M., Manecke, G. (1974) *Makromol. Chem.*, **175**, 2795.
325. Jpn Patent 55-106,814, 1979.
326. Tsuchida, E., Nishide, H. (1980) *Modification of Polymers*, edited by C.E.Carraher, Jr., M.Tsuda, (ACS Symp. Ser.) No. 147. Washington, DC: American Chemical Society.
327. Bied-Charreton ,C., Frostin-Rio, M., Pujol, D., *et al.* (1982) *J. Mol. Catal.*, **16**, 335.
328. Tsuchida, E., Kaneko, M., Kurimura, Y. (1970) *Makromol. Chem.*, **132**, 209; (1972) **155**, 45.
329. Korshak, V.V., Kasatochkin, V.I., Sladkov, A.M., *et al.* (1961) *Dokl. Akad. Nauk SSSR*, **136**, 1342.
330. Sorokin, M.F., Gershanova, E.L., Mikhitarova, Z.L., Lebedeva, A.A. (1982) *Lakokras. Mater. Ikh Primen.*, 5.
331. Shukla, A.K., Karkozov, V.G., Nikolaev, A.F. (1979) *Khimicheskaya Tekhnologiya: Svoistva i Primenenie Plastmass* (Chemical Technology: Properties and Utilization of Plastics), p. 59. Leningrad: Nauka.
332. Kurnoskin, A.V., Kanovich, M.Z. (1992) *Plast. Massy*, 28.
333. Shibalovich, V.G., Belgorodsskaya, K.V., Karakozov, V.G. (1989) *Plast. Massy*, 18.
334. Kurnoskin, A.V., Lapitskii, V.A. (1990) *Plast. Massy*, 65.
335. Maceil, G.E., Chuang, I.-S., Gollob, L. (1984) *Macromolecules*, **17**, 1081.
336. Uretskaya, E.A., Akutin, M.S., Pomogailo, A.D., *et al.* (1984) *Proceedings of the IV International Symposium on Homogeneous Catalysis*, (Leningrad 1984), vol. 1, p. 102. Chernogolovka: Institute of Chemical Physics, Russian Academy of Sciences.
337. Uretskaya, E.A., Fushman, E.A., Shupik, A.N., *et al.* (1984) *Proceedings of III All-Union Conference on Problems of Solvatation and Complex Formation in Solutions*, (Ivanovo 1983), vol. 2, p.330. Ivanovo: Ivanovo State University.
338. Khrustaleva, E.A., Kochneva, M.A., Fridman, L.I., *et al.* (1984) *Plast. Massy*, 6.
339. Yoneyama, M., Kakimoto, M.-A., Imai, Y. (1989) *J. Polym. Sci., Polym. Chem. Ed.*, **27**, 1985.
340. Morgan, P.U. (1970) *Polycondensation Processes for Synthesis of Polymers*, p. 242. Leningrad: Khimiya.
341. Barshtein, R.S., Sorokina, I.A. (1988) *Kataliticheskaya Polikondensatsiya* (Catalytical Polycondensation). Moscow: Khimiya.

342. Lipatova, T.E. (1974) *Kataliticheskaya Polimerizatsiya Oligomerov i Formirovanie Polimernykh Setok* (Catalytical Polymerization of Oligomers and Formation of Polymer Networks). Kiev: Naukova Dumka.

343. Jpn. Patent 58-144,360, 1972.

344. Solozhenko, E.G., Panova, G.V. (1986) *Zh. Obshc. Khim.*, **56**, 410.

345. Nizel'skii, Yu.N. (1983) *Kataliticheskie Svoistva β-Diketonatov Metallov* (Catalytic Properties of Metal β-Diketonates). Kiev: Naukova Dumka.

346. Nizel'skii, Yu.N., Ishchenko, S.S., Zapunnaya, K.V. (1983) *Ukr. Khim. Zh.*, **49**, 81; (1988) *Dokl. Akad. Nauk Ukr. SSR, Ser. B*, 51.

347. Voropaev, N.L., Nizel'skii, Yu.N., Lipatova, T.E. (1983) *Vysokomol. Soedin., Ser. A*, **25**, 2320.

348. Olezyk, W. (1981) *Polymery*, **26**, 247.

349. Berlin, P.A., Levina, M.A., Tiger, R.P., Entelis, S.G. (1989) *Vysokomol. Soedin., Ser. A*, **31**, 519.

350. Ptitsyna, N.V., Berlin, P.A., Tiger, R.P., Entelis, S.G. (1990) *Proceedings of the IV All-Union Conference on Chemistry and Physico-Chemistry of Oligomers*, (Nalchik, 1989), p. 63. Chernogolovka: Institute of Chemical Physics, Russian Academy of Sciences.

351. Berlin, P.A., Levina, M.A., Tiger, R.P., Entelis, S.G. (1989) *Vysokomol. Soedin., Ser. A*, **31**, 519.

352. Lipatov, Yu.S., Kosyanchuk, L.F., Kosenko, A.L., Nizel'skii, Yu.N., Rosovitskii, V.F. (1992) *Vysokomol. Soedin., Ser. A*, **34**, 129.

353. Arshady, R., Basato, M., Corain, B., *et al.* (1989) *J. Mol. Catal.*, **53**, 111.

354. Davletbaeva, I.M., Khramov, A.S., Frolov, E.N., Kuzaev, A.I., Tselikova, E.P., Delo, M. (1994) *Zh. Prikl. Khim.*, **67**, 254; 258.

355. Jpn. Patent 62-115,018, 1976.

356. Watanabe, K., Imazawa, A. (1982) *Bull. Chem. Soc. Jpn.*, **55**, 3208.

357. Zitsmanis, A.Kh. (1984) *Proceedings of the IV All-Union Conference on Organic Synthesis*, p. 122. Moscow: Nauka.

358. Mikaiyama, T., Iwakiri, H. (1985) *Chem. Lett.*, 1366.

359. Trumbo, D.L., Marvel, C.S. (1987) *J. Polym. Sci., Polym. Chem. Ed.*, **25**, 847.

360. Bassedas, M., Carreras, H., Castells, J. Loper-Calaharra, F. (1987) *React. Polym.*, **6**, 109.

361. Klyavin'sh, M.K., Raska, A.S., Skuin'sh, A.G., *et al.* (1990) *Zh. Prikl. Khim.*, **63**, 729.

362. Akutin, M.S., Tagil'tseva, R.B., Tarakhtunov, O.A. (1972) *Plast. Massy*, 4.

363. Jerabek, K., Stejber, J. (1979) *Proceedings of the IV International Symposium on Heterogeneous Catalysis*, (Sofia, 1978), Part 1, p. 481. Sofia: Akademia.

364. Pomogailo, A.D. (1989) *Advances in Interpenetrating Polymer Networks*, edited by. D. Klempner and K.C. Frish, p. 145. Detroit: Polymer Technology Inc.

365. USSR Inventor's Certificate no. 395407, 1973.

366. Baishiganov, E., Pomogailo, A.D., Raspopov, L.N. (1982) *Complex Organometallic Catalysts for Olefins Polymerization*, Part 8, p. 53. Chernogolovka: Institute of Chemical Physics, Russian Academy of Sciences.

367. Pomogailo, A.D., Torosyan, A.A., Raspopov, L.N., *et al.* (1980) *Dokl. Akad. Nauk SSSR*, **252**, 1148.

368. Pomogailo, A.D., Baishiganov, E., Torosyan, A.A., Raspopov, L.N. (1983) *Polymer-Polymernye Sinteticheskie Sistemy*, Prepr., 24. Chernogolovka: Institute of Chemical Physics Russian Academy of Sciences

369. Cansell, F., Siove, A., Belorgey, G. (1990) *J. Polym. Sci., Polym. Phys. Ed.*, **28**, 647.
370. Pomogailo, A.D., Dyachkovskii, F.S., Enikolopyan, N.S. (1985) *Polymerization Processes in Heterogeneous Systems*, ONPO "Plastpolimer", p. 108. Leningrad: Plastpolymer.
371. Lipatov, Yu.S. (1977) *Fizicheskaya Khimiya Napolnennykh Polimerov* (Physical Chemistry of Filled Polymers), p. 196. Moscow: Khimiya.
372. Lupinovich, L.N., Mamin, Kh.A., Orekhova, G.I. (1985) *Zh. Prikl. Khim.*, **58**, 1425.
373. Sun, L., Hsu, C.C., Bacon, D.W. (1994) *J. Polym. Sci., Polym. Chem. Ed.*, **32**, 2127; 2135.
374. Lee, D.-L., Yoon, K.-B. (1994) *Makromol. Rapid. Commun.*, **15**, 841.
375. Vizzini, J.C., Chien, J.C.W., *et al.* (1994) *J. Polym. Sci., Polym. Chem. Ed.*, **32**, 2049.
376. Yu, G., Li, Y., Qu, Y. (1993) *Macromolecules*, **26**, 6702.
377. Zhaothong, J., Aynsman, S. (1993) *Organometallics*, **12**, 1406.
378. Yang, M., Zheng M., Furlani, A., Russo, M.V. (1994) *J. Polym. Sci., Polym. Chem. Ed.*, **32**, 2709.
379. Ptitsyna, N.V., Tiger R.P., Entelis S.G., Berlin P.A. (1995) *Proceedings of the 3 International Symposium on Polymers Advanced Technologies* (Pisa, 1995), p.210. Pisa: University of Pisa.
380. Eisen, M.S., Marks, T.J. (1994) *J. Mol. Catal.*, **86**, 23.

CHAPTER 4

ACTIVATION OF SMALL STABLE MOLECULES AND STIMULATION OF CATALYTIC REACTIONS THROUGH THE PARTICIPATION OF IMMOBILIZED COMPLEXES

Some processes analyzed in this chapter are similar to those considered in previous chapters. For instance, CO hydrogenation is a kind of oxidative polymerization and hydroformylation (addition of CO and H_2 to multiple bonds of hydrocarbons) has many features in common with hydrogenation, etc. Introduction of oxygen to olefin molecules can be carried out not only by means of O_2 or H_2O_2, but also by means of other diverse reagents including water and carbon monoxide. Reactions of other types, for instance the activation of thermodynamically stable molecules N_2 and H_2O can proceed via untypical routes.

These processes are considered because of the prospect of using the products obtained as synthetic liquid fuel, organic half-products and cheap chemical compounds. These processes are applied to obtain alcohols (mainly methanol and ethanol) and gaseous hydrocarbons (paraffins C_1–C_4) from gas evolved on gasification of solid (coke, coal, biomass) or liquid (asphaltenes, heavy oils) fuels, from natural gas or oil by treatment with steam, or from waste gases of metallurgic plants (converters, coke ovens). Mechanisms of these reactions in the presence of homogeneous and heterogeneous catalysts are known rather well; the search for active catalysts continues.

In this chapter the effectiveness of two conventional methods of stimulation of reactions with participation of immobilized complexes, photoactivation and electrochemical stimulation, are compared.

4.1 Hydrogenation of Carbon Monoxide (the Fischer–Tropsch Synthesis)

Hydrogenation of CO is described by the equation

$$nCO + (2n + 1)H_2 \rightarrow C_nH_{2n+2} + nH_2O$$

and can be considered as hydrocondensation. Fe, Co and Ru complexes are the best catalysts of the reaction; the kinetics and mechanism of the reaction in the presence of homogeneous catalysts are rather well known [1–4]. The reaction proceeds by a two-step mechanism: addition of H^- to a CO molecule bound to a metal atom, then addition of H^+ to the intermediate; incorporation of CO into an M–H bond is also possible.

Alcohols and olefins are primary products of the reaction; the steps of the reaction are typical of homogeneous catalysis: coordination of CO and olefins, oxidative addition and reductive elimination, incorporation and β-transfer of hydrogen. The nature of the bond CO–metal ion, activation of this bond, migratory incorporation and dissociative addition are important factors.

Industrial catalysts consist of iron ions supported on Al_2O_3, SiO_2 or kieselgur promoted by salts of alkali metals; activity of industrial catalysts is 0.4–0.5 kg of CO/kg of Fe h. In recent years immobilized catalysts have also been applied to CO hydrogenation.

Data on selectivity and activity of supported transition metals of the first row in the Fischer–Tropsch reaction (Figure 4.1) [5] are important for optimization of catalysts. As a rule, only one parameter is optimized, sometimes at the expense of other parameters. Cyclopentadienyl cobalt dicarbonyl supported on macro- and microporous CSDVB (copolymer of styrene with divinylbenzene; 1–3% of DVB) with grafted cyclopentadiene (Cp) rings [6] seems to be the first catalyst of this type. At temperatures of 463–473 K this catalyst produces methane and higher hydrocarbons in the ratio 92/8; supported monocarbonyl complexes have the highest activity (up to 0.13 mol of CO/mol of Co h). The catalyst can be recycled many times; homogeneity and nuclearity of the catalyst are not changed in the course of the reaction. H–D exchange and deuteroexchange in polymer chains (products are CD_4, deuterated high hydrocarbons, D_2O and traces of CD_3H) testifies to formation of cobalt hydrides as active centers.

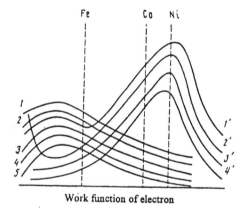

Figure 4.1 Main parameters for CO hydrogenation in the presence of supported transition metals of the first row. (1) Formation of CO_2, (2) mean molecular mass of products, (3) formation of coke, (4) yield of olefins, (5) formation of oxygen-containing products.
(1') Activity, (2') formation of CH_4, (3') yield of paraffins, (4') formation of H_2O.

Work function of electron

Probably, incorporation of a CO molecule into a formed Co–H bond is the first stage of chain initiation[1] (Scheme 4.1).

This process is a kind of polymerization, therefore macromolecular chains are involved in regular reactions of growing macromolecules:

$$L_n\text{--}M\text{--}(CH_2CH_2)_m\text{--}R \xrightarrow{CO + H_2} \begin{cases} \overset{a}{\longrightarrow} L_{n-1}\text{--}M\text{--}\overset{\|}{\underset{O}{C}}\text{--}(CH_2CH_2)_m\text{--}R \\ \overset{b}{\longrightarrow} L_n\text{--}M\text{--}H + CH_2\text{=}CH\text{--}R \\ \overset{c}{\longrightarrow} L_{n-1}\text{--}M\text{--}H + CH_3\text{--}CH_2\text{--}(CH_2\text{--}CH_2)_{m-1}\text{--}R \end{cases}$$

---

[1] A new mechanism of the Fischer–Tropsch polymerization in the presence of Ru complexes was proposed in review [7].

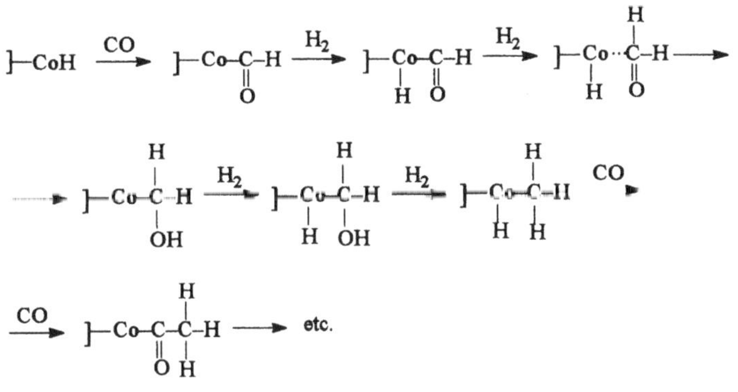

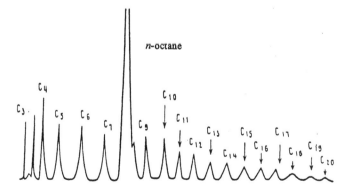

Scheme 4.1

Incorporation of CO increases the chain length by one link (a), proton elimination results in formation of α-olefin and metal hydride (probably the active center is regenerated in this reaction) (b), hydrogenation of metal–alkyl bond results in formation of a paraffin (c).

It was shown for model systems that cobalt crystallites and heterogeneous clusters are not responsible for catalyst activity [6]. It is interesting that soluble analogues of these systems are inactive because they are unstable under the conditions of the reaction. Supported complexes are more stable; probably, the specific surroundings of complexes in macroporous polymers leads to higher stability of supported complexes and prevents the loss of cobalt in the course of vacuum pyrolysis (458 K, 10^{-3} torr, 112 h). The primary products have the typical mass distribution of linear alkanes in accordance with the Schultz–Flory equation (Figure 4.2):

$$\log(m_P/P) = \log(\ln^2\alpha) + P\log\alpha$$

where m_P is the mass share proportional to the number of molecules with polymerization degree P and α is the probability of chain growth.

Figure 4.2 The product distribution of hydrocarbons obtained from CO and H_2 in the presence of cobalt carbonyl supported on CSDVB modified by Cp groups. Solvent n-octane; 100 h.

The Fischer–Tropsch process occurs at high temperatures (above 470 K), therefore cross-linked and thermostable polymers are used as supports; Al, Li or Na, Mg and Zn compounds, analogues of industrial catalysts, are often introduced as activators into these polymers.

For instance, catalysts of hydrocondensation of carbon monoxide are obtained [8] by immobilization of complexes of Co, Fe, Ni, Ru, Rh or Mo with activators and various donor–acceptor additions on porous CSDVB with the specific surface S_{sp}=60–250 m^2/g. Hydroconversion is carried out at 430–570 K and 0.5–40 MPa in decalin, tetralin or tetraglyme.

In addition to the nature of the catalyst, the content of products is affected by temperature, the basicity of support, the concentration of promoters and the composition of the gaseous mixture. These parameters affect also the selectivity of the process: the higher the basicity, the higher the mean molecular mass of products; an increase of pressure of synthesis gas leads to a rise in the share of hydrocarbons; the selectivity with respect to lighter hydrocarbons is increased with increasing temperature. For instance, CH_4 and CO_2 are major products of the reaction carried out at 570–620 K and catalyzed by Fe, Co and Ni salts supported on a poly(benzimidazole) matrix, reduced complexes of Ni being most active [9]. Methanol is produced under milder conditions (temperature up to 373 K, pressure up to 0.1 MPa) by the reaction of CO with H_2 in the presence of complexes of Fe, Zn, Mn, Cu, Co, Ni and Pt with porphyrins supported on poly(vinylpyrrolidone) [10].

Ruthenium catalysts carry out the process at relatively low temperatures (370–420K). For instance, surface hydrides of Ru^{1+} (or Rh^{1+}) are formed from Ru^{3+} (or Rh^{3+}) films supported on ionic regions of sulfurized linear polystyrene (PS) [11,12]. These regions containing Ru or Rh serve as isolated reactors which produce aliphatic hydrocarbons and promote decomposition of CH_3OH under mild conditions[2]. Linear carbonyls with bridge bonds and small particles of M^0 bound to copolymer are intermediates in these systems.

Synthesis gas can be converted to hydrocarbons also under the action of carbonyl clusters $Ru_3(CO)_{12}$ and $H_2FeOs_3(CO)_{13}$ supported on cross-linked macroporous chelating polymers with dipyridyl (Dipy), 2-aminopyridine, 2-aminophenol, 2-iminopyridine, sodium anthranilate and other groups[3] [22]. Ethylene homologation products are obtained in the presence of supported $Mo(CO)_6$ [23].

---

[2] It is supposed [13] that hydrogenation of synthesis gas or methanol proceeds via the same intermediate. The distribution of products depends on the conditions under which the hydrogenation is carried out.

[3] Nevertheless, clusters supported on inorganic oxides are used most often for hydroconversion of CO to hydrocarbons. In addition to papers considered in monograph [14], there are later studies. For instance, selective Fischer–Tropsch synthesis is carried out by Co clusters supported on SiO_2 or Al_2O_3 [15] and by $Os(CO)_{12}$ supported on Al_2O_3 [16]. Clusters $HFeCo_3(CO)_{12}$ grafted onto aminated silica gel catalyze the process at 470 K with high selectivity with respect to olefins C_4–C_8 [17]. Clusters $Co_2(CO)_8$, $CH_3CCo_3(CO)_9$, $Co_4(CO)_{12}$ and $Co_6(CO)_{15}$ supported on Al_2O_3, SiO_2, kieselgur and others catalyze, selectively, formation of low-molecular mass alkanes and alkenes [18]. High-molecular mass alkanes are produced in the presence of iron and nickel clusters (6–260 atoms) obtained by dispersion of the metals in flow of He [19]. Suspensions of Ru supported on mineral oxides are characterized by high selectivity to paraffins [20]. The process is characterized by higher selectivity, in comparison with gas-phase processes, with respect to long-chain hydrocarbons. Bimetal complexes $Cp_2Fe(CO)_2$–$Co(CO)_4$ on Al_2O_3 are also used [21].

Immobilized catalysts can be used in fluidized-bed and fixed-bed reactors to carry out the Fischer–Tropsch process in suspensions and in the gas phase. In principle, the composition of hydrocarbon mixtures formed depends on the reactor type: gasoline and light hydrocarbons are formed in fluidized-bed reactors, and mainly heavy hydrocarbons are formed in fixed-bed reactors [24]. Fibrous materials are promising supports for gas-phase hydrogenation; the use of fibrous materials enables the shape of catalysts to be varied (for instance, cobalt catalysts [25]) and to widen the range of operating temperatures. Bimetal complexes (M_1=W, V; M_2=Fe, Co, Ni), active catalysts for CO hydrogenation, are supported on carbon materials [26]. CO hydrogenation proceeds effectively in the gas phase in the presence of iron compounds [27]. The distribution of hydrocarbons formed depends on the method of obtaining the catalyst; the addition of $MgCl_2$ to the catalyst results in formation of predominantly high molecular mass products and olefins. It is important that the content of C_2–C_4 increases slightly with increasing temperature and exceeds the maximum yield calculated by the Shultz–Flory formula.

4.2 Carbonylation

The most interesting reaction of this kind is methanol carbonylation leading to formation of acetic acid and methyl acetate, products of etherification of acetic acid (in the presence of cobalt or rhodium catalysts promoted by CH_3I). This process, one of the few homogeneous processes realized on an industrial scale, proceeds under relatively mild conditions: $P<1.5$ MPa, T=450 K. Small (<1%) quantities of the side products dimethyl ether and acetaldehyde are formed in the process.

Perhaps the first immobilized catalysts for this process were metal complexes supported on ion-exchange resins with sulfo- or chelating groups [28], as well as Rh^{1+} complexes supported on phosphorylated CSDVB [29].

The selectivity of immobilized rhodium complexes (S=[CH_3COOCH_3]/[CH_3OCH_3]) supported on ion-exchange resins depends substantially on the nature of the polymer [30] (Table 4.1).

Table 4.1 Carbonylation of methanol[a] in different catalytic systems [30].

Catalyst	Content of Rh (mmol/g)	$w \times 10^{-3}$, (mol/mol min)	S
Rh/PEI[b]	0.317	6.03	0.31
Rh/PEI	0.317	9.73	0.51
Rh/NR$_2$[b]	0.234	14.30	0.38
Rh/amberlyst[b]	0.111	40.10	2.40
Rh/PSA[b]	0.103	54.80	1.90
Rh(0)/Al$_2$O$_3$	0.200	5.01	0.18

[a]Reaction conditions: P_{CO}=19.3 kPa, P_{CH_3I}=14.0 kPa, P_{CH_3OH}=8.65 kPa, T=403 K, 4 h

[b]Preparation of catalyst: P_{H_2}=P_{CO}=26.6 kPa, 403 K. Polymeric supports are ion-exchange resins on the basis of PEI; NR$_2$ is a product of interaction of chlormethylated CSDVB with pyrrolidone; PSA is the sodium salt of a sulfopolymer.

The activity and selectivity of Rh/PSA (PSA=poly(sulfobenzoic acid)) depend also on temperature (Figure 4.3). The initial rate of the reaction depends on the partial pressures of reactants in the following way:

$$w = P_{MeI}^{1.2}\, P_{MeOH}^{0.5}\, P_{CO}^{1.0}$$

The activation energy for the formation of methyl acetate is 14.5 kJ/mol for Rh/polyethyleneimine (PEI) and 50.2 kJ/mol for Rh/amberlyst.

Figure 4.3 Temperature dependence of the activity and selectivity of Rh/PSA in methanol carbonylation. (1) Yield of ether, (2) formation of methyl acetate, (3) selectivity. Pressures of CO, CH$_3$I and CH$_3$OH are 19.3, 14.0 and 8.65 kPa, respectively.

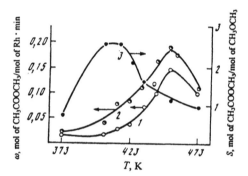

This process is similar to processes in homogeneous systems and includes consecutive reduction Rh^{3+}→Rh^{2+}→Rh^{1+} of bound ruthenium under the action of H$_2$, HCHO and CO, and oxidative addition of CH$_3$I to Rh^{1+} with the formation of active centers Rh^{3+}(CO)(CH$_3$)I (in the case of a polymeric chelate of PEI):

$$\text{]-C}_2\text{H}_4\text{NH-RhCl}_3 \xrightarrow[\text{HCHO}]{\text{H}_2\,,\,\text{CO}} \text{]-C}_2\text{H}_4\text{NH-Rh}^{2+}\text{Cl}_x \xrightarrow{\text{CO}} \text{]-C}_2\text{H}_4\text{NH-Rh}^{1+}\text{-CO} \xrightarrow{\text{CH}_3\text{I}}$$

$$\xrightarrow{\text{CH}_3\text{I}} \text{]-C}_2\text{H}_4\text{NH-Rh}^{3+}\text{(CO)(CH}_3\text{)I}$$

Scheme 4.2

For methanol carbonylation the activity of RhCl(CO)(PPh$_3$)$_2$ immobilized on phosphorylated CSDVB (2 and 20% DVB) is 60–70% of the activity of homogeneous catalyst (RhCl$_3$·3H$_2$O) [31]. The kinetics of the process are determined by swelling of the catalyst, steric hindrance and the diffusion of reactants. Deactivation of the catalyst is caused by washing out of the catalyst and by conversion of the active form of the catalyst, *trans*-(CO)ClRhP$_2$, into an inactive form *cis*-(CO)RhClP (P is the PPh$_2$ group of the macroligand).

Carbonylation of olefins, acetylenes and oxygen-containing compounds in the presence of immobilized complexes has been less well studied. Immobilization of complex PdCl$_2$(PPh$_3$)$_2$–SnCl$_2$ on phosphinated CSDVB increases the activity of the complex for etherification of 1-hexene by means of CO and methanol and does not change the selectivity with respect to the linear ester methyl heptanoate [32]. The activity of the immobilized complex depends on the content of polymer, temperature and CO pressure.

Carbonylation of hexene-1 proceeds effectively also in the presence of Pd-anionite catalysts [33]; the addition of $SnCl_2$ leads to selective synthesis of enanthic acid [34]. The use of polyglycols as cocatalysts for hydrocarboxylation of 1-octene or 1-decene leads to formation of carboxylic acids with high yield (>70%) and selectivity (>90%) [35].

A number of products including CO-containing oligomers are formed on carbonylation of diphenylacetylene in the presence of immobilized complexes of molybdenum (benzene, 520 K, P_{CO}=3.4 MPa) [36].

An interesting reaction is carried out in the presence of Pd^0 and Pt^{2+} supported on polystyrylphosphine, poly(2-vinylpyridine) (P2VP) or poly(vinylpyrrolidone) (PVPr): α-ketoamide is formed as a result of double carbonylation of iodobenzene under the action of CO in the presence of diethylamine at 358 K [37]. The catalyst is easily isolated from the reaction mixture and can be used many times. Anionic clusters $[HM_3(CO)_{11}]^-$ (M=Fe, Os) supported on CSDVB catalyze carbonylation of nitrobenzene with the formation of phenylisocyanate under mild conditions (P_{CO}<80 kPa, 373 K) [38].

Other compounds can be used as reactants of etherification. For instance, etherification of acrylic and methacrylic acids in the presence of Fe^{3+} or Cr^{3+} supported on mixed gels poly(4-vinylpyridine) + poly(ethylene oxide) (P4VP + PEO) proceeds with high efficiency and selectivity (96–100%) to alkylene oxides [39]. Palladium complexes supported on sulfocationite are active in carbonylation of allyl chloride; their activity is close to that of homogeneous catalysts [40].

4.3 Water-Gas Shift Reactions

Water-gas shift reactions are carried out to increase the yield of hydrogen in synthesis gas:

$$CO + H_2O \rightleftharpoons CO_2 + H$$

Data on homogeneous catalysts of this reaction were summarized in [41]. Only immobilized metal clusters are used as heterogenized catalysts; these promote the reaction under milder conditions. For instance, the clusters $Os_3(CO)_{12}$ supported on modified PS or P4VP are catalysts for the conversion, the content of Os decreasing from 1.3 to 0.6% at 383 K after 120 h [42].

Rhodium clusters supported on aminated PS are more effective [43]. The reaction rate depends on the nature of the polymeric amine (see Table 4.2), the sequence of activities being the same for supported catalysts and their homogeneous analogues: diethylenetriamine>propylenediamine>ethylenediamine>N,N-dimethylethylenediamine. The order of catalytic activity of rhodium clusters is $Rh_2CO)_4Cl_2$ (6.5) > $Rh_6(CO)_{16}$ (5.2) > $Rh_4(CO)_{12}$ (3.5) > $RhCl(PPh_3)_3$ (1.5) where the values in parentheses are the activity in mole of H_2 per g-atom of Rh for 24 h.

Heterogenized systems retain, in contrast to homogeneous systems, their activity for a long time, although the color and IR spectra of cluster catalysts are changed in the course of reaction and after regeneration. The cause of these changes is unknown.

Table 4.2 The dependence of the reaction rate on the nature of the polymeric amine.

Support[a]	Turnover number[b]
$]-CH_2-CH_2-NH_2$ (5)[c]	1.8
$]-CH_2-CH-NH_2$ (5) $\quad CH_3$	2.5
$]-CH_2-CH_2-N-CH_3$ (5) $\quad CH_3$	0.6
$]-(CH_2)_2 NH(CH_2)_2-NH_2$ (5)	5.2
$]-(CH_2)_2 NH(CH_2)_2-NH(CH_2)_2NH_2$ (5)	0.8
P4VP (5)	0.7
P4VP (10)	2.7
P4VP, 2% DVB (10)	1.4
P4VP, 7.8% DVB (10)	1.2
P4VP (0.1)	0.5
Amberlyst A-21 (0.1)	0.1
$NH_2-CH_2-CH_2-NH_2$ (5)	2.3

[a]Reaction conditions: P_{CO}=700 torr, 373 K, 24 h.
[b]In mole of H_2 per g-atom of Rh for 24 h.
[c]Values in parentheses are the nitrogen content in mg-atom per g of support.

The activity of supported clusters of $Rh_2(CO)_4Cl_2$ depends substantially on the H_2O content and is a maximum at $[H_2O]/[solvent]=0.3$, where the solvent is 2-ethoxyethanol [44]. It is supposed that cluster anions $[Rh_{14}(CO)_{25}]^{4-}$ and $[Rh_{14}H(CO)_{25}]^{3-}$ are the active centers.

4.4 Hydroformylation in the Presence of Immobilized Complexes

This reaction is the main method of incorporation of oxygen-containing functional groups into olefins and addition of formyl groups and hydrogen atoms to double bonds. This reaction is carried out for utilization of synthesis gas in the presence of catalyst compounds of Co or Rh (rhodium carbonyls are by 10^3-10^4 times more active than cobalt carbonyls) and sometimes Ru, Mn and Fe.

Industrial oxosynthesis is carried out in the presence of cobalt carbonyls (390–410 K, 20 MPa) for obtaining aldehydes (propion- and butyraldehydes are most valuable). The combination of hydroformylation with hydrogenation permits enables alcohols to be obtained including higher molecular mass compounds. However, in the system olefin–$CO-H_2$ there are many parallel side reactions (the proportion of these reactions reaches 10–15% for cobalt catalysts), the most important ones being hydrogenation and isomerization of olefins:

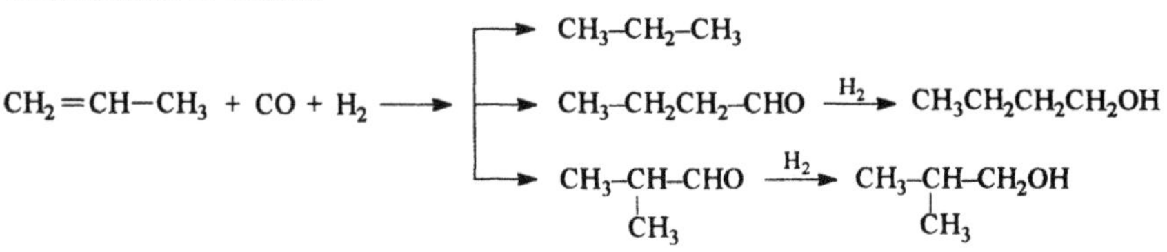

$$CH_2{=}CH{-}CH_3 + CO + H_2 \longrightarrow$$

$$\rightarrow CH_3-CH_2-CH_3$$

$$\rightarrow CH_3-CH_2CH_2-CHO \xrightarrow{H_2} CH_3CH_2CH_2CH_2OH$$

$$\rightarrow CH_3-CH-CHO \xrightarrow{H_2} CH_3-CH-CH_2OH$$
$$\qquad\quad CH_3 \qquad\qquad\qquad\quad CH_3$$

Scheme 4.3

Isoolefins are hydroformylated with formation of the same aldehydes as olefins with a terminal double bond. The most important characteristic of selectivity in this reaction is the ratio between n- and iso-forms of the aldehydes formed (in industrial conditions the ratio is 4–8). The ratio is determined by rates k_n and k_{iso} of incorporation of monooxide into n- and isometal–alkyl intermediates of catalysts:

$$S = k_n/k_{iso} = \exp[(F_a^n - F_a^{iso})/RT]$$

Values of k_n and k_{iso} depend on the nature of the transition metal.

The main shortcoming of rhodium catalysts is their high cost (rhodium is more costly than gold), therefore the loss of the ruthenium catalyst should be minimized. It is supposed [14] that the Fischer–Tropsch reaction and hydroformylation will be the first realized large-scale process with participation of supported metal complexes. This expectation stimulates the search for immobilized catalysts for these processes; the main requirement for catalysts is a technological one: a catalyst should be of fixed-bed type and should retain its activity and selectivity no less than homogeneous systems.

Let us consider features of these reactions. Olefins are the substrates most often used in hydroformylation reactions. Ethylene is hydroformylated at a high rate in the presence of rhodium triphenylphosphine supported on amberlite XAD-2 [45]. The degree of filling of the pores of the support affects the catalytic characteristics of the catalyst and its adsorption capability. Basic additions accelerate a side process, aldol condensation. In the case of gas-phase hydroformylation of ethylene ($100°C$, a layer of catalyst powder) in the presence of $RhCl(CO)(PPh_3)_{2-x}(PPh_2)_x-[$, the degree of conversion reaches 22.7% and after one pass through the catalytic reactor the selectivity is close to 100% [46].

Gas-phase hydroformylation of ethylene and propene proceeds at atmospheric pressure in the presence of bimetal carbonyl clusters $Rh_{4-x}Co_x(CO)_{12}$ (x=2,3) supported on coal, as well as in the presence of ruthenium–cobalt catalysts [47,48]. Bimetal catalysts are characterized by high selectivity to normal aldehydes (up to 90%) and activity substantially higher than that of monometal catalysts because of coordination of Rh with Co or Fe: formation of direct links Rh–Co–O or Rh–Fe–O prevents caking of Rh and, hence, stabilizes intermediate acyl derivatives of isolated Rh atoms[4].

Complexes of Co, Ir and Rh immobilized on supports of mixed type (SiO_2–phosphorus-containing oligomer) are used to hydroformylate propene with selectivity to n-BuOH > 90% [50]. Under milder conditions (348–363 K, total pressure 0.1 MPa, $C_3/H_2/CO$=1:1:1) gas-phase hydroformylation proceeds in the presence of $RhH(CO)(PPH_3)_3$ supported on the surface of phosphinated CSDVB (amberlite XAD-2 or XAD-4) [51]. The method of catalyst synthesis affects the activity: vacant phosphine groups of the catalyst affect exchange processes and decreases the activity of the catalyst. A catalyst obtained in H_2 atmosphere is much more active than that from an atmosphere of CO (Figure 4.4).

Propene is hydroformylated by $[Rh(AcAc)(CO)_2]$ supported on poly(trimethylol-propane) trimethacrylate and a grafted linear acrylate polymer bearing phosphines, proceeds with high rate (up to 1.1×10^{-4} mol butanol $s^{-1}(g\ Rh)^{-1}$ and stability (with the same rate for 215 h) [52]. The reaction rate is expressed by the equation:

---

[4] The promoting effect of Sn, Pb, Mo and W was found also for hydroformylation of propene in the presence of carbonyl complexes of Rh immobilized on inorganic supports [49].

Figure 4.4 Effect of treatment of catalyst obtained by interaction of $RhHCO(PPh_3)_3$ with amberlite XAD-2 on its activity. Catalyst was treated in an atmosphere of H_2 (1) and CO (2). $[C_3H_6]/[H_2]/[CO]=$ 1:1:1, 348 K.

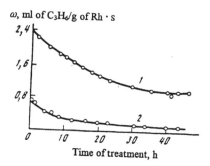

$$v=A_0\exp(-E_a/RT)\,(P_{propene})^{0.59}P_{H_2}(P_{CO})^{-0.75}$$

where $A_0=3.4\times10^9$, $E_a=90$ kJ/mol and P is in atm. The selectivity of these catalysts is $S=k_n/k_b=0.9-2.1$.

Impregnated catalysts formed from $RhH(CO)(PPH_3)_3$, especially monolayers at low degree of pore filling, are highly active in hydroformylation of butene (1.3 MPa, 333–373 K) [53]. A mixture of n- and isopentals is obtained by catalytic hydroformylation of butene-2 in the presence of supported Rh complexes [54]. In the course of this reaction the double bond is transferred not only in butene-2, but also in pentene-2, hexene-2, hexene-3 and others.

Metal polymers are widely used for oxosynthesis of olefins $C_5–C_7$. For instance, hydroformylation of pentene-1 in the presence of Rh complexes on phosphorylated CSDVB is more selective with respect to linear aldehyde ($S=12$) than hydroformylation with homogeneous systems ($S=3.3$) [55]. Perhaps, the higher selectivity is caused by the higher ratio P/Rh for the polymer-bound catalyst. However, immobilization of Rh complexes on the ion-exchange resin *Shirotherm* does not result in an increase of selectivity because at high temperatures most of the Rh complexes dissociate from the support and carry on the process in solution, whereas at decreasing temperature Rh complexes bind back to the support [56]. This subject was specially investigated for a composite that included polyamines and polycarboxylic acids impregnated by poly(vinyl alcohol) (PVAl) [57]. This composite separates metal ions in solution at high temperature (420 K) and binds them back at low temperature:

$$\vdash NH_2 + \vdash CO_2H + Na^+ + Cl^- \rightleftharpoons \vdash \overset{+}{N}H_2\cdot HCl + \vdash CO_2Na$$

high temperature   low temperature

$RhCl_3$ and $Fe(CO)_5$ supported on these polymers were used as catalysts for the Reppe reaction, hydroformylation of pentene-1. In this case active centers are hydrocarbonyl anions of Rh.

A number of polymers, including poly(vinylpyridine) (PVP), poly(vinylimidazole), polyvinylcarbazole, silicon-containing polymers and others, were used as macroligands of Rh complexes to obtain catalysts for this reaction [58–60]. Rhodium carbonyl supported on polyquinones, modified by phosphine groups and characterized by high thermal and chemical stability, is active at 373–383 K and selective to aldehydes (91%) [61]. Selectivities of complexes $Ru(CO)_3(PPh_3)_2$ and $Co_2(CO)_8$ supported on phosphinated CSDVB are close to those of the homogeneous systems.

Advantages of localization of reaction centers of heterophase catalytic systems on the surface or in a thin surface layer of polymer were emphasized many times in Chapters 1–3. The same conclusion is valid for hydroformylation of hexene in the presence of $Rh(AcAc)(CO)_2$ supported on the surface of polypropylene-gr-poly(styryldiphenylphosphine) [62–64]. Under mild conditions (338 K, 1.6 MPa) the ratio between n- and isoaldehydes for this system is higher by at least 3.5 times compared with that for the homogeneous analogue. The selectivity reaches 1.6 at P/Rh=8 with an insignificant decrease of activity.

Hydroformylation of hexene-1 in toluene at 80–120°C in the presence of $RhCl(CO)(PPh_3)_2$ supported on phosphinated PS proceeds with the same rate as in the homogeneous analogous systems [46]. However, in heterogenized systems only aldehydes are formed.

The chain loading affects the activity. For instance, for cyclooctene hydroformylation in the presence of Rh cyclooctadiene acetylacetone supported on copolymer of styrene and 2,2'-bis(4,6-di-t-butylphenyl)-p-styryl phosphite, the catalyst is most active at low chain loading [P]/[Rh]~10, because in this case active centers, monophosphite complexes, are easily accessible [65].

Rhodium indigo sulfonate on ion-exchange resin Vophatite-ES can be used many times without loss of activity; the major products of hydroformylation of hexene-1 in ethanol are the aldehyde acetals C_7 [66]. Interesting results were obtained by the direct synthesis of alcohols from hexene-1 in toluene at 120°C and syngas by rhodium supported on cross-linked acrylic resins [67]. The latter were obtained by dehydration of macromolecular isocyanides. The $[Rh(AcAc)(CO)_2]$ complex interacts with the support to form the macrocomplex $[(P-N≡C)_2Rh^{1+}(AcAc)(CO)]$. These molecular species do not survive the first catalytic run. Formation of the catalyst proceeds by reductive decomposition during the activation step, isocyanide hydrogenation, splitting the rhodium-acetylacetonate bond, and forming small metallic clusters (dimensions 1.6–7.2 nm):

$$(P-N≡C)_2 Rh^I L_n \xrightarrow{H_2} (P-N≡C)_2 Rh^{III}(H_2)L_n \longrightarrow (P-N=C)_2 Rh^{III} L_n \xrightarrow{\Delta/H_2}$$

$$Rh^0 + nL + P-N= \text{[hydrogenated isocyanide]}$$

It is likely that the hydroformylation reaction is catalyzed in the liquid phase by the soluble rhodium hydridocarbonyl species. The selective reduction of the aldehydes is probably catalyzed by the polymer-supported rhodium crystallites. Heptanol, 2-ethylpentanol, 2-methylhexene and aldehydes (at long times) are products of the reaction.

Complexes, $Rh_2Cl_2(CO)_4$, immobilized on supports of mixed type are highly effective and stable catalysts (the number of catalytic cycles occurring at one metal atom reaches 80 000) [68]. Sulfur-containing polymers (a product of interaction of poly(vinyl chloride) (PVC) and SiO_2 with RSH, or copolymer of vinyl butylsulfide with acrylonitrile and divinylbenzene in the presence of SiO_2) were used as supports. The catalysts are used to hydroformylate hexene or heptene in benzene at 313–423 K and 3–9 MPa, their selectivity with respect to heptanal being determined by the same factors as for the other rhodium catalysts.

Silica-grafted polymer-bound phosphite-modified rhodium complexes are able to hydroformylate styrene at moderate pressure (P_{CO/H_2} =30 bar) and temperature (100°C) [69]. The activity of the catalyst is low (an appreciable yield is obtained only over a period of at least 10 days). Higher activity is found in benzene as solvent and for a ligand-to-rhodium ratio of only 4. The highest activity (up to 50% after 12 h) was found in toluene. This was explained by swelling of the grafted layer and formation of a proper coordination environment. This could be attributed to contraction of the grafted polymer coils in this solvent, leading to incomplete coordination of rhodium.

Rhodium-containing polymers can hydroformylate hexene-1 by using the mixture CO–H$_2$O at 353 K [70]. In this case water serves simultaneously as solvent and the source of protons[5]. The yield of aldehydes is 53%, S=18; mononuclear complexes are less active than binuclear ones.

Co$_2$(CO)$_8$ on polyphosphazene is active and thermostable in the hydroformylation of 1-hexene (463–468 K, H$_2$/CO=2:1, total pressure 14 MPa) [72,73]. The yield of heptanols (84.7%) is close to the yield of alcohols C$_7$ in the homogeneous system. It is supposed that soluble species are formed as a result of cleavage of P–Co–P links and elimination of phosphine ligands leads to deactivation of the catalyst.

Anionic trinuclear-supported clusters of osmium and iron [HM$_3$(CO)$_{11}$]$^-$, where M=Fe or Os are more active in oxosynthesis with participation of hexene-1 [74]. These catalysts also promote isomerization of hexene-1; the iron cluster segregates and the reaction seems to occur both on clusters and the mononuclear centers. The immobilized osmium cluster remains unchanged and active; it was found that its polynuclear structure plays an important role in the selective formation of n-heptanal. Immobilized bimetal clusters, RuOs$_3$, are more active and selective in hydroformylation of hexene-1 than mononuclear Ru or Os clusters. The cause of this synergism is unclear. Synergism was also found for bimetal cluster Co$_2$Rh$_2$(CO)$_{12}$ supported on ion-exchange resin Dowex-1 [75].

Interesting features were found for the hydroformylation of hexene-1 in the presence of pentametallic clusters FeCo$_3$Cu and FeCo$_3$Au supported on phosphinated CSDVB [76]. These supported complexes are more stable than the unsupported ones, the Cu-containing cluster being more stable than the Au-containing analogue. Catalytic activity correlates with macrocluster stabilities. Polymeric complexes are more active than tetranuclear FeCo$_3$ analogues due to the synergistic effects of different metals. Only aldehyde is produced in the presence of supported complexes, and only alcohol is produced in the presence of unsupported complexes. The macrocomplexes facilitate the formation of n-isomers, because steric factors and phosphine groups facilitate addition reactions in spite of the Markovnikov rule.

Heptene is hydroformylated on the same catalysts, most effectively on complexes of Rh with phosphinated PS [77,78].

Hydroformylation is used in industry for the oxosynthesis of aldehydes with their subsequent transformation to alcohols and acids (C$_8$–C$_{16}$) . Immobilized catalysts can also be used in these processes. For instance, octene-1, decene-1, dodecene-1 and

---

[5] Palladium hydride complex HPdCl(PPh$_3$P)$_2$ in tetrahydrofuran (THF) in the presence of SnCl$_2$ hydrocarboxylates hexene with formation of n-C$_7$H$_{15}$COOH [71].

hexadecene-1 are hydroformylated to products containing up to 67% linear aldehydes in the presence of Ru–Co complexes supported on polyethylene (PE), PS and poly(methacrylamide) modified by $[(CH_2)_nPR_3]^+X^-$ groups [79].

Rhodium complexes supported on ion-exchange resins selectively hydroformylate octene-1 [80].

It was found [81] for hydroformylation of dodecene-1 in the presence of $Rh_4(CO)_{12}$ supported on phosphorus-containing polymers ($90°C$, $P_{H_2}=P_{CO}=0.5$ MPa, [Rh]=0.01%, [P]/[Rh]=10) that for effective hydroformylation, a macroligand should contain three aryl groups and no alkyl substituents near the phosphorus atom, phosphorus atoms should be localized far from backbone chain and the Rh atom should not be cis-chelated. The yield of normal aldehydes increases and the hydrofomylation rate decreases with the increase of the degree of functionalization. In the case of soluble macrocomplexes the increase of local concentration of phosphine groups facilitates their incorporation into the catalytic complex [82].

Carbonyl complexes of rhodium on polymers modified by cyclopentadienyl [83] or phosphor-containing groups [84] promote oxosynthesis of olefins C_{14} with high yield of linear aldehydes (up to 80%). Complexes $Rh_2(CO)_4Cl_2$ on CSDVB modified by N,O chelating units are effective catalysts for the hydroformylation of diisobutylene, their selectivity depending on the nature of the chelates [85].

Hydroformylation of styrene is catalyzed by Rh complexes on phosphinated polymers and proceeds by the Scheme 4.4:

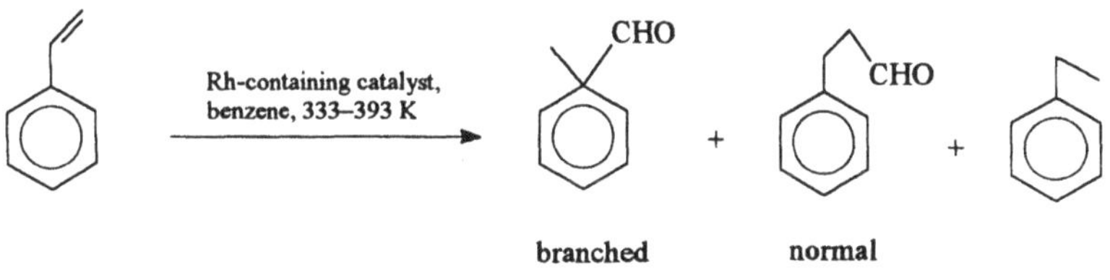

branched   normal

Scheme 4.4

Activities of the heterogenized system and its homogeneous analogue are given in Table 4.3. The influence of temperature, pressure of gas mixture and P/Rh ratio is the same in both the systems.

Table 4.3 Yields and selectivities of homogeneous and heterogenized rhodium catalysts in hydroformylation of styrene (353 K) [86].

Catalyst	P/Rh	Yield (%)	S
$RhH(C))_2(PPh_3)_3$	3.0	100	6.4
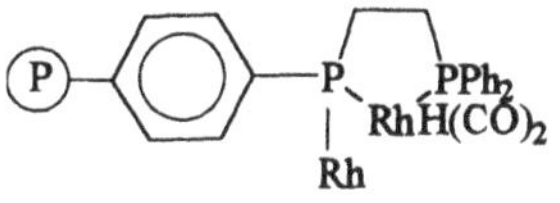	–	100	8.3
	3.3	100	7.2
	9.1	81	4.5
	16.7	100	3.2

Competition between Co and phopsphine ligands for coordination with Rh affects the properties of the immobilized complexes:

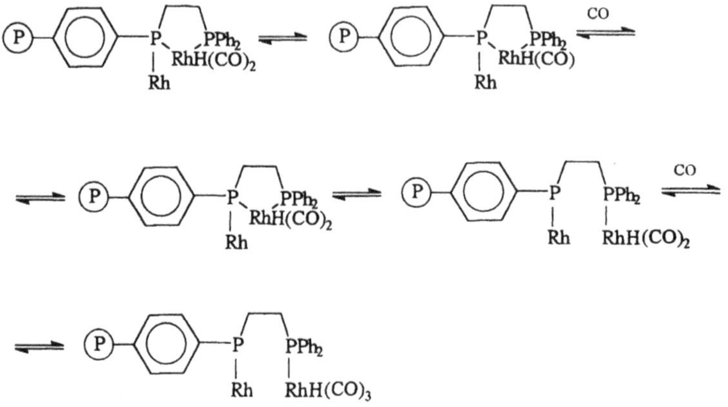

$$\text{(P)}-\bigcirc-\overset{\displaystyle P}{\underset{\displaystyle \overset{|}{Rh}}{\overset{\diagup}{\diagdown}}}\overset{PPh_2}{\underset{RhH(CO)_2}{}} \rightleftharpoons \text{(P)}-\bigcirc-\overset{\displaystyle P}{\underset{\displaystyle \overset{|}{Rh}}{\overset{\diagup}{\diagdown}}}\overset{PPh_2}{\underset{RhH(CO)}{}} \overset{CO}{\rightleftharpoons}$$

$$\rightleftharpoons \text{(P)}-\bigcirc-\overset{\displaystyle P}{\underset{\displaystyle \overset{|}{Rh}}{\overset{\diagup}{\diagdown}}}\overset{PPh_2}{\underset{RhH(CO)_2}{}} \rightleftharpoons \text{(P)}-\bigcirc-\underset{\overset{|}{Rh}}{P}\quad\underset{RhH(CO)_2}{PPh_2} \overset{CO}{\rightleftharpoons}$$

$$\rightleftharpoons \text{(P)}-\bigcirc-\underset{\overset{|}{Rh}}{P}\quad\underset{RhH(CO)_3}{PPh_2}$$

Scheme 4.5

Styrene is hydroformylated in the presence of cobalt clusters with bridged phosphine ligands [88]. In the presence of amines and clusters of $Fe_3(CO)_{12}$ the aldehydes formed are oxidized by the mixture CO/H_2O and are reduced by the mixture CO/H_2 [88].

Sparse data are available on hydroformylation of dienes. For instance, the cluster $Co_2Pt_2(CO)_8(PPh_3)_2$ is active in hydroformylation of 1,3-butadiene [89]. The activity of this cluster is higher than that of mononuclear Pt and Co complexes probably because of low electron density on the cobalt atom of the heterometallic cluster

Hydroformylation of dicyclopentadiene and the reduction of the products formed to corresponding diols in the presence of rhodium catalysts supported on microporous and gel amine-containing ion-exchange resins was studied in detail in [90]. The process proceeds by Scheme 4.6 which includes the formation of products of isomerization (I), dialdehydes (II), monoenes (III), oxoaldehydes (IV) and diols (V). Complexes on macroporous resins are characterized by high thermal and chemical stability even after 50 consecutive 2 h reaction cycles at 403 K (Figure 4.5). Catalysts on gel resins are deactivated due to substitution of supported Rh by reaction products. Deceleration of aldehyde reduction (Figure 4.5) is caused by deactivation of the catalyst.

. A number of valuable products can be obtained by direct hydroformylation of oxygen-containing substrates in the presence of immobilized catalysts. For instance, methylmethacrylate is transformed into linear $(CH_3OOCCH(CH_3)CH_2CHO)$ and branched $(CH_3OOCCH(CH_3)CH(CH_3)CHO)$ aldehydes in the presence of $Rh(CO)(PPh_3)_3$ supported on phosphinated CSDVB [91]. In comparison with the homogeneous analogue, the metal polymer is less active and more selective with respect to branched aldehydes (75% instead of 57%). At high temperatures competing polymerization decreases the yield of aldehydes, at the same time the share of n-aldehydes decreases. The mechanism of hydroformylation[6] includes incorporation of a double bond of MMAc into active bond Rh–H.

---

[6] Hydroformylation of methylmethacrylate (MMAc) by homogeneous rhodium–phosphine systems was studied in [92].

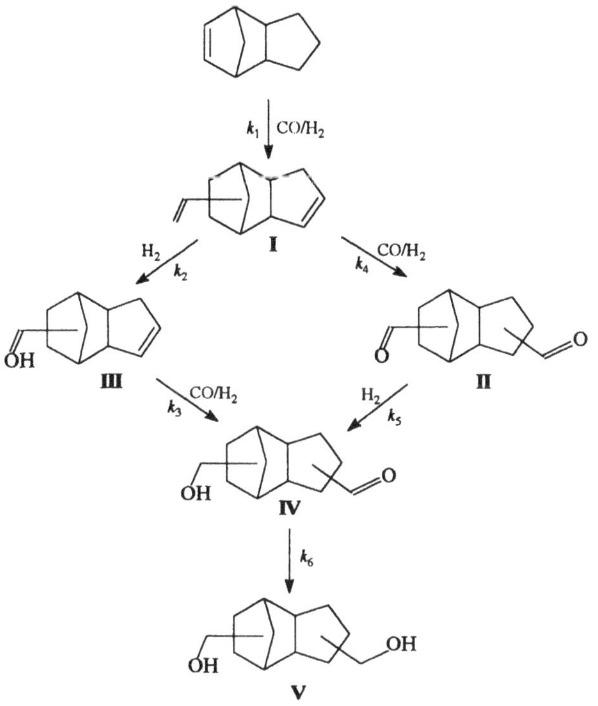

Scheme 4.6

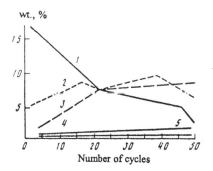

Figure 4.5 Variation of the content of the reaction mixture in 50 consecutive cycles for hydroformylation of dicyclopentadiene in the presence of rhodium complexes on CSDVB–P4VP. Curves (1)–(5) represent yields of products (see Scheme 4.6) **V** (1), **IV** (2), **II** (3), **III** (4) and **I** (5). [CO]/[H$_2$]=1:2, 403 K.

The main factors affecting regioselectivity in hydroformylation of vinyl esters ROXH=CH$_2$ in the presence of rhodium complexes are the ratio PPh/Rh, the temperature and the phenyl groups in α- and β-positions with respect to the double bond [93]. Esters of vicinal glycols (for example, ethylene glycol diacetate) can be obtained directly from synthesis gas with aliphatic carboxylic acids as solvents.

In my opinion, other reactions of interest are the hydroformylation of formaldehyde to glycols (yield up to 75%) by anionic Rh^{1+} complexes [94] and by heterometallic clusters $Co_2Rh_2(CO)_{12}$ at 383 K (selectivity to $HOCH_2CHO$ up to 65%) [92]; hydroformylation of allyl alcohol to 1,4-butanediol by rhodium catalysts with subsequent hydrogenation of the 4-oxybutyric aldehyde formed by skeleton nickel [95]; oxosynthesis of alkanes isolated from a biomass (for example, 3-(4-methoxyphenyl)-1-propene) in the presence of rhodium catalysts with a high yield of linear products [96].

The most interesting reaction is asymmetric hydroformylation with participation of immobilized chiral complexes. This reaction could be used for synthesis of large quantities of chiral products from small quantities of chiral material. The reaction proceeds by the scheme:

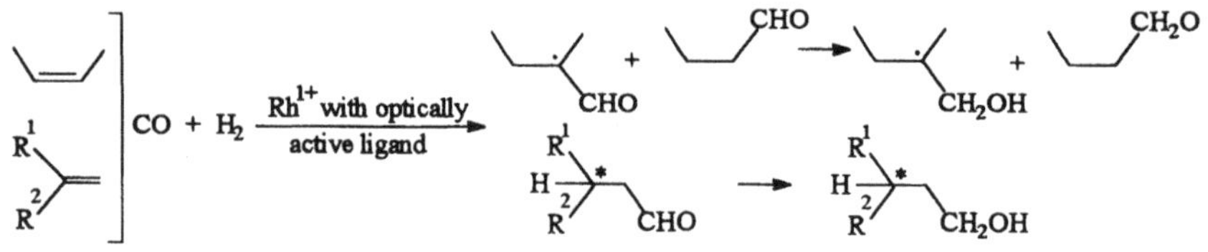

Scheme 4.7

Complexes $[Rh(CO)_2Cl]_2$ and $HRh(CO)(PPh_3)_3$ are immobilized by the same macroligands as those used for hydrogenation (see Section 1.10). Immobilization of Rh results in retention of both Rh and chiral ligand.

Hydroformylation of cis-2-butane under mild conditions [97] results in the formation of 2-methylbutanol-1 with the same yield (28%) as in analogous homogeneous systems. However, hydroformylation of styrene at 298–353 K is characterized by an enhanced ratio of branched to linear aldehydes (S=6–20 instead of 2 for homogeneous systems) and by a high yield of (–)(R)-aldehyde.

Complexes $PtCl_2/SnCl_2$ of the following structures

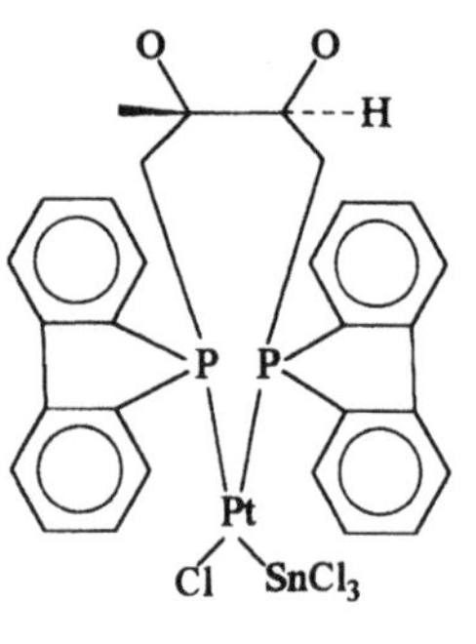

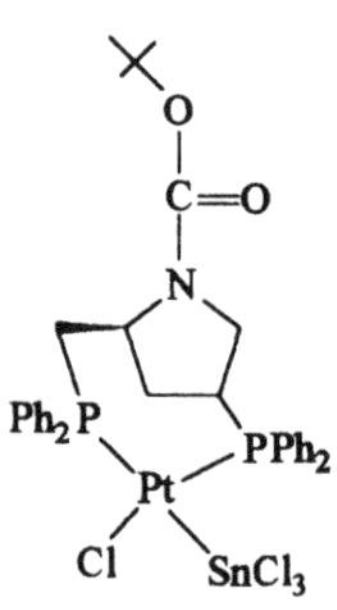

immobilized on phosphinated macrochelators are the best catalysts for asymmetric synthesis [98–101]. These catalysts promote asymmetric hydroformylation of vinyl acetate, N-vinylphthalamide, norbornadiene and styrene. In the case of styrene under certain conditions an enantiomer is formed, yield 98%, its hydrolysis proceeding also without racemization. The total selectivities to styrene and norbornadiene, vinyl acetate and N-vinylphthalamide are 40–100, 10–40 and 20–50%, respectively. An excess of the

$(+)(S)$-form of aldehydes is found for styrene and vinyl acetate and an excess of $(-)(S)$-form in other cases.

Hydroformylation on these catalysts is characterized by better parameters than hydroformylation in homogeneous systems; recycling of the catalysts leads to only a slight deterioration of the parameters. It is important that recycling does not decrease the reaction rate or the yield of the stereoisomer; however, racemization increases with increasing duration of reaction

Hydroformylation of norbornene[7] proceeds with a very high rate but low (10–26%) yield of asymmetric 2-formylnorbornane:

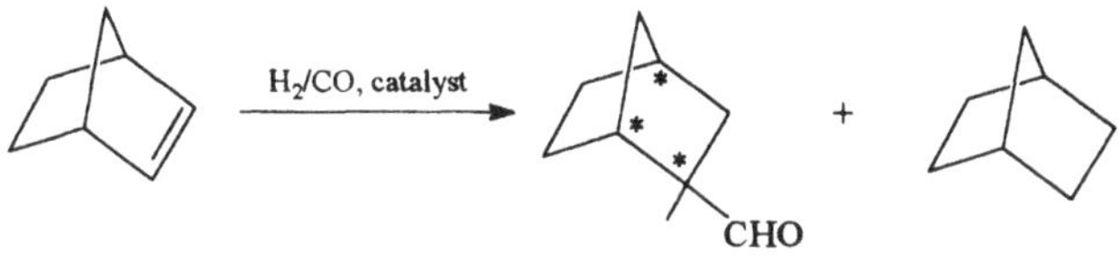

19–26%

N-vinyl phthalamide is hydroformylated with a lower rate and selectivity to *n*-aldehyde:

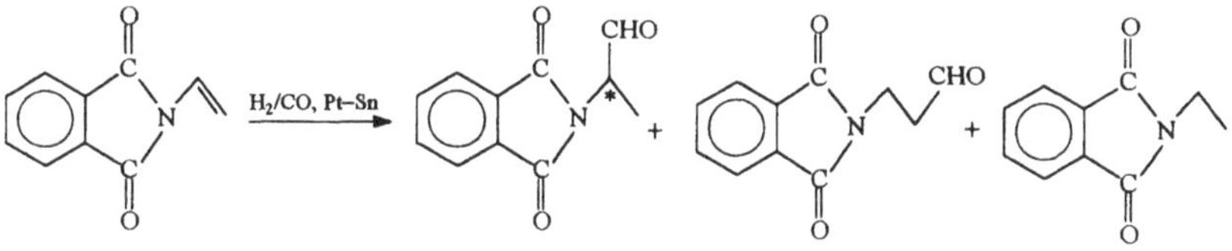

50–60%

Data for reactions with participation of CO in the presence of immobilized metal complexes can be summarized in the following way. Highly effective immobilized catalysts were found which are able to carry out processes under substantially milder conditions with substantially higher selectivity. In the case of immobilized catalysts, there are additional levers for process control: structure of support, distance between catalytic centers and polymer chain [103], nuclearity of catalytic complexes and others. Polynuclear centers seem to be more suitable for the activation of CO molecules. A number of bimetal systems (Co–Rh, Co–Ru, etc.) are more active than monometal systems, for instance in hydroformylation. Probably this synergism is explained by isolation of Rh or Ru atoms by means of the second metal atom, as well as by stabilization of intermediates and facilitation of the incorporation of CO into the metal–alkyl bond. In some cases the role of the second metal is to prevent carbonization of the catalyst [104].

An autoregulation effect was found in the heterogeneous catalytic reaction of CO with H_2 [105]. This effect is caused by generation of highly non-equilibrium concentrations of active states due to optimization of coupling of the catalytic reaction with destruction of the active centers (see Chapter 5).

---

[7] It is supposed [102] that hydroformylation and carbonylation of norbornene in the presence of tungsten-containing catalysts proceeds with formation of carbenes and ketones as intermediates.

Immobilized metal complexes could be used for carbonylation and hydroformylation of complex molecules with proper stereostructures (esters of geranic acid, half-products for synthesis of citronellal, vitamin A and others). At present this field is dominated by homogeneous catalysis.

One of the promising ways for chemical transformations of CO is its copolymerization, for instance alternating co-oligomerization of CO with olefins. In homogeneous systems such reactions are catalyzed by complexes of Rh, Pd, etc. [106].

Optimization of such processes including industrial processes [107,108] requires the search for new types of active immobilized catalysts. There is no literature on hydroformylation of polymeric supports, even those containing multiple bonds, under conditions of catalyzed reactions. However, recently published papers were concerned with the hydroformylation of polyisobutene [109] and hydroformylation and hydroxymethylation of styrene–butadiene copolymers with the aim of chemically modifing the copolymers [110].

4.5 Catalytic Hydrolysis and Dehydration

Hydrolysis of esters can be considered as activation of a $C=O$ group and the stable molecule H_2O. Many polymers catalyze this reaction, but a number of immobilized metal complexes have higher activity. For instance, hydrolysis of 2,4-dinitrophenyl acetate to 2,4-dinitrophenol in the presence of Cu^{2+} complexes on PVS proceeds via formation of a triple complex in which carbonyl group of the substrate and OH group are activated [111]:

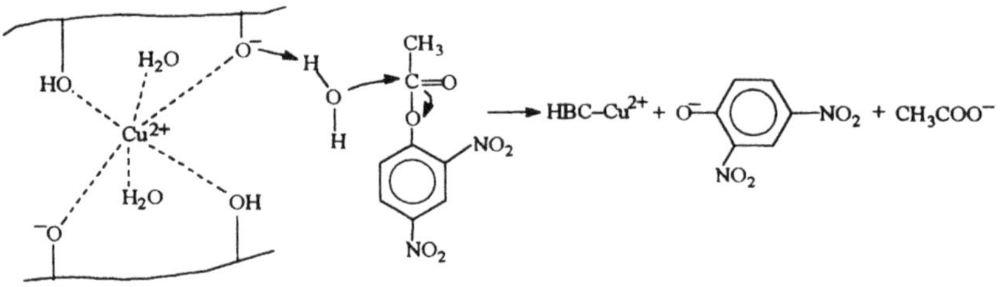

Scheme 4.8

The hydrophobic interaction between 2,4-dinitrophenyl acetate and metal polymer with formation of the triple complex (likewise in enzymatic catalysis) facilitates hydrolysis. However, the rate of basic hydrolysis of n-nitrophenyl acetate (aqueous solution of NaOH, 298 K) decreases from 12.3 to 0.12 s^{-1} after addition of a binuclear Cu^{2+} complex supported on α-cyclodextrin [112]. This inhibiting effect is explained by the formation of a strong complex of substrate and Cu^{2+} ion with α-cyclodextrin.

Co^{2+} ions in combination with humic acids inhibit hydrolysis of N-acetyl-L-tyrosine ethyl ester by α-chymotrypsin and trypsin [113]. A proportion of the Co^{2+} ions are bound with humic acids and do not inhibit the enzymatic process, whereas free Co^+ ions and unbound humic acids inhibit the enzymatic reaction by a mechanism of mutually independent two-component inhibition. Hydrolysis of this substrate in the presence of α-chymotrypsin is effected also by a polymeric analogue of salicylic acid and its macrocomplexes with biometals of Mn^{2+}, Co^{2+} and Ni^{2+} [114].

Complexes of Cu^{2+} ions with P4VP, partially quaternized copolymer of ethylenediamide and N-vinylimidazole, or copolymer of 4-VP and styrene, catalyze hydrolysis of esters of phosphoric acid, for instance diisopropyl fluorophosphate [115]. The time of substrate half-transformation is 18–28 min in the case of non-metal catalysts and 150 min in the case of Cu^{2+}-PVI.

Complexes of Cu^{2+} with quaternized P4VP were used in [116] for hydrolysis of toxic organophosphorus compounds DFP and soman:

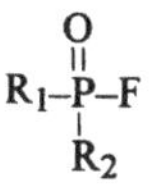

$$R_1 - \overset{\overset{\displaystyle O}{\|}}{\underset{\underset{\displaystyle R_2}{|}}{P}} - F$$

DFP: $R_1=R_2=i\text{-}C_3H_7O$; soman: $R_1=(CH_3)_3CCH(CH_3)O$, $R_2=CH_3$

which are hazardous as potent acetylcholinesterase inhibitors. For instance, for soman this represents a 14-fold increase over the uncatalyzed rate in the same buffer. DFP hydrolyzes with the rate constant 0.011 min^{-1} ($t_{1/2}=63.8$ min). It can be expected that immobilized complexes of Cu^{2+}, as well as low molecular mass analogues, are able to catalyze hydrolysis of the chemical warfare agent ethyl–N,N-dimethyl-phosphoramidocyamidate (tabun). This route could be used for the destruction of chemical weapons.

Catalysis of enantioselective hydrolysis of mixtures of esters of D- and L-amino acids by immobilized complexes is especially interesting as a method of isolation of more valuable L-amino acid from the mixture. Most often catalysts are obtained by supporting MX_n on CSDVB or polyacrylamide modified by (S)-hydroxyproline groups (up to 4 mmol/g) [117–119]. The catalyst has the following structure:

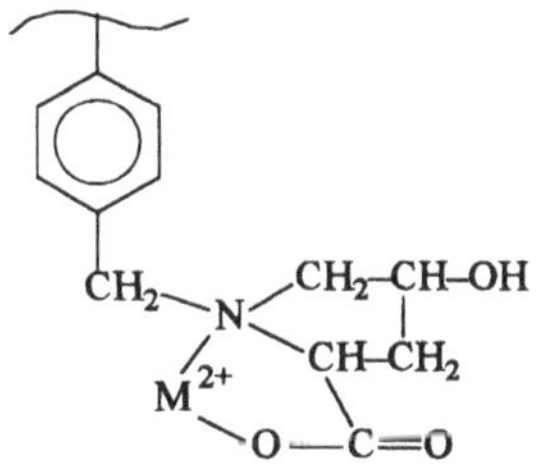

The process proceeds at 348 K and pH 9.4 by a two-step mechanism: formation of catalyst–substrate complex:

$$[Pr–Cu]^+ + Am \rightarrow [Pr–Cu–Am]^+$$

where Am is an amino group of the polymer, Pr is proline, which then hydrolyses the coordinated ester to amino acid (Aa):

$$[Pr–Cu–Am]^+ + OH^- \rightarrow Pr–Cu–Aa.$$

The ratio of rate constants of hydrolysis of (R)- and (S)-forms of phenylalanine and leucine esters is $K_R/K_S =1.16$ and 1.39, respectively, because of the higher stability of the complex $[(S)-Pr–Cu(R)–Am]^+$ in comparison with the complex $[(S)–Pr–Cu(S)–Am]^+$ [118].

The hydrolysis of enantiomers of methyl esters of other α-amino acids in a borate buffer at pH 8.3 and 308 K is characterized by a higher ratio K_R/K_S: 1.66 for histidine and 0.54 for leucine. Values of E_a are 55.9 and 50.0 kJ/mol for R- and S-forms of the methyl ester of leucine, respectively. Amide groups of polyacrylamide, participating in coordination of Cu^{2+} or Ni^{2+}, affect the values of K_R and K_S.

Thus, hydrolysis of esters could be used as a simple method of enantioselective enrichment of racemic mixtures[8].

The selectivity of hydrolysis of methyl esters of D- and L-histidine in the presence of Ni^{2+} complexes supported on polymers with chiral residues of L-histidine rises with an increase of content of hydrophilic links of 2-hydroxymethyl methacrylate in the polymer, the ratio [substrate]/[catalyst] and the temperature [121].

Under optimal conditions the activity and selectivity of these catalysts are higher than those of the homogeneous analogues. The highest enantioselective excess (at pH 9.0 the ratio L/D is equal to 6 in the case of leucine ester) was found for $Ru^{3+}(NH_3)_5X$ supported on oligomeric surface-active substance (SAS) where X is a derivative of histidine [122]. The same features were found for stereoselective hydrolysis of n-nitrophenyl esters of L- and D-aminoacids at pH 8.0 and 298 K in the presence of Cu^{2+} complexes supported on PEI containing covalently bound optically active L-histidines [123]. The cooperative action of a ligand and Zn^{2+} or Co^{2+} ions was found in the hydrolysis on n-nitrophenylpicoline [124]; these catalysts can be considered as models of enzymes.

Zn^{2+}, Co^{2+} and Cu^{2+} ions activate aminopeptidase, a kind of Zn-containing hydrolytic enzyme, which catalyzes hydrolysis of L-leucine-n-nitroanilide [125]. Equilibrium constants for the formation of the enzyme–substrate complex at 293 K are 8.56, 8.40, 8.68 and 9.09 mmol/l and rate constants are 10.7, 10.7, 208 and 363 s^{-1} for aminopetidase pristine and activated by Zn^{2+}, Co^{2+} and Cu^{2+} ions, respectively, i.e. the Zn-containing enzyme is characterized by the best binding and the Cu-containing enzyme is characterized by the highest activity.

Macroligands do not always remain unchanged in hydrolysis by immobilized complexes. For instance, the copolymer of acrylamide and acrylic acid decompose rapidly in the presence of small quantities of Fe^{2+} ions in aqueous solutions at pH 5–13 and 333–373 K even under anaerobic conditions [126].

It was found [127] that unactivated peptide bonds are selectively hydrolysed in the presence of Pt^{2+} complexes anchored to an amino acid side chain.

Immobilized complexes are used relatively often to catalyze dehydration processes. For instance, metal porphyrins ($M=Cr^{3+}$, Ti^{4+}, Zr^{4+}) immobilized in a matrix of poly-

---

[8] Optically active polymers containing no metals carry out hydrolysis at substantially lower rates. For instance, substituted PEI (containing optically active imidazole residues) catalyzes hydrolysis of derivatives of phenyl esters [120]. The rate constant is proportional to the concentration of non-protonated imidazole residues, the active species in this reaction. Cooperative effects were found for this reaction. Enantioselectivity K_D/K_L is equal to 1.1 and does not depend on the nature of the chiral substrate and the pH of the medium.

acrylamide gel cross-linked by N,N'-methylenebisacrylamide, are effective catalysts for dehydration of aldoses [128,129]. To prevent hydrolysis of metal–polymer in aqueous medium in the course of hydration of xylose and glucose at 353–393 K, acetic acid is used as solvent. These catalysts are also active in dehydration of pentose to furfural [130]. Sulfurized CSDVB in which a part of the sulfogroups are substituted by Al, Fe, Zn, Cu and Ni ions are used for dehydration of alcohols at 403–423 K, hydration of olefins at 373–403 K and decomposition of ethers to olefins and alcohols at 433–443 K [131]. The activity of membrane-type catalysts based on sulforized CSDVB in the gas-phase dehydration of C_2H_5OH at 393 K depends on the distribution of sulfo groups: the rate of deactivation of the catalyst (due to inhibition by the reaction product, H_2O) is higher for catalysts with inhomogeneous and high local concentrations of acid centers than that for catalysts with homogeneous and low local concentrations [132]. Macroporous sulfurized CSDVB with different porosity and degree of cross-linking are active in the dehydration of 2-butanol [133]. The parameters mentioned above determine the ratio cis/trans-butene-2, butene-1 not being formed. Hydration of aldehydes C_4–C_6 in the presence of Cu^{2+} or Pd^{2+} complexes supported on modified CSDVB results in the formation of dienes (butadiene, isoprene, 1,3-hexadiene and others) [134].

Macrocomplexes of Cu^{2+} with synthetic fluorated tetrasilylcellulose selectively catalyze the dehydration of methanol to methylformate at 450–490 K [135]; immobilized complexes are used very rarely to catalyze this process. Active centers of these immobilized catalysts are coordinationally unsaturated Cu^{2+} ions having three vacant coordinational sites.

Optimization of these reactions seems to require great effort to make them technologically acceptable.

4.6 Acid Catalysis by Metal Polymers

Solid acids are widely used as catalysts for the cracking of hydrocarbons, for the isomerization, oligomerization and polymerization of olefins, for the alkylation of aromatic hydrocarbons and for the dehydration of alcohols. [136].

Recently a new kind of solid acids, Lewis acids on polymeric supports, has been recognized. These are sometimes called solid superacids [137] and are classified by the state of aggregation (liquid or solid), number of components (mono- and multicomponent), character of acidity of components (protic acids (Brønsted acids), aprotic acids (Lewis acids), and combined acids (mixtures of protic and aprotic acids)).

A number of studies on metal–polymer acid catalysts have been carried out to find new polymeric supports and new methods of immobilization of regular Lewis acids ($AlCl_3$, $AlBr_3$, BCl_3, SbF_5 and others), to increase the acidity of regular protic polyacids (for example, cation-exchange sulforized resins) and to increase activity, selectivity and thermal stability by means of modification of CSDVB; modification of CSDVB is by sulfonation, nitration, chlorination, fluorination or sulfoalkylation (a review of these studies is given in [138]).

Cation-exchange polymers[9] with partial or complete substitution of functional groups by metal ions, as well as cation-exchange resins in which a metal ion is bound to the polymer via another ligand or sulfate group, are promising types of acid catalysts [140]. In these catalysts we can control the number of exchange acid centers and thus vary the degree of bifunctional catalysis.

Let us consider typical monometal polymeric Lewis acids and their catalytic properties. EtMgBr bound to phenol groups of PS modified by methylvinylphenol, is active in reactions proceeding by an acid–base mechanism (self-condensation of linear aliphatic aldehydes into unsaturated α- and β-adducts, isomerization of epoxides into simple carbonyl products with high yield of ketones and low yield of aldehydes) [141]. In the presence of this catalyst side reactions, for instance self-condensation of aldehydes and their co-condensation with ketones, proceeds more slowly than side reactions in the presence of the homogeneous analogue ArOMgBr.

CSDVB in chloroform interacts with BBr_3 with the formation of a stable product containing $\sim 8 \times 10^{-5}$ mol of BBr_3 g^{-1} of CSDVB [142]. This product catalyzes acetylation, etherification, formation of ketals and Friedel–Crafts alkylation. The catalyst can be recycled many times.

Polymeric acid catalysts containing aluminium, $AlCl_3$ on gel polymers (1.7–2.6% of Al) and macroporous resins (3.3–6.1%) [143], $RAlCl_2$ on sulfocationite (active in cationic polymerization of isobutylene) [144] and others, are most frequently used. These catalysts are obtained mainly by interaction of CSDVB (2–8% DVB) with a solution of $AlCl_3$ in CS_2, 1,2-dichlorethane, CH_2Cl_2, nitrobenzene or ether followed by filtration, washing with water and ether, and drying [145]. The content of bound Al decreases from 0.75 to 0.02% with increasing degree of cross-linking of CSDVB. The catalyst is effective in etherification of propionic acid by BuOH. Immobilization of a Lewis acid on a modified support, for instance the immobilization of $AlCl_3$ on CSDVB modified by Dipy [146], increases the activity of the catalyst.

The yield of p-hexylpropionate in esterification of carboxylic acids by alcohol in benzene at 298–343 K in the presence of the "mild" catalyst $AlPO_4$ bound to CSDVB is 65%, whereas in the uncatalyzed reaction under the same conditions the yield is only 8% [147]. A polymer-bound Lewis acid, AlH_3 on aminated CSDVB, reduces aldehydes, ketones and esters [148]. The catalytic activity of $AlBr_3$ on PS is often higher than the activity of supported $AlCl_3$ [138].

The reaction of anhydrous $GaCl_3$ with CSDVB in CS_2 at 293 K for 4 h results in the formation of a stable complex containing 0.287 mmol of $GaCl_3$ g^{-1} [149]. The complex is active in etherification and esterification, acetylation, synthesis of ketals and Friedel–Crafts alkylation. The catalyst can be recycled; however, in water/acetone solution at pH 7–13 immobilized $GaCl_3$ is hydrolyzed in ~ 1 h. As mentioned in Section 3.4, such complexes are initiators of cationic polymerization of α-methylstyrene [150].

$SnCl_4$ is used rather often as a Lewis acid, immobilized on various polymeric supports [151, 152]. $TiCl_4$ is easily immobilized on polymers, both from the gas phase and from solutions in CS_2, chloroform, hydrocarbons and others, with formation of a strong solid Lewis acid; main characteristics of these Lewis acids are given in Table 4.4.

---

[9] Sulforized ion-exchange resins are substantially better solid acids than alumosilicates and other oxides [139].

Table 4.4 Polymeric Lewis acids containing immobilized TiCl$_4$ and reactions catalyzed by these acids.

Polymeric support	Conditions of immobilization	[Ti] (mmol/g)	Catalyzed reaction	Ref.
CSDVB (4% DVB)	CS$_2$, TiCl$_4$	2.03	Esterification, alkylation, ketogenesis, polymerization of α-methylstyrene	[153]
Weak-base anion-exchange resin, PS modified by–NH$_2$ and =NH groups (exchange capacity is 5 mmol/g)	TiCl$_4$ in chloroform	2.59	Esterification, synthesis of ketals, acetylation	[154]
Copolymer of styrene and vinyl pirrolidone (10%) cross-linked by DVB	TiCl$_4$ in chloroform	0.994	Esterification, acetylation, alkylation	[155]
Copolymer of styrene and vinylcarbozole (17.3%) and DVB (3%)			Esterification, acetylation, alkylation, ketalation	[156]
Copolymer styrene, DVB and methacrylacetone	TiCl$_4$ in chloroform	1.4	Esterification, acetylation, alkylation, ketalation	[151]
Copolymer of styrene with DVB (3.6%) and acrylonitrile (41%).		1.03	Esterification, acetylation, alkylation	[157]
Polymer containing OH groups	Ti(OPr-i)$_2$Cl$_2$		Diels–Alder reaction (cyclopentadiene + methacrolein)	[158]

TiCl$_4$ on these supports is stable in the presence of water, can be kept for 1 year and can be recycled more than 6 times without marked loss of activity. The catalytic properties of these catalysts are close to those of supported AlCl$_3$. TiCl$_4$ and EtAlCl$_2$ complexes supported on a polymer containing hydroxyl groups are active in the Diels–Alder reaction [158]; however, regeneration of the catalysts, especially Al-containing ones, leads to a partial loss of activity. SnCl$_4$, VCl$_4$, VOCl$_3$ and others are immobilized on polymers with benzene rings due to formation of π-complexes [159]. Interaction of TeCl$_4$ with CSDVB (4% DVB) in CHCl$_3$ produces a polymeric Lewis acid stable in air and capable of catalyzing all the reactions given in Table 4.4 [160].

Immobilized bimetal chloride complexes are the strongest solid superacids. These complexes are obtained in the following way. Interaction of CSDVB (4% DVB) with an equimolar mixture of anhydrous AlCl$_3$ and TiCl$_4$ in CS$_2$ results in formation of a macrocomplex, a dibasic acid, with the ratio Al/Ti=2:1 [161]. This complex is stable in air, has higher activity (for the reactions given in Table 4.4) than TiCl$_4$/PS or AlCl$_3$/PS and can be recycled more than 7 times. Double salt PS–AlCl$_3$/SnCl$_4$ with a ratio Al/Sn=2:1 has the same features [162].

Products of interaction of anhydrous AlCl$_3$ and FeCl$_3$ with anion- and cation-exchange gel and porous resins [163] are characterized by the ratio FeCl$_3$/AlCl$_3$=4–6, Al and Fe being distributed randomly in the copolymers. These metal–polymer Lewis acids catalyze effectively esterification of dibasic acids. In the case of unsaturated acids *trans*-isomers react at a higher rate; double salts on ion-exchange resins are more active than on gel resins.

PVC is used relatively rarely as a support, although catalysts on PVC are promising. For instance, complex PVC/AlCl$_3$/Fe$_2$(SO$_4$)$_3$ catalyzes the reaction of carboxylic acids with epoxides with an efficiency of up to 90% [164].

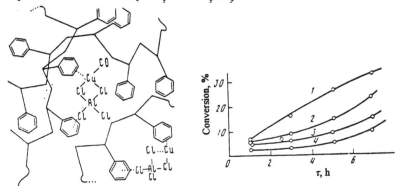

Figure 4.6 Structure of polymeric Al–Cu Lewis acids (see text).

Figure 4.7 The effect of the ratio [AlCl$_3$]/[CuSO$_4$] on the conversion of *n*-hexane. Al/Cu (mass%): 1:5.5 (1), 1:11 (2), 1:16.5 (3), 1:22 (4). T=288 K.

Complexes of AlCuCl$_4$ with PS are obtained by mixing AlCl$_3$, CuCl and PS at 323 K in 1,2-dichloroethane [165–167]. In the binuclear complexes formed Al and Cu ions bridged by chlorine atoms are bound to aromatic rings of PS with formation of charge-transfer complex (CTC) (Figure 4.6). Localization of AlCuCl$_4$ in hydrophobic domains of PS results in an interesting phenomenon, separation of CO from H$_2$O and other admixtures because of higher stability of the complexes with respect to CO, and prevents irreversible inactivation of the complexes by water vapors[10]. Moreover, homogeneous films of such polymeric double salts are stable against steam [166]. Immobilized acids PS/AlAgCl$_4$ are obtained by the same method (by keeping AlCl$_3$, AgCl and PS in benzene at 323 K for 4 h) [166,169,170]. Films made of these immobilized acids are selective adsorbents of ethylene; electroconductivity of these films is 10^8–1.4×10^9 times higher than that of pristine PS. In my opinion, polymeric double salts (Lewis acids) offer great promise as new catalysts[11].

Immobilized superacids have a relatively great ability to activate alkanes under mild conditions. It is known (see, for example, [173]) that σ-bonds of alkanes are strong, and the donor ability of alkanes is so low that they are protonated only in the presence of

---

[10] In homogeneous systems AlCuCl$_4$ decomposes in the presence of H$_2$O by the scheme [168]:

$$2AlCuCl_4 + H_2O \longrightarrow AlCuCl_4Al(OH)Cl_2 + CuCl + HCl$$
$$\downarrow$$
$$AlCuCl_4AlOCl + HCl$$

[11] For instance, it was found recently that the new superacid HGeCl$_3$–AlCl$_3$ (unsupported) is active in hydrogermalation of benzene and toluene [171]. The structure of PdCl$_2$–CuCl$_2$ complexes supported on SiO$_2$ was clarified [172].

superacids[12]. Therefore, activation of such substrates could be carried out also by superacids supported on polymers. Indeed, isomerization and dehydrogenation of *n*-butane occurs at 333–393 K in the presence of Lewis acids supported on macroporous sulfurized CSDVB [176]. Catalysts with $SbCl_5$ and BF_3 are substantially more active in isomerization than catalysts with $SnCl_4$ and $TiCl_4$.

Interaction of $AlCl_3$ and $CuSO_4$ with PVC in 1,2-dichloroethane results in formation of an immobilized superacid protected by poly(vinyl chloride) (PVC) [177]. This acid catalyzes low-temperature (at 288 K) transformation of *n*-hexane. The share of transformed *n*-hexane depends substantially on the ratio Al/Cl which determines the acidity of the catalyst (Figure 4.7). The constituent products of isomerization also depend on the ratio Al/Cu and the time of reaction (Table 4.5.).

Table 4.5 The dependence of product composition from the low-temperature isomerization of *n*-hexane in the presence of PVC/$AlCl_3$–$CuSO_4$ on the ratio Al/Cu and the time of reaction [177].

Products	Product composition (%)					
	Al/Cu=					
	1:5.5, 3 h	1:11, 3 h	1:11, 6 h	1:11, 24 h	1:16.5, 3 h	1:22, 3 h
n-Hexane	85.3	74.8	65.1	38.2	89.1	92.2
2-Methylpentane	6.0	9.7	13.1	15.8	4.8	3.7
3-Methylpentane	1.7	3.3	5.6	4.9	1.6	1.2
2,2-Dimethylbutane	0.3	1.6	2.6	6.6	1.0	0.4
Isopentane	3.5	5.6	8.1	16.2	1.7	1.3
2,2-Dimethylpentane	3.1	4.8	5.2	15.8	1.7	1.1
n-Pentane	0.1	0.2	0.3	2.5	0.1	0.1
Isoheptane	–	–	–	Traces	–	–
Total conversion (%)	14.7	25.2	34.9	61.8	10.9	7.8

$AlCl_3$ on sulfurized CSDVB is active in isomerization and cracking of *n*-hexane [178]. $AlCl_3$/$CuSO_4$ is active in low-temperature conversion of pentane (conversion up to 21% at 301 K for 3 h) [179].

The examples given above show that the study of immobilization of superacids and their use for catalysis of numerous reactions has developed intensively (see also [180,181]).

4.7 Binding and Activation of Molecular Nitrogen by Immobilized Complexes

After the discovery of the reduction of molecular nitrogen in aprotic media in the presence of metal complexes [182], many complexes of N_2 have been found. Analysis of these studies shows that effective catalysts interact with N_2 with formation of binuclear

---

[12] For instance, cracking of *n*-alkanes occurs at room temperature under the action of "aprotic superacids" RCOX·$AlBr_3$ [174]. Isomerization of *n*-hexane occurs at 343 K in the presence of the catalytic system $AlCl_3$ (30 mol %)–$SbCl_5$ (65 mol%)–KCl (5 mol%) [175].

complexes MNNM, where metals are in d^2 or d^3 states, with subsequent reduction of bound N_2 to NH_3 or N_2H_4 [183–185].

The first step of nitrogen reduction is binding of dinitrogen by d-orbitals of metals and vacant π_g^*-orbitals of nitrogen; as a rule, complexes of metals with dinitrogen have a linear structure. Diazocomplexes, such as $(Cp_2TiR)_2N_2$, where R is aryl, are stable because of the donation of π-electrons from phenyl groups via the Ti atom to N_2.

A number of complexes of dinitrogen with metal polymers have been found. The first adduct of this type was a complex of dinitrogen with cyclopentadienyl compounds of Ti^{4+} supported on modified CSDVB (up to 20% DVB) [186].

Polymeric benzcarbonyl complexes of chromium, obtained by radical copolymerization of chromium benzylacrylatetetricarbonyl or by polymeranalogous transformations of poly(styrene-co-acrylic acid) with chromium (3,5-dimethylaminobenzyl)tricarbonyl, bind N_2 reversibly [187]. UV irradiation of the Cr-containing polymers (273 K, 1 h, THF) in an atmosphere of N_2 results in elimination of CO and formation of the bond $Cr-N_2$ (for different polymers $v_{N\equiv N}=2140$ or 2120 cm^{-1}, time of reaction = 9.1 or 21 h). This process is reversible: flow of CO regenerates benzyltricarbonyl complexes of Cr. The process is described by Scheme 4.9.

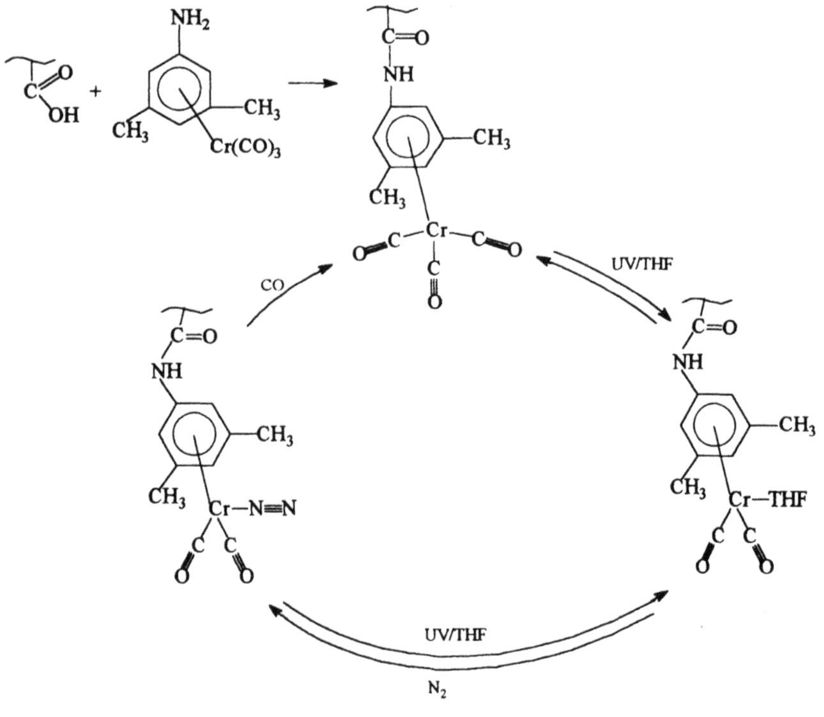

Scheme 4.9

The polymer matrix shields dinitrogen complexes from water molecules, the shielding results in an increase of the lifetime of the complexes.

A direct reaction of N_2 with Mn^{2+}-containing polymers (obtained by copolymerization of N-vinylpyrrolidone or styrene with 1,2- or 1,3-isomers of vinylmethyl(η^5-C_5H_5)Mn(CO)$_3$ in the mixture THF/benzene) results in the formation of dinitrogen complexes (via substitution of CO by THF under UV irradiation) [188]:

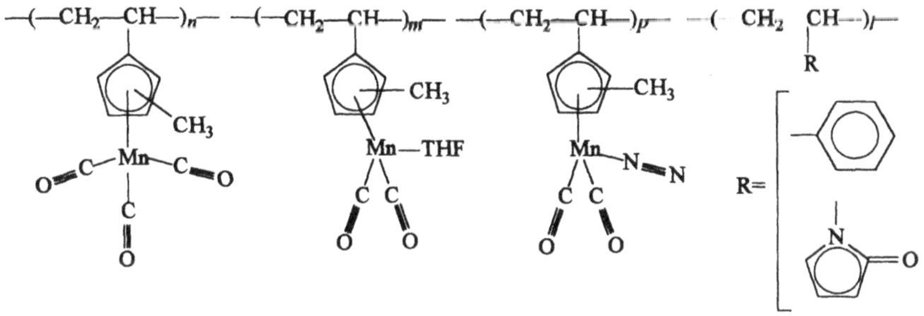

Only one CO group is substituted by THF after UV-irradiation for 2 h. The linear structure Mn–N≡N was confirmed by IR spectroscopy ($\nu_{N\equiv N}$=2160 cm^{-1}). It is interesting that the monomeric complex $CH_3(\eta^5$-$C_5H_5)Mn(CO)_3$ does not bind dinitrogen.

Polymeric complexes of Mn^{2+} with dinitrogen are obtained also by oxidation (in the presence of Cu^{2+}–H_2O_2) of polymeric complexes of Mn^{2+} with hydrazine, hydrazine being oxidized to dinitrogen. These dinitrogen complexes are stable for several days at room temperature and 6 h at 363 K; removal of N_2 is not accompanied by destruction of the polymeric complex.

Membranes, made of Mn^{2+} macrocomplexes cross-linked by DVB, bind N_2 [189,190]. UV irradiation of these membranes for 30–90 min at 273 K results in substitution of CO by N_2. In the course of this reaction λ_{max} shifts to 320 nm and an isobestic point is observed at 350 nm (Figure 4.8). The reaction proceeds on the solid–gas boundary, 14% of the CO groups being substituted by N_2.

Dinitrogen–molybdenum complexes bound to polymers by covalent links are obtained by the immobilization of $Mo^0(N_2)_2(PPh_2Me)_4$ on a methacrylamide macroligand modified by mono-, di- and tridentate ligands, tetrakis(diphenylphosphine)bis(dinitrogen) (THF, 273 K, 48 h) [191,192]. These complexes have the structure:

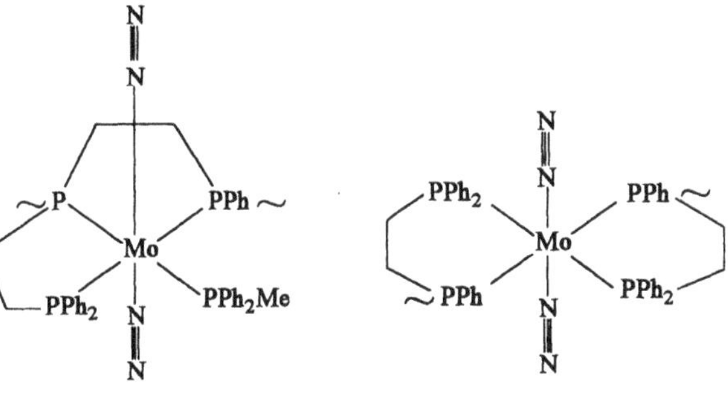

These dinitrogen–molybdenum complexes are very stable and do not decompose with release of ammonia or hydrazine; it is one of the rare cases when binding does not result in activation.

The authors of [193] believe that at present catalytic reduction of coordinated molecules of dinitrogen under mild conditions is impossible; however, several examples of N_2 activation by macrocomplexes are known. For instance, reduction of Cp_2TiCl_2, supported on brominated CSDVB or CSDVB modified by acrylic, 4-vinylbenzoic or maleic acids (content of Ti^{4+}=0.19–2.5 mass%), by lithium naphthamide (LiNp) in an atmosphere of N_2 results in formation of mononuclear (λ_{max}=564–574 nm in THF) and binuclear (λ_{max}=620–660 nm) dinitrogen complexes [194–196]. The share of dinuclear complexes increases with increasing content of supported Ti, and at 2.5 mass% predominantly binuclear dinitrogen complexes are formed, especially for polymeric supports with flexible chains. The dinitrogen–titanium complexes decompose under the action of isopropanol with formation of ammonia and hydrazine (Table 4.6).

Table 4.6 Reduction of dinitrogen on various titanium-containing polymers.

Macroligand	Degree of cross-linking (%)	[Ti] (mol%)	Yield of NH$_3$ (mol/mol of Ti)
Homogeneous system	–	–	50–65
PS (M_r 13 500)	0	0.60	41.7
CSDVB	8	0.30	67.8
Copolymer of styrene and acrylic acid (M_r 11 800)	0	0.14	150
Copolymer of styrene and maleic acid cross-linked by derivatives of bis(acrylamide)	1	0.20	123
Copolymer of styrene and 4-vinyl-benzoic acid (M_r 13 500)	0	0.58	55.0
Copolymer of styrene and 4-vinyl-benzoic acid (M_r 13 500) cross-linked by divinylbenzene	10	0.30	0.37

The yield of NH_3 (as well as the share of mono- and binuclear titanium complexes) depends substantially on the rigidity of the polymer chain (Figure 4.9). The yield increases with decreasing content of Ti in the copolymer, i.e. when Ti is dispersed randomly in the polymer matrix. The presence of residues of acrylic and maleic acids in the polymer backbone chain facilitates reduction of dinitrogen because the carboxyl groups are donors of electrons. The yield of NH_3 increases and the yield of N_2H_2 decreases with increasing temperature (195–298 K). These features are explained satisfactorily by Scheme 4.10 for the binding and activation of N_2.

Immobilized compounds of cobalt also bind dinitrogen. For instance, pre-synthesized dinitrogen complexes $CoH(N_2)(PPh_3)_3$ or $CoH(N_2)(EtPPh_2)_3$ were supported on phosphinated PS (M_r 1400–140 000) or CSDVB by the method of ligand exchange [197]. Dinitrogen complexes of this type can also be obtained by direct interaction of $Co(AcAc)_3$ with PPh_3 and $Al(isoBu)_3$ in an atmosphere of N_2 in the presence of the polymers mentioned above. The content of fixed N_2 (determined by the intensity of the band at $\nu_{N\equiv N}$ =2088 cm^{-1}) depends on many factors including the molecular mass of polymer. Hydrolysis of dinitrogen complexes (in the system

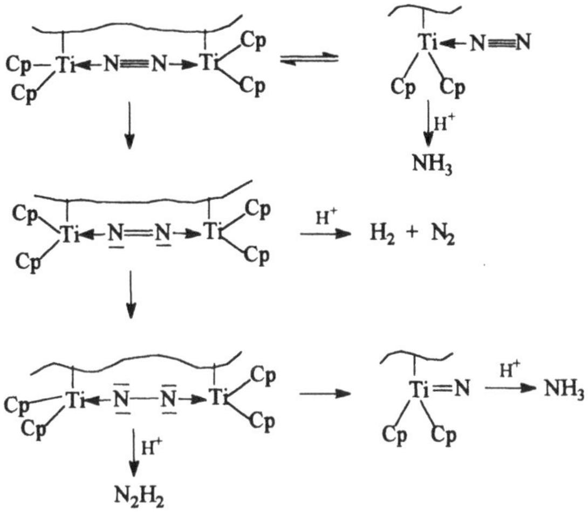

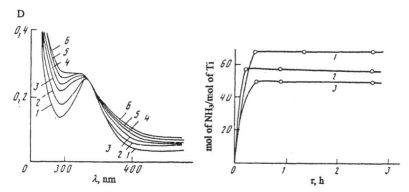

Scheme 4.10

Figure 4.8 Changes in optical spectra of the Mn^{2+}-containing polymer in the course of UV illumination in an atmosphere of N_2. Time of illumination (min): 0.2 (1), 5 (2), 10 (3), 15 (4), 20 (5), 30 (6).

Figure 4.9 Dependence of the rate of reduction of dinitrogen–titanium polymeric complex (support CSDVB) in the presence of LiNp in THF and isopropanol on the degree of cross-linking of the metal–polymer. Degrees of cross-linking: 8% (1), 2% (2) and 1% (3).

$C_{10}H_8/Li/TiCl_4/Co=20:60:2:1$) results in the formation of NH_3 with a yield of 2 mol% per complex formed.

The enzyme nitrogenase consists of two components, an iron-containing protein and a protein containing 32–34 Fe atoms and two Mo atoms [183,198]. Nitrogenase reduces dinitrogen in the presence of ATP and natural (ferredoxins, flavodoxins) or artificial ($Na_2S_2O_4$, reduced methylviologen) donors of electrons:

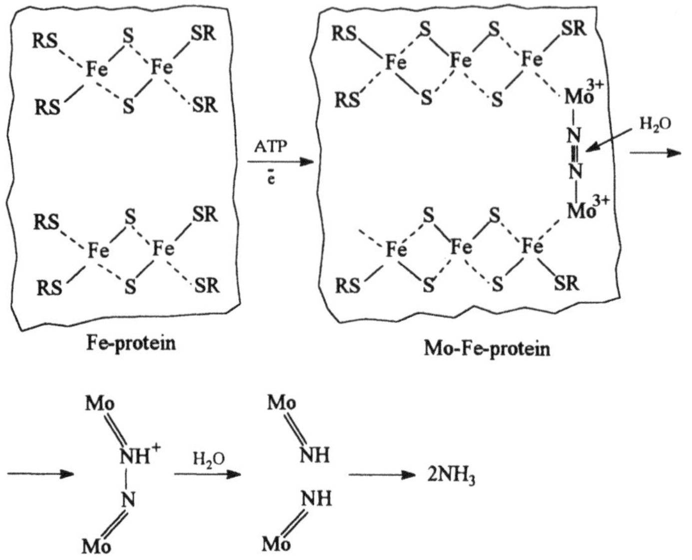

Scheme 4.11

Chemical systems able to reduce N_2 in protic media contain polynuclear metal complexes and a macroligand modeling a protein molecule. For instance, mononuclear Mo^{3+} complexes immobilized on p-mercapto-substituted PS are stabilized by ~20 phenyl rings; this stabilization prevents the formation of inactive Mo–Mo complexes [199]:

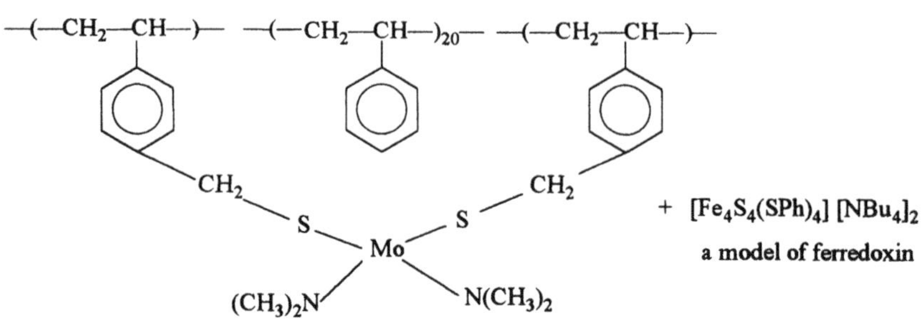

Scheme 4.12

Yields of hydrazine and ammonia in homogeneous and heterogenized systems are given in Table 4.7. Ammonia is a major product in all the systems. Activity of the immobilized catalyst is substantially higher than activity of the homogeneous analogue, and approaches activity of a polypeptide containing –Cys–Mo(NMe$_2$)$_4$ fragment [200].

Table 4.7 Reduction of dinitrogen[a] by various Mo^{4+} complexes in the presence of $NaBH_4$ and $[Fe_4S_4(SPh)_4]^{2-}$ [199].

Catalytic system	Yield (mol/mol of Mo)	
	Hydrazine	Ammonia
$Mo(NMe_2)_4 + [Fe_4S_4(SPh)_4]^{2-}$	0.001	0.007
Marcapto-substituted PS/Mo	0.013	0.15
$(NMe_2)_4 + [Fe_4S_4(SPh)_4]^{2-}$		
Random polypeptide	0.004	0.37
$[(Cys-\gamma-Glu_6)_{4.5}]/Mo(NMe_2)_4 + [Fe_4S(SPh)_4]^{2-}$		

[a]Conditions of reaction: volume of nitrogen 2000 ml, [Catalyst]=7.35×10^{-5} mol of Mo, [NaBH$_4$]=5.88×10^{-3} mol, THF/EtOH/MeOH=16:8:8 ml, T=303 K, 24 h, Fe/Mo=4. Cys=cysteine, γ-Glu=γ-glutamic acid.

Thus, there are serious reasons to expect creation of effective catalysts on the basis of immobilized metal complexes for activation of thermodynamically stable molecule N_2.

4.8 Immobilized Complexes in Photochemical Decomposition of H₂O

Over the years photocatalytic systems capable of converting solar into chemical energy have been found [183,201,202]. Photocatalytic decomposition of water is the most important reaction in these systems; it proceeds by the following hypothetical scheme:

$$2H_2O \xrightarrow[\text{Catalyst-1 Catalyst-2}]{h\nu,\ PC,\ D,\ A} 2H_2 + O_2$$

Scheme 4.13

where PC=photocatalyst, D=donor, A=acceptor.

The conversion of solar energy into a chemical fuel (creation of the complete fuel couple hydrogen–oxygen) would have great social consequences. However, this problem is far from solution.

The main steps of photochemical decomposition of water are as follows. A PC is needed because water is absolutely transparent for most of the solar spectrum; the PC, like chlorophyll, absorbs solar light and then causes chemical transformations. The PC separates the electric charges of an acceptor (A) and a donor (D) under the action of visible light:

$$A + D \xrightarrow[PC]{h\nu} A^- + D^+$$

The photoseparation results in formation of reagents for water decomposition, a strong reductant A^- and an oxidant D^+. A sensitizer S excited by quantum light supplies the energy required to carry out this endothermic reaction (a part of the photon energy transforms into energy of the separated charges):

$$A + S^* \longrightarrow S^+ + A^-$$

$$S^+ + D \longrightarrow S + D^+$$

Scheme 4.14

Catalysts Ct_1 and Ct_2 accelerate further "dark" multielectron reactions in the reduction of water by the reductant formed to H_2 and oxidation of water by the formed oxidant to O_2:

$$2A^- + 2H_2O \xrightarrow{\;Ct_1\;} 2A + 2H_2 + 2OH^-$$

$$4D^+ + 2H_2O \xrightarrow{\;Ct_2\;} 4D + O_2 + 4H^+$$

Scheme 4.15

These reactions are called reductive and oxidative half-reactions.

Porphyrins, phthalocyanines and their metal derivatives (analogues of natural photosynthetic pigments chlorophylls), dyes, and complexes of transition metals with rigid nitrogen-containing ligands Dipy and phenantroline (Phen) are used as PCs. $Ru(Dipy)_3^{2+}$ is used most often as a PC, it is chemically stable in various media, posseses strong luminescence in the visible region at room temperature and it can be reversibly reduced[13].

Complexes of V^{3+} and Eu^{3+} and most often MV^{2+} (this organic bication has the following structure[14]:

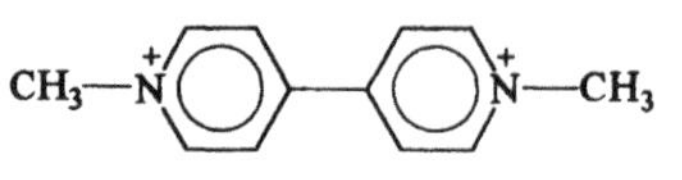

$$CH_3-\overset{+}{N}\!\!\left\langle\!\!\bigcirc\!\!\right\rangle\!\!-\!\!\left\langle\!\!\bigcirc\!\!\right\rangle\!\!\overset{+}{N}-CH_3$$

are used as acceptors. Reduced forms of these acceptors are able to reduce H_2O to H_2 at an appropriate pH. Ethylenediamine, ethylenediaminetetraacetate (EDTA), cysteine, ethanol and triethylamine are used as donors. In the course of a photoinduced reaction in dimethyl formamide (DMFA)/H_2O the reductant triethanolamine is split into $NH(CH_2CH_2OH)_2$ and $CHO-CH_2OH$, MV^{2+} and Ru^{2+} being transformed to MV^+ and Ru^{3+}.

Finely dispersed noble metals Pt and Rh and iso- and heteropolycompounds containing several (up to 12) cations of transition metals Mo, W and others, are used most often as catalysts of the oxidative half-reaction (Ct_1), although homogeneous

---

[13] The maximum of its absorption in water (λ_{max}=452 nm) is close to the maximum of solar light (λ_{max}=500 nm), its extinction coefficient is very high (1.38×10^4 l/mol cm^2) and the redox potential Ru^{3+}/Ru^{2+} (1.27 V) is sufficient to oxidize water. The rate of electron transfer from $Ru(Dipy)_3^{2+*}$ to MV^{2+} is very high, about 10^8–10^9 l/(mol s) and the lifetime of the excited state is 650 ns.

[14] A shortcoming of MV^{2+} is its irreversible hydrogenation in the presence of most catalysts for the reduction of water to hydrogen.

catalysts are used in some cases. These catalysts participate in multielectron redox reactions, like the reaction centers of hydrogenases.

Water-soluble complexes of Co and Cu with porphyrins and phthalocyanines, heterogeneous catalysts based on the hydroxides of transition metals (Co, Fe, Mn and others) and oxides of spinel type RuO_2, RuO_4 and PtO_2, colloidal platinum and others, catalyze the oxidative half-reaction.

Thus, the two-component system for the photoreduction of water by visible light can be represented as the following:

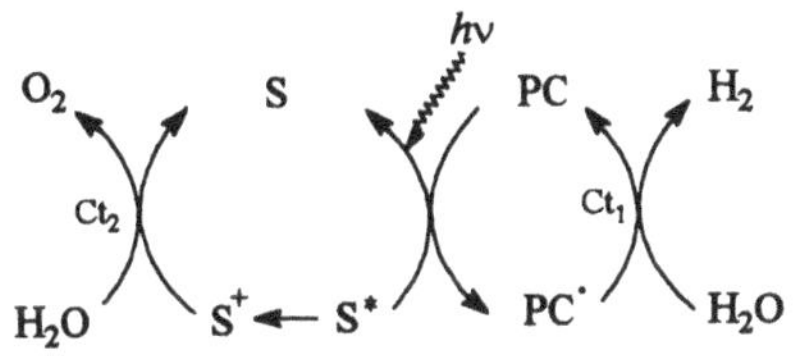

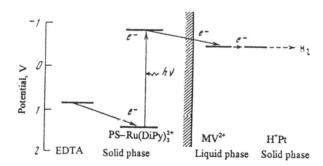

Scheme 4.16

The main problem is the stabilization of the photoseparated charges D^+ and A^-. In the homogeneous phase these charges recombine very rapidly, dispersing the stored energy (Figure 4.10). Only heterogeneous and microheterogeneous systems are really promising systems from the viewpoint of stabilization of photoseparated charges because in these systems charges are stabilized not only by chemical but also by physicochemical means. By analogy with natural photosynthesis, where lipid membranes and vesicles of micellar solutions, etc. participate in stabilization, the charges must be separated by a distance of 1–2 nm to suppress substantially their recombination.

Figure 4.10 Photoinduced separation of charges on the boundary between the solid phase [EDTA + PS–Ru(Dipy)$_3$$^{2+}$] and the liquid phase [MV^{2+}–MeOH] and evolution of hydrogen on solid platinum in the liquid phase (see text).

Immobilized metal complexes and macroligands (both soluble and insoluble) are used in many ways for solving the problem of charge stabilization [203,204]: as matrices for migration of energy, as catalysts of multielectron reactions, as membranes, as synthetic concentrators of solar light, and as *p*- and *n*-semiconductors for construction of solar batteries, etc.

Polymeric viologens are used most often [205–207]. These are obtained by polymerization and copolymerization of suitable monomers (for example, by radical polymerization of N-vinylbenzyl-N'-n-propyl-4,4'-dipyridyl bromide [208]) or by the method of polymeranalogous transformations (for example, by interaction of chloromethylated CSDVB with 4-(4'-pyridyl)-N-hexadecyl pyridyl bromide). The polymeric reagent affects the rate constants for electron transfer and the lifetime of the excited state. The electron transport toward colloidal platinum in these systems with participation of polymeric viologens is shown in Figure 4.11. The rate of luminescence quenching[15] in the system Ru(Dipy)$_3^{2+}$–viologen-containing polymer is half that for the system Ru(Dipy)$_3^{2+}$–MV^{2+} because of a lower probability of encounters of Ru(Dipy)$_3^{2+}$ with viologen links of the polymer chain. The rate of H$_2$ release in these systems is lower than in homogeneous systems for the same reason [210].

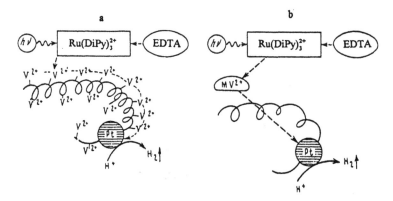

Figure 4.11 Model of electron transfer along immobilized viologen (V^{2+}) groups on poly(N-vinylbenzyl-N'-n-propyl 4,4'-dipyridylbromide) chloride (a) and in colloidal platinum on PVPr (b).

Photosensitizers are immobilized more often than electron carriers. They are immobilized by various methods. Bengal Pink was bound covalently at 333–353 K to beads of chloromethylated CSDVB in DMFA [211]. Bengal Pink was also grafted onto the surface of hydrophilic polymers (for instance, the copolymer of chloromethylstyrene and the monomethacrylic ester of ethylene glycol) [212]. Polymeric pigments (models of light-harvesting antennas) were obtained by radical copolymerization of 1-vinyl-1,2,4-triazole with 1,4-DVB in the presence of a fluorescent pigment [213]. Photosensitive groups were introduced into polymeric chelates, most often derived from azoaromatic polyureas [214]. The most promising systems are those in which macromolecular complexes of Ru(Dipy)$_3^{2+}$ were used as PCs. These macrocomplexes were obtained in two ways: by copolymerization of Ru-containing monomers, for instance [Ru(Dipy)$_2$(4VP)]$^{2+}$, [Ru(Dipy)$_2$(4VP)Cl]$^+$ or [Ru(Dipy)$_2$(4VP)(CO)]$^{2+}$, with styrene [215],

---

[15] Note for comparison that rate constants for the oxidation of viologens in solution are close to the diffusion rates (10^8–10^{10} l/(mol s)); incorporation of viologen molecules into a bilayer lipid membrane results in a decrease of the rate constants to 4.6×10^6–2.7×10^3 l/(mol s) whereas the rate constant for electron transfer from a radical localized on the outer surface of the membrane is ~ 20 s^{-1} [209].

4VP [216] and others [217], as well as by supporting Ru(Dipy)$_3$$^{2+}$ on polymers containing usually Py and Dipy groups (see, for example [218–221]). In comparison with monomeric analogues, the macrocomplexes are usually characterized by almost the same sensitivity with respect to photoreduction of MV^{2+} and longer lifetimes (τ^*) of photoexcited states (500–600 ns), whereas for quaternized polymeric complexes the values of τ^* are substantially shorter (for instance, in the case of bromine octyl τ^* is shorter by 10 ns due to the effect of the heavy atom Br [220]).

The intensity of the luminescence of the Ru(Dipy)$_3$$^{2+}$ complex on a polymer is 4 times less than that of monomeric analogues (Figure 4.12) because of a specific interaction of the complex with the polymer chain and the solvent[16] [216]. The decrease of intensity of luminescence of Ru(Dipy)$_3$$^{2+}$ immobilized on poly(p-sulfurous acid) or its sodium salt was explained by formation of polymer regions with a high local density of negatively charged hydrophobic groups which adsorb cations of complexes [223]. It is assumed [218] that the macrocomplex creates conditions facilitating photoinduced charge separation between liquid and solid phases. Intramolecular interactions (MV^{2+} as PC) do not decrease the rate of quenching of excited dimeric Ru^{2+} complexes, supported on polypyridine and divided by two or three methylene groups, because of slow migration of electrons through this sequence of groups separating Ru^{2+} complexes [221].

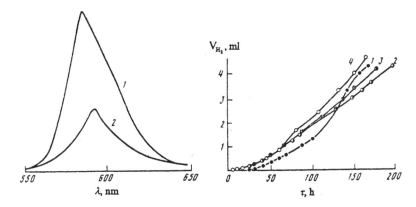

Figure 4.12 Luminescence spectra of Ru^{2+} complexes in water at room temperature. (1) [Ru(Dipy)$_2$(Py)$_2$]Cl$_2$, (2) [Ru(Dipy)$_2$(P4VP)(Py)]Cl$_2$; [Ru]=1.3×10^{-5} mol/l.

Figure 4.13. The rate of H$_2$ evolution under the action of Ru complexes immobilized on gels of PS (1) or a gel of saccharose–MMAc (2–4).

Let us consider only a few examples of the effect of immobilization of metal complexes on the reductive half-reaction of water photodecomposition (two-electron reduction of water to H$_2$). Evolution of hydrogen from water at a rate 0.037 ml/g was observed in the

---

[16] The excited state of Ru(Dipy)$_3$$^{2+}$ supported on a copolymer of 4-methyl-4'-vinyl-2,2-dipyridine and acrylic acid is quenched by MV^{2+} in CH$_3$OH; various additions affect the quenching [222]. The role of CH$_3$OH is to stimulate sequential binding of MV^{2+} to domains of Ru complexes.

presence of $Ru(Dipy)_3^{2+}$ immobilized on hydrophobic supports, poly(4-aminostyrene) or gels of saccharose-modified methacrylate (Figure 4.13) [224]. The catalytic systems retain their activity for 8 days (Table 4.8).

Table 4.8 Photocatalytic formation of hydrogen in the presence of $Ru(Dipy)_3^{2+}$ immobilized on a gel[a] of saccharose-modified MMAc [224].

Molar ratio polymeric support/dipyridyl	[Ru] ($\mu mol/g$)	W (ml of H_2/g h)	Total amount of formed H_2 (ml)
1:3	–	–	–
1:5	153	0.025	4.35
1:3	258	0.029	4.27
1:0.6	282	0.036	5.50
1:2	630	0.037	6.20
1:2	524	0.023	0.90[b]
1:5	286	0.015	0.60

[a]Support is

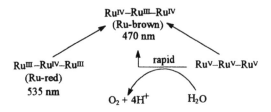

Conditions of reaction: concentrations of Ru, MV^{2+}, EDTA and Pt are 8.8×10^{-4}, 2.56×10^{-2}, 2.0×10^{-2} and 5.5×10^{-4} mol/l, respectively; total volume of solution is 25 ml.

[b]Illumination for 40 h; in the other cases for 160 h.

Complexes of Rh^{3+} with PEI were the first catalysts for the reduction of water to hydrogen by such one-electron reductants as cation-radicals of methylviologen and aqua-ions of V^{2+} and Cr^{2+} [225–229]. The catalytic properties of these metal polymers depend on the ratio [Rh]/[N] (Figure 4.14). Coordinationally unsaturated Rh^{3+} complexes form multinuclear associates inside polymer globules. These associates are formed by reduction of the complexes and are chemically bound to the polymer. Probably, these associations are active centers for the catalytic reduction of water to hydrogen.

Recently [230] trinuclear ruthenium complexes were identified which were used as water oxidation catalysts in a homogeneous aqueous solution as well as in a Nafion membrane-incorporated system. It was shown that Ru-red complexes adsorbed into a Nafion photosynthetic membrane underwent oxidation to form Ru-brown initially and further oxidation to give a higher oxidation state complex. The latter was readily reduced by water molecules and turned back to Ru-brown without decomposition:

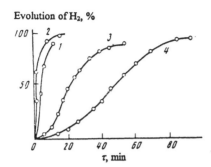

Evolution of H_2, %

τ, min

Figure 4.14 Kinetics of evolution of H_2 (percentage of stoichiometric quantity) in the presence of various Rh complexes. (1) $Rh(en)Cl_3$, (2,3) Rh–PEI at $[Rh]/[N]=1:2$ and $1:4$, respectively, (4) $Rh(en)_2Cl_2(ClO_4)$, $[Rh]=5\times10^{-4}$ mol/l, 298 K. en=ethylenediamine.

The highest value of the SCA (640 mol of H_2/g-atom of Rh h) was found for Rh^{3+} supported on nitrogen-containing anionites [225]. Polymeric rhodium catalysts have high stability; their activity does not diminish even after 6000 cycles. Moreover, in many cases the rate of the reaction is limited by mass transfer in the polymer pores, and not by the catalytic reaction itself.

Reduced forms of 12-silicon–tungsten acid reduce H_2O to H_2 in the presence of immobilized rhodium complexes at a high rate (SCA~5 mol of H_2/g-atom of Rh h) [227,228].

Polyphthalocyanines of Co^{2+}, Ni^{2+} and Cu^{2+} obtained by polymer-analogous transformation are active in photodecomposition of water and are able to reduce MV^{2+} to MV^+ (the initial rate is $(1.5–3.7)\times10^{-7}$ mol/(l min)) [231]. Polymeric macrocyclic N_4 and N_2O_2 complexes of with Zn, Mg, Mn, Fe and Ru are also very effective in the sensitized photodecomposition of water in the presence of a regular electron-transferring system, Zn^{2+} macrocomplexes being the most active [232].

The catalytic activity of noble metals for the photodecomposition of H_2O depends on their degree of dispersion and is substantially higher for colloidal dispersions. Natural and synthetic polymers are used widely for the stabilization of colloids (see Section 1.8). For instance, the activity of colloidal Pt depends mainly not on dispersity but on the protective ability of the polymer support (PAP) to stabilize the colloid. PAP is measured by the volume of polymer required for stabilization of 1 g of Au in a 1% NaCl solution (Table 4.9). To obtain colloidal catalysts, noble metals are sometimes confined in microemulsions, vesicles and polymerized vesicles of SAS [234]. In this case an optimal concentration of colloidal Pt should be maintained because colloidal Pt can catalyze also hydrogenation of MV^{2+} and, thus, deactivate it. Vesicular systems for reduction of water to hydrogen contain platinum metal particles[17] of size 1.4 ± 0.3 nm [235]. The oxidative half-reaction (four-electron oxidation of H_2O) is also catalyzed by colloidal catalysts for O_2 evolution:

---

[17] Particles of such a size should be considered not as metal, but multinuclear associations of Pt atoms.

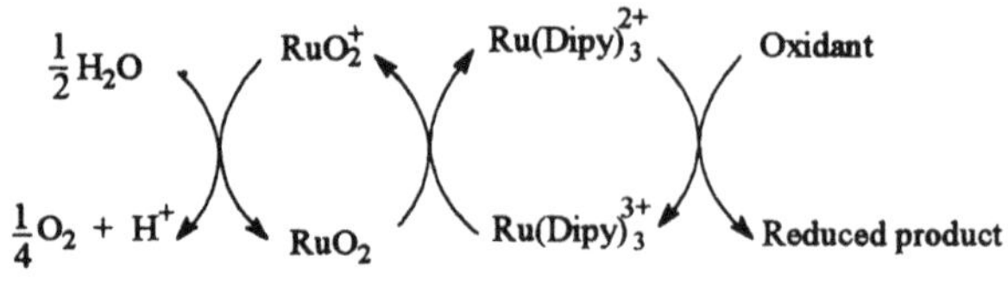

$$\frac{1}{2}H_2O \quad \diagdown \quad RuO_2^+ \quad \diagup \quad Ru(Dipy)_3^{2+} \quad \diagdown \quad \text{Oxidant}$$

$$\frac{1}{4}O_2 + H^+ \quad \diagup \quad RuO_2 \quad \diagdown \quad Ru(Dipy)_3^{3+} \quad \diagup \quad \text{Reduced product}$$

Scheme 4.17

In addition to RuO_2, stabilized RuO_4 and PbO_2, colloidal particles and multinuclear complexes of a number of metals can be used to oxidize water. These carry out multielectron oxidation of water to O_2, functioning as a reservoir of holes. New catalysts, Co^{3+} complexes immobilized on ion-exchange resins with additions of Al^{3+}, Fe^{3+} or Ce^{4+} ions, have been proposed for the oxidation of H_2O to O_2 at pH 9 [236].

Table 4.9 Photoinduced evolution of H_2 from water in the presence of platinum catalysts stabilized by various polymers[a] [233].

Polymer	PAP	Catalyst		Rate of H_2 evolution
		Type	Mean diameter (nm)	(l/day)
PEI	0.04	Complex	–	0.1
Poly(allyl amine)	1.3	Particles	–	–
Poly(vinyl methyl ester)	–	Colloid[b]	–	0.1
PVAl (*PD*=500)	5	Colloid	3.4	2.1
Poly(acrylic acid)	0.07	Colloid	2.9	2.5
Hydroxyethylated cellulose	–	Colloid	2.9	3.1
PVPr (*PD*=3250)	50	Colloid	2.9	4.0
Gum arabic	–	Colloid	2.0	3.3
Gelatin	90	Colloid	1.6	3.2

[a]Concentrations of $Ru(Dipy)_3^{2+}$, MV^2, EDTA–Na and Pt are 5×10^{-5}, 5×10^{-3}, 5×10^{-2} and 6.6×10^{-5} mol/l, respectively.
[b]Platinum formed after illumination

In the presence of an ion-exchange resin with iminodiacetate (IDA) chelating groups, photoreduction of oxygen proceeds by the scheme [237]:

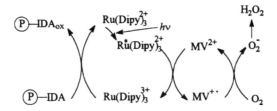

Scheme 4.18

The effectiveness of H_2O_2 formation in polymeric systems is 10–20 times higher than with homogeneous analogues because of inhibition of the back reaction due to binding of the Ru^{3+} formed, as well as due to facilitation of the interaction of two positively charged reactants and more effective use of the polymeric donor.

Let us compare the activities of natural and synthetic systems for photoconversion. The most promising metal polymer systems (for instance, Ru^{3+}/PEI) produce up to 1×10^3 mol of H_2/(g-atom of M h). Hydrogenases, the best natural catalysts of H_2 evolution, have activities of 10^4–10^7 mol of H_2/(mol of enzyme h). These activities, represented as activities per g of catalyst, are as follows: 1 mol of H_2 /(g of M h) for metal–polymer systems and 0.2–200 mol of H_2/(g of enzyme h). However, hydrogenases are effectively deactivated instantaneously in acidic media.

We can expect rapid progress in modeling of oxygen-evolving photosystem II of plants. These contain polynuclear metal complexes as elements of electron-transporting chains or catalytic centers carrying out redox transformations of substrates [198]. Natural polypeptide polymers bind metal complexes and create an optimal micro-environment for the effective functioning of the metal complexes.

A plausible pattern for the binding of a tetranuclear manganese complex to polypeptides of oxygen-evolving center in photosystem II of plants was proposed in [238]. It is assumed that this complex catalyzes the oxidation of water to oxygen, the most important natural process. The exact structure of the complex is unknown but there are serious reasons to think that the complex is a cubane-like four-manganese-atom cluster [238–241] immobilized on protein by carboxylated groups of polypeptides D_1 and D_2 of the reaction center.

Carboxylated groups are most suitable for this binding because they form stable complexes with manganese in oxidation states 3+ and 4+ in aqueous media, in contrast to the amide groups of side polypeptide chains. The amide groups can be hydrolyzed to carboxylates on coordination with Lewis acids. The weak acidity of the medium of the inner surface of the thylakoid membrane, where the manganese complex is localized, facilitates the hydrolysis. Therefore, they assume that the most probable sites of the manganese cluster binding are sequences Asp-308-Val-Val-Ser-Gln-304 (where Asp=aspartic acid, Val=valin, Ser=serine and Gln=glutamine) in polypeptide D_1 and 304-Se-Val-Phe-Asp-300 (where Phe=phenylalanine) in chain I_2 [238]. Both the sequences have two carboxylated ligands, Asp and glutamic acid (Glu) (hydrolyzed glutaminate Gln), as shown in Figure 4.15. The binding of two terminal carboxyl groups of a 33 kDa apoprotein complete the coordination sphere of the manganese ions.

In each catalytic cycle of water oxidation the manganese complex, transferring electrons to a primary acceptor generated by light, accepts four one-electron oxidative equivalents which are used for oxidation of two coordinated oxoligands with formation of the O–O bond of dioxygen [239]. Probably, the carboxylated groups of the side chains of the polypeptide, which bind the complex, prevent its destruction after release of dioxygen from the manganese coordination sphere. This was shown for a number of model polynuclear manganese complexes [242]. Immobilization of a manganese catalyst for water oxidation on the surface of lipid vesicles increases substantially the yield of oxygen [243].

As seen from the analysis given above, conversion of water is a difficult problem, therefore researchers have investigated intensively molecular photocatalytic systems (including systems of immobilized complexes) capable of storing solar energy in

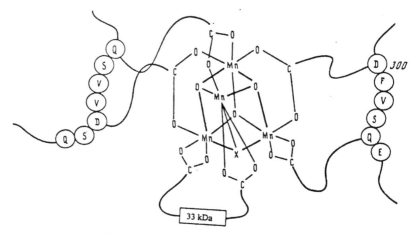

Figure 4.15 Scheme of binding of four-nuclear manganese complex with peptides of the oxygen-evolving center of photosystem II of plants (see text).

reactions different from water decomposition. Transformations of strained, cyclic, unconjugated hydrocarbons into the corresponding conjugated compounds are the most extensively studied reactions of this type, for example isomerization of norbornadiene to quadricyclane in the liquid phase in combination with catalytic isothermic (ΔH=1090 kJ/kg) back transformation under the action of various catalysts:

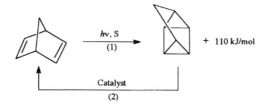

Scheme 4.19

Carbonyl compounds (acetophenone, benzophenone and others) and Cu^{1+} complexes and other transition metals are sensitizers for this reaction [244].

This reaction is interesting from the viewpoint of practical applications because of the high value of the stored energy (chemical not thermal energy) and the stability of quadricyclane (at room temperature reaction 2 is forbidden by the law of conservation of orbital symmetry, whereas at 413 K the period of half-decay of quadricyclane is longer than 14 h). The high stability of quadricyclane enables unstable and intermittent solar energy to be stored.

Immobilization of the sensitizer or catalyst of reaction 2 improves substantially the effectiveness of the process, because contamination of reactants by sensitizer and catalyst decreases the effectiveness. For instance, immobilized sensitizer 4(N,N-dimethyl-amino)benzophenone is more effective than its monomeric analogue [245]. Co^{2+}–tetraphenylporphine (TPP) supported on CSDVB [246], immobilized phthalocyanines of Mg, Mn, Fe and Ru [247], metal porphyrins supported on ion-exchange resins [248], complexes of Co, Rh and Pd supported on modified CSDVB [249], were used as catalysts of the back reaction 2. They are characterized by high effectiveness, ease of isolation and reductive regeneration, and effective concentration of substrate.

Co²⁺ chelates immobilized on CSDVB are the most promising catalysts for this reaction [250,251]. The heterogenized complexes are more active and stable than their low molecular mass analogues (Figure 4.16), however a decrease in size of the pores of the support leads to a decrease of catalyst activity because of diffusion limitations. Fragmentation of catalyst particles increases their activity by one order of magnitude.

Figure 4.16 Kinetics of isomerization of norbornadiene by low molecular mass (1,2,3) and macromolecular (1',2',3') Co²⁺ chelates with different structures of the chelating units.

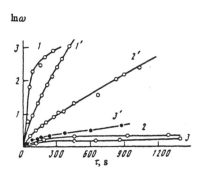

As in many other catalytic processes, the transfer of catalytic centers from the bulk of the polymeric support to its surface increases substantially their activity in isomerization of quadricyclane to norbornadiene. This was shown for immobilized Co-containing catalysts [252] obtained as follows. Firstly chloromethylstyrene from the gas phase was grafted by γ-irradiation to the surface of powders or granules of low density poly-ethylene, high density polyethylene, glass or cellulose fibers (grafting degree 1.7–76.4 mass%). The grafted polymers were then treated with cobaltocene or cobalt tetra(p-carboxylmethyl)porphyrin. These catalysts are easily regenerated and can be used many times without loss of activity.

Other simple methods of conversion of solar into chemical energy, for instance direct photocatalytic dehydration of alcohol to aldehyde and hydrogen, or conversion of sulfur dioxide and water (hydrated form of sulfuric anhydride) to hydrogen and sulfuric acid in molecular photocatalytic systems proposed by the Westinghouse firm:

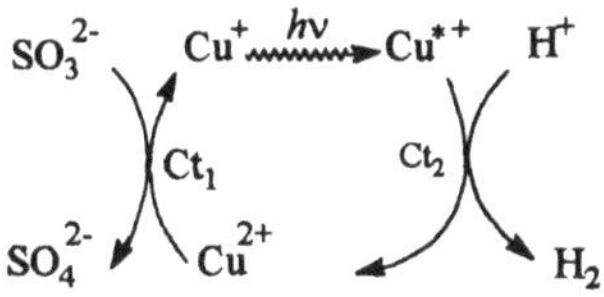

$$SO_3^{2-} \diagdown \quad Cu^+ \overset{h\nu}{\rightsquigarrow} Cu^{*+} \quad H^+ \diagup$$
$$\diagup Ct_1 \qquad Ct_2 \diagdown$$
$$SO_4^{2-} \diagup \quad Cu^{2+} \qquad \diagup H_2 \diagdown$$

Scheme 4.20

are used very rarely.

Macroligands facilitate various phototransformations of immobilized catalysts (compare with the influence of catalyzed reactions on immobilized catalysts considered in other sections). Phototransformations of complexes MX_n on polymeric matrices occur most often. These numerous reactions are beyond the scope of the present book, and only two examples are presented here. Illumination of the system Cu^{2+}–PVS by visible light causes d–d transitions in Cu^{2+} complexes; in the dark the system returns back to its initial state [253]. Illumination of the Cp_2TiCl_2 supported on PMMA results in reduction of Ti^{4+} [254]:

$$Cp_2TiCl_2 \xrightarrow{\ h\nu\ } \dot{C}p + Cp\dot{T}iCl_2$$

Catalytic systems based on finely dispersed noble metals can also hydrogenate macroligands. This results in gradual destruction of the photosystems. Processes of photoreduction and photooxidation of macroligands, including those resulting in chain destruction (see, for example [255,256]), were studied in detail and should be taken into account when designing systems for conversion of solar energy. Immobilized metal complexes are promising chemical materials for such systems.

4.9 Electrochemical Stimulation of Catalytic Processes in the Presence of Immobilized Complexes

Metal complexes supported on polymers are used relatively often for electrocatalytic activation of kinetically and thermodynamically stable small molecules. As a rule, metal polymers are supported on electrodes, most often based on pyrolyzed graphite. These electrodes are called chemically immobilized electrodes and were made primarily to accelerate electrode reactions[18] [258].

Polymer films on electrodes possess an electroconductivity due to inclusion of metal complexes or organic compounds (for example, methylviologen); electroconductive coats can be obtained also from conducting polymers (PA, poly(p-phenylene), polythiophene, polythiazole SN_x, polyaniline, polypyrrole and others) [259]. Coats can by obtained by adsorption, immersion of electrode into polymer solution, electro- and photocoating, casting and different variants of polymerization, including polymerization *in situ* [204, 260–262]. There are four mechanisms of action of electrodes coated by polymer films: (i) interphase reactions between polymer film and solution due to the elecroconductivity of the film, (ii) diffusion through pores and channels and then reaction on electrodes, (iii) electron transfer through the polymer film (electrocatalysis), and (iv) diffusion into films (considered as a highly viscous medium).

The thicknesses of such films are 100–1000 molecular layers and the concentrations of electroactive centers are 10^{-10}–5×10^{-6} mol/cm^2; therefore their concentration in volume is high and reaches 0.1–0.5 M. To increase the adhesion of films to the electrode, its surface is modified in appropriate way. Electron transport in poly[Ru(Dipy)$_3$]$^{2+}$ film[19] can be represented by the following scheme:

---

[18] Moreover, even the simple introduction of polymers into a system can affect substantially electrode processes. For instance, the mechanism of electrode charging in ferrocene–ferricene system changes after adding P4VP due to formation of the complex ferricene–P4VP [257].

[19] Synthesis and coordination chemistry of films of poly(4-vinyl-4'-methyl-2,2'-Dipy) on the surface of electrodes are considered in [263].

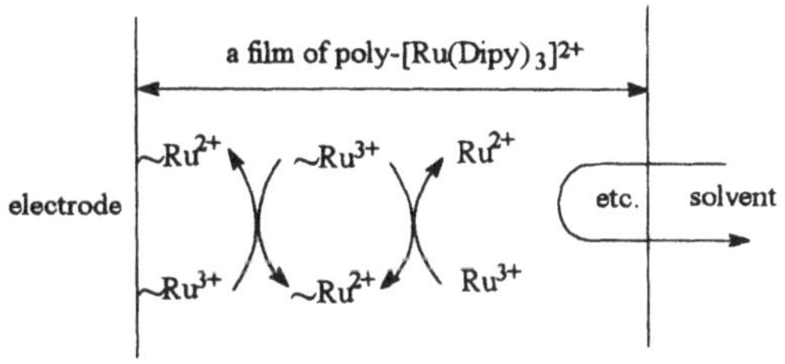

a film of poly-[Ru(Dipy)$_3$]$^{2+}$

electrode

$\sim$Ru^{2+} $\sim$Ru^{3+} Ru^{2+}

$\sim$Ru^{3+} $\sim$Ru^{2+} Ru^{3+}

etc. solvent

Scheme 4.21

A polymeric film with electrochemically active hydroquinone groups supported on glass-like carbon is considered in [264] as a multilayer system in which the internal layer undergoes electrochemical transformations, then further charge transfer to the boundary with the electrolyte proceeds by a redox reaction. For sufficiently thick films, oxidation of hydroquinone to quinone proceeds relatively rapidly. In some cases, for instance in oxidation of bis(hydroxymethyl)ferrocene on an electrode, it is believed that the a catalytic effect is caused by the influence of the redox system on charge transfer to the boundary with the electrolyte. Addition of organic solvent to the electrolyte, on the one hand, facilitates charge transfer due to swelling of the polymer, and on the other hand, results in the destruction of external layers of polymer or desorption of active groups from these layers.

Bilayer electrodes are known which consist of two types of electroactive films with different redox potentials. In such composites the external layer is influenced by the solution, and the internal layer comes into direct contact with the electrode surface. Typical bilayer electrodes have the following structure: the internal layer is made of poly[Ru(Dipy)$_3^{2+}$–vinylpyridine], the external layer is made of poly(vinylferrocene) (PVFer). The energy levels of this bilayer electrode [265] are shown in Scheme 4.22.

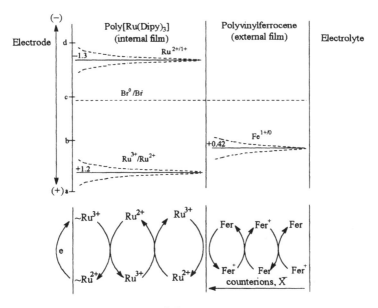

Scheme 4.22

At electrode potential a the outer film participates in the reaction:

$$poly[Ru(Dipy)_3]^{2+} + PVFer \rightarrow poly[Ru(Dipy)_3]^+ + PVFer^+ .$$

The outer film is reduced at potential c, the internal film is oxidized at potential d:

$$poly[Ru(Dipy)_3]^+ + PVFer^+ \rightarrow poly[Ru(Dipy)_3]^{2+} + PVFer .$$

The most convenient method of obtaining electroconductive films on electrodes is electrochemical polymerization of pyrrole monomers, polypyridyl complexes of Ru^{2+}, Fe^{2+}, Ru^{1+}, Cu^{2+}, Co^{2+}, Ni^{2+} and other metals [266], and organic cations [267] (see below).

Electrochemical polymerization of pyrrole on an electrode proceeds by the following mechanism:

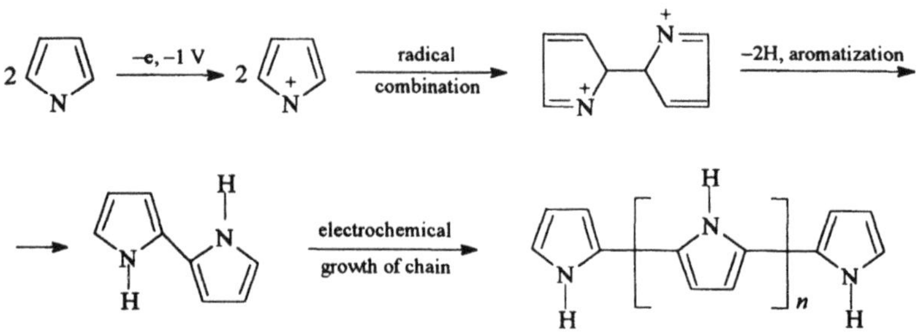

Scheme 4.23

For electrochemical purposes metal-containing monomers have also been used, for example pyrrole–tris(dipyridyl ruthenium(2+)):

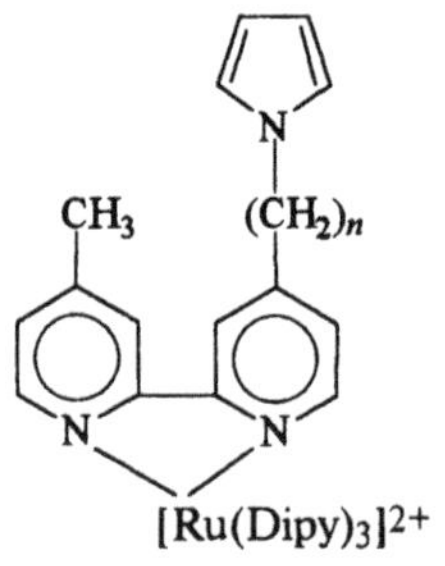

$[Ru(Dipy)_3]^{2+}$

Particles of various metals, for instance platinum, can be introduced into the polypyrrole films in the course of this polymerization [268]. Particles can also be formed by electrochemical reduction of platinum tetrachloride ion, supported on oxidized polypyrrole, to metallic Pt [269]. Moreover, bulky dianions, Fe–S clusters, can also be introduced into the films in the course of the polymerization [270].

Polymeric film can be deposited on electrodes made of pyrolyzed graphite by electrochemical oxidation of o-phenylenediamide or its N-methyl derivatives (aqueous solutions of Na_2SO_4/H_2SO_4 or $NaClO_4/HClO_4$, pH 1) [271]. Film thickness and concentration of redox centers on the surface of the film are proportional to the value of the electric charge passed through the solution.

Polymeric multifunctional films were deposited on Pt or glass-like carbon by anode oxidation of aniline and its derivatives [272]. Depending on the nature of the monomer, films reveal (o-phenylenediamide, N-methylalanine and others) or do not reveal (chloranil, m- and p-phenylenediamide and others) a redox transition characteristic of monomers. Polymeric derivatives of sulfoaniline acid, p-aminophenol and others are soluble and not deposited on electrodes. Doped polyacetylene can be used as a material for the electrodes of batteries and accumulators with aqueous electrolytes [272].

Electrochemical stimulation of immobilized metal complexes are used most often for catalysis of numerous redox reactions. For instance, typical heterogeneous catalysts deposited on electrodes are used for hydrogenation, hydrogen required for this reaction is evolved in the course of water electrolysis.

A carbon electrode ($20\times20\times10$ mm) modified by poly(pyrrole viologen) film ($2-5\times10^{-5}$ mol of MV^{2+}) with microparticles of Pt ($3-7\times10^{-5}$ g-atom of Pt) reduces various organic substrates [273] (Table 4.10).

Table 4.10 Electrochemical hydrogenation in the presence of Pt–poly(pyrrole viologen) deposited on carbon electrode[a] [273].

Substrate	Amount (mmole)	Consumption of charge (electron/ molecule)	Product	Yield (%)
	15	2		100
CH=CHCOOH	7	4	CH₂CH₂COOH	95
—C≡C—	3	4	CH₂CH₂—	98
CHO OCH₃	4	2	CH₂O OCH₃	100
NO₂	8	6	NH₂	86
	4	4	1% 17% 60%	78

[a]Current 100–300 mA, potential –(0.4–0.5) V.

This catalyst is very stable: 5000 ketone molecules are hydrogenated on each Pt atom without a marked decrease of activity. If the film does not contain viologen, catalytic current decreases by 2.5 times and activity decreases by 75% after hydration of 120 substrate molecules on 1 Pt atom. This shows that viologen is important for conductivity and stability of catalytic films. In some cases these catalysts are selective. For instance, predominantly vicinal and cyclic double bonds of carvone are hydrogenated (see Table 4.10).

Electrochemical hydrogenation of CO is used to obtain methanol[20] (293 K, electrode is coated by a conducting polymer containing $Ru(Dipy)_3^{2+}$) [274]. This electrochemical reaction is carried out at the potential −0.1 to 0.5 V and the current density 1–100 mA/cm^2.

Electrochemical oxidation with participation of immobilized complexes is no less widespread than hydrogenation, for instance electrooxidation of carbinol [275], epoxidation of cis-cyclooctadiene by molecular oxygen on an electrode modified by poly(pyrrole–manganese-porphyrin) [276]. RuO_2 particles included in polypyrrole–$Ru(Dipy)_3^{2+}$ films oxidizes selectively benzyl alcohol to benzaldehyde with TN>5000 [277].

Not only the modified electrode, but also the polypyrrole itself containing manganese and porphyrin[21] has high activity in the oxidation of 2,6-di-tert-butylphenol by dioxygen to 2,2,6,6-tetrabutyldiphenoquinone [280]. At the same time an electrode coated with Cu^{2+} macrocomplex carries out only oxidative polymerization of 2,6-dimethylphenol [281]:

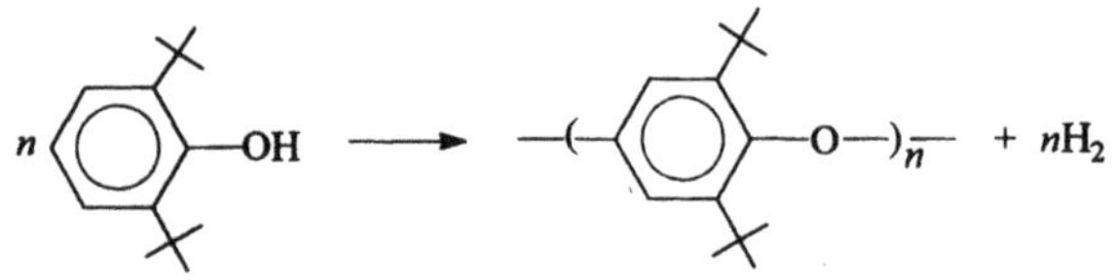

Scheme 4.24

The amount of poly(2,6-dimethylphenylene oxide) formed with molecular mass ~10^4 depends on the charge passed, the share of oligomers decreasing with an increase of charge (Figure 4.17). The reaction proceeds by the scheme (compare with Section 3.6.1):

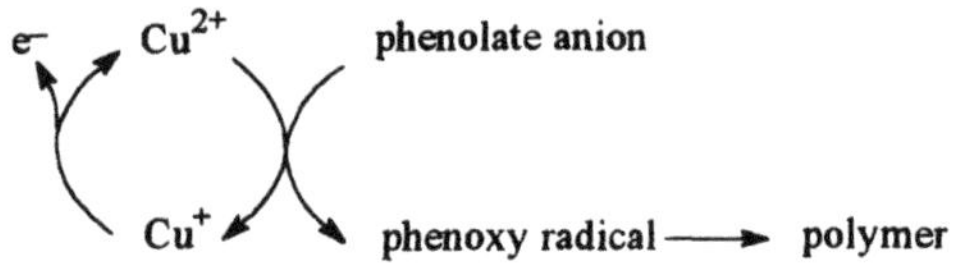

Scheme 4.25

---

[20] The reaction includes the following steps: CO adsorbed on the electrode is oxidized to CO^+, then CO^+ transforms to radical COH which is oxidized further to COH^+; then COH^+ adds H_2 with formation of CH_3O^+ which is reduced to the CH_3O radical. This radical attaches a proton with formation of a cation-radical $CH_3OH^{+\cdot}$ which is reduced to MeOH: $CO \rightarrow CO^+ \rightarrow COH$ $\rightarrow COH^+ \rightarrow CH_3O^+ \rightarrow CH_3O \rightarrow CH_3OH^{+\cdot} \rightarrow CH_3OH$.

[21] Electrocatalytic reactions with the participation of metal porphyrins (cathode reduction of O_2, H_2O_2 and CO_2, anode oxidation of organic and inorganic compounds) were considered in a review [278]. Photophysical, electrochemical and luminescent properties of complexes Ru–polypyridine were also considered in a recent review [279].

A kinetic model of electrochemical reactions catalyzed by polymeric redox electrodes was proposed in [282]. This model takes into account, as limiting factors, electron transfer to the surface of the film and diffusion of substrate through the film to the electrode. The catalytic effect is at a maximum at a certain thickness of the film. This feature of Ru–polymer complexes on electrodes (Pt, SnO₂, glass) is caused by two factors [283]: spatial separation of centers along the polymer chain and variation of center surroundings. It was shown theoretically [284] that under certain conditions the limiting step is a charge transfer through the polymer film, not a charge transfer between electrode and catalyst.

Figure 4.17 Dependence of the yield of polyphenylene oxide (in %) in electrochemical polymerization of 2,6-dimethylphenol on the charge passed. (1) high molecular mass (10^4) polymer, (2) monomer + oligomer.

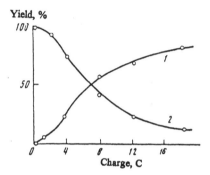

Some electrochemical catalysts carry out selective oxidation and reduction of other organic and inorganic substrates, for instance $IrCl_6^{2-}$ and $Fe(CN)_6^{3+}$. Electroreduction of dioxygen is the most important reaction of this type [285]. Most of the known electrocatalysts reduce O_2 to H_2O_2, and only some are able to reduce O_2 to H_2O, for instance complexes of Cu^{2+} with P4VP on pyrolyzed graphite [286], polymeric cobalt phthalocyanines [287]. Synthesis of H_2O from O_2 and H_2 under the action of Pt compounds, for instance $H_2PtCl_6 \cdot 6H_2O$ or Pt supported on modified soot or CSDVB with their subsequent reduction, is carried out in an elecrolysis apparatus of a closed type [288]. Such catalysts are hydrophobic and retain their activity in aqueous media. However, electrocatalysts effective in reduction of H_2O_2 to water are often ineffective in the reduction of O_2. Several different complexes of transition metals can be supported on one electrode [289]. For instance, two catalysts, the first reducing O_2 to H_2O_2 and the second disproportionating H_2O_2 to H_2O, were supported on the surface of the same electrode [290]. This electrode carries out the complete four-electron reduction[22] of dioxygen. The main step in obtaining this electrode is the binding of chloro(2-histamyl)tetraaminoruthenium(3+) dichloride to a chlorinated copolymer of methacrylic acid and methyl methacrylate:

---

[22] Double catalysts for complete four-electron reduction of dioxygen are used in fuel cells and batteries [291]. Reduction of O_2 and decomposition of H_2O_2 in fuel cells can be carried out by iron polyphthalocyanine on the surface of the electrode [292].

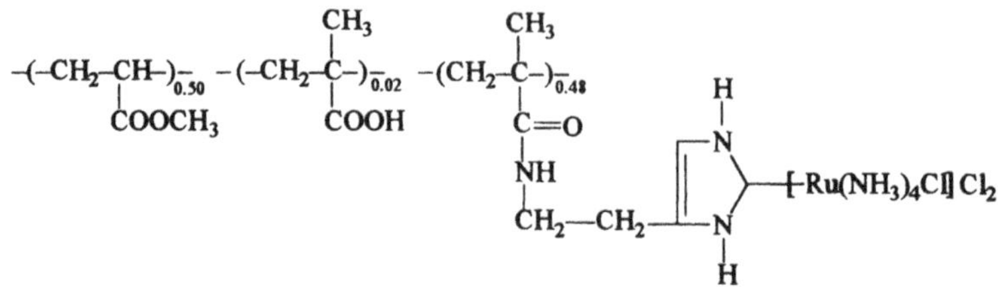

$$-(-CH_2-CH-)\overline{)_{0.50}} \quad -(-CH_2-\underset{\underset{COOH}{|}}{\overset{\overset{CH_3}{|}}{C}}-)\overline{)_{0.02}} \quad -(CH_2-\underset{\underset{C=O}{|}}{\overset{\overset{CH_3}{|}}{C}}-)\overline{)_{0.48}}$$

A solution of $CuPhen_2(ClO_4)_2$ in CH_3CN is then added to the above product to produce a coating for graphite electrodes. This coating includes complexes of two transition metals: Ru^{3+} reduces O_2 to H_2O_2 and Cu^{2+} disproportionates H_2O_2 to H_2O. This coating is stable for several days, although Cu^{2+} ions are partially washed out from the electrode. Four-electron reduction of O_2 on this electrode proceeds with high rate; catalysts of this type are effective in acid electrolytes at lower positive potentials than for other catalysts.

Modified electrodes are used also for solving the problem of dinitrogen fixation. For instance, metal polymers including $Mo(N_2)(PPh_2Me)_4$ were deposited on silicon, platinum or glass electrodes ($1 \times 10^{-7} - 1 \times 10^{-9}$ mol/cm^2) [191]. N_2 was released from these electrodes at half-potential $-(0.37 - 0.40)$ V (this voltage corresponds to the transition Mo^0/Mo^{+1}). Unfortunately, it is not possible to convert N_2 to NH_3 by this electrode.

Photoelectrochemical transformations under the action of immobilized complexes are promising processes. For instance, in combination with Pt counter-electrode, n-type semiconductors TiO_2, ZnO, Si, GaAs and CdS decompose water to O_2 (on a TiO_2 electrode) and H_2 (on a Pt electrode). Unfortunately, TiO_2 and ZnO are stable only in the near-UV region, whereas Si, GaAs and GdS are unstable in water under illumination. These shortcomings can be overcome by putting inorganic semiconductors into capsules of polymers including dispersed RuO_2. PS, PMMA, PVC and most often polyacetylene (PA) and polypyrrole are used as polymers. In these systems photoinduced charges are separated by a phase boundary. A CdS electrode modified in this way enables the use of up to 68% of photogenerated holes (h^+) for oxidation of water to O_2, and only 32% of holes is consumed for corrosion of CdS (Figure 4.18) [293]. A CdS monocrystal, the surface of which, was stabilized by PS films modified by $Ru(Dipy)_3^{2+}$ and RuO_2, was studied in detail in [294].

Films of metal polymers, especially those derived from doped PA, are interesting also as elements of solar batteries (see, for example [203]; the efficiency of conversion reaches 4–6%) and photogalvanic cells (a photochemical process in solution causes a photoresponce of electrode). Metal polymers are used also in dry and semidry photochemical cells.

Finally metal polymers are used also to create photosensitive electrodes, a new type of photodiodes. If photoexcitation is transferred by a donor or acceptor to the electrode, it works as a photodiode:

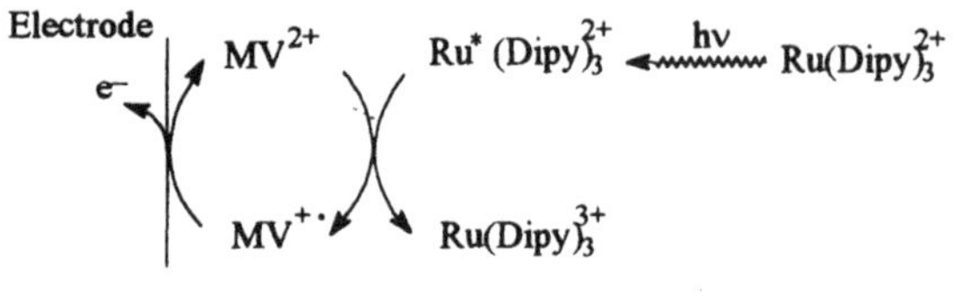

Scheme 4.26

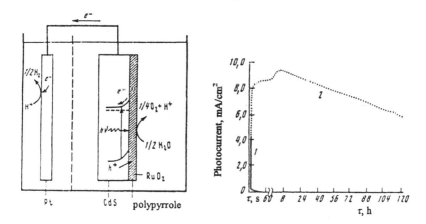

Figure 4.18 Scheme of electrochemical decomposition of water by electrodes with a narrow semiconduction zone coated by polypyrrole.

Figure 4.19 Photocurrents for bare (1) and polypyrrole-coated (2) electrodes made of polycrystalline Si. Conditions: aqueous solution, pH 1.0, 0.15 M $FeSO_4$, 0.15 M $FeNH_4(SO_4)_2$, 0.1 M Na_2SO_4.

To obtain such a photodiode from photochemical reaction, the surface of a carbon electrode was coated with a thin film of anionic polymer, then $Ru(Dipy)_3^{2+}$ was supported on the film [295]. These modified electrodes give, in combination with a Pt electrode, a photoresponse in MV^{2+} solution. Photodiodes can be obtained also by coating the electrodes with bilayer membranes. For instance, the surface of a graphite electrode was first coated with a polymeric complex $poly[Ru-(Dipy)_3]^{2+}$, then the outer surface of the polymeric film was grafted with MV^{2+} [296]. In a 0.2 M solution of CF_3COONa in water at pH 2 this photodiode generates current under illumination by visible light. The potential of the electrode is –0.2 V. In the case of bilayer membranes the value of the cathode current is 4 times higher than in the case of monolayer membranes.

Thus, the photosensitivity of electrodes coated with metal polymers is caused by the introduction of photoreactive groups into multilayer metal polymer composites.

The main advantage of coated electrodes is their long-term stability. For instance, for an uncoated n-Si electrode the photocurrent is diminished after several seconds to zero, whereas for an electrode coated with polypyrrole the photocurrent 6–7 mA/cm^{2+} was constant during tens of hours (Figure 4.19).

Many electrochemical processes proceed with the participation of polymers which do not contain metals, for instance water-soluble and solid polyelectrolytes, bioelectrocatalysts of H_2 oxidation on an electrode coated with hydrogenase [297], polypeptide membranes with covalently bound viologen [298], and others. However, these systems are far beyond the scope of the present book.

4.10 Conclusions

Research considered in this chapter shows that immobilized metal complexes can be used as catalysts for diverse catalytic processes. These can be applied to the large-scale production of a number of valuable products, many technological stages being improved substantially. For instance, they can be used in processes with participation of carbon monoxide (including the Fischer–Tropsch process), carbonylation of alcohols, hydroformylation of olefins and other reactions. Immobilized complexes offer additional control levers in determining product content, in carring out processes under milder conditions (especially with respect to temperature and pressure), in increasing the degree of conversion of metal ions to active centers, in permitting control of the nuclearity of complexes. Probably, a polynuclear center is preferable for activation of CO by supported metal complexes, likewise in catalysis by free complexes.

Asymmetric hydroformylation and hydrolysis with the participation of immobilized chiral complexes shows the possibility of generation of a large quantity of stereoasymmetric products from small quantities of chiral material (under optimal conditions the optical excess reaches 55–70%). It is important that the recycling of catalysts does not decrease the rate and selectivity of the reaction and the optical yield. Perhaps, we can expect the wide use of heterogenized systems in asymmetric reactions, at present applied in the field of homogeneous catalysts. The same is valid for metal–polymer acid catalysis which has developed only in the last few years.

In the near future immobilized catalysts will be expected to be used for activation of thermodynamically stable small molecules (N_2, H_2O, CO, CO_2); immobilized catalysts carry out the activation in the same way as biocatalysts. Metal complexes on polymeric supports are used for photo- and electrochemical stimulation of chemical reactions. In many cases (separation of photoinduced charges, realization of complete cycle of H_2O conversion, oxidation of O_2 and other processes) immobilized metal complexes have no homogeneous analogues. It is also probable that immobilized complexes could be used in photoelectrochemical catalysis (see, for example, recent papers [299–303]).

References

1. Henrici-Olive, G. and Olive, S. (1984) *The Chemistry of the Catalyzed Hydrogenation of Carbon Monoxide.* Berlin: Springer-Verlag.
2. Wojciechowski, B.W. (1988) *Catal. Rev. Sci. Eng.*, **30**, 629.
3. Lapidus, A.L. and Savel'yev, M.M. (1988) *Usp. Khim.*, **53**, 925.
4. Sheldon, A.R. (1983) *Chemicals from Synthesis Gas.* Dordrecht: Reidel.
5. Snel, R. (1986) *Mol. Chem.*, **1**, 427; (1988) *Appl. Catal.*, **37**, 35.
6. Perkins, P. and Vollhardt, K.P.C. (1979) *J. Am. Chem. Soc.*, **101**, 1395.
7. Maitlis, P.M. (1989) *Pure Appl. Chem.*, **61**, 1747.
8. Ger. Patent 2564748, 2563750.
9. Kondrat'yeva, L.T., Muranova, L.M., Belyi, A.A., *et al.* (1985) *Ref. Zh. Khim.*, **16**, 291
10. Jpn. Patent 59–88436.
11. Matera, V.D. (1984) *Inorg. Chem.*, **23**, 3597.
12. Shim, I.W., Materra, V.D. and Risen, W.M. (1985) *J. Catal.*, **94**, 531.
13. Blansher, M. (1985) *Proceedings of the III All-Union Conference on the Mechanism of Catalysis,* Vol. 1, p. 175. Novosibirsk: Nauka.

14. Hartley, F.R. (1985) *Supported Metal Complexes. A New Generation of Catalysts.* Dordrecht: Reidel.
15. Vanhove, D., Zhuyong, Z., Makambo, L., Branchord, M. (1984) *Appl. Catal.*, **9**, 327.
16. Lapidus, A.L., Savel'yev, M.M., Muranova, L.M., et al. (1985) *Kinet. Katal.*, **26**, 248.
17. Roper, M. (1984) *Ger. Chem. Eng.*, **7**, 329.
18. Meyers, G.F., Hall, M.B. (1987) *Inorg. Chim. Acta*, **129**, 153.
19. Parks, E.K., Nieman, G.G., Pobo, L.G., Rili, S.I. (1987) *J. Phys. Chem.*, **91**, 2671.
20. Fujimoto, K., Kajioka, M. (1987) *Bull. Chem. Soc. Jpn.*, **60**, 2237.
21. Khomenko, T.I., Kadushin, A.A., Khandozhko, V.N., et al. (1988) *Kinet. Katal.*, **29**, 987.
22. U.S. Patent 4596831.
23. Nakamura, T., Yamadu, M., Yamaguchi, T. (1992) *Appl. Catal. A. General.*, **87**, 69.
24. Dry, M.E. (1982) *J. Mol. Catal.*, **17**, 133.
25. Malykh, O.A., Krylova, A.Yu., Emel'yanova, G.I., Lapidus, A.L. (1988) *Kinet. Katal.*, **29**, 1362.
26. U.S. Patent 4605751.
27. Kikuchi, E., Koizumi, A., Akinasa, A. (1982) *J. Jpn. Pet. Inst.*, **25**, 360.
28. Uematsu, T., Tsukada, K., Fjisia, M., Hashimoto, H. (1974) *J. Catal.*, **32**, 369.
29. Jarrel, M.S., Gates, B.C. (1975)., *J. Catal.*, **40**, 255.
30. Shimazu, S., Ishibashi, Y., Miura, M., Uematsu, T. (1987) *Appl. Catal.*, **35**, 279; (1987) *React. Kinet. Catal. Lett.*, **33**, 35.
31. Ki, S. R., Seong, I. W. (1990) *J. Mol. Catal.*, **59**, 353.
32. Zhao, Y.-I., Pan, C.-Y., Chen, Y. (1987) *Tsuikhua syuebao* (J. Catal.), **8**, 288.
33. Lapidus, A.L., Pirozhkova, S.D., Buiya,M.A., et al., (1985) *Izv. Akad. Nauk SSSR, Ser. Khim.,* **12**, 2816.
34. Lapidus, A.L., Krylova, A.Yu., Petrachkova, N.E., et al. (1985) *Izv. Akad. Nauk SSSR, Ser. Khim,* **6**, 1257.
35. Hagen, Y., Bauermann, P. (1986) *Chem.-Ztg.*, **110**, 151.
36. Vantanatham, S., Farona, M.F. (1980) *J. Catal.*, **61**, 540.
37. Feng, Z., Chen, B., Liu, H. (1987) *J. Macromol. Sci. A*, **24**, 289.
38. Marrakchi, H., Nguini, E.J.-B., Haimeur, M., et al. (1985) *Bull. Soc. Chim. Fr.*, **3**, 390.
39. Stamenova, R., Ivanova, P., Tsvetanov, Ch. (1987) *Prog. Prague Meeting on Macromolecules, XXX Microsymposium: Polymer Supported Organic Reagents and Catalysts* (Prague, 1986), p. 44.
40. Ger. Patent 1800379.
41. Laine, R.M., Crawford, E.J. (1988) *J. Mol. Catal.*, **44**, 357.
42. Bhaduri, S., Khwaja, H., Naravayanam, B.A. (1984) *J. Chem. Soc., Dalton Trans.*, **10**, 2327.
43. Kaneda, K., Kobayashi, M., Imanaka, T., Tetranishi, S. (1984) *Chem. Lett.*, **9**, 1483.
44. Kaneda, K., Kitaona, K., Imanaka, T., Teranishi, S. (1985) *Sekubai* (Catalyst), **7**, 419.
45. Pelt, N.L., Gerritsen, L.A., van der Lee, G. (1983) *Preprints of the Catalysis, III International Symposium* (Louvain-la-Neuve, Amsterdam, 1982), p. 369.

46. Gao, G., Temure, Y.Y. (1990) *J. Mol. Catal.* (in Chinese), **4**, 200.
47. Fruoka, A., Ichikawa, M. (1986) *Sekubai* (Catalyst), **28**, 461.
48. Ichikawa, M. (1986) *V International Symposium on Homogeneous and Heterogeneous Catalysis*, (Novosibirsk, 1985), Vol. 3, p. 37.
49. Chalganov, E.M., Kuznetsov, B.N., Ermakov, Yu.I. (1980) *Proceedings of the Symposium: Catalysts containing Supported Complexes* (in Russian) (Novosibirsk, 1979), Vol. 2, p. 103.
50. U.S. Patent 4504684, 1985.
51. De Munk, N.A., Verbruggen, M.W., Scholten, J.J.F. (1981) *J. Mol. Catal.*, **10**, 313.
52. Heindrich, B., Hjorkjaer, J., Nikitidis, A., Andersson, C. (1993) *J. Mol. Catal.*, **81**, 333.
53. Pelt, H.L., Verburg, R.P.J., Scholten, J.J.F. (1985) *J. Mol. Catal.*, **32**, 77.
54. Br. Patent 2028793, 1980.
55. Pittman, C.U. (Jr.), Hanes, R.M., Jacobson, S.E., Hirao, A. (1976) *Polym. Prepr., Am. Chem. Soc.*, **17**, 246.
56. Kowabata, Y., Pittman, C.U. Jr. and Kobayashi, R. (1981) *J. Mol. Catal.*, **12**, 113.
57. Pittman, C.U. (Jr.), Kanabata, Y., Liang, Y.F. (1980) *Abstracts of the Papers of the 179th ACS National Meeting*, (Huston, TX, 1979), Washington DC: American Chemical Society.
58. Sun, I., Weng, I., Li, H., He, B. (1987) *Chem. J. Chin. Univ.*, **8**, 653.
59. U.S. Patent 4189448, 1980.
60. Hiang, H.-Y., Ren, C.-Y., Zhou, Y.-S., *et al.* (1985) *Organosilicon and Bioorganosilicon Chemistry: Proceedings of the VII International Symposium on Organosilicon Chemistry* (Kyoto; 1984), p. 275. New York.
61. Ding, M., Stille, J.K. (1983) *Macromolecules*, **16**, 839.
62. Hartley, F.R., Murray, S.G., Nicholson, P.N. (1982) *J. Mol. Catal.*, **16**, 363.
63. Hartley, F.R., Murray, S.G., Sayer, A.T. (1986) *J. Mol. Catal.*, **38**, 295.
64. Hartley, F.R. (1984) *Br. Polym. J.*, **16**, 199.
65. Jongsma, T., Fossen, M., Challa, G., Leeuwen, P.W.N.M. (1993) *J. Mol. Catal.*, **83**, 17.
66. Avilov, V.A., Chepaikin, E.G., Khidekel', M.L. (1974) *Izv. Akad. Nauk SSSR, Ser. Khim.*, **11**, 2559.
67. Corain, B., Basato, M., Zecca, M., Braca, G., Raspolli, Galletti, A.M., Lora, S., Palma, G., Guglielminotti, E. (1992) *J. Mol. Catal.*, **73**, 23.
68. Zong, H.J., Quo, X.Y., Tang, Q., *et al.* (1987) *J. Macromol. Sci. A*, **24**, 277.
69. Jongsma, T., Aert, H., Fossen, M., Challa, G., Leeuwen, P.W.N.M. (1993) *J. Mol. Catal.*, **83**, 37.
70. Escaffre, P., Thorez, A., Klack, P. (1987) *J. Chem. Soc., Chem. Commun.*, **3**, 142.
71. Terekhova, M.I., Sigalov, A.B., Petrova, N.E., Petrov, E.S. (1985) *Zh. Obshch. Khim.*, **55**, 944.
72. Dubois, R.A., Garrou, P.E., Lavin, K.D., Allcock, N.R. (1984) *Organometallics*, **3**, 649; (1986) **5**, 460; 466.
73. Dubois, R.A., Garrou, P.E. (1985) *Appl. Catal.*, **19**, 259; 275.
74. Marrakchni, H., Efta, N.J.B., Haimeur, M., Lieto, J. (1985) *J. Mol. Catal.*, **30**, 101.
75. U.S. Patent 4144191, 1979.
76. Jia, C.-G., Wang, Y.-P., Feng, H.-Y. (1992) *React. Polym.*, **18**, 203.
77. Capka, M., Svoboda, P., Cerny, M., Hetflejs, J. (1984) *J. Mol. Catal.*, **26**, 355.
78. Terreros, P., Pastor, E., Fierro, J.L.G. (1989) *J. Mol. Catal.*, **53**, 359.
79. U.S. Patent 4511740, 1985.

80. Ford, M.E., Premecz, J.E. (1983) *J. Mol. Catal.*, **19**, 99.
81. Imyanitov, N.S., Rybakov, V.A., Tupitsin, S.B. (1992) *Neftekhimiya*, **32**, 446.
82. Imyanitov, N.S., Tupitsin, S.B. (1993) *Neftekhimiya*, **33**, 438.
83. U.S. Patent 4173575, 1980.
84. U.S. Patents 4178313, 4625069, 1980.
85. He, B., Zheng, F., Sun, J., Li, H., *et al.* (1985) *Tsuikhua Suebao* (J Catal.), **6**, 296.
86. Pitman, C.U. Jr. and Lin, C.-C. (1978) *J. Org. Chem.*, **43**, 4928.
87. Collin, J., Jossart, C., Balavoine, G. (1985) *Organometallics*, **5**, 203.
88. Marko, L., Radhi, M., Palagyi, I., Palyi, G. (1986) *Kem. Kozl.*, **65**, 144.
89. Li, D.-G., Zhai, W.-X., Chen, Z.-S., *et al.* (1986) *Khuansue Suebao* (Acta Chim. Sin.), **44**, 990.
90. Hunter, D.L., Moore, S.E., Dubois, R.A., Garrou, P.E. (1985) *Appl. Catal.*, **19**, 275.
91. Pittman, C.U. (Jr.), Honnick, W.D., Yang, Y.Y. (1980) *J. Org. Chem.*, **45**, 684.
92. Marchionna, M., Longoni, G. (1986) *J. Mol. Catal.*, **35**, 107.
93. Lazzaroui, R., Rertozzi, S., Pocai, P., *et al.* (1985) *J. Organomet. Chem.*, **295**, 371.
94. Jacobson, S.E. (1987) *J. Mol. Catal.*, **44**, 163.
95. U.S. Patent 4465873.
96. Kalek, P., Park, D.C., Serein, F., Thores, A. (1986) *J. Mol. Catal.*, **36**, 349.
97. Futschel, S.J., Ackerman, J.J., Keyser, T., Stille, J.K. (1979) *J. Org. Chem.*, **44**, 3152.
98. Stille, K. (1984) *J. Macromol. Sci.*, **21**, 1689.
99. Stille, J.K. (1985) *Polym. Prepr., Am. Chem. Soc., Div. Polym. Chem.*, **27**, 9.
100. Parrinello, G., Deschenaux, R., Stille, J.K. (1986) *J. Org. Chem.*, **51**, 4189.
101. Parrinello, G., Stille, J.K. (1986) *Polym. Mater. Sci. Eng.*, p. 889.
102. Bencze, L., Prokai, L. (1985) *J. Organometal. Chem.*, **294**, C5.
103. Fan, C., Zhao, Y. (1987) *J. Chin. Univ. Sci. Technol.*, **17**, 160.
104. Guczi, L. (1981) *Catal. Rev.*, **23**, 329.
105. Karan, Yu. B., Rozovskii, A.Ya. (1987) *Kinet. Katal.*, **28**, 1508.
106. Sen, A., Brubbaugh, J.S., Lin, M. (1992) *J. Mol. Catal.*, **73**, 297.
107. Lisichkin, G.V., Yuffa, A.Ya. (1983) *Khim. Prom-st. (Moscow)*, **5**, 293.
108. Tyukova, O.A. (1983) *Khim. Prom. Za Rubezhom*, **8**, 25.
109. Serio, M., Garatta, R., Santacesaria, E. (1991) *J. Mol. Catal.*, **69**, 1.
110. Sibtain, F., Kempel, G.L. (1991) *J. Polym. Sci., Part A: Polym. Chem.*, **29**, 629.
111. Takeishi, M., Ueda, M., Hayama, S. (1981) *Macromol. Chem. Rapid Commun.*, **2**, 167.
112. Matsui, Y., Suemrisu, D. (1985) *Bull. Chem. Soc. Jpn.*, **6**, 1658.
113. Zhorobekova, Sh.Zh., Karabaev, S.O. (1986) *Problems of Modern Bioinorganic Chemistry: Proceedings of Session* (in Russian), p. 96, Novosibirsk.
114. Semenets, O.P.(1994) A new inhibitor of α-chymotrypsin and effect of metal ions. *Cand. Sci. (Chem.) Dissertation* (in Russian), Bishkek.
115. Bao, Y.T., Saltzman, H., Pitt C,. (1985) *Am. Chem. Soc., Polym. Prepr.*, **26**, 81.
116. Hammond, P.S., Forster, J.S. (1991) *J. Appl. Polym. Sci.*, **43**, 1925.
117. Yamskov, I.A., Berezin, B.B., Belchich, L.A., Davankov, V.A. (1979) *Eur. Polym. J.*, **15**, 1067.
118. Yamskov, I.A., Berezin, B.B., Davankov, V.A. (1980) *Macromol. Chem. Rapid Commun.*, **1**, 125.

119. Yamskov, I.A., Berezin, B.B., Davankov, V.A. (1980) *Katalizatory Soderzhashchiye Nanesennye Compleksy* (Catalysts containing supported complexes), p. 51, Novosibirsk.
120. Eijik, J.M., Nolte, R.J., Richters, V.E.M., Drenth, W. (1981) *Rec. Trav. Chim. Pays-Bas.*, **100**, 222.
121. Spassky, N., Reix, M., Sapulchre, M.-O. (1983) *Makromol. Chem.*, **184**, 17.
122. Ohkubo, K., Nahano, Y., Sasaki, S. (1985) *J. Mol. Catal.*, **32**, 5.
123. Mamoru, N., Yoshiharu, K., Yasuji, I., *et al.* (1986) *Chem. Express*, **1**, 232.
124. Broun, R.S., Zamkanei, M., Cocho, J.L. (1984) *J. Am. Chem. Soc.*, **106**, 5222.
125. Vanni, A., Gastaldi, D. (1986) *Ann. Chim.*, **76**, 75.
126. Francois, J., Kheadmand, H., Plazanet, V., *et al.* (1986) *Polym. Mater. Sci. Eng.*, **55**, 694.
127. Burgeson, I.E., Kostic, N.M. (1991) *Inorg. Chim.*, **30**, 4299.
128. Potapov, G.P., Krupenskii, V.I., Alieva, M.I. (1985) *React. Kinet. Catal. Lett.*, **28**, 331.
129. Potapov, G.P., Krupenskii, V.I., Alieva, M.I. (1987) *Kinet. Katal.*, **28**, 205.
130. Inventor's Certificate no. 1351649 (1986); no. 1155600, USSR.
131. U.S. Patent 4551567.
132. Chee, Y.C., Ihm, S.K. (1986) *J. Catal.*, **102**, 180.
133. Sivanand, S.P., Singh, R.S., Chakrabarty, D.K. (1983) *Proc. Indian Acad. Sci., Chem. Sci.*, **92**, 227.
134. U.S. Patent 4665266.
135. Takagy, K., Morikava, J., Ikava, C. (1986) *Sekubai* (J. Catal.), **28**, 78.
136. Tanabe, K. (1987) *Solid Acids and Bases*. Moscow: Mir.
137. Bogoczek, R., Surowiec, J. (1987) *Progress of the Prague Meeting on Macromolecules, XXX Microsymposium: Polymer Supported Organic Reagents and Catalysis* (Prague, 1986), p. 46.
138. Widdecke, H. (1984) *Br. Polym. J.*, **16**, 188.
139. Vit, Z. (1986) *React. Kinet. Catal. Lett.*, **32**, 205.
140. Waller, F.J. (1986) *Catal. Rev.*, **28**, 1.
141. Dochini, A., Ungaro, R. (1980), *IUPAC Macro Florence, International Symposium on Macromoleculecules* (Pisa, 1979), Vol. 4, p. 170.
142. Ran, R., Rei, W.W., Jia, X., Wu, X. (1987) *Youtszi Huasyue* (Org. Chem.), **4**, 268.
143. Naik, K.M., Sivaram, S. (1977) *Proceedings of the International Symposium on macromolecules*, (Dublin, 1976), Vol. **2**, p. 481.
144. Sangalov, Yu. A., Gladkikh, I.F., Minsker, K.S. (1982) *Vysokomol. Soedin.*, **26**, 356.
145. Klein, J., Widdecke, H. (1982) *Chem.-Ing.-Techn.*, **54**, 595.
146. Neckers, D.S. (1987) *J. Macromol. Sci.* A, **24**, 431.
147. Fernandez-Sanchez, C., Marinas, J.M. (1987) *Angew. Makromol.*, **149**, 197.
148. Boda, C., Contento, M., Manescalchi, F. (1989) *Macromol. Chem. Rapid Commun.*, **10**, 303.
149. Reicheng, R., Shuojian, J. (1987) *J. Macromol. Sci.* A, **24**, 669.
150. Ran, R.-C., Jia, X.-R., Li, M.-Q., Jiang, S.-J. (1988) *J. Macromol. Sci.* A, **25**, 907.
151. Ran, R., Fu, D. (1990) *Macromol. Sci.* A, **27**, 625.
152. Ran, R.C., Mao, G.P. (1990) *J. Macromol. Sci.* A, **27**, 125.
153. Ran, R., Jiang, S.-J. (1985) *Gaofentszy Tyasyun'* (Polymer Commun.), **5**, 376.

154. Ran, R., Huang, J., Shen, J. (1987) *Gaofentszun Syuebao* (Acta Polym. Sin.), **4**, 312.
155. Ran, R., Huang, J., Shen J. (1988) *Gaofentszun Syuebao* (Acta Polym. Sin.), **6**, 476.
156. Ran, R., Huang, J., Jia, X.-R., Shen, J. (1987) *Tsyuhua Syuebao* (J. Catal.), **8**, 440.
157. Ran, R., Hang, J., Shen, J. (1987) *Inyun Huasyue* (Chin. J. Appl Chem.), **4**, 49.
158. Luis, S.V., Burguete, M., Ramirez, N., *et al.* (1992) *React. Polym.*, **18**, 237.
159. Pomogailo, A.D. (1988) *Polimernye Immobilizovannye Metallokompleksnye Katalizatory* (Polymeric Immobilized Metal Complex Catalysts), Moscow: Nauka.
160. Ran, R., Wu, X.-H., Jia, X., Pei, W. (1988) *Gaofentszy Syuebao* (Acta Polym. Sin.), **1**, 67.
161. Ran, R., Pei, W., Jia, X., *et al.* (1986) *Gaofentszy Tynsyun'* (Polym. Commun.), **6**, 453.
162. Ran, R., Jia, X., Wu, X., *et al.* (1986) *Gaoden Syuesyao Huasyue Syuebao* (Chem. J. Chin. Univ.), **7**, 646.
163. Whang, K.-J., Lee, K.-I., Lee, Y.-K. (1984) *Bull. Chem. Soc. Jpn.*, **57**, 2341.
164. Liu, F., Lin, Y., Yi, W., Huang, H. (1987) *Tszilin' Dasyue Tszyzhan' Kesue Syuebao* (Acta Sci. Natur. Univ. Jilinensis) **4**, 79.
165. Hirai, H., Hara, S., Komiyama, M. (1982) *Chem. Lett.*, **10**, 1685.
166. Hirai, H., Shigekura, M., Komiyama, M. (1986) *Macromol. Chem. Rapid Commun.*, **7**, 351.
167. Toshima, N., Kanaka, K., Komiyama, M., Hirai, H. (1988) *Macromol. Sci.*, **25**, 1349.
168. Haase, D.J. (1975) *Chem. Eng.*, August 4th, p. 52.
169. Hirai, H., Hara, S., Komiyama, M. (1985) *Angew. Makromol. Chem.*, **130**, 207.
170. Hirai, H., Hara, S., Komiyama, M. (1986) *Chem. Lett.*, **2**, 257; *Bull. Chem. Soc. Jpn.*, **59**, 109.
171. Kolesnikov, S.P., Povarov, S.L., Nefedov, O.M. (1989) *Izv. Akad. Nauk SSSR, Ser. Khim.*, **7**, 1708.
172. Utarov, N.P., Mikheikin, I. D., Bakshin, Yu. M., Gel'bshtein, A.I. (1990) *Koord. Khim.*, **16**, 42.
173. Shilov, A.E. (1977) *Zh. Vses. Khim. Obshch.*, **22**, 521.
174. Akhrem, I.S., Orlikov, A.V., Vol'pin, M.E., *et al.* (1988) *Dokl. Akad. Nauk SSSR*, **298**, 107.
175. Ryabov, V.G., Uglev, N.P., Trofimov, S.V. (1985) *Proceedings of the IV Ural Conference on High-temperature physicochemistry and electrochemistry* (in Russian), (Perm',1985), p. 58. Sverdlovsk.
176. Dooley, K.M., Gates, B.C. (1985) *J. Catal.*, **96**, 347.
177. Jiang, H., Qiongfren, L., Yingyan, J. (1982) *Fundamental Research on Organometallic Chemistry, Proceedings of the China–Jpn.–US Trilateral Seminar*, (Peking; 1980), p. 637. New York.
178. Magnotta, V.L., Gates, B.C., Schuit, G.C.A. (1976) *J. Chem. Soc., Chem. Commun.*, **10**, 342.
179. Ono, Y., Tanabe, T., Kitajima, N. (1978) *Chem. Lett.*, **6**, 625; (1979) *J. Catal.*, **56**, 47.
180. Minsker, K.S., Babkin, V.A., Zaikov, G.E. (1994) *Usp. Khim.*, **63**, 289.
181. Biglova, R.Z., Malinskaya, V.P., Minsker, K.S. (1994) *Vysomol. Soedin., Ser. A*, **36**, 1276.
182. Vol'pin, M.E., Shur, V.B. (1964) *Dokl. Akad. Nauk SSSR*, **156**, 1102.

183. Shilov, A.E. (1980) *New Trends in the Chemistry of Nitrogen Fixation*, London: Academic Press.
184. Shilov, A.E. (1987) *J. Mol. Catal.*, **41**, 221.
185. Shilov, A.E. (1977) *Kinet. Katal.*, **18**, 90.
186. Bonds, W.D. (Jr.), Brubaker, C.H. (Jr.), Chandrasekaran, E.S., *et al.* (1975) *J. Am. Chem. Soc.*, **97**, 2128.
187. Nishide, H., Kawakami, H., Kurimura, Y. (1988) *J. Macromol. Sci.*, **25**, 1339.
188. Kurimura, Y., Uchino, Y., Ohta, F., *et al.* (1981) *Polym.*, **13**, 247.
189. Kirimura, Y., Ohta, F., Gohda, J., *et al.* (1982) *Macromol. Chem.*, **183**, 2889.
190. Kurimura, Y., Takagi, Y., Suka, M., *et al.* (1985) *J. Chem. Soc., Jpn. Chem. Ind. Chem.*, **3**, 373.
191. Dubois, D.L. (1984) *Inorg. Chem.*, **23**, 2047.
192. George, T.A., Kaul, B.B. (1990) *Inorg. Chem.*, **29**, 4969.
193. Sellmann, D., Weiss, W. (1978) *Angew. Chem., Int. Ed. Engl.*, **17**, 269.
194. Koide, M., Masubuchi, T., Kurimura, Y., Tsuchida, E. (1980) *Polym. J.*, **12**, 793.
195. Tsuchida, E., Nishide, H., Tomiyama, H., Kurimura, Y. (1981) *Macromol. Chem. Rapid Commun.*, **2**, 627.
196. Koide, M., Tsuchida, E., Kurimura, Y. (1981) *Makromol. Chem.*, **182**, 749.
197. Koide, M., Kato, M., Tsuchida, E., Kurimura, Y. (1981) *Makromol. Chem.*, **182**, 283.
198. Likhtenstein, G.I. (1988) *Chemical Physics of Redox Metalloenzyme Catalysis.* Heidelberg: Springer-Verlag.
199. Oguni, N., Shimazu, S., Iwamoto, Y., Nakamura, A. (1981) *Polym. J.*, **13**, 845.
200. Oguni, N., Shimazu, S., Nakamura, A. (1980) *Polym J.*, **12**, 891.
201. Semenov, N.N. (1973) *Nauka I Obshchestvo* (Science and Society). Moscow: Nauka.
202. Zamaraev, K.I., Parmon, V.N. (1980) *Usp. Khim.*, **49**, 1457; (1983) *Makromol. Chem.*, **52**, 1433.
203. Kaneko, M., Yamada, A. (1984) *Adv. Polym. Sci.*, **55**, 1.
204. Skorobogaty, A., Swith T.D. (1984) *Coord. Chem. Rev.*, **53**, 55.
205. Vannikov, A.V., Grishina, A.D. (1984) *Fotokhimiya Polymernykh donorno-aktseptornykh Kompleksov* (Photochemistry of Polymeric Donor–Acceptor Complexes). Moscow: Nauka.
206. Ivanov, V., Tverskoi, V.A., Timofeeva, T.V., Pravednikov, A.N. (1985) *Vysomol. Soedin.*, **27**, 392.
207. Saotome, Y., Endo, T., Okamato, M. (1983) *Macromol.*, **16**, 881.
208. Nishijama, T., Nagamura, T., Matsuo, T. (1981) *J. Polym. Sci., Polym. Lett.*, **19**, 65.
209. Shafirovich, V.Ya., Levin, P.P., Khannanov, N.K., Kuz'min, V.A. (1986) *Izv. Akad. Nauk SSSR, Ser. Khim.*, **4**, 801.
210. Ageishi, K., Endo, T., Okawara, M. (1981) *J. Polym. Sci., Polym. Chem. Ed.*, **19**, 1085.
211. Neckers, D.C. (1985) *React. Polym.*, **3**, 277.
212. Shaap, A.P., Thayer, A.L., Zaklika, K.A., Nalenti, P.C. (1979) *J. Am. Chem. Soc.*, **10**, 4016.
213. Pshezhetskii, V.S., Yaroslavov, A.A., Rakhnyanskaya, A.A., *et al.* (1983) *Dokl. Akad. Nauk SSSR*, **271**, 466.

214. Kumar, G.S., De Pra, P., Zhang, K., Neckers, D.S. (1984) *Macromolecules*, **17**, 2463.
215. Clear, J.M., Kelly, J.M., O'Connell, C.M., Vos, J.G. (1981) *J. Chem. Res. Microfishe.*, **30**, 3039.
216. Shimidzu, I., Izaki, K., Akai, Y., Iyoda, T. (1981) *Polym. J.*, **13**, 889.
217. Ennis, P.M., Kelly, J.M., O'Connell, C.M. (1986) *J. Chem. Soc. Dalton Trans.*, **11**, 2485.
218. Kaneko, M., Ochiai, M., Kinosito, K., Yamada, A. (1982) *J. Polym. Sci., Polym. Chem. Ed.*, **20**, 1011.
219. Changkai, H., Qiwen, M., Yungi, L., Xiuchang, Z. (1982) *Fundamental Research on Organometallic Chemistry, Proceedings of the China–Jpn–US Trilateral Seminar, (Peking, 1980)*, p. 643. New York.
220. Hou, X.-H., Kaneko, M., Yamada, A. (1984) *Kobunshu Ronbunshu*, **41**, 311.
221. Furue, M., Kuroda, N., Sano, S. (1988) *J. Macromol. Sci.*, **25**, 1263.
222. Kaneko, M., Jou, X.-H., Yamada, A. (1987) *Bull. Chem. Soc. Jpn.*, **60**, 2523.
223. Kurimura, Y., Yokota, H., Shipgehara, K., Tsuchida, E. (1988) *Bull. Chem. Soc. Jpn.*, **55**, 55.
224. Nussbaumer, W., Gruber, H., Greber, G. (1988) *Macromol. Chem.*, **189**, 1027.
225. Parmon, V.N. (1985) Physicochemical basis of Conversion of Solar Energy by Water Decomposition in Molecular Photocatalytic Systems, *Dr. Sci. (Chem.) Dissertation*, Novosibirsk.
226. Savinova, E.R., Kokorin, A.I., Shepelin, A.P., et al. (1985) *J. Mol. Catal.*, **32**, 149.
227. Saidkhanov, S.S., Savinov, E.N., Kokorin, A.I., et al. (1984) *Izv. Akad. Nauk SSSR, Ser. Khim.*, **11**, 2447.
228. Saidkhanov, S.S., Kokorin, A.I., Savinov, E.N., et al. (1983) *J. Mol. Catal.*, **21**, 365.
229. Buyanova, E.R., Matvienko, L.G., Kokorin, A.I., et al. (1981) *React. Kinet. Catal. Lett.*, **16**, 309.
230. Ramaraj, R., Kaneko, M. (1993) *J. Mol. Catal.*, **81**, 319.
231. Yamaguchi, H., Fujiwara, R., Kusuda, K. (1986) *Macromol. Chem. Rapid Commun.*, **7**, 225.
232. Tanno, T., Kaneko, M., Yamada, A., et al. (1980) *IUPAC Macro Florence, International Symposium on Macromolecules.*, (Pisa, 1979), Vol. **4**, p. 103.
233. Toshima, N., Kaneko, M. (1981) *Kagakuto Siebutsu*, **19**, 80.
234. Fendler, J., Kurihara, K. (1985) *Metal Containing Polymer Systems*, p. 341-353.New York: Plenum Press,
235. Mayer, V.E., Levchenko, L.A., Shafirovich, V.Ya. (1986) *Kinet. Katal.*, **27**, 1378.
236. Elizarova, G.L., Kim, T.V., Matvienko, L.G., et al. (1986) *React. Kinet. Catal. Lett.*, **31**, 455.
237. Kurimura, Y., Tsuchida, E., Kaneko, M., Yamada, A. (1982) *International Union of Pure and Applied Chemistry, XXVIII Macromolecules Symposium*, (Amherst, 1981), Vol. **1**, p. 141.
238. Dismukes, G.C. (1988) *Chem. Scr. A*, **28**, 99.
239. Khramov, A.V., Moravsky, A.P., Shilov, A.E. (1984) *Abstracts of the IV International Symposium on Homogeneous Catalysis*, (Leningrad, 1983), Vol. **2**, p. 219.
240. Brudvig, G.W., Crabtree, R.H. (1986) *Proc. Natl. Acad. Sci. U.S.A.*, **83**, 4586.
241. Sheats, J.E., Unni, N., Petrouleas, B.C., et al. (1986) *Progress in Photosynthesis Research.* Vol. **1**, p. 721. Amsterdam: Nijhoff.

242. Sheats, J.E., Czernuszewicz, R.S., Dismukes, G.G., *et al.* (1987) *J. Am. Chem. Soc.*, **109**, 1435.
243. Knerel'man, E.I., Luneva, N.P., Shafirovich, V.Ya., Shilov, A.E. (1986) *Dokl. Akad. Nauk SSSR*, **291**, 632.
244. Kutal, Ch., Schwendiman, D.P., Grutsch, P. (1977) *Sol. energy*, **19**, 651.
245. Kaneko,M., Nemoto, S., Yamada, A., Kurimura, Y. (1980) *Inorg. Chim. Acta*, **44**, L289.
246. King, R.B., Sweet, E.M. (1979) *J. Org. Chem.*, **44**, 385.
247. Sikube, H. (1979) *Kagau To Koge Kagaku To Kogyo* (Chem. Chem. Ind.), **32**, 361.
248. Smierciak, R.C., Giardano, P.J. (1985) *Appl. Catal.*, **18**, 353.
249. U.S. Patent 4606326.
250. Wöhrle, D., Bohlen, H. (1983) *Makromol. Chem.*, **184**, 763.
251. Wöhrle, D., Buttner, P. (1985) *Polym. Bull.*, **13**, 57.
252. Taoda, H., Hayakawa, K., Kawase, K., *et al.* (1987) *Kobunsi Rombunsu*, **44**, 437.
253. Ignat'yev, N.G., Karasev, A.L., Savel'yev, V.V., Vannikov, A.V. (1986) *Izv. Akad. Nauk SSSR, Ser. Khim.*, **3**, 676.
254. Kopietz, M., Lechneg, D.M., Steinmeier, D.G. (1986) *Makromol. Chem.*, **187**, 2787.
255. Treushnikov, V.M., Frolova, N.V. (1983) *Vysokomol. Soedin.*, **25**, 1400.
256. Pasinkov, L.M., Margolin, A.L., Vichutinskaya, E.N. (1981) *Natur. und Kunst. Alter. Kunstst. Vortr. Munchen-Wien*, p. 33.
257. Turbanov, K.Yu (1993) *Zh. Obshch. Khim.*, **63**, 2036.
258. Snell, K.D., Keenan, A.G. (1979) *Chem. Soc. Rev.*, **8**, 259.
259. Wegner, G. (1981) *Angew. Chem. Intern. Ed.*, **20**, 361.
260. Murray, R.W. (1984) *Electroanalytical Chemistry*, Vol. 13, p. 191. New York: Dekker.
261. Peerce, P.J., Bard, A.J. (1980) *J. Electroanal. Chem.*, **112**, 97.
262. Kaneko, M., Wöhrle, D. (1988) *Adv. Polym. Sci.*, **84**, 141.
263. Meyer, T.J., Sullivan, B.P. (1987) *Inorg. Chem.*, **26**, 4145.
264. Fukui, M., Kitani, A., Degrand, C., Miller, L.L. (1982) *J. Am. Chem. Soc.*, **104**, 28.
265. Denisevich, P., Willman, K.W., Murray, R.W. (1981) *J. Am. Chem. Soc.*, **103**, 4727.
266. Deronzier, A., Moutet, J.-C. (1989) *Acc. Chem. Res.*, **22**, 249.
267. Coche, L., Moutet, J.-C. (1988) *J. Electroanal. Chem.*, **224**, 112.
268. Vork, F.T.A., Janssen, L.J.J., Barendrecht, E. (1986) *Electrochim. Acta*, **31**, 1569.
269. Holdcroft, S., Funt, B.L. (1988) *J. Electroanal. Chem.*, **240**, 89.
270. Moutet, J.-C., Pickett, C.J. (1989) *J. Chem. Soc. Chem. Commun.*, **3**, 188.
271. Oyama, N., Chiba, K., Ohuuki, Y., *et al.* (1985) *J. Chem. Soc., Jpn. Chem. Ind. Chem.*, **6**, 1172; 1124.
272. Mommone, R.J., Macoliarmid, A.G. (1984) *Synth. Met.*, **9**, 143.
273. Coche, L., Motel, J.-C. (1987) *J. Am. Chem. Soc.*, **109**, 6887.
274. Fr. Patent 2548177.
275. Deronzier, A., Limasin, D., Moutet, J.-C. (1987) *Electrochim. Acta*, **32**, 1643.
276. Moisy, P., Bedioui, F., Pobin, Y., Devynck, J. (1988) *J. Electroanal. Chem.*, **250**, 191.
277. Cosnier, S., Deronzier, A., Moutet, J.-C. (1988) *Inorg. Chem.*, **27**, 2389.
278. Tarasevich, M.P., Radyushkina, K.A. (1980) *Usp. Khim.*, **49**, 1498.

279. Juris, A., Balzani, V., Barigelletti, F., *et al.* (1984) *Chem. Rev.*, **84**, 85.

280. Bedioui, F., Bongars, C., Devynck, J., *et al.* (1986) *J. Electroanal. Chem.*, **207**, 87.

281. Tsichida, E., Nishide, H. (1983) *Initial Polymer 183rd ACS National Meeting*, (Las Vegas, 1982), p. 49. Washington DC: American Chemical Society.

282. Andrieux, C.P., Dumas-Bouchiat, J.M., Saveanz, J.M. (1980) *Electroanal. Chem.*, **114**, 159.

283. Calver, J.M., Meyer, T.J. (1982) *Inorg. Chem.*, **21**, 3978.

284. Anson, F.C. (1980) *J. Phys. Chem.*, **84**, 3336.

285. Kublanovskii, V.S., Novikova, E.M., Maletin, Yu.A., Strizhakova, N.G. (1986) *Elektrokataliz i Elektrokataliticheskiye Protsessy* (Electrocatalys and Electrocatalytic Processes). p. 29. Kiev: Naukova Dumka.

286. Orlov, S.B., Tarasevich, M.R., Bogdanovskaya, V.A., Pshezhetskii, V.S. (1986) *Dokl. Akad. Nauk SSSR*, **298**, 1167.

287. Kalacheva, V.V., Bazanov, M.M., Smirnov, R.P., Al'yamov, M.I. (1983) *Sovershenstvovaniye Protsessov Krasheniya I Methodov Sinteza Krasitelei* (Improvement of Processes of Dyeing and Methods of Dye Synthesis), p. 74. Ivanovo.

288. Jpn. Patent 57-27136.

289. Oyama, N., Anson, F.C. (1979) *J. Am. Chem. Soc.*, **101**, 3450.

290. Shigehara, K., Anson, F.C. (1982) *J. Electroanal. Chem.*, **132**, 107.

291. Kordesch, K. (196) *Fuel Cells*, Chapter 2. New York: Reinhold.

292. Kreja, L. (1987) *Monatsh. Chem.*, **118**, 717.

293. Frank, A.J., Honda K. (1982) *J. Phys. Chem.*, **86**, 1933.

294. Rajeshwark, K., Kaneko, M., Yamada, A., Noufi, R.N. (1985) *J. Phys. Chem.*, **89**, 806.

295. Kaneko, M., Ochiai, M., Yamada, A. (1982) *Macromol. Chem. Rapid Commun.*, **3**, 769.

296. Kaneko, M., Moriya, S. (1984) *Electrochim. Acta*, **29**, 115.

297. Yaropolov, A.I., Karyakin, A.A., Gogotov, I.N., *et al.* (1984) *Dokl. Akad. Nauk SSSR*, **274**, 1434.

298. Ishikawa, K., Nambu, Y., Endo, T. (1989) *J. Polym. Sci. A., Polym. Chem.*, **27**, 1625.

299. Fujinami, T., Nikawahara, N., Sakai, S., Ogita, M. (1993) *Chem. Lett.*, **6**, 585.

300. Sakamoto, K., Aramaki, K., Nishihara, H. (1993) *Chem. Lett.*, **6**, 659.

301. Wöhrle, D., Wendt, A., Wetemcycr, A., *et al.* (1994) *Izv. Akad. Nauk SSSR, Ser. Khim.*, **11**, 2071.

302. Orlova, I.A., Timonov, A.M., Shagisultanova, G.A. (1995) *Zh. Prikl. Khim.*, **68**, 468.

303. Bedioni, F., Devynck, J., Bied-Charreton, E.C. (1995) *Acc. Chem. Res.*, **28**, 30.

CHAPTER 5

SPECIFIC FEATURES OF CATALYSIS BY IMMOBILIZED METAL COMPLEXES

In Chapters 1–4 special attention was paid to both general features and specific effects of immobilized catalysts having polymer character in each catalytic reaction considered. It is clear that only comparative analysis of the actions of homogeneous and heterogenized catalysts in each particular system and catalyzed reaction can reveal the specificity of immobilized catalysts. To generalize this effect is very difficult because one must take into account a great number of catalyst parameters. In addition, the activity of particular immobilized catalysts arises from their molecular nature and so can hardly be totally defined. For example, as shown earlier, the formation of finely dispersed metal particles promotes catalyst activity in hydrogenation, but inhibits polymerization.

Nevertheless, all observed experimental data allow identification of the most impotant effects of the polymer chain. Macromolecular effects include the roles of the microenvironment (local polarity, the neighbor effect, the effect of the size of the polymer globule, the degree of substitution of functional groups, L, etc.), the electrostatic properties of the ionic groups bound to the polymer chain (polyelectrolytic catalysis), isolation of active centers (dilution effect), local concentration effect (changes in both local active site concentrations due to specific curdling of polymer chains and that of substrate), and the effect of ligand coordination number, etc. When insoluble polymers are used as supports, it is necessary to take into account polymer swelling ability, the heterogeneity of the surface, migration of the attached complexes during the course of the reaction, etc. The situation becomes much more complicated if the immobilized complex undergoes partial dissociation during the reaction and catalysis is achieved simultaneously with mobile and bound particles.

In this chapter we shall try to summarize the most important features of catalysis by immobilized complexes (see also [1]).

The activity of immobilized catalysts can be affected by local polarity, "neighbor effect", the condition of both the polymer chain and the whole polymer mass, chain loading with metal complexes, electrostatic properties of ionic groups bound to the polymer chain (electrolytic catalysis), active site isolation (the effect of dilution), coordination number, etc.

Almost all studies have demonstrated long-term stability of immobilized metal complexes. Let us consider the most probable reasons for this.

5.1 Ligand Exchange in Metal–Polymeric Systems

Since the interaction of MX_n with a macroligand proceeds in an enclosed microvolume and quite a large number of macromolecules, functional groups and other different types of particles (MX_n, electrons, protons, radicals, oligomers, etc.) are involved, the state of the metal-containing polymeric system as a whole can be characterized by a combination of factors: the concentrations of free functional groups, and the complexes, which may contain different numbers of ligands of each type which may occupy various positions in the macromolecule.

There are two main reasons for a decrease of catalyst stability and even disruption of the immobilized complexes.The first is thermodynamic instability. After the system reaches an equilibrium state, one can observe a decrease in content of bound metal ions as some part of the complexes pass into solution (due to their reversible dissociation):

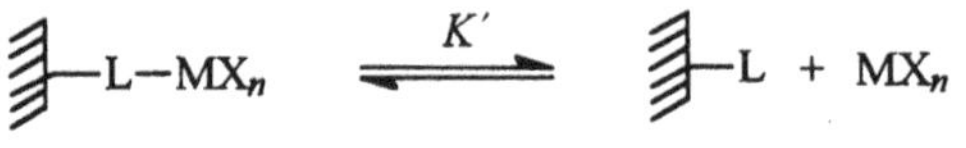

$$\text{} -\text{L}-\text{MX}_n \quad \xrightleftharpoons{K'} \quad \text{} -\text{L} + \text{MX}_n$$

Scheme 5.1

However, the contribution of chemical instability of macrocomplexes is usually higher than that arising for thermodynamic reasons. For example, the substitution of metal complexes with some reactive components (say with solvent molecules) may go so far as to cause their decomposition:

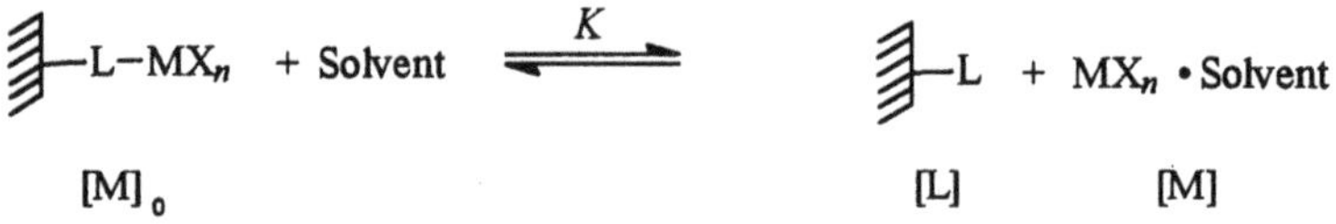

$$\text{} -\text{L}-\text{MX}_n + \text{Solvent} \quad \xrightleftharpoons{K} \quad \text{} -\text{L} + \text{MX}_n \cdot \text{Solvent}$$

$$[M]_0 \qquad\qquad\qquad\qquad\qquad\qquad [L] \qquad [M]$$

Scheme 5.2

It is rather difficult to estimate equilibrium constant values (which indicate metal ion interactions with the macroligand [2]) for these systems because there is usually no data on complex stoichiometry (both bound and mobile), nuclearity and the number of interacting particles. Formally, one can take the relation which is usually used for calculation of the equilibrium constant for low molecular mass complexes[1] [5,6]: $K=[L][M]/[M_0]$, where L=ligand and M=metal.

The specific contribution of each reaction is reflected as a decrease in content of the metal on the support or in the concentration of the metal passing into the solution. These processes have been analyzed[2] in detail by the use of tracer techniques for systems which are used in olefin polymerization [8]:

$$\text{MX}_n \cdot \text{D} + \text{AlR}_x\text{Cl}_{3-x} \quad \rightleftharpoons \quad \text{MX}_n + \text{D} \cdot \text{AlR}_x\text{Cl}_{3-x}$$

Scheme 5.3

---

[1] Strict approaches have been developed for the calculation of equilibrium constants for metal complexes bound to inorganic supports [3,4].

[2] Radiochemical studies (as well as technique details) of exchange rates in ion-exchange resins by means of a strontium tracer are to be found in [7].

When $MX_n=VCl_4$, D=methyl methacrylate and $AlR_xCl_{3-x}=AlEt_3$ (benzene, 293 K), equilibrium constant values were shown to be equal $K'=1.8\times10^{-3}$ mol/l, $K=8.7\times10^2$. Replacement of the ligand environment by the polymeric environment leads to an increase in both thermodynamic and chemical stability of the complexes. Thus by using the stopped-flow technique the kinetics of a pseudo-first-order exchange reaction (pH 9.3–11.0; 298 K, excess of ethylendiaminetetraacetate (EDTA) has been studied [9]:

$$Cu^{2+}-PVAl + EDTA \longrightarrow Cu^{2+}-EDTA + PVAl$$

Scheme 5.4

The kinetic equation for this reaction has the following form:

$$-\frac{d[Cu^{2+}-PVAl]}{dt} = k_1\frac{[Cu^{2+}-PVAl]}{[PVAl]} = k_2\frac{[Cu^{2+}-H-PVAl]}{[PVAl]}$$

where PVAl=poly(vinyl alcohol). Under these conditions $k_1=3.62\times10^{-3}$ mol/s and $k_2=3.68\times10^4$ mol/s. Ligand exchange occurs by two mechanisms, either H^+ attack (from EDTA) with formation of the intermediate complex $Cu^{2+}-H-PVAl$ and its subsequent dissociation, or solvolysis involving water.

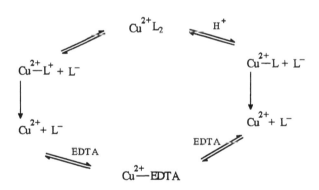

L is PVAl unit

Scheme 5.5

One of the characteristic examples of exchange reactions involving low molecular mass and macromolecular ligands is the exchange of the axial ligand in benzylcobaltoxime and in cobaltoxime $(Co(DH)_2)$ covalently bound to chloromethylated polystyrene (PS) [10]:

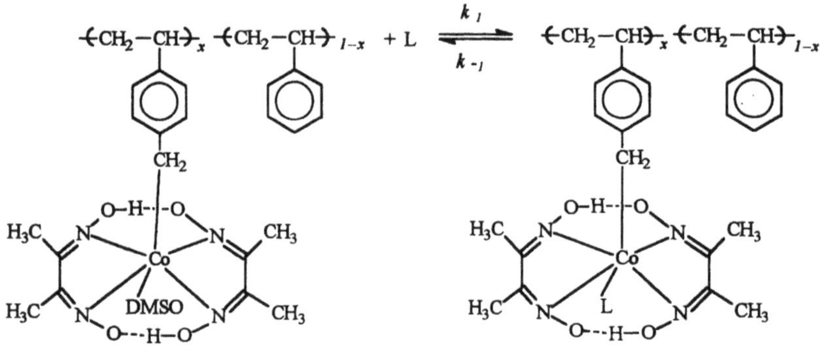

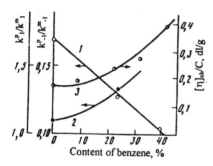

K = k_1 / k_{-1},　　L = Py, SCN⁻

Scheme 5.6

Although Co^{3+} complexes are comparatively resistant to substitution, the presence of the macromolecular ligand promotes a more rapid substitution of the axial ligand in the *trans*-position (solvent dimethyl sulfoxide (DMSO) or DMSO/benzene). It is apparent from Table 5.1 that rate constants for the forward reactions in the systems compared are similar, although the rate constants of the reverse reactions for dissociation of the complex, which contains the low molecular mass *trans*-ligand, are approximately 3 (L=SCN⁻) and 9 (L=pyridine) times higher.

Figure 5.1 Variation in the relative rate constants of pyridine coordination with cobaltoxime k_1 (1) and k_{-1} (2) and the characteristic viscosity (η) of the macrocomplex (3) as a function of the composition of the DMSO/benzene mixture.

Thus, the higher values of the equilibrium constants for ligand exchange in metal-containing polymer systems in comparison with the low molecular mass benzyl complexes may be attributed to the smaller values of the rate constants for the reverse reactions. This may be caused by both the higher local concentration of cobaltoxime in the polymer matrix in comparison with its concentration in the volume of DMSO solvent and also by steric hindrances which the macrochain makes, hampering ligand transport from the polymer "cells" and promoting its recombination. In the mixture of solvents, DMSO/benzene, with the proportion of the latter increased, the polymer globule expands, as a result of which the effect of these factors decreasing. The increase of characteristic viscosity is accompanied by a decrease in the rate constant of the reverse reaction (Figure 5.1). The higher value of the rate constant of ligand exchange (L–pyridine) in polymer-bound cobaltoxime (in DMSO) is the result of large changes in entropy (Table 5.2). The higher values of ΔS and ΔH for polymer systems are caused by higher pyridine lability.

Table 5.1 Thermodynamical parameters of pyridine coordination by cobaltoxime at 298 K.

R	Solvent	ΔF (kJ/mol)	ΔH (kJ/mol)	ΔS (J/mol deg)
Polymer	DMSO	–17.6	–38.6	–67
Benzyl	DMSO	–16.4	–67.2	–167
Polymer	DMSO+benzene (3:1)	–24.8	–14.7	33
Benzyl	DMSO+benzene (3:1)	–18.9	–14.7	13

Table 5.2 Rate constants of ligand exchange in $R–Co(DH)_2DMSO$ [10].

R	Ligand	k_1 l/(mol·s)	k_{-1} l/(mol·s)	$K \times 10^{-2}$
Polymer	SCN–	1×10^2	4.8	2.1
Benzyl	SCN–	1.2×10^3	15	0.8
Polymer	Py	23.6	2×10^{-3}	118
Benzyl	Py	16.5	1.7×10^{-2}	9.8

Similar results [11] have been obtained in the studies of outersphere electron exchange involving Co^{2+} and Fe^{3+} ions. Both calculation of the standard potentials and measurements of rate constants of exchange reactions, with allowance made for changes in structural parameters of the macrocomplexes, indicate that the systems examined are in many ways similar to metalloprotein complexes.

Hence, the replacement of a low molecular mass ligand by a polymeric environment hinders the entry of reagents into the coordination sphere of the transition metal and causes its additional polarization in the non-uniform electrostatic field of the ligands, which also confers stability upon the macrocomplex. The ability of the coordinatively saturated metal ion in the macrocomplex to give coordinative vacancies is also significant for ligand exchange. They may arise during immobilization owing to the prevention of metal complex dimerization or on account of the facile disruption of M–M bonds, as a result of the transition metal ion reduction due to effective adsorption of excess ligands by the polymer matrix, etc.

5.2 Characteristics of Electron Transfer Reactions in Polymer–Metal Complexes

Electron transport in metal-containing polymer systems determines the rate of many processes, and is almost always the slowest stage of the catalytic act. Such an intramolecular process is very sensitive to the active center environment being dependent on the nature of both ligand and solvent, the ionic strength, the pH, etc. It is accompanied by rearrangement of the complex and by conformational changes in the polymer.

The primary structure of the support, the distribution of the redox-centers, the charge density, and the polarity of the polymer domains have a strong effect on electron transport in macromolecules. Redox transformations in macrocomplexes of a series of transition metal ions, such as $Cu^{2+} \leftrightarrow Cu^+$, $Co^{2+} \leftrightarrow Co^+$, $Fe^{2+} \leftrightarrow Fe^{3+}$, $Mo^{6+} \leftrightarrow Mo^{5+}$ are most widely exemplified. Redox processes sometimes occur just at the stage of metal complex immobilization [12] and they most often take place in oxidation reactions. Thus, Cu^{2+} chelation with poly[N-(acryloylamino)methyl]-mercaptoacetamide cross-linked with 30% N,N'-methylenebisacrylamide is accompanied by formation of bound Cu^+ complexes [13] which are reoxidized in air to Cu^{2+}; their concentration increases progressively for several months (Figure 5.2). The more probable structure of these complexes is planar with some rhombic distortion provided by S and O atoms.

Similar redox processes can accompany catalytic cycles. Thus, chelate complexes of Cu^{2+} are active in oxidation of aldehydes to acids or cyclohexanol to cyclohexanone. In this case Cu^{2+} is reduced to Cu^+ [14].

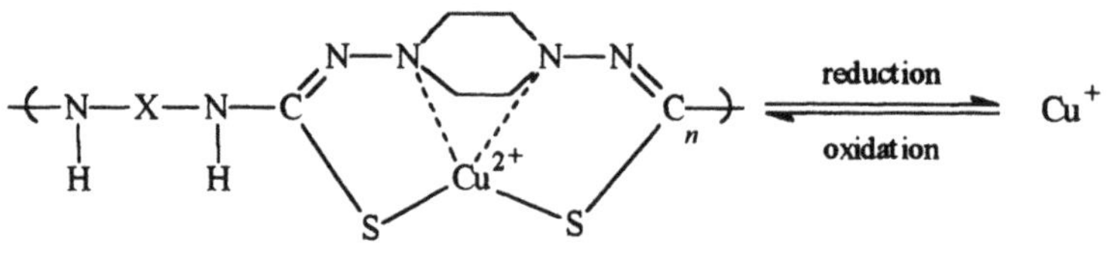

Scheme 5.7

In turn the Cu^+ ion in the polymeric complex is oxidized to Cu^{2+} in reduction reactions (nitrobenzene to azoxybenzene, nitrocyclohexane to cyclohexanoxime, 1,4-benzoquinone to hydroquinone, and $KMnO_4$ to MnO_2). It has been observed that no copper leaching occurs in the course of these reactions, with the exception of reactions with 1,4-benzoquinone and hydroquinone. Such a conjugated process is referred to as the "ping-pong" mechanism. The mechanism of cyclic redox reactions of Cu^+ chelates with poly(acroleinoxime)

includes the formation of an intermediate complex with O_2 [15].

Table 5.3 Kinetic and activation parameters for the oxidation of Cu^+ complexes at 298 K [16].

Ligand	pH	$K_{eff} \times 10^{-2}$ (l/mol·s)	$\Delta F^{\neq}$ (kJ/mol)	$\Delta H^{\neq}$ (kJ/mol)	$\Delta S^{\neq}$ (J/mol·deg)
Poly-1-vinylimidazole	4.7	2.7	61.3	48.3	−44.1
	4.3	14.0	57.1	29.4	~92.4
Poly-4(5)-vinylimidazole	6.0	0.10	66.0	61.8	−14.4
	4.7	0.26	64.7	57.5	−24.0
Imidazole	8.0	4.9	60.5	13.4	−154.6

To determine the specific characteristics of the polymer environment, the oxidation of Cu^+ complexes with polyvinylimidazoles by molecular oxygen was compared with that of imidazole [16]. This study was made by using the stopped-flow technique in buffer systems. Both kinetic parameters of the reaction and dependence on pH values are presented in Table 5.3. The higher enthalpy and lower entropic factor are typical of polymer complex oxidation. The discrepancies in the kinetic and activation parameters are associated with conformational transformations of the macroligand in the course of the redox reaction leading to a change in bond lengths and angles between them in Cu^+ and Cu^{2+} complexes. The occurrence of conformational transformations during oxidation was demonstrated by viscometric measurements (Figure 5.3).

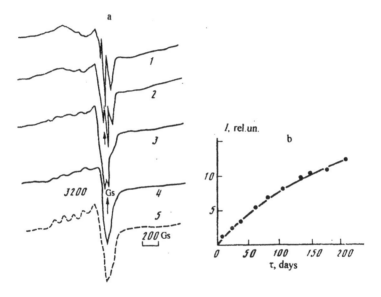

Figure 5.2 ESR spectra of regenerated Cu^{2+} complexes in macrochelates (*a*) and variation in their intensity with time (*b*) (room temperature). Exposure time (in air), days: (1) 1, (2) 7, (3) 23, (4) 38; intensity, *I*, in related units: (1) 1, (2) 1.4, (3) 2.6, (4) 3.6; (5) calculated spectrum.

Similar patterns have also been found in the redox cycles $Co^{3+} \leftrightarrow Co^{2+}$. For example, the outersphere reactions of Co^{3+} macrocomplexes with poly(4-vinylpyridine) (P4VP), poly(ethyleneimine) (PEIm), poly(vinylimidazole) (PVIm) and Fe^{2+} complexes with electron transport occur at lower rates than the reactions of Co^{3+} complexes with low molecular mass analogues [17,18]. The opposite situation is found when hydrophobic interactions[3] occur between the polymer ligand and Fe^{2+}. This is associated with an

---

[3] The hydration of tris(oxalato)cobalt ions accompanied by Co^{3+} reduction is considerably accelerated by the polyelectrolyte, poly(*N*-laurylethyleneimine bromide) [19]. Vice versa, the polarogram of Co^{3+}–P4VP complex reduction shows considerably lower saturation current than that of the monomeric complex which is caused by differences in diffusion coefficients of the systems compared [20]. This effect was attributed to partial blocking of the electrode process.

increase in the local concentration of the reducing agent in the macrocomplex due to electrostatic attraction between metal ion and macrocomplex as polycation. Thus addition of styrene sulfonate increases the rate of reduction of the Co^{3+} ion in P4VP complexes by 10–30 times in comparison with pyridine complex. In both cases bridging ligands (Cl^-, Br^-) decrease the rate of reduction. An increase of polymer chain loading (coordination degree f) is followed by a decrease of activation parameters (Table 5.4). This may be associated with the state of the polymer chain, and particularly with elongation (as indicated by an increase in the characteristic viscosity) and hence with a change in the charge density on the polymer chain.

The reverse oxidation reaction $Co^{2+} \rightarrow Co^{3+}$ may alternatively be either more rapid or slower than that with the low molecular mass reagents. Thus, this reaction acceleration in the system $Co(ClO_4)_2$–polystyrene potassium sulfonate 2-(2-pyridylazo)-1-naphthol (very valuable for studying biological processes as the polyelectrolyte here acts as the protein part of an enzyme) is caused [21] by intermediate chelate stabilization due to coordination with 2-(2-pyridylazo)-1-naphthol. The reaction may follow both electrostatic and coordination routes. Structural changes in the macrocomplex are of great interest. The initial complex of Co^{2+} shows positive dichroism caused by changes in extinction coefficient due to chromophore orientation in an electrical field. Dichroism changes its sign in the course of cobalt oxidation which is the result of chelate complex rearrangement with respect to the polymer chain due to coordination of additional chain units.

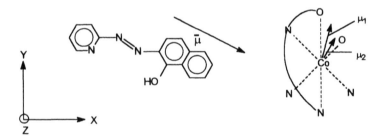

Scheme 5.8

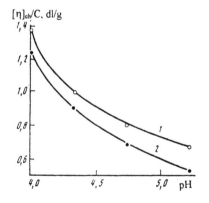

Figure 5.3 The dependence of characteristic viscosity of Cu^+ and Cu^{2+} complexes with poly(1-vinylimidazole) in water solution on pH values: (1) Cu^+, (2) Cu^{2+}; buffer CH_3COONa/CH_3COOH, 298 K.

Of the other comparatively numerous examples that confirm the concept being developed, redox reactions involving molybdenum macrocomplexes should be mentioned. Peptide Mo^{5+} complexes immobilized on chloromethylated PS oxidize to Mo^{6+} by one-electron transfer from NO_3^- [22]. The cycle of reaction consists of polymer-bound molybdenum reduction under the action of PPh_3 and subsequent oxidation of the reduced form with the nitrate ion. This mechanism is similar to the cyclic action of reductase and oxidase. Another analogy (molybdo-oxidase model) is the redox reaction of cysteine (cys)thiolato dioxomolybdenum(6+) centers immobilized to polyethyleneglycol (PEG), $MoO_2(Cys$-O-PEG$)_2$ [23]. Under the action of water the cleavage of one of the Mo=O bonds proceeds and S–S fragment oxidation is accompanied by $Mo^{6+} \rightarrow Mo^{5+}$ reduction:

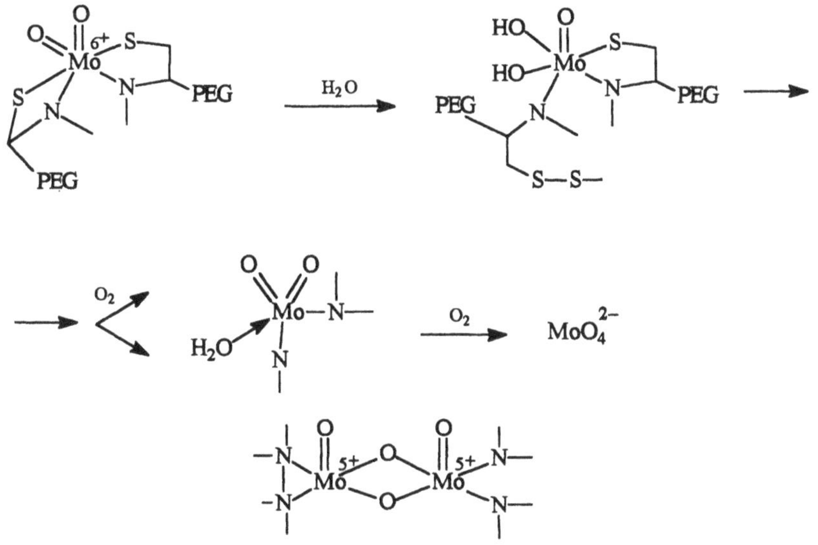

Scheme 5.9

These complex reductions proceed in the system dimethylformamide (DMFA)/H_2O/P(OMe)$_3$)and also follow a one-electron transfer mechanism yielding mononuclear complexes of Mo^{5+}. At the same time an excess of H_2O activates the Mo=O bond. However the hydrophobic environment of the complex regulates its contact with water molecules thus stabilizing the Mo^{6+} chelate complex in the redox cycle.

Redox steps, such as $Pd^{2+} \rightarrow Pd^+ \rightarrow Pd^0$; $Ti^{4+} \rightarrow Ti^{3+} \rightarrow Ti^{2+}$; $V^{5+} \rightarrow V^{4+} \rightarrow V^{3+}$; $Ni^{2+} \rightarrow Ni^+ \rightarrow Ni^0$; $Co^{2+} \rightarrow Co^+ \rightarrow Co^0$, etc. often accompany catalytic reactions, as already mentioned above (Chapters 1–4). The formation of polynuclear complexes of the reduced forms frequently causes deactivation of homogeneous metal complexes. It is important that multiple recycling of electron transport does not cause changes in the nuclearity of Cu, Co, and Mo bound complexes, whereas macroligand reorganization relates to certain of the most important factors influencing the reaction with electron transfer in the polymer matrix [24].

Table 5.4 Activation parameters for the electron transfer reaction and the change in viscosity in the $Co^{3+} \cdot PVIm-Fe(EDTA)^{2+}$ system [18] at pH 5 and 303 K (buffer CH_3COONa/CH_3COOH).

Coordination degree (f)	[η] (dl/mmol)	ΔF≠ (kJ/mol)	ΔH≠ (kJ/mol)	ΔS≠ (J/mol deg)
0.10	2.26	51.7	144.5	72.8
0.12	2.33	50.4	136.1	67.4
0.15	2.86	49.1	106.3	45.0
0.25	3.25	46.6	97.4	39.3
0.32	3.34	45.3	94.5	30.3
N-Vinyl-2-methylimidazole	–	52.0	31.9	15.6

5.3 Macromolecular Effects in Catalysis by Immobilized Complexes

The influences of polymer support properties on catalyzed reactions have been numerously exemplified below for different reactions, such as oxidation, hydrogenation, n-merization, hydroformylation, and especially for asymmetric synthesis. This effect is greatest for soluble metal–polymer catalysts although is may also be considerable for insoluble promotors.

One of the protective functions of the macroligand is to prevent diffusion within the support, which impedes deactivation of the active centers or interaction of the substrate with reagents or microadditives in the reaction medium (for example H_2O, O_2, etc.). Thus to protect high-spin Fe^{2+}–porphyrin complexes, which are oxidized in solution to low-spin Fe^{3+} complexes, the former have been immobilized (see Section 2.2) on macroligands (polymers and copolymers based on 1-vinylimidazole, 1-vinyl-2-methylimidazole, 1-vinylpyrrolidone, water-soluble copolymers based on quaternizied 4-vinylpyridine, etc.). This is an effective example of hydrophobic protection. Mo^{5+} complexes bound to chloromethylated CSDVB (copolymer of styrene with divinylbenzene) modified with peptides became stable within several days in the presence of O_2 and H_2O [22].

Another mechanism for the acceleration of catalytic reactions is based on water molecules binding to hydrophilic polymers [19]. Thus the rate of polyelectrolytic catalysis of tris(oxalato)cobalt hydration in a binary mixture of water with DMSO or DMFA is increased 10^5 times due to the polymer solvation effect (poly(ethylene)iminopropionate, poly(4-vinyl-N-ethylpyridinium bromide, etc.). This effect is caused by selective water sorption from the solvent mixtures and gives a double advantage due to localization of the anionic reagent in the vicinity of the polymer chain and activating repulsion of H_3O^+.

Hydrophobic protection plays a major role in reactions catalyzed by bound organometallic compounds or those formed at the stage of catalyst synthesis (see Chapter 2). Polar media are also used for these purposes. This is possible because in many reactions the globule functionalized macromolecules with bound catalytic centers surrounded by solvents fulfill the function of isolated microreactors [25,26]. The increased activity of immobilized catalysts may be caused by an increase in both the local concentration of substrate and the active forms in such microreactors, despite possible steric hindrances as the reagent approaches the catalysts and during removal of the reaction products. The hyperequilibrium substrate concentration in domains situated adjacent to the active centers is achieved by electrostatic, donor–acceptor, and π-interactions involving active regions and functional groups of the polymer. The macrochain provides a favorable microenvironment as it does in enzymatic catalysis.

The polymer chain conformation also considerably affects catalytic properties of polymer-immobilized systems. There are numerous examples of such effects, especially for macroligands showing α-helical structure with specified distances between bound metal ions. This has already been mentioned above for the oxidation of $L(+)$-ascorbic acid, catalyzed by $trans$-[Fe(2,2′,2″,2‴-tetrapyridyl)(OH)$_2$]$^+$ bound to Na-poly(L-glutamate) [27].

"Renaturation" of the polymer chain caused by conformational asymmetry in metal–polymer chains is observed following metal complex binding to peptides. This effect is favorable for asymmetric induction (stereospecificity factor defined as K_{FeD}/K_{FeL}) increasing with a rise of the conformational asymmetry due to an increase in the numbers of α-helix fragments in polypeptide matrices. The observed regularities are associated with changes in standard thermodynamic potentials of diastereomeric transition states, $G^{\neq}_{LL}$ and $G^{\neq}_{DL}$ which differ with steric factors; steric hindrances are less for the DL diastereomers than with the LL. As a result of the more effective electron transfer caused by favorable changes in activation parameters ($G^{\neq}_{LL} - G^{\neq}_{DL}$=1250–3400 J/mol, K_{FeD}/K_{FeL}=1.5–4.0, with the changes in chain loading f=0.01–0.20) an increase of relative conformational mobility of intermediates takes place.

Differences in polymer support configurations can influence not only composition, structure and stability of the metal complexes formed but also their catalytic properties. There are only a limited number of data on these questions. Thus a specific feature of metal ion interactions with stereoregular poly(methacrylic acid) (PMAA) is that two neighboring carboxylic groups are involved in metal binding, ΔH always being higher for complexes with isotactic PMAA than that for complexes with syndiotactic PMAA [28]. Isotactic PMAA behaves as a weaker acid with respect to the syndiotactic form [29], for conformational reasons. Isotactic PMAA shows local helical structure in solution, whereas the syndiotactic form is planar zig-zag; atactic PMAA takes on a compact conformation in an ionized or partially ionized state. Rh and Ru complexes bound to isotactic or atactic phosphorylated polystyrene exhibit different activities in oxidation, hydrolysis and isomerization, etc. [26], although it is difficult to reach a clear conclusion as to the causes of this phenomenon.

A feature of the soluble macrocomplexes is that not only the structure and conformation of the chain but also the chain length have significant effects on their activity and selectivity. Thus the rate of tetralin oxidation catalyzed by CoCl$_2$ bound to PEG increases by ten times with an increase of the number average molecular mass of PEG from 4000 to 40000 [30]. This is an example of a new effect in catalysis: an increase in rate of the process owing to the energies of non-equilibrium configurations of the polymer chains arising in the course of the reactions. The energy increases with an increase of chain length, and under certain conditions (for example when the rate of the catalytic step is greater than the rate of relaxation of the polymer non-equilibrium) may be transmitted to the reacting components, thereby increasing the rate of their transformation.

This effect is similar to the energy recuperation effect in enzymatic catalysis in which the protein molecules assume the role of an energy reservoir. The energy released in the course of the reaction that is not dissipated as heat is utilized to activate the reagents. During the enzyme reactions the protein macromolecule may be transformed from an equilibrium to a non-equilibrium state. The energy of this transition is utilized to overcome the substrate transformation activation barrier, whereas the energy released as a result of the reaction is utilized to return the protein molecule to its original equilibrium state [31].

Other considerations [32] suggest that the small value of the activation barrier for the enzymatic reaction may be attributed to the fact that enzyme–substrate complex formation proceeds in the presence of some functional groups with a high potential energy. For example the zinc ion within carboxypeptidase or carbonic anhydrase is much more reactive than zinc occurring in the normal ionic form or in a complex with a low molecular mass ligand[4].

Another hypothesis [33] suggests that conformational enzyme transformation is the result of the elementary step of the enzymatic reaction: before the stage of enzyme–substrate complex formation, the macromolecule is in conformational equilibrium.

Thus the high activity and selectivity of enzymes in contrast to metal complexes immobilized on synthetic polymers is probably caused by the conformational specificity of these macromolecules. Changes in conformational state are equivalent in strength to a redistribution of chemical bonds in the substrate molecules.

Thus, immobilized macrocomplexes can be assumed only to be the simplest model of enzymatic catalysts with the aim of demonstrating improvements of catalytic properties in the course of the binding of active centers to macromolecules.

5.4 Main Factors Regulating the Activity of Immobilized Metal Complexes

Generally any of the usual means of regulating homogeneous catalyst activity (pH, ionic strength of solvent, additives, temperature, pressure, reagent concentrations, the ratio [substrate]/[catalyst], etc.) results in a specific change for immobilized catalysts. Moreover, these catalysts provide a number of additional factors which have no analogies in homogeneous catalysis which enable their activity, selectivity, and sometimes asymmetric induction to be regulated. These are both the nature and structure of the polymer matrix, cross-linking degree, the presence of different functional groups, including inert ones, the nature of the chemical bond of MX_n with the macroligand, the degree of occupied functional groups, the profile of metal ion distribution along the polymer chain or in the mass of the polymer, etc. The effects of each factor are specified for each catalyzed reaction.

Polyacids, polymers based on primary and secondary amines, thiols and some other compound are usually used for σ- or ionic metal complex binding. Catalyst supports most widely used are phosphorylated PS or CSDVB, as well as polymers and copolymers based on vinylpyridines, polyesters, etc.

The frequently higher specific catalytic activity of the immobilized complexes may be attributed to structural defects in the coordination centers and to their unsaturation, which promotes substrate coordination, and also to cooperative effects. Many of these factors may be taken into account during the construction of immobilized catalysts. For

---

[4] The reason is that in the active center the zinc atom is only partially associated with protein ligands forming the stressed configuration of a deformed tetrahedron. Such formations are a specific feature of protein macromolecules with secondary and tertiary structures.

example, the composition of bound complexes may be controlled to a certain extent by the L:M ratio. Simultaneous formation of mono- and disubstituted Ni^{2+} complexes during their fixation on PE-gr-poly(acrylic acid) (PAAc) is well illustrated in Figure 5.4 [34]. In most reactions between the macroligand and MX_n equilibrium is established almost immediately after the components are mixed, but sometimes a prolonged period is required for this to be attained, particularly if the macroligand is polyfunctional. Thus, in the course of bonding of cis-Ru(Dipy)$_2$Cl$_2$ (Dipy=dipyridyl) to a copolymer of 4-vinyl-pyridine (4VP) with 4-methyl-4'-vinyldipyridyl, 4VP groups react first followed by prolonged (tens or hundreds of hours, Figure 5.5) substantial complex rearrangement with Dipy unit participation [35]:

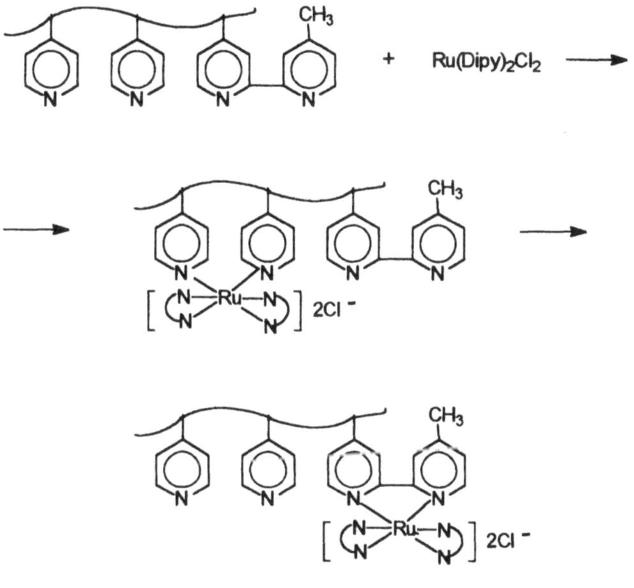

Scheme 5.10

In certain cases the total rearrangement of the complex occurs more slowly than the catalyzed reaction. Since the structural homogeneity of the complex is important for ensuring high and stable activity, it becomes essential to take into account the rate of such a rearrangement.

The formation of the required structure, determined by the electronic configuration of the transition metal ion, is also dependent on the conformational exchange in the macroligand which leads to the adoption of the appropriate orientation. Such an exchange is associated with energy losses which are greater the more the chain conformation differs from that required. In this sense the "pliability" of the macromolecule is important, and therefore the macroligand can not be regarded as a simple set of independent functional groups[5].

---

[5] The pattern is complicated if one takes into account the mutual influence of functional groups, which is conducted by such electronic effects as induction effect, field induction,

Furthermore the polymer matrix may bind and stabilize a coordinatively unsaturated metal complex [36]. The rate of rearrangement (relaxation) of the active centers is low because of the large size of such formations, especially in the case of strongly cross-linked matrices. These processes become complicated on going from homogeneous to heterogeneous systems and especially to enzymatic catalysis [37]. Therefore in a catalyzed reaction a non-equilibrium configuration of active centers (and their hyperequilibrium concentration) becomes possible. In this case in heterogeneous catalysis non-equilibrium phases may arise. Asymmetric distribution of transition metal ions on the support surface (the result of both metal ion binding and the catalytic reaction) has been observed in main studies. This particularly concerns the immobilized complexes of Pd, Pt, Rh, Cu, Ti, V, Ni, etc. For example a non-uniform distribution of palladium ions[6] on CSDVB has been demonstrated by different methods. Three different centers of H_2PtCl_4 adsorption (and consequently distribution) have been revealed (π-binding units of carbonic support) which were characterized by different strengths of Pt^{2+} bonding and hence of different reactivity with respect to oxidants [39].

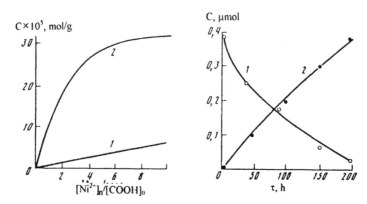

Figure 5.4 The dependence of mono- (1) and disubstituted (2) Ni^{2+} contents on the initial ratio of $[Ni^{2+}]_0/[COOH]_0$. Support PE-gr-PAAc; grafting degree 7.2%.

Figure 5.5 The kinetics of pyridine–Ru^{2+} complex: (1) rearrangement into dipyridyl complexes; (2) on the polymer support.

During catalysis by immobilized complexes not only the total content (concentration) of bound MX_n but also its distribution profile in the polymer matrix (topography) is of great importance. Numerous observations indicate the statistical nature of the distribution of the transition metal ions on the support. Analysis and calculations of the topography of immobilized complexes opens avenues for increasing the efficiency of these catalysts. Optimization is achieved by moving the active sites to the surface (or within a thin surface layer) of the polymer support. When such catalysts are constructed

---

conjugation, dispersive interactions, hydrogen bond formation, the formation of supramolecular structures, etc.

[6] An interesting method for active component determination (for instance Pd) in polymer granules has been suggested [38]. This is based on catalytic decomposition of copper formate on granules. The deposition of the cooper metal formed reflects the Pd ion distribution, whereas the rate of the reaction correlates with catalyst activity.

their main role is attributed to the topography of the functional groups of the macroligand and their accessibility. The latter is determined by the contribution of the polymer chain to the changes in microenvironment of functional groups, viscosity and density of cross-linkages in gel and microgel particles and by the thermodynamic compatibility of the polymer matrix and reaction medium as well as the polymer support pore structure and pore size distribution (including in the swollen state). Due to the mobility of the polymer chains in the surface and at the boundary of the polymer–medium phases, reorientation and rearrangement of the macromolecular fragments occurs [40], as a result of which the polar groups may be turned to the side of the polar phases or vice versa. Moreover the nature of their distribution may also affect even the properties of functional groups. For example the sulfogroups located on the surface possess lower acidity than groups situated within the polymer mass [41]. Interactions between adjacent dicarboxylic groups (a copolymer of maleic acid with methyl vinyl ether) leads to the emergence of alternating singly and doubly charged dicarboxylic acid groups [42], etc.

When the catalytic systems under examination are used one should remember that the temperature usually affects the rate of the reaction in a complicated manner. The dependence of reaction rate (it has been shown already for ethylene polymerization and dimerization, in some hydrogenation reactions, oxidation, etc.) catalyzed by immobilized polymer complexes often has a bell-shaped relationship. With an increase in temperature the reaction rates at first increase in accordance with Arrhenius law, and then after the second-order transition temperature (T_g) is reached, or after softening of the polymer support, the rates of these reactions fall. Thus in the temperature range in which the chain segments possess adequate mobility and in which the relaxation processes associated with the reorientation of the entire macromolecule become controlling, there is a change in the structure of the active centers (a change in the coordination unsaturation state of the transition metal, aggregation of active centers, reduction of the metal, etc.).

In particular the rate of gas–phase ethylene polymerization ($P4VP \cdot VCl_4 - AlEt_2Cl$ system) is dependent on the mobility restoration temperature (rate of the relaxation processes) [43]. In the Arrhenius region the process is reversible and E_a has a normal anionic-coordination polymerization value (16 kJ/mol). In the T_m range and above the polymerization temperature coefficient becomes negative (–130 kJ/mol).

This phenomenon was studied in greatest detail in the case of gas–phase ethylene hydrogenation by Rh complexes bound to triple copolymers consisting of styrene, diphenyl-p-styrylphosphine (DPSP) (5–10%) cross-linked with DVB (2%) [44]. The temperature dependence is readily reproducible over a broad range. At T_g a discontinuity in the Arrhenius curve is found (Figure 5.6). The activation barrier of the reactions at $T>T_g$ remains essentially the same, i.e. all changes in the rate constant are associated with the entropy factor. This phenomenon arises as a result of the restoration of the mobility of the polymer segments and the creation of the most suitable configuration of the active center for the catalytic action. Also one cannot rule out contributions from other factors. When $T>T_g$, relaxation of micro-Brownian movement promotes diffusion and increases the probability of substrate collisions with the active centers. Furthermore movements of adjacent groups and ligands may increase steric hindrances for the coordination of ethylene and may reduce the stability of the bond of the coordinated substrate. Finally a change in the nature of the movement in the polymer chain also leads to structural transformations of the active centers, since in this case there are changes in the coordinating capacity of the functional groups, their configuration and coordination unsaturation of the transition metal, etc. Both the reaction rate and the content of absorbed substrate n decreases with olefin elongation at temperatures below T_g (Table 5.5).

The anomalous character of the temperature dependence of the reaction rate can be associated neither with an increase in ethylene diffusion into small pores of the catalyst (which would become limiting close to T_g) nor with the fact that at a certain temperature the overall rate of the process begins to be controlled by another mechanism (since the slope of the Arrhenius plot remains virtually constant). A comparison of adsorbed olefin content at several temperatures showed that the complex with ethylene is more stable than say that with cis-2-butene, which is in accord with the rates of hydrogenation of these monomers.

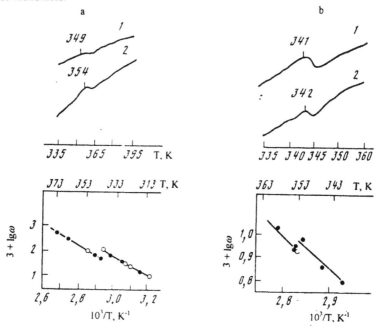

Figure 5.6   Differential scanning calorimetry curves of polymer supports (1) and macrocomplexes of Rh (2) and the Arrhenius dependence of the rate of ethylene hydrogenation in the gas phase on rhodium macrocomplex (a) CSDVB–DPSP (5%); (b) CSDVB–DPSP (10 %).

The reversible discontinuity in the Arrhenius dependence may be explained if one assumes the existence of two different coordination states differing in catalytic activity in the T_g region. The somewhat non-uniform slope of the curve on different sides of T_g (33 kJ/mol at higher and 35 kJ/mol at lower temperatures) may probably be attributed to the fact that the enthalpy and entropy of ethylene coordinated at various Rh centers differs on account of movements of the polymer network.

It is interesting that similar effects have also been found in soluble metal–polymer catalysts. Thus the effective rate constant (k_{eff}) for cholesterol oxidation by O_2 in the presence of soluble porphyrin-containing metal polymers was shown to be temperature dependent [45] with a maximum at room temperature, and there were no analogies with normal homogeneous catalysis. The non-monotonous nature of $k_{eff}(T)$ dependence indicates that conformational transitions possibly occur in such systems. One such transition may be caused by a disruption of the ordering of polymer fragments associated with hydrogen bonds, coordinate bonds, and other bonds during their concomitant rupture. In this temperature range the temperature dependence of the characteristic viscosity has the form $\eta = \eta_0 \exp(-E/RT)$, as in any activation process which is also associated with conformational transitions in metalloporphyrin-containing macromolecules. In the transition region a significant fluctuation of the conformations of the chains in the solution is found, and each conformation corresponds to the value of the reaction rate contant, because of the difference in the pre-exponential factor, which reflects steric factors, or because of the multiple effects of the active centers. Since there is a set of different rate constant corresponding to metastable conformational states in the transition region, the dependence $k_{eff}(T)$ also has a bell-shape non-Arrhenius character.

In principle such effects may influence not only the activity but also the selectivity of the immobilized catalysts. Only one example will be presented, i.e. a reduction in temperature during the synthesis of cyclopropane ethers leads to an increase in the proportion of the cis-isomer, probably as a consequence of the limitation in the mobility of the macromolecular chain [46].

The examples of temperature effects on the reaction rates discussed show the unusual action of this usual factor for reaction control in the case of polymer-immobilized catalysts. However the action of other factors for regulation of both homogeneous and heterogeneous catalysis also result inspecific effects for immobilized polymer systems.

Table 5.5 Olefin adsorption by CSDVB–DPSP (10%) at 313 K [44].

Olefin	Adsorbed olefin, $n \times 10^5$ (mol/g)	Adsorption rate $\times 10^5$ (mol/g·min)
C_2H_4	1.0	0.19
C_3H_6	2.0	1.1
1-Butene	3.6	2.6
cis-2-Butene	7.9	4.7
trans-2-Butene	8.1	7.5

5.5 The Effects of Cluster Formation and Cooperative Stabilization in Immobilized Systems

Isolation of transition metal ions by binding them to the polymer support, one of the basic ideas in the creation of immobilized metal complexes, is difficult to achieve in practice. There are several reasons for this. Firstly, regions with higher local concentration of transition metal ions may be formed immediately during the stage of their fixation. Different cooperative interactions (including exchange processes between paramagnetic ions) between catalyst components are observed in these regions. The proportion of such interactions increases with the rise of binding metal-complex surface density. If one assumes non-uniform catalytic activity of different forms, the bell-shaped dependence of catalytic activity (as mentioned above for hydrogenation, n-merization reactions, etc.) on the surface density of bound metal complexes[7] becomes clear. Particle enlargement is one of the underlying reasons for metal complex catalyst deactivation in the process of ethylene polymerization and even dimerization. At the same time the seat of activity may well be the particles located at the boundaries of cluster associations, stabilized by the electronic system of the cluster[8].

Cooperative interactions of this type increase the stability and activity of immobilized catalysts. Thus in polymers based on Rh^+–bis–(4,4′-diisocyanodiphenyl) chloride (the catalyst for hydrogenation of olefins) [47] in which the Rh is arranged regularly in a tetragonal or hexagonal environment where each rhodium atom is associated with two or four isocyanate ligands, there are three types of active center (Figure 5.7a). The first of these (a) incorporates Rh at the end of the chain, the second (b) incorporates rhodium atoms in the backbone of the macroligand, and the third (c) carries Rh on the outer side (001) of the material. The number of centers active in catalysis does not exceed 1% of the number of Rh atoms, but cooperative interactions are observed between the centers (a) arranged at close spacings, which increases the activity of the catalysts in the isomerization of the substrate, whereas the cooperative effect created by interaction between centers (b) and (c) or adjacent centers (b) is the reason for the increase in activity in the hydrogenation reactions (in this case the Rh–Rh interactions have the characteristics of a metal bond). The relationship between the different centers is dependent on the previous history, the method of catalyst preparation, its topography, ρ_s value, temperature, as well as on the nature of the polymer support. Thus a more uniform distribution of $RhCl(PPh_3)_3$ on CSDVB modified with phosphyne groups [48] is reached by its binding to a triple copolymer of styrene, DVB and 4-diphenyl-p-styrylphosphyne (in comparison with the case when CSDVB was phosphynated by polymer analogous reactions) (Figure 5.8). Certainly Rh^0 particles of different size and activity in olefin isomerization[9] are formed by reduction (i.e. by hydrogen) of immobilized metal complexes.

The processes of cluster formation and particle enlargement are influenced both by the method of catalyst preparation and the catalytic process usually in the direction of increased nuclearity. Due to the small size of the particles formed (1–2 nm) they may scarcely be metal-like, rather being in the form of multinuclear associates. The dynamics of their formation is a complex process. This is only one example [32]. Cu^+ and Fe^{2+} ions reduced in the absence of O_2 are located far from each other in the enzyme molecule. Meanwhile O_2 binding proceeds as a result of the approach of an ion pair, and is accompanied by conformational transformations of the enzyme but does not lead to strong M–M bond formation to preserve coordination vacancies after activation.

If one takes into account that metal complex association can involve the interaction of several segments of the same polymer chain, characterized by different mobilities and

---

[7] The specificity of catalyst concentration action on reaction parameters for the case of immobilized polymer–metal complexes is demonstrated by this.

[8] It has been mentioned above that the macroligand can stabilize copper complex dimers impeding them from monomerization.

[9] The process of reduction of bound metal complexes is also dependent on the nature of the support, metal ion concentration and distribution profile. Thus in [49] the activation energy of Ru crystallite reduction in catalysts 1% Ru/Al_2O_3 and 1% Ru/SiO_2 are 128 and 96 kJ/mol, respectively. The activation energy increased to 274.6 and 147.8 kJ/mol with a rise of rhutenium concentration of up to 5%.

even different macromolecules each with their own conformational transformation, the complexity of the process becomes evident.

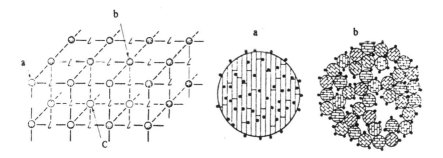

Figure 5.7 Types of active center in Rh^+-bis-(4,4'-diisocyanodiphenyl) chloride polymers [47].

Figure 5.8 Models of rhodium complexes bound to CSDVB–DPSP (a) and to phosphynated CSDVB (b).

A simpler way of creating cluster-containing polymer catalysts is immobilization of preformed cluster complexes, although the approach has been developed mainly for inorganic supports [50,51]. Immobilization of cluster complexes may proceed either with preservation of the cluster structure or with their rearrangement and even segregation to mononuclear species (which may provide coordination vacancies in the coordination sphere of the transition metal). The route of the reaction depends on the nature of the cluster complex, the reaction conditions, and mainly on the properties of the polymer support. These effects however require systematic studies.

There is another interesting example of reactions with the common stage of active cluster formation and catalyst deactivation due to formation of other clusters of different sizes [52]. Thus olefin hydroformylation proceeds at a constant rate in the presence of $Rh_4(CO)_{12}$ clusters (including those formed from mononuclear species), whereas catalyst deactivation occurs via multinuclear cluster $Rh_x(CO)_y$ formation. Consequently under optimal conditions (for example favorable for active clusters stabilization), a hyperequilibrium concentration of active clusters can be realized, otherwise irreversible catalyst deactivation may proceed. The effect of heterogeneous catalyst self-regulation is associated with a combination of these parallel reactions [53].

Thus one of the principal transformations of the immobilized complexes may be associated with the evolutionary transformations accompanied by an increase (more rarely a decrease) in nuclearity. The following sequence of such transformations is observed: immobilized monomer complex→formation of small (bi-, tri-, etc.) particle (clusters)→formation of multinuclear associates (diameter of 10–20 Å and more)→appearance of metal phase. Small-sized particles interact very intensively with the polymer matrices. Usually as a result of electron transfer from the metal atoms a positive charge arises promoting substrate coordination, which also affects the activity of the catalyst. The deep interaction of such particles with the support facilitates the conservation in them of a developed π-electron system. The situation becomes complicated because the systems examined are dynamic structures. The concentration and structure of the metal complexes may be altered in the course of the catalytic reaction through the sequence of cluster formation, aggregation, crystallization and displacement of the metal onto the surface from the body of the matrix, which is especially intensive at the stage of activation of immobilized metal complexes. The study of the relation between all these reactions and their consequences for catalysis are fundamental problems.

5.6 The Outlook for Polyfunctional Catalysis

Systems consisting of several transition metal compounds (including those bound by a uniting ligand) may be used in different variants in catalysis. They may act either in parallel in unbound catalytic cycles or independently in sequential stages or in reactions, or as combination in one reaction [54]. Polyfunctional catalysts, by means of which it is possible to initiate successive stages of one multi-stage reaction, are widely used commercial systems [55]. These particularly include the Pd–Cu catalyst for CH_3CHO synthesis from ethylene, Co–Rh complexes for 2-ethylcyclohexanol production from C_3H_6, CO and H_2O, bismuthate molybdate used in the oxidative ammonolysis of propylene, etc[10].

The role of additionally introduced metal (M_1) in catalytic hydrogenation, oxidation, (co)polymerization and other reactions has already been mentioned above. On the one hand M_1 (usually non-transition metal) may serve as a "trap" for active centers, binding them more firmly with the support and preventing migration and aggregation. On the other hand these effects can be achieved as a result of the insertion of this metal ion between attached complexes ($-M-M_1-M-$ type) promoting their matrix isolation. However, it has been shown that there is a strong inductive influence between Zn^{2+} ions additionally introduced into PMAA–Fe^{3+} systems although they were separated by several monomer units [57]. Strong electron exchange between paramagnetic ions M and M_1 bound to PE-gr-PAAc has been also detected [58].

In addition, the chemical interactions (formation of weak $M-M_1$ bonds, co-crystallization, formation of intermetallic compounds, and alloys, etc. proceeding on the support) may promote an increase of catalyst productivity. This effect has been clearly demonstrated by example of titanium–magnesium catalysts in olefin polymerization (see Section 3.1.5). Synergism of bifunctional catalyst action appeared in ethylene polymerization by $VOCl_3$ and $Ti(OEt)_4$ [59], etc.

However, those cases in which each of the bound complexes contributes (according to the coordinated mechanism in the optimum variant) its own step to a multistage process are of more interest. The cases in point are the "relay-race" (matching) copolymerization in the presence of immobilized complexes (Ni^{2+} and V^{4+}, see Section 3.3.3) of two types, one responsible for ethylene dimerization and the second for copolymerization of the butene formed with ethylene, as well as joint polymerization of di- and monoolefins ($Co^{2+}-V^{4+}$ system) yielding "chemical hybrids" (see Section 3.2.6).

The most complicated task in construction of polyfunctional systems is the coordination of the action of the diverse centers and prevention of their autonomous

---

[10] These catalysts are most widely used in the petrochemical industry for reforming and hydrocarbons hydroisomerization, such as reforming of petroleum distillers (dehydrogenation of paraffins, olefin hydrogenation, olefin isomerization by $Pt/Re/\gamma$-Al_2O_3), hydrodesulfurization and hydrogenation ($NiO/MoO_3/\gamma$-Al_2O_3) and some others. Details concerning a bifunctional catalyst for benzene alkylation with ethylene are to be found in [56].

operation. The main approach to resolving this problem is through control of the spatial arrangement of the active centers, which determines the extension of the transfer pathways of reagents between them, regulates diffusional limitations and promotes the existence of the bifunctional catalysis itself [60]. The size of the areas of the monofunctional action of the active centers has been estimated to be 12 Å [61]. Both types of active center should react with the substrate molecule, so that the matching mechanism is realized.

This may be shown in the case of the three-stage synthesis of 2-ethylhexanol from propylene and synthesis gas [62]. The process includes the formation of butyric aldehyde (stage 1), its aldol condensation (stage 2), and hydrogenation (stage 3).

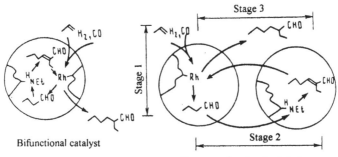

Scheme 5.11

It may be achieved by the use of two monofunctional catalysts (hydroformylation and hydrogenation) and one bifunctional catalyst (chloromethylated CSDVB, modified by secondary amino groups with bound RhCl(PPh$_3$)$_2$(CO)) (Figure 5.9). The rates of the first, second and third stages of the bifunctional catalyst are 5-, 15- and 30-fold higher respectively than those of the two monofunctional immobilized catalysts.

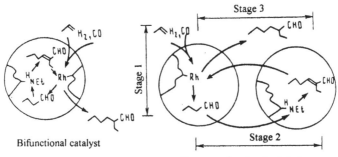

Figure 5.9 Comparison of diffusional inhibition in reactions catalyzed by one bifunctional and two monofunctional catalysts in the case of three-stage synthesis of 2-ethylhexanol [62].

The range of such examples has been extended significantly in recent years. We shall refer only to oligomerization of butadiene with the hydroformylation of vinyl cyclohexene by Ni–Rh complexes bound to phosphorylated CSDVB and the preparation of 4-methyl-2-pentanone on composites of polymers consisting of polyamines and polycarboxylic acids impregnated with PVC and incorporating two transition metal ions [63], etc.

All examples mentioned above relate to immobilized heterometallic clusters, which have a higher stability than that of the monometallic clusters under the conditions of the catalyzed reactions. A search for methods of constructing such systems is a promising direction for metallocomplex catalysis [64].

Yet another important example [65] relates to H_2O decomposition to H_2 and O_2. Berlin blue $KFe^{3+}[Fe^{2+}(CN)_6]^{4-}$ (colloidal suspension in water) in combination with $Ru(Dipy)_3^{2+}$ catalyze water decomposition under exposure to visible light yielding H_2 and O_2. The polymer-like structure of Berlin blue and the total cycle of water decomposition by this catalyst are presented in Figure 5.10.

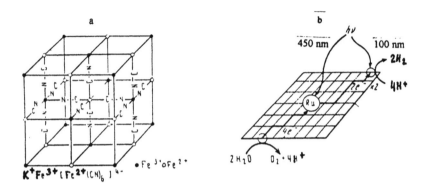

Figure 5.10 Photoconversion of water molecules catalyzed by the bifunctional catalyst composed of Berlin blue and $Ru(Dipy)_3^{2+}$. (*a*) The structure of a Berlin blue cell; (*b*) the total cycle of H_2O decomposition by a colloidal suspension of Berlin blue.

Finally let us note the example of unequal ability of several metal complexes binding to phosphoryl oxygen of phosphonic acid groups, which gives the ability of the progressive loading of this group of ampholytes by different metal ions [66]:

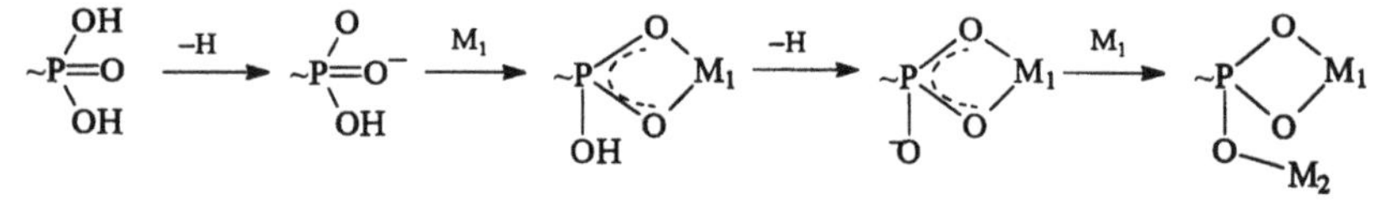

Scheme 5.12

In a similar way a $Mo^{6+}-Fe^{3+}$ pair can be attached within a single elementary unit of phosphorus-containing ampholyte, which is the model of nitrogenase composed of Mo–Fe proteins. Such analogies are extremely deep-rooted and they can be extended also to polynuclear metalloenzymes. In the latter the role of multifunctionality consists of the creation of favorable energy and probability factors owing to the optimal usage on the respective active centers at each stage of the reaction. In other words in the catalyst–substrate complex there is a restriction on the number of inefficient diffusional movements from the standpoint of the catalyzed reaction which do not lead to the intended products [67]. An important condition for the realization of synchronous catalytic processes is the possibility of the dynamic adaptation of the form of the catalytic matrix to the structure of the substrate and the changing of the active centers in the course of the reaction (fine-tuning). Rapid and reversible conformational transitions both in enzymes and in macroligands and also fluctuations in the matrix structure ensure the necessary structural adaptation. The specific characteristics of enzyme systems reside in the presence of secondary and tertiary structures, which limit the spatial orientation of metal ions and the possibility of side reactions occurring. By using synthetic metal-containing polymers it has still not been possible to create a model which under similar conditions (pH, temperature, catalyst and substrate concentrations) would approach enzyme systems in the levels of activity and selectivity.

5.7 Technological Aspects of Catalysis by Immobilized Metal Complexes

The application of metal complexes immobilized on polymers is of technological value if only due to the fact that such a difficult stage as catalyst recycling can be excluded from the technological cycle (particularly valuable in the case of expensive catalysts containing either noble metals or optically active ligands, etc.).

The data on technological aspects of catalysis by immobilized complexes are mainly concentrated in patents. Some of them have been already realized on a pilot plant scale. However their wide commercial use is restricted by two main reasons: (i) the absence of industrial technologies for catalyst preparation; (ii) the task of elaboration of both technologies and apparatus for their exploitation needs additional work because one should take into account a great number of specific features of catalysis by these systems.

It is most probable that they will be primarily used in production of medicinal compounds, dyes, perfumes, and foodstuffs, i.e. in those processes that require total catalyst separation. The processes of industrial chemistry which are in prospect for the application of these catalysts are hydrogenation, especially of nitrocompounds, oxidation of thiosalts, hydroformylation, polymerization (e.g. polyethylene (PE) and its copolymer production in the gas phase or suspension, and probably in solutions).

Industrial applications may be conducted either batchwise or by flow method. The technological value also consists of the possibility of catalyst preparation in the form of solution, suspension or as a fixed bed.

Generally almost all immobilized complexes can be constructed in such a way that thay can be exploited in standard technological processes without complications. Thus the system PE-gr-PAAl–VCl$_4$–AlEt$_2$Cl (PAAl=poly(allyl alcohol)) has been used without considerable modification in a slurry high-density PE manufacturing process [68]. The scheme of the process is presented in Figure 5.11. It consists of the following operations: transport of immobilized components into the unit for preparation of the catalytic complex; catalyst pumping into the polymerization reactor; treatment and drying of the PE formed. To regulate properties of the PE manufactured all conventional means such as hydrogen, comonomers and several additives have been used. The process showed high stability with respect to the main technological parameters. A similar process can be used for high-density linear PE manufacturing by bifunctional catalysts.

A fixed bed of catalyst may be composed of powder, granules, braids and even Raschig rings (as mentioned for reactions of deuterium/hydrogen exchange).

Preparation of catalysts in the form of microporous metal–polymer membranes seems to be valuable. The catalyst support can be prepared from cellulose ethers, polyamide, poly(tetrafluoroethylene) (PTFE) and organosilica polymers. Metal complexes can be bound either to the outside surface or throughout the total volume of the polymer support. The membrane catalysis manufacturing process is one of the most promising technologies in prospect [69] especially for reactions proceeding at the membrane–gas interface. At the optimal ratio between reactor volume and the width of active particle layer a considerable increase in catalyst efficiency can be achieved.

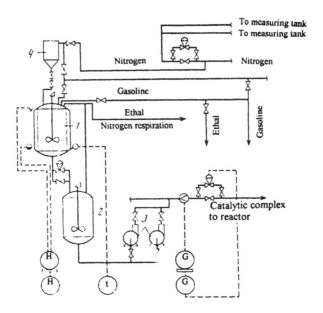

Figure 5.11 A technological process for the preparation of a catalytic complex based on immobilized catalysts and its transport to the polymerization reactor: (1) unit for slurry diluent; (2) vessel for catalyst slurry preparation; (3) pumps; (4) the vessel contaning the immobilized catalyst.

Methods of metal complex immobilization on cord fibers made of polyamide, PTFE, etc. [70] are also of practical engineering importance. Ethylene polymerization in the presence of these catalysts yields a hydrophobic covering on the fibers. This is an example of a flowing catalyst. The technological challenge consists of coordination of the rates of the catalytic process and the flow of the catalyst (the rate of thread winding is about 600 m/min). In a similar way polymer composites on fibers (PVAl, PA, cotton, flax, viscose, etc.) as well as several interesting (e.g. conductive) polymer–polymer composites *in situ* are obtained.

There is another important practical feature of immobilized catalysts. On the one hand strong binding of metal ions gives the opportunity of avoiding the stage from catalyst separation of the reaction products thus promoting a reduction in the production of waste and on the other hand they can be used for neutralization of waste waters in the processes of oxidative destruction of phenol and pure production of ethanol and amino acids, etc.

To have widespread practical use it is necessary to develop special technologies for the use of immobilized catalysts which will require an effort just as great as that directed toward studying their synthesis and catalytic activity.

References

1. Pomogailo, A.D. (1992) *Russ. Chem. Rev.*, **61**, 133.
2. Hartley, F.R., Burgess, C. and Alcock, R.M., (1980) *Solution Equilibria*, New York: John Wiley.
3. Fillipov, A.P. (1983) *Teor. Eksp. Khim.*, **19**, 463.
4. Lisichkin, G.V. and Kudryavtsev, G.V. (1986) *Immobilized Metal Complexes in Catalysis*, Preprint, Vol.2, Part 1, p.84, Novosibirsk: Institute of Catalysis Siberian Division, Russian Academy of Sciences.
5. Semikolenov ,V.A., Likholobov ,V.A.and Ermakov, Yu.I. (1977) *Kinet. Katal.*, **18**, 1294.
6. Pomogailo, A.D. (1985) *Khimiya Geterogennykh Soedinenii* (Chemistry of Heterogeneous Compounds), edited by A.Ya.Yuffa, p.1134, Tumen: Tumen University.
7. Leiser, K.H. and Zuber, G. (1977) *Angew. Makromol. Chem.*, **57**, 183.
8. Pomogailo, A.D., Baishiganov, E. and Khvostik, G.M. (1980) *Kinet. Katal.*, **21**, 1535.
9. Suzuki, T., Shirai, F. and Hojo, N. (1983) *Polym. J.*, **15**, 409.
10. Nishikawa, H., Terada, E.-I., Tsuchida, E. and Kirumura, Y. (1978) *J. Polym. Sci.; Polym. Chem. Ed.*, **16**, 2453.
11. Rorabacher, D.B., Martin, M.J., Malik, M., et al. (1980) *Abstracts of the Papers of the 179th ACS National Meeting*, (Huston, TX, 1979), p.98, Washington, DC: American Chemical Society.
12. Pomogailo, A.D. (1988) *Polymer-Immobilized Metal Complex Catalysts*. Moscow: Nauka.
13. Deratani, A., Sebille, W., Hommel, H. and Legrand, A.P. (1983) *React. Polym.*, **B1**, 261.
14. Donaruna, L.G., Kiton, S., Walsworth, G., *et al.* (1979) *Macromolecules*, **13**, 435.
15. Masuda S., Nakabayashi, I., Ota, T. and Takemoto, K. (1979) *Polym. J.*, **11**, 641.
16. Sato, M., Kondo, K. and Takemoto, K. (1979) *J. Polym. Sci.; Polym. Chem. Ed.*, **17**, 2729.
17. Tsuchida, E., Karini, Y. and Nishide, H. (1974) *Makromol. Chem.*, **175**, 161; 171.
18. Tsuchida. E., Kiyotaka.S. and Yoshimi. K. (1974) *J. Polym. Sci., Polym. Chem. Ed.*, **12**, 2207; **13**, 1457.
19. Sugimura, M., Okudo, T. and Ise, M. (1979) *Proceedings of the International Symposium on Macromolecules*, (Pisa, 1979), Vol.4, p.174. Pisa: University of Pisa; (1981) *Macromolecules*, **14**, 124.
20. Oyama, N., Shimomura, T. and Anson, F.C. (1980) *Rev. Polarogr.*, **26**, 25.
21. Yamagishi, A. (1982) *J. Phys. Chem.*, **86**, 229.
22. Topich, J. (1980) *Inorg. Chim. Acta*, **46**, L97.
23. Ueyama, N., Kamada, E. and Nakamura, A. (1983) *Polym. J.*, **15**, 519.
24. Vannikov, A.V. and Grishina, A.D. (1989) *Usp. Khim.*, **58**, 2056.
25. Challa, G. (1980) *Preprints of IUPAC Macro*, Florence: International Symposium on Macromolecules (Pisa, 1980), Vol.1, p.101. Pisa: University of Pisa.
26. Carlinii, C. and Sbrana, G. (1981) *J. Macromol. Sci.*, **16**, 323.
27. Barteri, M. and Pispisa, B. (1980) *Preprints of IUPAC Macro*, Florence: International Symposium on Macromolecules (Pisa, 1980), Vol.4, p.155.
28. Kalawole, E.G. and Bello, M.A. (1980) *Eur. Polym. J.*, **16**, 325.
29. Barone, G., Oresenzi, V. and Qwadrifoglio, F. (1965) *Ric. Sci.*, **35**, 393; 1067.

30. Seleznev, V.A., Artemov, A.V., Vainshtein, E.F. and Tyulenin, Yu.P. (1981)
    Neftekhimiya, **21**, 114.
31. Purmal, A.P. and Nikolaev, L.A. (1985) *Usp. Khim.*, **54**, 786.
32. Willams, R.J. (1971) *Inorg. Chim. Acta, Rev.*, **5**, 137.
33. Blyumenfel'd, L.A. (1974) *Problemy Biologicheskoi Fiziki: (Fizika Zhiznennykh
    Protsessov)* (Some Aspects of Biological Physics (Physical Aspects of Living
    Processes)). Moscow: Nauka.
34. Bravaya, N.M., Pomogailo, A.D. and Vainshtein, E.F. (1984) *Kinet. Katal.*, **25**,
    1140.
35. Hou, X.-H., Kaneko, M. and Yamada, H. (1984) *Kobunshi Ronbunshu*, **41**, 311.
36. Braca, G. and Carlini, C. (1978) *Inf. Chim.*, 149.
37. Krylov, O.V. (1984) *Mekhanizm Kataliza* (Mechanism of Catalysis), Part 1, p.51.
    Novosibirsk: Institute of Catalysis Siberian Division Russian Academy of Sciences.
38. Cerveny, L., Marhoul, A., Cervinka, K. and Ruzicka, V. (1980) *J. Catal.*, **63**, 491.
39. Semenov, P.A., Semikolenov, V.A., Likholobov, V.A., *et al.* (1988) *Izv. Akad. Nauk
    SSSR, Ser. Khim.*, 2714.
40. Andrale, J.D. and Chem, W.Y. (1986) *Surf. Interface Anal.*, **8**, 253.
41. Widdecke, H. (1987) *Progress of the Prague Meeting Macromolecules. XXX.
    Microsymposium: Polymer Supported Organic Reagents and Catalysis*, (Prague,
    1986), p.14. Prague: Institute of Macromolecular Chemistry Czecoslovak Academy
    of Sciences.
42. Schults, A.W. and Strauss, P. (1972) *J. Phys. Chem.*, **76**, 1767.
43. Pomogailo, A.D., Irzhak, V.I., Burikov, V.I., *et al.* (1982) *Dokl. Akad.Nauk SSSR*,
    266, 1160.
44. Uematsu, T., Nakazawa, Y., Akatsu, F., *et al.* (1987) *Makromol. Chem.*, **188**, 1085.
45. Solovieva, A.B. (1990) Metal Porphyrines as Catalysts of Non-Chain Olefin
    Oxydation, *Doctoral Dissertation*, Moscow.
46. Viout, P. and Artaud, I. (1984) *J. Chem. Soc., Perkin. Trans. I*, 1351.
47. Lawrence, S.A., Serman, P.A. and Feinstein-Jaffe, I. (1989) *J. Mol. Catal.*, **51**, 117.
48. Uematsu, T., Kawakami, T., Saitho, F., et al. (1981) *J. Mol. Catal.*, **12**, 11.
49. Liu, J., Yang, L., Gao, S., *et al.* (1987) *Thermochim. Acta*, **123**, 121.
50. *Metal Clusters in Catalysis* (1986) edited by. B.C.Gates, L.Guczi and H. Knözinger.
    Amsterdam: Elsevier.
51. Gates, B.C. and Lamb, H.H. (1989) *J. Mol. Catal.*, **52**, 1.
52. Rozovskii, A.Ya. (1988) *Katalizatory i Reaktsionnaya Sreda* (Catalysts and Reaction
    Media). Moscow: Nauka.
53. Kagan, Yu.B. and Rozovskii, A.Ya. (1987) *Kinet. Katal.*, **27**, 1508.
54. Imyanitov, N.S. (1984) *Koord. Khim.*, **10**, 1443.
55. Schult, G.C.A. and Gates, B.C. (1983) *Chem. Tech.*, **13**, 556.
56. Isakov, Ya.I. and Minachev, Kh.M. (1981) *Neftekhimiya*, **21**, 773.
57. Tripathi, K.C., Sarma, P.R. and Gupta, N.M. (1978) *J. Polym. Sci., Polym. Lett.
    Ed.*, **16**, 629.
58. Pomogailo, A.D., Uflyand, I.E. and Golubeva, N.D. (1985) *Kinet. Katal.*, **26**, 245.
59. Hseih, H.L., McDaniel, M.P., Martin, J.L., *et al.* (1985) *Polymer Materials Science
    Engineering, Proceedings of the ACS Departement of Polymer Materials Science
    and Engineering*, Vol.53, p.105. Washington,DC: American Chemical Society.
60. Surkov, N.F., Davtyan, S.P., Pomogailo, A.D. and Dyachkovskii, F.S. (1986) *Kinet.
    Katal.*, **27**, 714.

61. Ione, K.G. (1982) *Polifunktsionalnyi Kataliz na Tseolitakh* (Polyfunctional Catalysis on Zeolites). Novosibirsk: Nauka.

62. Batchelder, R.F., Gates, B.C. and Kuijpers, F.J.P. (1976) *Preprints of the VI International Congress on Catalysis*, L.

63. Pittmann, C.U., Jr., Kawabata, Y., Liang, Y.F. (1980) *Abstract of the Papers of the 179th ACS National Meeting (Huston, TX, 1979)*. Washington, D.C.: American Chemical Society.

64. Moiseev, I.I. (1984) *Itogi Nauki Tekh., Kinet. Katal.*, **13**, 147.

65. Kaneko, M., Takabayashi, N., Yamada, A. (1982) *Chem. Lett.*, 1647.

66. Kopylova, V.D. (1986) *Teoriya i Praktika Sorbtsionnukh Protsessov* (Theory and Practice of Sorbtion Processes), p.3. Voronezh: Voronezh. University.

67. Likhtenshtein, G.I. (1979) *Mnogoyadernye Okislitel'no-Vosstanovitel'nye Metallofermenty* (Polynuclear Redox Metal Enzymes). Moscow: Nauka.

68. Pomogailo, A.D., Kolesnikov, Yu.N., Shishlov, S.S. (1977) *Plast. Massy*, 30.

69. *Membranes and Membrane Separation Processes: Proceedings International Symposium*, (Torun, Poland, 1989) p.357. Torun: Nicolaus Copernicus University.

70. Dobbs, B., Jackson, R., Gaiske,l J.N., *et al.* (1976) *J. Polym. Sci., Polym. Chem. Ed.*, **14**, 1429.

CHAPTER 6

CONCLUSIONS

A significant number of segments of the polymer chain are usually involved in the MX_n binding (and active centers based on these structures). The rupture of one of the bonds between the metal ion and the macroligand unit requires the expenditure of approximately the same amount of energy as for the rupture of the bond in the analogous low molecular mass complex. However these losses in the first case are not compensated for by the appearance of new progressive degrees of freedom (i.e. they are not accompanied by an increase in the configurational entropy). Thus the probability that the active center in the immobilized system is disrupted under equally stable conditions sharply declines, whereas the stability of the intermediates may be increased owing to the rapid spontaneous conformational transformations in adjacent units in the chain, fluctuations in its structure, etc. The polymeric structure of ligand environment hampers ligand exchange and introduces specific properties to the electron transfer reactions and redox processes. Non-valent interactions affect reactivity, especially the geometry of the transition state. The differences in the kinetic and activation parameters for the reactions of polymeric and monomeric complexes are associated with conformational changes in the macroligands.

Catalysis by immobilized metal complexes is accompanied by a whole series of macromolecular effects, i.e. an increased local concentration of active centers and substrate in the macromolecular globules or in domains close to the active centers, the effect of the spatial structure and polymer chain length, etc. It is also important that immobilized systems provide additional means of regulating catalyst activity and selectivity and sometimes asymmetric induction. The most significant features are the nature and concentration of functional groups, the distribution of transition metal ions, the temperature, etc. Evolutionary transformations of multinuclear complexes into associations in the immobilized systems occur to a lesser extent than in homogeneous systems.

Immobilization of a metal complex catalyst on a polymer support is an effective method for synthesizing polyfunctional catalysts, including matching catalysts. The macroligand serves as a uniting node binding different active centers and controlling their localization.

The realization of the potentials of these interesting systems will help to explain in detail both homogeneous and heterogeneous catalysis, and the information obtained will help to explain the basis of the catalytic action and promote the creation of new industrial technologies for their exploitation.

Here we have tried to summarize the more widespread combinations of examined macrocomplexes and macroligands which are used in different catalytic reactions (Table 6.1).

Table 6.1 The summary of combinations of metal complexes and macroligands, as well as catalyzed reactions most commonly used in practice.

Polymer support	Functional group	MX_n	Catalyzed reaction
Phosphynated CSDVB	$-PPh_2$	$RhCl(PPh_3)$	Hydrogenation
		$[CODRhCl]_2$	
		$Rh(CO)_2(PPh_3)_3Cl$	
		$PdCl_2(PPh_3)_2$	
		$RuCl_2(CO)_2(PPh_3)_2$	
		$RhHCO(PPh_3)_3$	Hydroformylation
		$[CODRhCl]_2$	
		$CoCl_2(PPh_3)_2$	
		$Mo(CO)_2(PPh_3)_2$	
		$Fe(CO)_4PPh_3$	
		$Ir_4(CO)_{12}$	
		$RuCl_2(PPh_3)_3$	Deuterium/hydrogen exchange
		$PdCl_2PPh_3CH_2CH_2PPh_2$	Isomerization
		$Ni(CO)_2(PPh_3)_2$	Oligomerization
		$NiCl_2(PPh_3)_2$	Cyclooligomerization
CSDVB	Dipy	$Pd(OCOCH_3)_2$	Hydrogenation
	Cp	$CpTiCl_3$	
	$NH_2CH_2CH_2NH_2$	$Pt(PhCN)_2Cl$	
PA	–	$Ni(napht)_2$	PhA hydrogenation
PEI/SiO$_2$	$-NH$	Pd^0	Nitrobenzene hydrogenation
	—		
PE-gr-P4VP	Py	$PdCl_2(PhCN)_2$	p-Nitrochlorobenzene hydrogenation
PMMA	$-COOCH_3$	Pd^0	Nitrocompound hydrogenation
PTFE	–	Au^0	Deuterium/hydrogen exchange
CSDVB	$-NMe_2$	MCl_2 (M=Pt, Ru)	Hydrosilylation
PE-gr-PAAc	$-COOH$	$Co(AcAc)_2$	Cyclohexene oxidation
PAN	$-C\equiv N$	$M(AcAc)_2$ (M=Mn, Co)	Ethylbenzene
Poly(styrene-co-divinyl-benzene-2-Me-5-VP)	N—	Cu^{2+}	Isopropylbenzene oxidation
CSDVB		Co, Ni, VO, CO, Fe, Mn, phthalocyanines	Cyclohehene oxidation
PEG	$-OCH_2CH_2-$	MCl_2(M=Co, Mn, Cu)	Tetralin oxidation
	–	Co^{2+}, Cu^{2+} polyacrylates	H_2O_2 decomposition
Poly(2-vinylpyri-dine-co-styrene)		Fe^{3+}, Co^{2+}, phthalocyanines	H_2O_2 decomposition

Polymer support	Functional group	MX_n	Catalyzed reaction
Phosphate cellulose	$-PO_2OH$	V^{5+}, Mo^{5+}	Cyclohexene epoxydation
Chloromethy-lated CSDVB		Dithiocarbamate Mo^{5+}	Olefin epoxydation
P4VP	$-\overset{\mid}{\underset{\mid}{N}}-$	Cu^{2+}	Dialkylphenol oxidation
Polyvinylamine	$-NH_2$	Co^{2+}, Fe^{3+}, Cu^{2+}, phthalocyanines	Thiol oxidation
P4VP	$\overset{\mid}{\underset{\mid}{N}}-$	Cu^{2+}	Ascorbic acid oxidation
PMMA	$>C=O$	$TiCl_4$, VCl_4	Ethylene polymerization
PE-gr-PAAl	$-OH$	$Ti(OBu)_4$	Ethylene polymerization
PE-gr-P4VP	$\overset{\mid}{\underset{\mid}{N}}-$	$TiCl_3$	Stereospecific propylene polymerization
	$\overset{\mid}{\underset{\mid}{N}}-$	$CoCl_2$	1,4-cis-butadiene polymerization
Copolymer of styrene and acrylic acid	$-COOH$	$Ni(napht)_2$	1,4-cis-Butadiene polymerization
PE-gr-PAAc	$-COOH$	$Ni(CH_3COO)_2$	Ethylene dimerization
	$-COOH$	Ni^{2+}, V^{4+}	"Relay-race" ethylene copolymerization
Triple copolymer of ethylene propylene and P4VP	$\overset{\mid}{\underset{\mid}{N}}-$	$CoCl_2$, $NiCl_2$	Butadiene di- and oligomerization
PS	$-$	$AlCl_3$, $TiCl_4$	Styrene and α-methylstyrene polymerization
PE-gr-PAAl	$-OH$	$MoCl_5$, WCl_6	PhA polymerization
CSDVB-P4VP	$\overset{\mid}{\underset{\mid}{N}}-$	Cu^{2+}	Oxidative polycondensation of phenols
PE-gr-PAAc	$-COOH$	$Co(OCOCH_3)_2$	Phenolformaldehyde oligomers solidification
CSDVB	Cp	$CpCo(CO)_2$	Fischer–Tropsch synthesis
PEI	$-NH$	$RhCl_3$	Metanol carbonylation
PVAl	$-OH$	Cu^{2+}	2,4-Dinitrophenyl-acetate hydrolysis
CSDVB		$-TiCl_4$	Etherification, alkylation, acetalization

Polymer support	Functional group	MX_n	Catalyzed reaction
Copolymer of styrene and AAc	–COOH	Cp_2TiCl_2	Reduction of molecular nitrogen to ammonia
Thioacetal derivatives of	–CH$_2$	–Mo(NMe$_2$)$_4$	Nitrogen reduction
poly(4-amino-styrene)	–NH$_2$	–	Photocatalytic hydrogen formation
PEI	–NH	Rh^{3+}	Formation of H$_2$ from H$_2$O
CSDVB	–	Mg, Mn, Fe, Ru, phthalocyanines	Isomerization of quadricyclane to norbornadiene
Poly[Ru(Dipy)$_3^{2+}$]	–	–	Chemically modified electrodes
P4VP/Carbon	$\overset{\mid}{\underset{\mid}{N}}$—	Cu^{2+}	Electrochemical reduction of O$_2$ to H$_2$O
Poly[Ru(Dipy)$_3^{2+}$ - 4VP] /Pt	–	–	Photodiodes (chemically immobilized photoelectrodes)

6.1 The Relations Between Polymer-Immobilized, Homogeneous, Heterogeneous and Enzymatic Catalysts

The data considered in this monograph may be reviewed through the relationship between homogeneous, heterogeneous, enzymatic catalytic systems and those immobilized on polymers.

Homogeneous catalysts may be considered as an individual particle (complex or combination of complexes) located in the same media as reagents (usually solution). The reaction occurs in the transition metal ion coordination sphere by spatial and electronic rearrangement of complex ligands via the formation of active intermediates. All active centers are equally accessible to the reagents. Their activity and selectivity may be regulated by changes in the metal–complex ligand enviroment.

In heterogeneous catalysis the reaction occurs on the interphase boundary. Adsorption of reagents on the catalyst surface is followed by their activation. However all these centers located on the catalyst surface are not equivalent. It is possible that only one of the set is capable of catalyzing the desired reaction, while others are inactive and even responsible for some side reactions. It should be noted that there is also evidence of a primary role for the individual active center in heterogeneous catalysis but not of collective electronic properties of the surface.

The main advantage of homogeneous catalysts is that they provide the opportunity to study the structure of the active center and the mechanism of catalytic action. However due to lower activity in comparison with heterogeneous catalysts, they are usually less attractive for technological applications. In addition, heterogeneous catalysts are easily separated from reaction products and sometimes may be regenerated and recycled. This is especially important in the case of expansive or toxic metal complexes. This is the reason why these catalysts are scarcely used in the chemical industry with the exception of such important processes as Ziegler–Natta manufacturing processes (polymerization of olefins and dienes), the Wacker process (ethylene oxidation to acetaldehyde), the Monsanto process (methanol carbonylation to acetic acid), and some others. Conventional methods of catalyst separation from reaction products (effective distillation, ultrafiltration, ion exchange, etc.) are energy-consuming and unreliable.

In the creation of polymer-immobilized catalysts there were two main goals, both of which are useful for operation and theory. These were (i) synthesis of stable and reproducible catalysts showing high activity and selectivity (like homogeneous processes) and easily separated from reaction products (like heterogeneous processes); (ii) synthesis of heterogeneous-like catalysts but carrying active centers of definite structure which would provide the possibility of studying the mechanism of catalytic action and lead toward the solution of fundamental problems. This is not an empirical but a scientific method of catalyst selection and regulation of their properties.

However, polymer-immobilized catalysts are closer to homogeneous catalysys in chemical character. They are synthesized from soluble metal complexes by ligand exchange with polymer functional groups often extending to the preservation of one or more ligands. Their operating temperatures are close to those for homogeneous catalysts (usually below 100°C). Their chemical behavior is also similar to soluble metal complexes.

Immobilization of metal complexes usually increases their efficiency and stability for two main reasons. Firstly, their concentration on the polymer support is not limited by solubility (under certain conditions one can reach the state of an infinitely dilute system). Secondly, binding of metal complexes promotes their stability protecting active centers from deactivation processes. Although the idea was to create catalysts that would combine the advantages of both homogeneous and heterogeneous catalysts and were deprived of their drawbacks, the problem has proved to be more complicated, as one can see from analysis of the data discussed in this monograph. Thus, this approach is still far from its potential realization because almost every concrete case is accompanied by a number of complications. It is difficult to create bound complexes of single-type on the polymer support and overcome the problem of low stability of the metal–polymer bond under the conditions of the catalyzed reaction.

A brief comparative summary of catalysts of different type can be found in Table 6.2.

As already mentioned above, there is another type of catalyst which has many similar features to immobilized metal complexes. These are enzymatic catalysts. It should be noted that all enzymes (e.g. myoglobin and hemoglobin, active in the binding and activation of oxygen; nitrogenase in the activation and binding of molecular nitrogen, cytochrome P-450 in direct oxidation of paraffins; ferredoxin in electron transfer, etc.) are composed of two parts having different chemical natures and purposes. There are apoenzymes (the high molecular mass protein part) and prosthetic groups (the low molecular mass parts which include the transition metal ion or its complex). The last group is the active center of enzymatic catalyst binding to substrate yielding the reaction products. However it is not the prosthetic group but the apoenzyme part that specifies the enzymatic process. It is the apoenzyme that separates and orientates the substrate with respect to the active center providing extremely high selectivity. This involves only selected bonds in the reaction. The protein carrier plays another important role in enzymatic catalysis by preventing the prosthetic groups from aggregation and oxidation (which occurs with unstable active centers). Thus the protein part of a biocatalyst ensures perfection to the process, attracting the attention of investigators for a long time with the aim of using these main principles for the development of synthetic enzymes.

Table 6.2 The formal comparison of catalysts of different types.

Properties	Catalyst		
	Homogeneous	Heterogeneous	Immobilized
Specific catalytic activity	Moderate	High	High
Degree of transition metal employment	Low	High	High
Selectivity	High	Low	High
Ability in activity and selectivity regulation by changes in ligand surroundings	Possible	Impossible	Possible
Indentification of the nature of the active forms	Possible	Difficult	Possible
Bimolecular active center deactivation	High	Low	Low
Asymmetric induction	Yes	No	Yes
Activity in peptide synthesis	Low	Modest	Modest
Catalyst separation of reaction products, regeneration, recycling	Difficult	Easy	Easy
Application in flow manner	Difficult	Easy	Easy
Catalytic reaction conditions	Mild	Hard	Mild/Hard
Corrosion-resistance, deposition on reactor walls	High	Low	Low
Mechanical strength	Absent	High	High
Thermal and chemical stability in catalyzed reaction	Low	High	High

However the main drawbacks (chemical and thermal instability) of biocatalysts lie originally in the protein nature of the apoenzyme. All known enzymatic reactions occur within narrow ranges of temperature and pH. This reason clarifies the general direction for developments. Synthetic catalysts operating as enzymes but stable to the action of external factors should be developed.

Since in metal enzymes metal ions are incorporated into the structure of biomembranes they may be considered as immobilized complexes incorporated into swollen polymer matrices (gels).

Such an approach has proved to be successful. In the course of the above discussions we have often pointed out some analogies in the action of the catalysts examined. Let us summarize only some of them. The high catalytic activity of enzymes is specified by a number of controling factors, such as structural conformity (some parts of biopolymers possess enhanced affinity to substrate and form a specific spatial network in the vicinity of the active center whose form and size correspond to those of the substrate; this increases substrate concentration, so-called substrate enrichment, and provides favorable substrate orientation), multicenter catalysis (functional groups of biopolymer coordinatvely additionally activate substrate molecules), etc. The same effect can be realized for immobilized catalysts.

We have already concentrated on the contribution of the energy of non-equilibrium configurations of polymer chains to catalysis. This effect is also similar to the energy recuperation effect in enzymatic catalysis in which the protein molecule assumes the role of an energy reservoir. The energy released is utilized to activate reagents. The analogy between polyfunctional catalysis by immobilized metal complexes and polynuclear metal enzymes should also be mentioned.

Consequently immobilized catalysts combine the main features of homogeneous, heterogeneous and enzymatic catalysts which can be represented as follows:

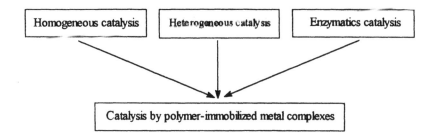

Scheme 6.1

So at present polymer-immobilized metal complexes can be assumed only as a simple model of a catalysts, which combines both enhanced catalytic activity and selectivity as a result of the binding of a low molecular mass active group with a high molecular mass support.

However even a formal comparison of these catalysts is not a simple task. It is often difficult to find unbiased parameters for catalyst activity and selectivity evaluation. This particularly concerns comparative models of the reactions, specific catalytic activity and turnover numbers. It is also difficult to evaluate the part of the transition metal involved in the active centers for each compared catalyst; the ranges of operating temperatures do not often overlap, etc. Under strict comparison of the action of heterogeneous and immobilized catalysts one should be sure of low molecular mass analogue similarity with immobilized particles at least with respect to its electronic systems and stereochemistry. However the problem is also complicated by chain effects and the presence of non-valent interactions.

6.2 Perspectives on Future Developments of Catalysis by Immobilized Metal Complexes

Although the main conclusions on the nature and mechanism of immobilized catalyst action are mostly based on the logical analysis of available experimental data and general scientific concepts, special theories of both the mechanism of catalytic action and the nature of active centers will be generated in the future. This statement is based on the additional advantages of immobilized catalysts and the ability to study intermediate products by the use of analytical and especially spectroscopic techniques: NMR (including techniques for solid samples), ESR, UV, Raman, IR, Mössbauer, Auger, ESCA, photoacoustic spectroscopy; electron and field-emission microscopy; slow electron diffraction; magnetic methods; mass spectrometry of secondary ions, etc. Valuable information on coordination number, the nature and length of bonds between polymer support and transition metal ion has been obtained by EXAFS (extended X-ray absorption fine structure); some dynamic characteristics have been determined by spin label and probe technique.

As mentioned above, the combination of several methods sometimes gives the ability to detect the oxidation state of the transition metal ion in the active center, characterize metal-ligand interactions, and reveal the polyhedral structure of an active center.

In the course of studies, the general features and phenomena common for different catalytic systems have been revealed. It has been shown that macroligands stabilize the mononuclear complexes incorporated into unidimensional chains, two-dimensional networks, and three-dimensional frameworks. Non-valent interactions affect reactivity, especially the geometry of the transition state. Numerous examples show that the polymer support posesses a dynamic ability to adapt to reaction conditions (medium, temperature, etc.) by reversible changes in its conformational structure. Cooperative effects of the polymer chain promote interactions of bound active centers:

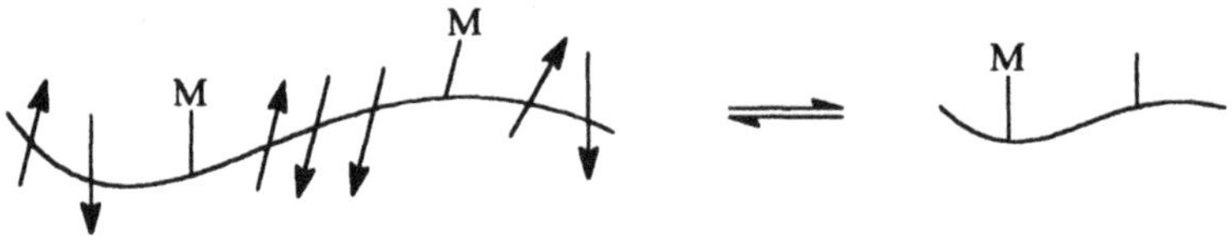

Scheme 6.2

Active center rearrangements (relaxation) proceed at higher rates in immobilized systems than in homogeneous processes and reach the values of catalytic reaction rates. This yields an excess of active center concentration (above the equilibrium value). With respect to all these features immobilized metal complexes resemble, to a certain extent, enzymatic catalysts.

The author considers that by analyzing the problems considered in this monograph it is possible to follow the growing scientific interest in catalysis by immobilized complexes and the characteristics of their operation. During the study of immobilized systems new patterns and phenomena have been revealed which are of fundamental importance. Here we shall only point once more to the most important factors which one should be taken into account when study catalysis by immobilized metal complexes. One is metal complex binding. The most important factors are the nature of the bond between the transition metal ion and the polymer support, the profile of the metal ion distribution (topography) on the support, possible interactions between immobilized metal complexes and cluster formation. As immobilization conditions and subsequent modifications affect to a great extent evolutionary transformations of bound mononuclear complexes to associates and thus their activity, they should also be taken into consideration. Special attention should be paid to dynamic factors such as metal complexes and macroligand transformations in the course of catalyzed reactions; methods for their regulation should be elaborated. Diffusion factors, effect of medium, operating temperature (below the T_g of the polymer support) are very important for catalysis by immobilized metal complexes.

Bi- and polyfunctional catalysts are very promising approaches for the future. Even the present rather scarce studies in this field show that ensembles of several transition metal ions promote synchronous "relay-race" catalysis. Extra side products with unusual properties may be formed when the product formed by the first active center serves as the substrate for the second. A development of this approach is the case of "optimal movements" (the number of participants in the catalytic action should be high enough to provide optimal energy to the reaction) proposed for enzymatic catalysis that can be realized in catalysis by immobilized metal complexes. The systems examined are multi-parameter systems, and it is necessary to take into account the effect of many factors in order to optimize their activity and selectivity. Total optimization with respect to all variables would be an incredibly complicated experimental objective. Theoretically based, economically viable methods for its resolution in immobilized systems are only beginning to be developed.

Specific features of the most widespread organic reactions catalyzed by immobilized metal complexes have been examined in this monograph. The range of catalyzed reactions becomes wider with each year. Recently these systems have found their usage in carbon monoxide utilization; telomerization of dienes with amines; polymerization by organometallic compounds in water; in cracking process; processes connected with C–C and C–H bond activation; ethylene transformation to butadiene; cyclopropane group formation; dehydrogalogenation, etc. Certainly in the near future these developments will promote new technological approaches and applications of these catalysts on a large industrial scale.

INDEX

titanium-containing polymers, reduction
  of dinitrogen, 334
titanium–magnesium catalysts, 366
topochemistry, 9
transfer hydrogenation, 73
transformations of immobilized complexes, 385
transition metal complexes with oligomer
  ligands, 8
transparent polymer–gel-like membranes, 7
trinuclear Pd clusters bound to polymers, 56
turnover cycles, 75, 80
turnover number, 15, 18, 56, 75, 80, 136,
  240, 248, 300, 316
  of hydrogenation, 8

— V —

Vaska's complex, 16
vitamin A, 323
vitamin B_{12}, 169

— W —

Wacker-type catalysts, 87, 104, 116,
  379
water decomposition, 388
water photodecomposition, 341
water-gas shift reactions, 312
water-soluble polymers, 49
Westinghouse process, 327
Wilkinson catalyst, 3, 4, 18, 48, 50, 60,
  70, 128

— X —

X-ray analysis, 28

— Z —

Ziegler–Natta catalysts, 69, 201, 232,
  242, 245, 249, 379
zirconium-containing polymers, 217
zirconocene 137

Milton Keynes UK
Ingram Content Group UK Ltd.
UKHW031125141024
449569UK00006B/428